STRIPLINE CIRCULATORS • *Joseph Helszajn*

THE STRIPLINE CIRCULATOR: THEORY AND PRACTICE • *Joseph Helszajn*

LOCALIZED WAVES • *Hugo E. Hernández-Figueroa, Michel Zamboni-Rached, and Erasmo Recami (eds.)*

MICROSTRIP FILTERS FOR RF/MICROWAVE APPLICATIONS • *Jia-Sheng Hong and M. J. Lancaster*

MICROWAVE APPROACH TO HIGHLY IRREGULAR FIBER OPTICS • *Huang Hung-Chia*

NONLINEAR OPTICAL COMMUNICATION NETWORKS • *Eugenio Iannone, Francesco Matera, Antonio Mecozzi, and Marina Settembre*

FINITE ELEMENT SOFTWARE FOR MICROWAVE ENGINEERING • *Tatsuo Itoh, Giuseppe Pelosi, and Peter P. Silvester (eds.)*

INFRARED TECHNOLOGY: APPLICATIONS TO ELECTROOPTICS, PHOTONIC DEVICES, AND SENSORS • *A. R. Jha*

SUPERCONDUCTOR TECHNOLOGY: APPLICATIONS TO MICROWAVE, ELECTRO-OPTICS, ELECTRICAL MACHINES, AND PROPULSION SYSTEMS • *A. R. Jha*

OPTICAL COMPUTING: AN INTRODUCTION • *M. A. Karim and A. S. S. Awwal*

INTRODUCTION TO ELECTROMAGNETIC AND MICROWAVE ENGINEERING • *Paul R. Karmel, Gabriel D. Colef, and Raymond L. Camisa*

MILLIMETER WAVE OPTICAL DIELECTRIC INTEGRATED GUIDES AND CIRCUITS • *Shiban K. Koul*

MICROWAVE DEVICES, CIRCUITS AND THEIR INTERACTION • *Charles A. Lee and G. Conrad Dalman*

ADVANCES IN MICROSTRIP AND PRINTED ANTENNAS • *Kai-Fong Lee and Wei Chen (eds.)*

SPHEROIDAL WAVE FUNCTIONS IN ELECTROMAGNETIC THEORY • *Le-Wei Li, Xiao-Kang Kang, and Mook-Seng Leong*

ARITHMETIC AND LOGIC IN COMPUTER SYSTEMS • *Mi Lu*

OPTICAL FILTER DESIGN AND ANALYSIS: A SIGNAL PROCESSING APPROACH • *Christi K. Madsen and Jian H. Zhao*

THEORY AND PRACTICE OF INFRARED TECHNOLOGY FOR NONDESTRUCTIVE TESTING • *Xavier P. V. Maldague*

METAMATERIALS WITH NEGATIVE PARAMETERS: THEORY, DESIGN, AND MICROWAVE APPLICATIONS • Ricardo Marqués, Ferran Martín, and Mario Sorolla

OPTOELECTRONIC PACKAGING • *A. R. Mickelson, N. R. Basavanhally, and Y. C. Lee (eds.)*

OPTICAL CHARACTER RECOGNITION • *Shunji Mori, Hirobumi Nishida, and Hiromitsu Yamada*

ANTENNAS FOR RADAR AND COMMUNICATIONS: A POLARIMETRIC APPROACH • *Harold Mott*

INTEGRATED ACTIVE ANTENNAS AND SPATIAL POWER COMBINING • *Julio A. Navarro and Kai Chang*

ANALYSIS METHODS FOR RF, MICROWAVE, AND MILLIMETER-WAVE PLANAR TRANSMISSION LINE STRUCTURES • *Cam Nguyen*

FREQUENCY CONTROL OF SEMICONDUCTOR LASERS • *Motoichi Ohtsu (ed.)*

WAVELETS IN ELECTROMAGNETICS AND DEVICE MODELING • *George W. Pan*

OPTICAL SWITCHING • Georgios Papadimitriou, Chrisoula Papazoglou, and Andreas S. Pomportsis

SOLAR CELLS AND THEIR APPLICATIONS • *Larry D. Partain (ed.)*

ANALYSIS OF MULTICONDUCTOR TRANSMISSION LINES • *Clayton R. Paul*

INTRODUCTION TO ELECTROMAGNETIC COMPATIBILITY, Second Edition • *Clayton R. Paul*

ADAPTIVE OPTICS FOR VISION SCIENCE: PRINCIPLES, PRACTICES, DESIGN AND APPLICATIONS • *Jason Porter, Hope Queener, Julianna Lin, Karen Thorn, and Abdul Awwal (eds.)*

ELECTROMAGNETIC OPTIMIZATION BY GENETIC ALGORITHMS • *Yahya Rahmat-Samii and Eric Michielssen (eds.)*

INTRODUCTION TO HIGH-SPEED ELECTRONICS AND OPTOELECTRONICS • *Leonard M. Riaziat*

NEW FRONTIERS IN MEDICAL DEVICE TECHNOLOGY • *Arye Rosen and Harel Rosen (eds.)*

ELECTROMAGNETIC PROPAGATION IN MULTI-MODE RANDOM MEDIA • *Harrison E. Rowe*

ELECTROMAGNETIC PROPAGATION IN ONE-DIMENSIONAL RANDOM MEDIA • *Harrison E. Rowe*

HISTORY OF WIRELESS • *Tapan K. Sarkar, Robert J. Mailloux, Arthur A. Oliner, Magdalena Salazar-Palma, and Dipak L. Sengupta*

PHYSICS OF MULTIANTENNA SYSTEMS AND BROADBAND PROCESSING • *Tapan K. Sarkar, Magdalena Salazar-Palma, and Eric L. Mokole*

SMART ANTENNAS • *Tapan K. Sarkar, Michael C. Wicks, Magdalena Salazar-Palma, and Robert J. Bonneau*

NONLINEAR OPTICS • *E. G. Sauter*

APPLIED ELECTROMAGNETICS AND ELECTROMAGNETIC COMPATIBILITY • *Dipak L. Sengupta and Valdis V. Liepa*

COPLANAR WAVEGUIDE CIRCUITS, COMPONENTS, AND SYSTEMS • *Rainee N. Simons*

ELECTROMAGNETIC FIELDS IN UNCONVENTIONAL MATERIALS AND STRUCTURES • *Onkar N. Singh and Akhlesh Lakhtakia (eds.)*

ELECTRON BEAMS AND MICROWAVE VACUUM ELECTRONICS • *Shulim E. Tsimring*

FUNDAMENTALS OF GLOBAL POSITIONING SYSTEM RECEIVERS: A SOFTWARE APPROACH, Second Edition • *James Bao-yen Tsui*

RF/MICROWAVE INTERACTION WITH BIOLOGICAL TISSUES • *André Vander Vorst, Arye Rosen, and Youji Kotsuka*

InP-BASED MATERIALS AND DEVICES: PHYSICS AND TECHNOLOGY • *Osamu Wada and Hideki Hasegawa (eds.)*

COMPACT AND BROADBAND MICROSTRIP ANTENNAS • *Kin-Lu Wong*

DESIGN OF NONPLANAR MICROSTRIP ANTENNAS AND TRANSMISSION LINES • *Kin-Lu Wong*

PLANAR ANTENNAS FOR WIRELESS COMMUNICATIONS • *Kin-Lu Wong*

FREQUENCY SELECTIVE SURFACE AND GRID ARRAY • *T. K. Wu (ed.)*

ACTIVE AND QUASI-OPTICAL ARRAYS FOR SOLID-STATE POWER COMBINING • *Robert A. York and Zoya B. Popović (eds.)*

OPTICAL SIGNAL PROCESSING, COMPUTING AND NEURAL NETWORKS • *Francis T. S. Yu and Suganda Jutamulia*

SiGe, GaAs, AND InP HETEROJUNCTION BIPOLAR TRANSISTORS • *Jiann Yuan*

ELECTRODYNAMICS OF SOLIDS AND MICROWAVE SUPERCONDUCTIVITY • *Shu-Ang Zhou*

The Stripline Circulator

The Stripline Circulator

Theory and Practice

By

J. HELSZAJN

WILEY

A JOHN WILEY & SONS, INC., PUBLICATION

Published by John Wiley & Sons, Inc.
Published simultaneously in Canada

Library of Congress Cataloging-in-Publication Data:

Helszajn, J. (Joseph).
Stripline circulators : theory and practice/by Joseph Helszajn.
p. cm.
ISBN 978-0-470-25878-1 (cloth)
1. Circulators, Wave-guide—Design and construction. I. Title.
TK7871.65.H46 2008
621.381'331—dc22

2007039340

Printed in the United States of America
10 9 8 7 6 5 4 3 2 1

Contents

Preface

The stripline junction circulator is a unique nonreciprocal device, which is embodied in many pieces of microwave equipment. The text is devoted to the theory and practice of this class of circulator. It starts with a chapter on the architecture of stripline circulators, and chapters on the tensor permeability in a magnetic insulator, and on the spatial shape demagnetizing factors of magnetic insulators. It continues with chapters on the scattering, eigenvalues, and admittance descriptions of the circulator as well as on its degree-1 and degree-2 one-port circuits. These chapters embody various classic experimental procedures for the characterization of the classic circulator. It proceeds with a block of chapters dealing with properties of gyromagnetic planar cloverleaf, wye, irregular hexagonal, and triangular resonators and the use of magnetic walls. The lumped element single junction circulator is dealt with as a preamble to dealing with the distributed circulator in that it embodies all the theoretical considerations of the general problem. Synthesis of the classic junction using a disk resonator is dealt with separately, as is the important Green's function and the finite element method. Special attention is devoted thoughout to bridging the gap between its circuit and electromagnetic descriptions. These chapters are followed by one that deals with circulators employing triangular and irregular hexagonal gyromagnetic resonators. A separate chapter provides a detailed investigation of the frequency responses of the classic circulator using very weakly, weakly, strongly, and very strongly magnetized disk resonators. Still another chapter is devoted to the theory of the negative permeability circulator. The text continues with two chapters on circulators using wye resonators and a chapter on the little understood four-port single junction. A block of three chapters deals with the synthesis problem and the frequency responses of reciprocal and nonreciprocal junctions. The last two chapters but one are devoted to the fabrication of UHF and microstrip circulators. The last chapter deals with some discrepancies between idealized or theoretical models and practice. A number of important topics such as spinwave instabilities and nonlinear effects in magnetic insulators have been omitted from the text in order to keep what is already a large volume in check. These topics are in every case already in place in a number of classic textbooks. Inevitably, some works, which have appeared elsewhere, have been duplicated for the sake of understanding.

CHAPTER ONE

Architecture of Symmetrical Stripline Junction Circulators

1.1 INTRODUCTION

The three-port circulator is a unique nonreciprocal symmetrical junction having one typical input port, one output port, and one decoupled port. The fundamental definition of the junction circulator has its origin in energy conservation. It states that the only matched symmetrical three-port junction corresponds to the definition of the circulator. A wave incident in such a junction at port 1 is emergent at port 2, one incident at port 2 is emergent at port 3, and so on in a cyclic manner. One possible model of a circulator is a magnetized ferrite or garnet gyromagnetic resonator having three-fold symmetry connected or coupled to three transmission lines or waveguides. The purpose of this introductory chapter is to provide one phenomenological description of the operation of this sort of device, to summarize some of the more common resonator geometries met in its construction, and to indicate some of its uses. The introduction of any such resonator at the junction of three striplines produces a degree-1 circulation solution. In practice, the gyromagnetic resonator is embedded in a filter circuit in order to produce a degree-2 or degree-3 frequency response. The possibility of realizing a single junction circulator with more than three ports is understood.

1.2 PHENOMENOLOGICAL DESCRIPTION OF STRIPLINE CIRCULATOR

The geometry of the stripline circulator geometry is depicted in Fig. 1.1. It consists of two ferrite planar disk resonators separated by a disk center conductor symmetrically coupled by three transmission lines. The gyromagnetic material is magnetized perpendicularly to the plane of the device by a static magnetic field. An important property of

The Stripline Circulator: Theory and Practice. By J. Helszajn

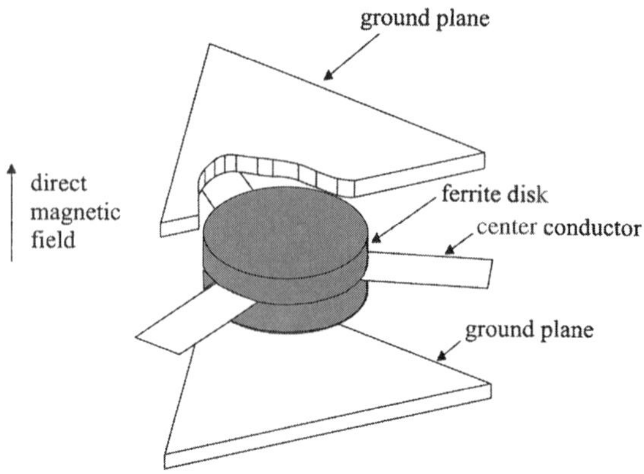

FIGURE 1.1 Schematic diagram of three-port stripline circulator.

this device is that a circulator condition is met whenever all three ports are matched. For a three-port junction this requires two independent variables. Under certain simplifying conditions its adjustment can be described in terms of a figure-eight standing wave pattern within the disk due to the interference of a pair of degenerate field patterns rotating in opposite directions. When the gyromagnetic junction is unmagnetized, the resonant frequencies of the two field patterns are identical. When it is magnetized, the degeneracy is removed, and the standing wave pattern within the resonator is rotated. One circulation condition is established by operating between the two split frequencies. This requirement essentially fixes the radius of the gyromagnetic resonator. The second condition is met by adjusting the splitting, until the standing wave pattern is rotated through 30°. From symmetry, port 3 is then situated at a null of the standing wave pattern and is therefore isolated and the junction displays properties akin to that of a two-port transmission line resonator between the other two ports. This condition fixes the magnitude of the gyrotropy or the direct magnetic field. Figure 1.2 depicts the two standing wave patterns under discussion. A third, in-phase mode, also strictly speaking enters into the description of this type of junction. It has, however, an electric wall at the periphery of the resonator so that it does not affect the total field pattern there.

The rotation of the standing wave pattern in a gyromagnetic resonator under the application of a direct magnetic field may be understood by decomposing the linearly polarized radiofrequency (rf) magnetic field on its axis into counterrotating ones, which are then split by its gyrotropy. The direction in which the pattern in such a resonator is rotated is fixed by that of the direct magnetic field so that it may be utilized to realize an electrically actuated waveguide switch.

1.3 ADJUSTMENT OF JUNCTION CIRCULATOR

The operation of any junction may be understood by having recourse to superposition. It starts by decomposing a single input wave at port 1 (say) into a linear

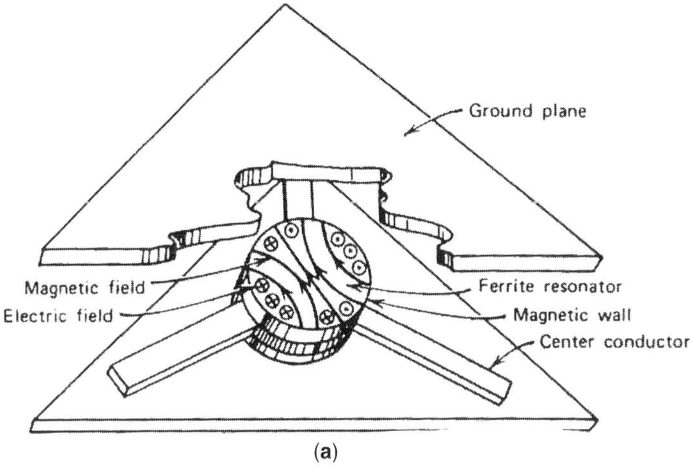

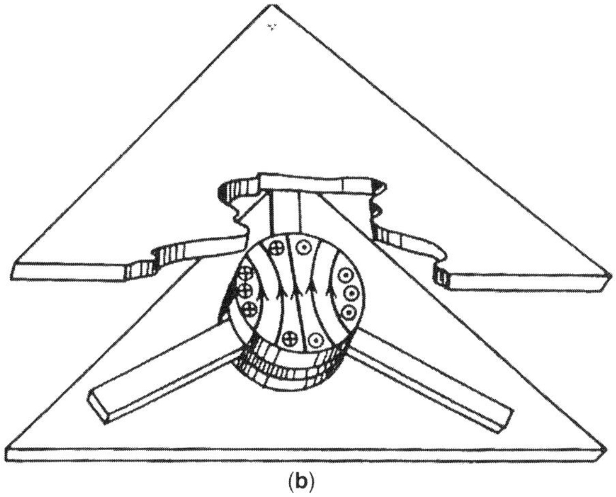

FIGURE 1.2 (a) Standing wave patterns in (a) demagnetized stripline junction and (b) magnetized stripline junction. (Reproduced with permission from C. E. Fay and R. L. Comstock, Operation of the ferrite junction circulator, *IEEE Trans. Microwave Theory Tech.*, Vol. MTT-13, pp. 15–27, January 1965.)

combination of voltage settings at each port:

$$\begin{bmatrix} 1 \\ 0 \\ 0 \end{bmatrix} = \frac{1}{3}\begin{bmatrix} 1 \\ 1 \\ 1 \end{bmatrix} + \frac{1}{3}\begin{bmatrix} 1 \\ \alpha \\ \alpha^2 \end{bmatrix} + \frac{1}{3}\begin{bmatrix} 1 \\ \alpha^2 \\ \alpha \end{bmatrix} \tag{1.1}$$

where

$$\alpha = \exp(j120) \quad \text{and} \quad \alpha^2 = \exp(j240)$$

A scrutiny of the first, so-called in-phase generator settings indicates that it produces an electric field along the axis of the junction. The reflected waves at the three ports of the junction are therefore in this instance unaffected by the details of the gyrotropy. A scrutiny of the second and third, so-called counterrotating generator settings indicates, however, that these establish counterrotating circularly polarized alternating magnetic fields in the plane of the junction. A characteristic of a suitably magnetized gyromagnetic insulator is that it has different scalar permeabilities under the two arrangements. It therefore provides one practical means of removing the degeneracy between the reflected waves associated with these two generator settings. The fields produced at the axis of the junction by each of these three possible generator settings are illustrated in Fig. 1.3.

A typical reflected wave at any port is constructed by adding the individual ones due to each possible generator setting. A typical term is realized by taking the product of a typical incident wave and a typical reflection coefficient.

$$\begin{bmatrix} b_1 \\ b_2 \\ b_3 \end{bmatrix} = \frac{\rho_0}{3}\begin{bmatrix} 1 \\ 1 \\ 1 \end{bmatrix} + \frac{\rho_-}{3}\begin{bmatrix} 1 \\ \alpha \\ \alpha^2 \end{bmatrix} + \frac{\rho_+}{3}\begin{bmatrix} 1 \\ \alpha^2 \\ \alpha \end{bmatrix} \tag{1.2}$$

An ideal circulator is now defined as

$$\frac{\rho_0 + \rho_- + \rho_+}{3} = 0 \tag{1.3a}$$

$$\frac{\rho_0 + \rho_-\alpha + \rho_+\alpha^2}{3} = -1 \tag{1.3b}$$

$$\frac{\rho_0 + \rho_-\alpha^2 + \rho_+\alpha}{3} = 0 \tag{1.3c}$$

To adjust this, and other circulators, requires a 120° phase difference between the reflection coefficients of the three different ways in which it is possible to excite the three ports of the junction. One solution is

$$\rho_+ = \exp\left[-j2\left(\phi_1 + \phi_+ + \pi/2\right)\right] \tag{1.4a}$$

$$\rho_{-1} = \exp[-j2(\phi_1 + \phi_- + \pi/2)] \tag{1.4b}$$

$$\rho_0 = \exp[-j(2\phi_0)] \tag{1.4c}$$

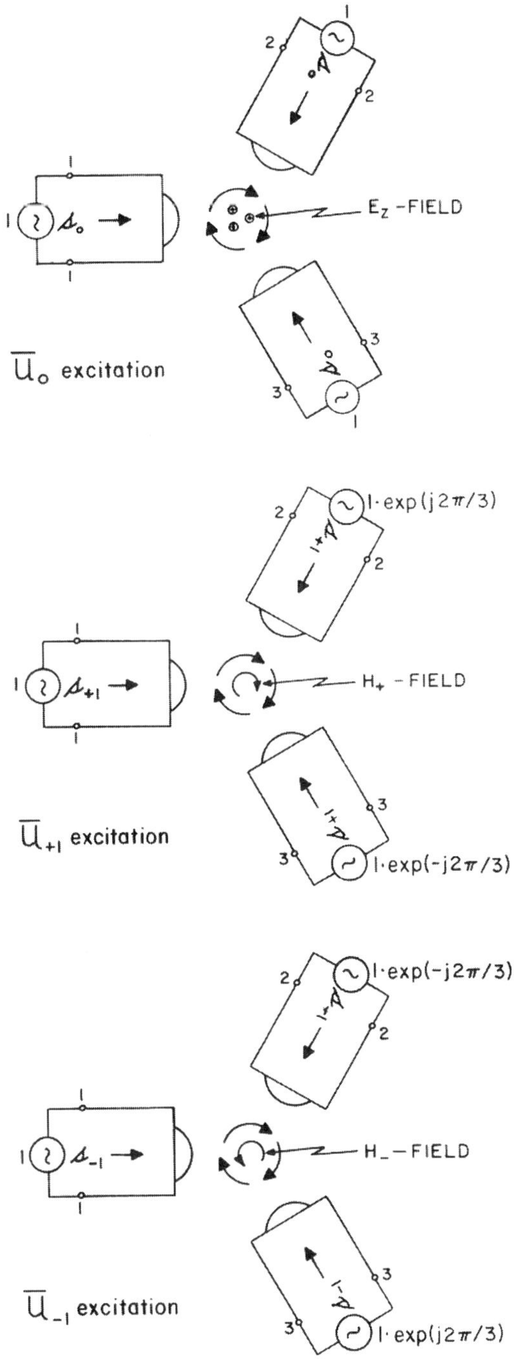

FIGURE 1.3 Voltage settings on three-port circulator.

provided that

$$\phi_1 = \phi_0 = \pi/2 \tag{1.5a}$$

$$\phi_+ = -\phi_- = -\pi/6 \tag{1.5b}$$

The in-phase and degenerate counterrotating reflection angles are established by adjusting the details of the corresponding one-port eigen-networks of the demagnetized ferrite section so that the angle between the two is initially 180°. The degenerate phase angles of the counterrotating reflection coefficients are thereafter separated by 120° by the gyrotropy of the gyromagnetic region, thereby producing the ideal phase angles of the circulator. These two steps represent the necessary and sufficient conditions for the adjustment of this class of circulator.

The relationship between the incident and reflected waves at the terminals of a network or junction is often described in terms of a scattering matrix. It is therefore appropriate to reduce the result established here to that notation. The nomenclature entering into the definition of this matrix is indicated in Fig. 1.4. Its entries relate

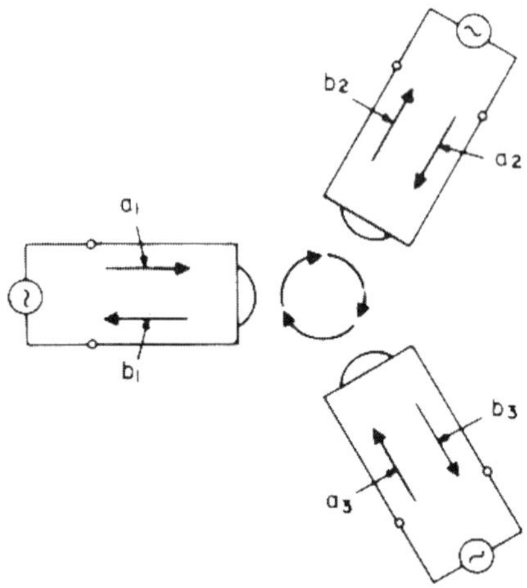

$$\begin{bmatrix} b_1 \\ b_2 \\ b_3 \end{bmatrix} = \begin{bmatrix} S_{11} & S_{21} & S_{31} \\ S_{31} & S_{11} & S_{21} \\ S_{21} & S_{31} & S_{11} \end{bmatrix} \begin{bmatrix} a_1 \\ a_2 \\ a_3 \end{bmatrix}$$

FIGURE 1.4 Scattering variables in three-port junction.

incident and reflected waves at suitable terminal planes of the circuit:

$$S_{11} = \frac{b_1}{a_1}\bigg|_{a_2 = a_3 = 0} \tag{1.6a}$$

$$S_{21} = \frac{b_2}{a_1}\bigg|_{a_2 = a_3 = 0} \tag{1.6b}$$

$$S_{31} = \frac{b_3}{a_1}\bigg|_{a_2 = a_3 = 0} \tag{1.6c}$$

A scrutiny of these definitions indicates that the entries of the scattering matrix may readily be evaluated once the reflected waves at all the ports due to an incident wave at a typical port are established. Taking a_1 as unity and making use of the results for b_1, b_2, and b_3 gives the required parameters without ado.

$$S_{11} = \frac{\rho_0 + \rho_+ + \rho_-}{3} \tag{1.7a}$$

$$S_{21} = \frac{\rho_0 + \alpha\rho_+ + \alpha^2\rho_-}{3} \tag{1.7b}$$

$$S_{31} = \frac{\rho_0 + \alpha^2\rho_+ + \alpha\rho_-}{3} \tag{1.7c}$$

The entries of the scattering matrix are therefore linear combinations of the reflection variables at any port associated with each possible family of generator settings. One definition of an ideal circulator, which is on keeping with the description of the junction circulator, is therefore

$$S_{11} = 0 \tag{1.8a}$$

$$S_{21} = -1 \tag{1.8b}$$

$$S_{31} = 0 \tag{1.8c}$$

This solution may be established separately by having recourse to the unitary condition and may therefore be taken as a universal definition of a three-port lossless junction circulator.

1.4 GYROTROPY IN MAGNETIC INSULATORS

One means of removing the degeneracy between a pair of counterrotating field patterns is by having resource to a suitably magnetized magnetic insulator. The

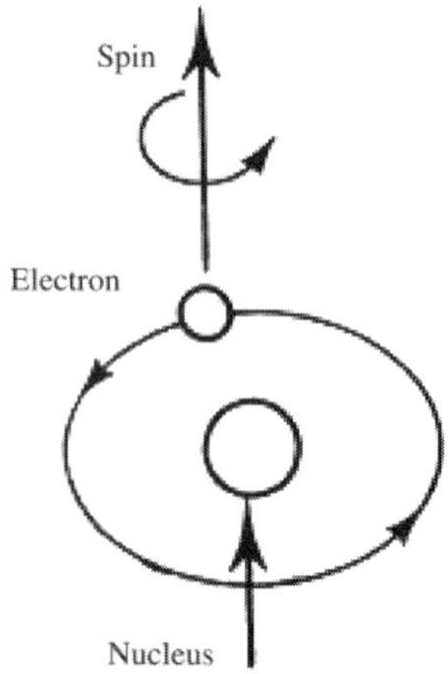

FIGURE 1.5 Atomic orbit.

origins of the magnetic effects or magnetization in magnetic insulators are due to the effective current loops of electrons in atomic orbits and the effects of electron spin and atomic nuclei (Fig. 1.5). Each of these features produces a magnetic field that is equivalent to that arising from a magnetic dipole—the total magnetic moment being the vector sum of the individual moments. In ferromagnetic insulators the predominant effect is due to the electron spin. A property of this sort of medium is that while it has, in general, a tensor permeability, it displays scalar permeabilities under one of three specific arrangements. One solution is a circularly polarized magnetic field in the plane transverse to the direct magnetic field, which rotates in the same sense as that of the electron spin; another is one that rotates in the opposite direction. The scalar permeabilities ($\mu \pm \kappa$) displayed by the medium under these two situations are simple linear combinations of the diagonal (μ) and off-diagonal (κ) entries of the tensor permeability. The absolute values of these quantities are essentially fixed by the frequency of the alternating radio magnetic field and the direct magnetization of the magnetic insulator and its direct magnetic field. The third normal mode coincides with a linearly polarized alternating magnetic field along the axis of the electron spin. It involves no gyromagnetic interaction.

1.5 PLANAR RESONATORS

In the design of any directly or transformer coupled planar circulator, it is essential to simultaneously reconcile physical, magnetic, and network conditions. It is necessary in order to do so with acceptable microwave characteristics to adjust either the substrate thickness or the resonator shape of the junction. The substrate thickness is often specified by the system rather than by the junction design, so that an ideal synthesis procedure is one where the resonator shape can be varied. If this is the case, a quarter-wave coupled triangular resonator coupled at its corners is best at low frequencies, a disk resonator is best at intermediate frequencies, and a triangular

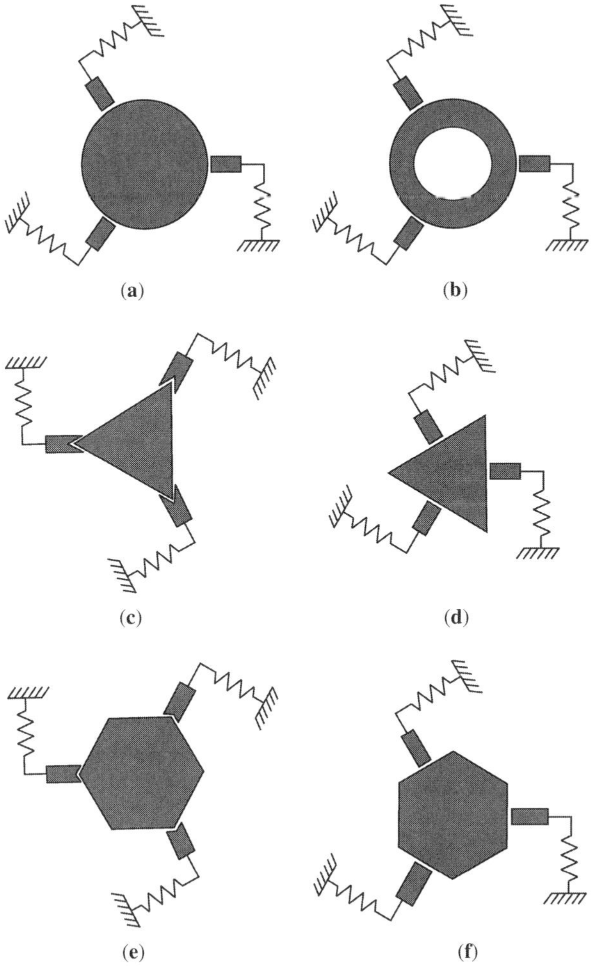

FIGURE 1.6 Some planar resonators that are suitable for the construction of three-port circulators.

resonator fed midway along its side is most suitable at high frequencies. Figure 1.6 depicts some possibilities. Figure 1.7 shows the construction of circulation solutions using triangular and wye resonators in terms of the field patterns of the demagnetized resonators.

1.6 PARALLEL PLATE WAVEGUIDE MODEL OF MICROSTRIP CIRCULATORS

The usual approach to the design of microstrip passive circuits and circulators using weakly magnetized resonators is to replace the problem region with imperfect

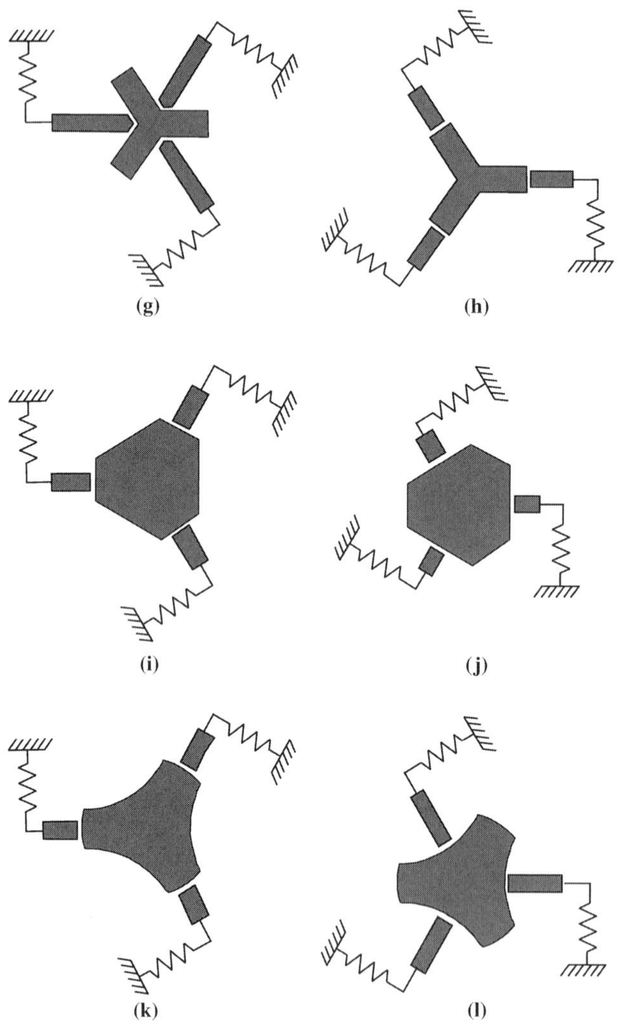

FIGURE 1.6 *Continued.*

magnetic side walls by an equivalent waveguide model. Figure 1.8 indicates a possible equivalence for a microstrip circulator using a circular resonator. The concepts entering into this sort of model are well established and need not be dwelt upon. Once any design is complete, in terms of the equivalent parallel plate waveguide approximation, it is necessary to invoke the relationship between the actual and effective parameters of the problem region. Another matter of concern in the design of such circuits is that if the fringing fields on a typical contour are excessive then it becomes difficult to preserve the definitions of both the coupling angle at the terminals of the resonator and its shape. One way to partially avoid both difficulties is to impose a lower bound on the aspect ratio (r/H) of the resonator.

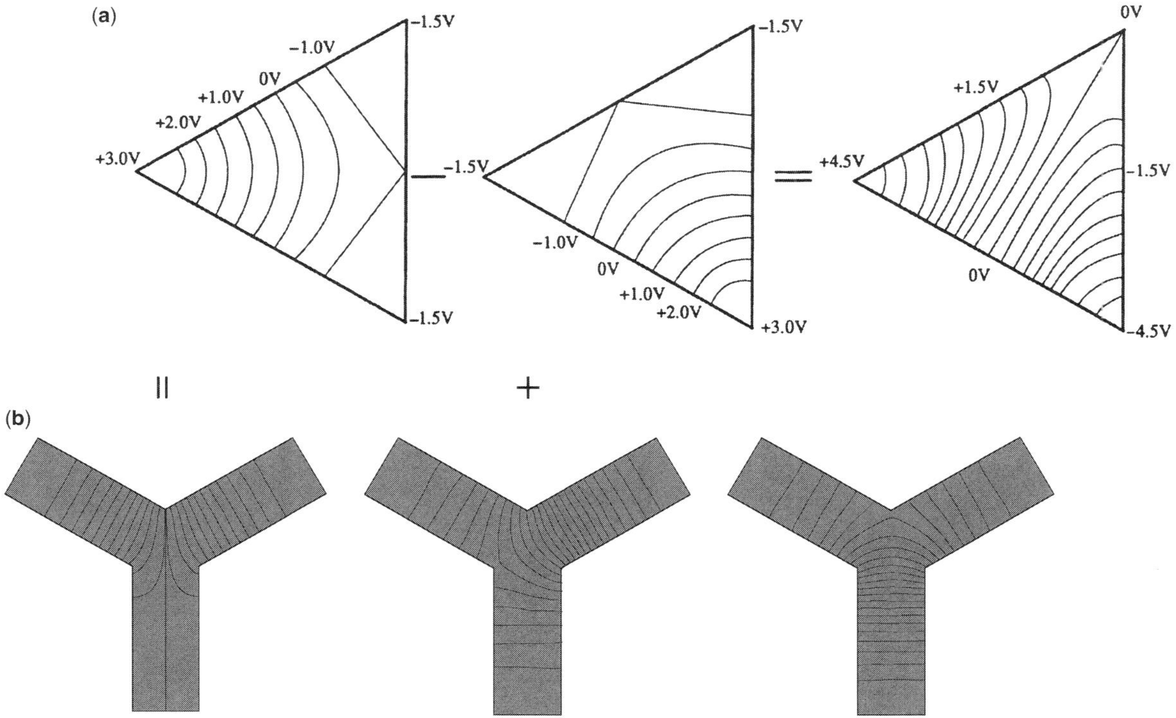

FIGURE 1.7 Standing wave solution of 3-port circulator using (a) triangular resonator and (b) wye resonator. (Reproduced with permission from J. Helszajn, D. S. James, and W. T. Nisbet, Circulators using planar triangular resonators, *IEEE Trans. Microwave Theory Tech.*, Vol. MTT-27, February 1979.)

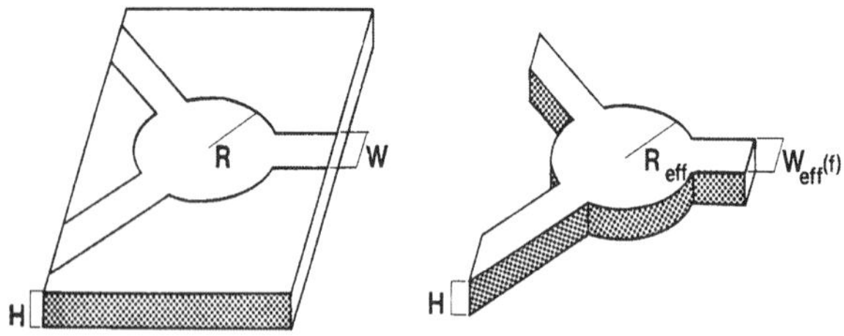

FIGURE 1.8 Equivalent waveguide model of planar disk.

1.7 DROP-IN AND PACKAGING TECHNIQUES

An ongoing activity in the area of microstrip devices is that of integration of the microstrip circulator or isolator in the subsystem. Some possibilities include an all-ferrite substrate, a suitable magnetic ferrite region inserted into a nonmagnetic ferrite plate, a ferrite puck inserted into a hole in a dielectric substrate, and a ferromagnetic film deposited on a dielectric sheet. The choice of technique used in any given application will be at least partly determined by cost and size considerations as well as performance. A number of drop-in techniques have also evolved over the years. One problem with this sort of fabrication is the transition between the microstrip circuit and the transmission line. Figure 1.9 shows one possibility that permits a planar resonator to be mounted onto an existing alumina substrate. Figure 1.10 illustrates a ferrite or garnet resonator embedded in an alumina substrate.

1.8 SWITCHED RESONATORS

The direction of circulation of a circulator is determined by that of the direct magnetic field. It may therefore be employed to switch an input signal at one port to either one of the other two. This may be done by replacing the permanent magnet by an

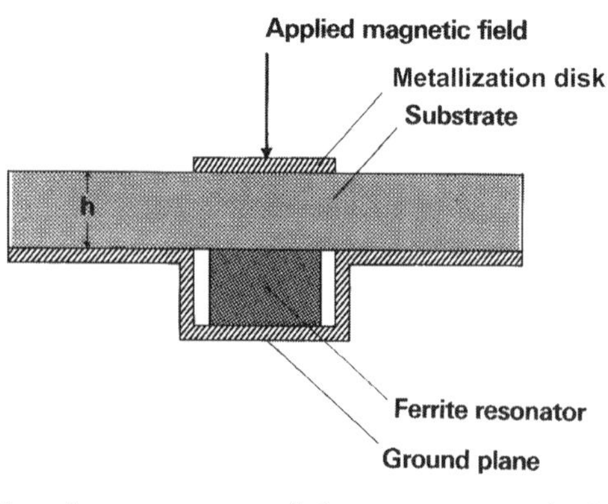

FIGURE 1.9 Mounting arrangement of planar resonator on back of microstripline.

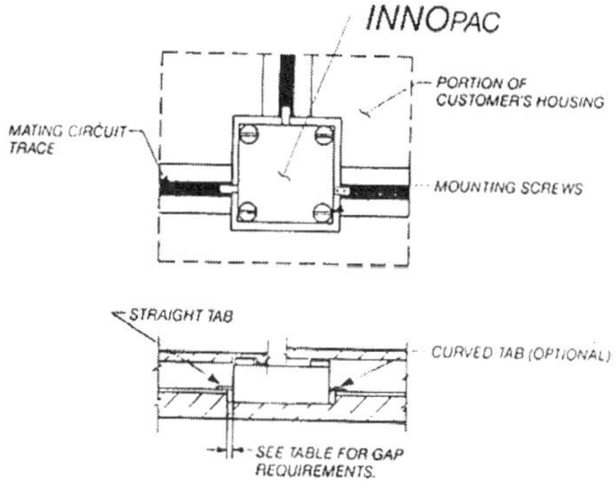

Microstrip circulator on alumina substrate. (Courtesy INNOWAVE Inc., CA, USA)

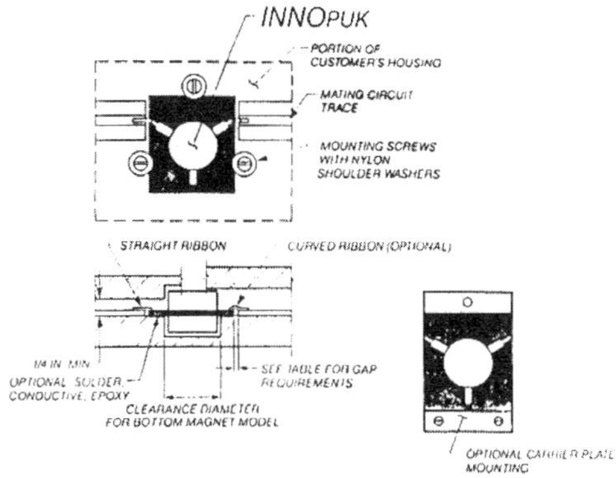

: Microstrip circulator on garnet substrate (Courtesy IN-NOWAVE INC., CA, USA)

FIGURE 1.10 Microstrip circulator on alumina substrate.

electromagnet or by latching the microwave ferrite resonator directly by embedding a current-carrying wire loop within the resonator. One practical arrangement is illustrated in Fig. 1.11. This sort of switch is particularly useful in the construction of Butler type matrices in phased array systems. The switching power necessary to actuate this sort of circuit is determined by the stored energy in the magnetic circuit and the switching time. Switching times between 10 ns and 1 ms are achievable depending on whether the gyromagnetic circuit is internally or externally magnetized.

1.9 COMPOSITE RESONATORS

The conventional stripline junction relies for one of its two circulation conditions on an open dielectric resonance in a demagnetized ferrite or garnet geometry. The

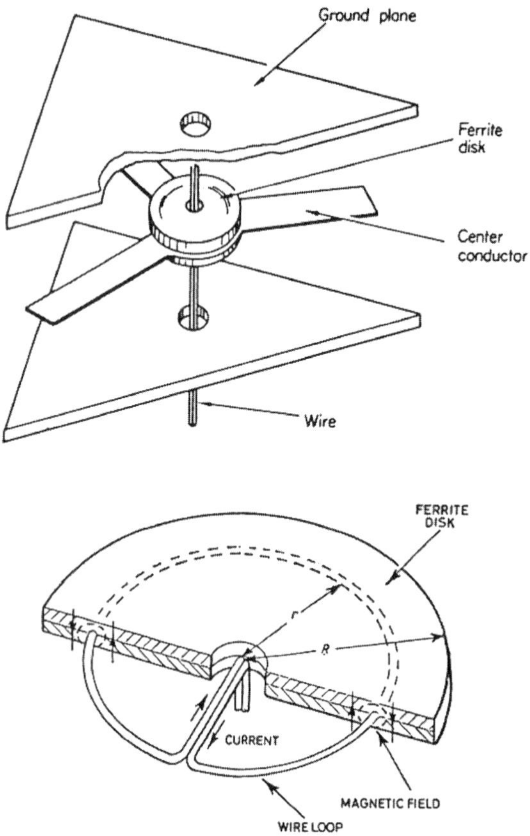

FIGURE 1.11 Details of wire loop in switched resonator.

maximum average power that such a circulator can handle is determined by the temperature drop across the thin dimension of the structure and its surface area. The thermal conductivity of the ferrite material is relatively low. For instance, the thermal conductivity of WESGO AL-995 ceramic is 29.31 W/m · °C (70×10^{-3} cal · m/s · °C) and that of beryllia oxide is 219.81 W/m · °C (525×10^{-3} cal · cm/s · °C). For ferrites it is 2.09 W/m · °C (5×10^{-3} cal · cm/s · °C). One way to overcome this difficulty is to employ composite resonators in the design. Such resonators have mainly been utilized in the construction of devices capable of handling hundreds of watts of mean power. Figure 1.12 depicts one planar and one radial configuration. A further advantage of this class of resonator is that the temperature stability is improved because many dielectric materials are temperature stable.

The second circulation relates the gyrator conductance, the susceptance slope parameter, and the split frequencies of the junction to the specification of the device—a large separation between the split frequencies being essential for the realization of high-quality circulators. Substituting a dielectric for part of the ferrite material

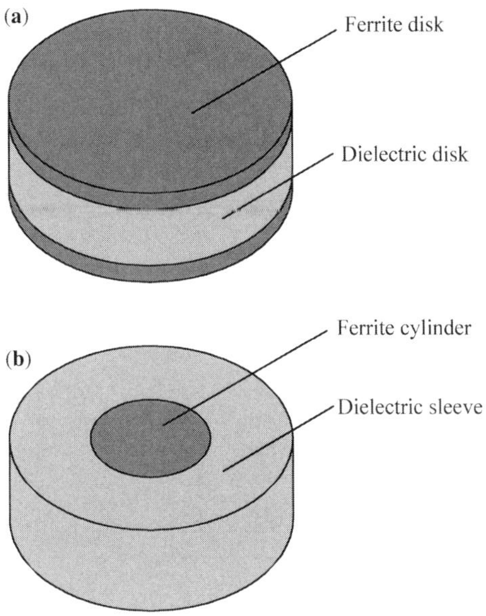

FIGURE 1.12 (a) Axial composite resonator and (b) radial composite resonator.

reduces the difference between the split frequencies of the junction; a compromise is therefore necessary between the microwave specification and the power rating of the device.

1.10 POWER RATING OF GYROMAGNETIC RESONATOR

Important aspects of microwave components are its peak and average power ratings. The peak power rating is usually fixed by arcing and by nonlinear effects in magnetic insulators due to spinwave instabilities. The average power rating is restricted by the temperature rise of the device, which may to some extent be mitigated by cooling it by forced air or water. The power dissipated in the device has its origin in dielectric losses and linear and nonlinear magnetic ones and dissipation in the striplines. The choice of resonator in any particular situation is dictated by one or both of these difficulties. The family of planar resonators is in general suitable for the construction of devices with modest mean power ratings. Composite resonators are used where a compromise between mean and peak power is necessary. Insertion losses on the order of 0.06 dB are not unusual in practice. The power rating of a typical circulator is dependent on the choice of resonator, the connector size, the frequency, the bandwidth, the available ferrite material, the cooling arrangement, the temperature range, pressurization, and the outline drawing to mention but some considerations that enter into any design. It is therefore difficult to compile precise recommendations in any particular situation.

1.11 QUARTER-WAVE COUPLED CIRCULATOR

Any gyromagnetic resonator with threefold symmetry suitably placed at a junction of three striplines may be adjusted to produce an ideal three-port circulator at a single frequency. Practical circulators, however, have to operate over finite frequency intervals with a specified ripple level or return loss and isolation. One way to realize a classical frequency response is to absorb each port of the gyromagnetic resonator into a two-port filter or matching network. A knowledge of the one-port complex gyrator circuit of the gyromagnetic junction at a typical port is sufficient for this purpose. One typical topology is indicated in Fig. 1.13. It consists of a quarter-wave unit element (UE) in shunt with the gyrator conductance of the junction. Figure 1.14

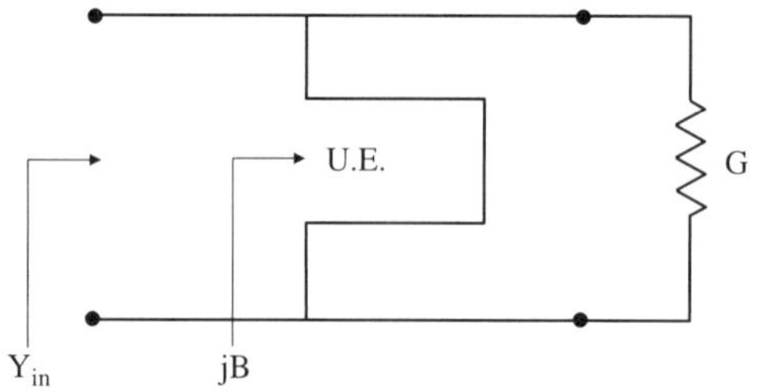

FIGURE 1.13 Complex gyrator circuit of three-port circulator.

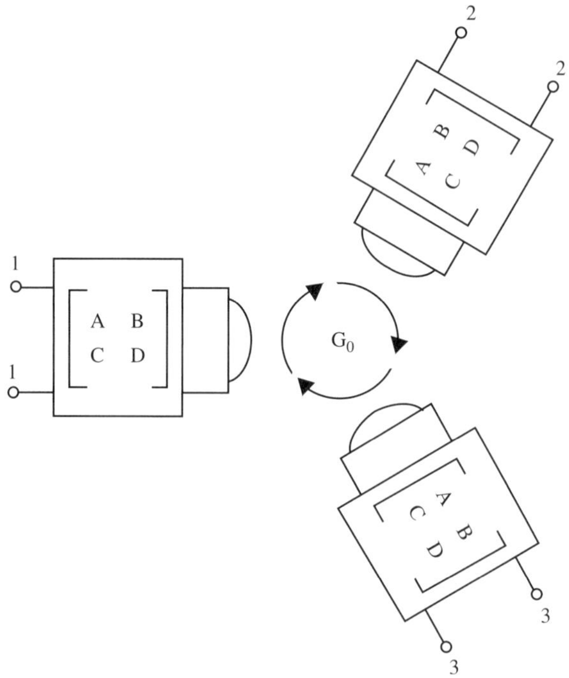

FIGURE 1.14 Topology of three-port circulators using two-port filter circuits.

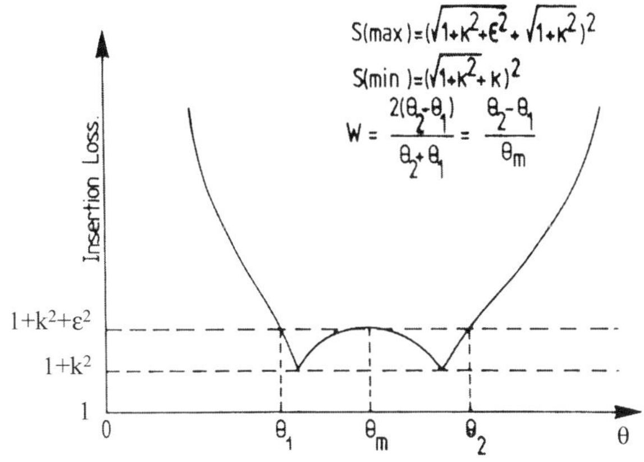

FIGURE 1.15 Frequency response of quarter-wave coupled junction circulation.

depicts the overall schematic arrangement. While a host of network solutions are possible in practice, the most common arrangement consists of one or more quarter-wave long impedance transformers. A typical frequency response of a circulator using a single quarter-wave long transformer at each port of the gyromagnetic resonator is indicated in Fig. 1.15. This sort of topology is a classic network problem in the literature and much of modern circulator practice rests on an understanding of this topic. The actual gain bandwidth of any circulator is fixed in practice by the quality factor of the gyromagnetic resonator and the topology of the filter circuit. It fixes the relationship between the isolation or return loss in decibels of the circulator and its bandwidth. A return loss or isolation of 23 dB (say) and a bandwidth of typically 25% is realizable in practice in conjunction with a single quarter-wave long impedance transformer. A similar return or loss or isolation specification is readily realizable over a bandwidth of between 40% and 66% using two such transformers.

1.12 FOUR-PORT SINGLE JUNCTION CIRCULATOR

While the three-port single junction circulator is the most common arrangement met in practice, four-port ones may also be realized without too much difficulty. Such junctions, in common with three-port devices, have some of the properties of a transmission line cavity resonator between ports 1 and 2, and a definite standing wave pattern exists within the junction with nulls at ports 3 and 4 also. An important difference between the two, however, is that the four-port device cannot be adjusted with external tuning elements only. This remark may be understood by recognizing that a four-port device can be matched without being a circulator. Scrutiny of its eigenvalue problem indicates that its adjustment requires three independent variables. These may be established in a systematic way by perturbing the scattering matrix

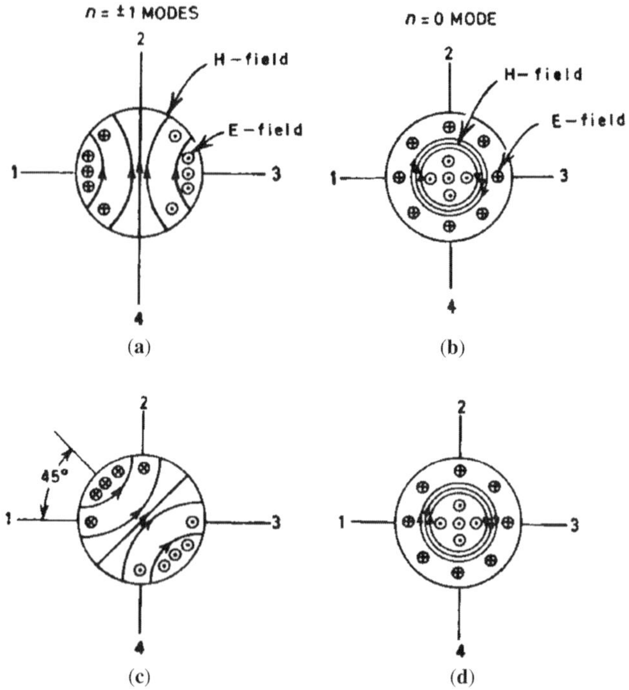

FIGURE 1.16 Standing wave solution of four-port single junction circulator: (a) $n = \pm 1$ for unmagnetized ferrite post, (b) $n = 0$ field patterns for unmagnetized ferrite post, (c) $n = \pm 1$ field patterns for magnetized ferrite post, and (d) $n = 0$ field patterns for magnetized ferrite post. (Reproduced with permission from C. E. Fay and R. L. Comstock, Operation of the ferrite junction circulator, *IEEE Trans. Microwave Theory Tech.*, Vol. MTT-13, pp. 15–27, January 1965.)

eigenvalues one at a time on the unit circle until these coincide with those of an ideal circulator.

The four-port junction may be realized by having recourse to a linear combination of radial $TM_{\pm 1,1,0}$ and $TM_{0,1,0}$ modes. The field patterns in the demagnetized junction are depicted in Fig. 1.16a,b. The $TM_{0,1,0}$ mode is tuned to the frequency of the $TM_{\pm 1,1,0}$ modes with the help of a thin nonresonant capacitive post at the center of the junction instead of a resonant one. Figure 1.16c,d indicates the field patterns of the magnetized junction.

1.13 EDGE MODE CIRCULATOR

A junction that does not rely on a resonant effect in a gyromagnetic resonator is the edge mode arrangement. This effect is manifested by a suitably magnetized wide strip in a microstrip or stripline gyromagnetic line. The substrate is strongly magnetized perpendicular to the direction of propagation. The structure supports a TE-type solution in the transverse plane of the form

$$E_y = A\exp(-\alpha x)\exp(-j\beta z) \tag{1.9a}$$

$$H_x = \zeta_0 E_y \tag{1.9b}$$

$$H_z = 0 \tag{1.9c}$$

This solution indicates that the fields decay exponentially across the strip with no attenuation along the direction of propagation. The line also has the property that its field pattern is displaced to one side of the strip for one typical direction of propagation. It is displaced to the other side for the opposite direction of propagation. Once such a mode is established on one edge of the strip at one port, it may be wrapped at a second port by suitably bending the strip. Figure 1.17 illustrates one three-port structure. In principle, any number of ports may be connected in this way.

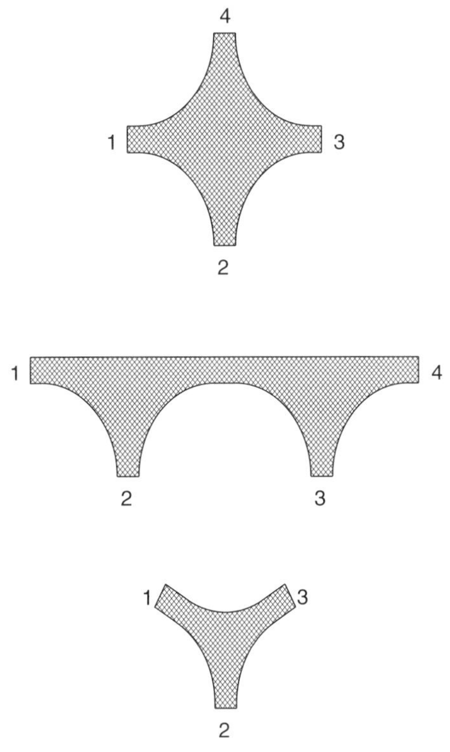

FIGURE 1.17 Topologies of edge mode circulators.

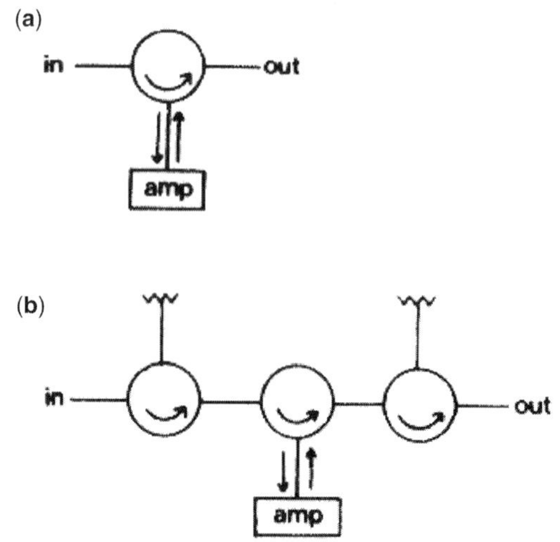

FIGURE 1.18 Single port amplifier using (a) three-port circulator and (b) five-port circulator.

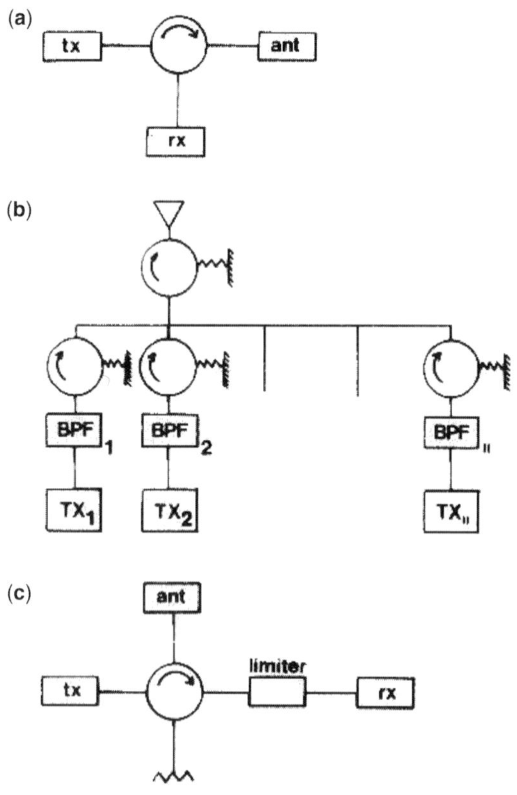

FIGURE 1.19 (a) Duplexing using single three-port circulator, (b) duplexing between closely spaced transmitters, and (c) high-power duplexer using four-port circulators.

1.14 SINGLE-PORT AMPLIFIERS USING JUNCTION CIRCULATORS

A simplified schematic of a reflection tunnel diode amplifier (TDA) that utilizes a circulator to separate the input signal from the amplified one is shown in Fig. 1.18a. When the gain of the amplifier is comparable to the isolation of the junction, it is often used in conjunction with five-port circulators. One such arrangement is depicted in Fig. 1.18b. The input and output junctions in this arrangement are connected as isolators in order to minimize gain variations due to source and load impedance variations. Here the magnetic field for the input circulator is sometimes supplied by an electromagnet. Reversal of the magnetic field in this circuit allows the TDA to be protected from radiofrequency leakage during the transmitting period by reversing the direction of circulation during this interval.

1.15 DUPLEXING USING JUNCTION CIRCULATORS

The ferrite circulator may also be used in duplexing systems for simultaneous transmission and reception of microwave energy with a single antenna. Here ferrite circulators replace conventional types of duplexing and are suited to both high- and low-power systems. Circulators are also employed in communication systems to eliminate mutual interference between closely separated transmitters. Figure 1.19b gives an example of a single antenna being shared by a number of

FIGURE 1.20 Microwave duplexer employing power splitters, isolators, and filters. (Courtesy of Dynatech Microwave Technology.)

transmitters with the help of circulators and bandpass filters (BPFs). Figure 1.19c illustrates a four-port high-port duplexer that utilizes a reflection limiter to protect the receiver during the transmission interval. Figure 1.20 illustrates a commercial duplexer using a power divider, ferrite isolators, and filters.

CHAPTER TWO

Tensor Permeability in a Magnetic Insulator

2.1 INTRODUCTION

The nature of the tensor permeability in a magnetic insulator is a classic problem in the literature. Its existence has its origin in the interaction between a spinning electron and an alternating radiofrequency magnetic field in a suitably magnetized magnetic insulator. An important feature of such a gyromagnetic insulator is that the entries of its permeability tensor exhibit an absorption line when the frequency of the alternating magnetic field intensity coincides with that of the electron spin. This sort of material may be biased below the absorption line, at the line, or above it. It is of particular interest in the construction of gyromagnetic resonators encountered in the design of nonreciprocal circulators and other devices. The diagonal and off-diagonal elements of this tensor involve, at a specific frequency, the direct magnetization of the magnetic insulator and the external applied direct magnetic field intensity. A scrutiny of the different connections between these various quantities indicates the possibility of imaging six possible relationships between the physical variables under consideration. Of some separate interest in the design of practical devices employing this sort of material is the temperature stability of the entries of the permeability tensor. This aspect is considered in some detail in Chapter 26. The occurrence of so-called low-field loss in a partially magnetized magnetic insulator is also given some consideration. It has its origin in the effective shape demagnetizing fields in individual magnetic dipoles. Throughout the text the shape demagnetizing factor has been specialized to that of an infinitely thin plate. Its effect can always be recovered in any practical situation by scaling the magnetization (M_0) by the shape demagnetizing factor (N_z) in the manner dealt with in Chapter 3.

The Stripline Circulator: Theory and Practice. By J. Helszajn

2.2 TENSOR PERMEABILITY

A characteristic property of a magnetized magnetic insulator, at microwave frequencies, is that while it has in general a scalar dielectric constant, it has a tensor permeability. It has its origin in a magnetic moment precessing about a direct magnetic field in the manner indicated in Fig. 2.1. Inside a ferrite medium Maxwell's equations must therefore be solved in conjunction with this tensor permeability.

$$[\mu_r] = [I] + [\chi] \tag{2.1}$$

The relative permeability tensor with the direct magnetic field intensity along the z-coordinate is given by the linearized equation of motion of the magnetization vector in a number of classic books. Its detailed derivation will not be reproduced here. The result is

$$[\mu_r] = \begin{bmatrix} \mu & -j\kappa & 0 \\ j\kappa & \mu & 0 \\ 0 & 0 & 1 \end{bmatrix} \tag{2.2}$$

The presence of imaginary off-diagonal components having opposite signs in this tensor is the basis for a number of important nonreciprocal effects not usually encountered in a medium with a scalar permeability.

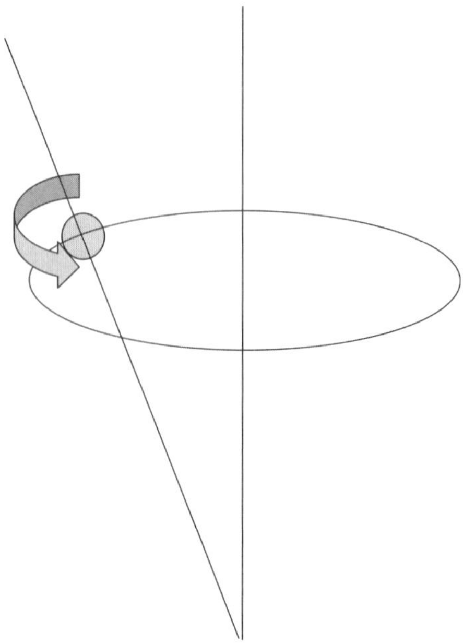

FIGURE 2.1 Spin motion in magnetic insulator.

$[\chi]$ is known as the susceptibility tensor. It is defined separately as

$$[\chi] = \begin{bmatrix} \chi_{xx} & \chi_{xy} & 0 \\ \chi_{yx} & \chi_{yy} & 0 \\ 0 & 0 & 0 \end{bmatrix} \tag{2.3}$$

where

$$\chi_{xx} = \chi_{yy} = \frac{\omega_m \omega_i}{-\omega^2 + \omega_i^2} \tag{2.4a}$$

$$-\chi_{yx} = \chi_{xy} = \frac{j\omega_m \omega}{-\omega^2 + \omega_i^2} \tag{2.4b}$$

ω is the radian frequency of the alternating radio magnetic field and

$$\omega_m = \gamma M_0, \quad \text{rad/s} \tag{2.5}$$

$$\omega_i = \gamma(H_0 - N_z M_0), \quad \text{rad/s} \tag{2.6}$$

M_0 is the saturation magnetization (A/m), γ is the gyromagnetic ratio (2.21×10^5 rad/s per A/m), ω is the radian frequency (rad/s), μ_0 is the free space permeability ($4\pi \times 10^{-7}$ H/m), and H_0 is the applied direct magnetic field intensity (A/m).

N_z is a z-directed shape demagnetizing factor of the geometry under consideration, which relates to the external and internal direct magnetic field intensity. It is dealt with in some detail in Chapter 3.

The entries of the permeability tensor are related separately to those of the susceptibility ones by

$$\mu = 1 + \chi_{xx} \tag{2.7a}$$

$$-j\kappa = \chi_{xy} \tag{2.7b}$$

The elements of the susceptibility and permeability tensors have a singularity at $\omega = \omega_i$. This singularity is determined by the product of the gyromagnetic ratio and the amplitude of the internal direct magnetic field. It denotes a resonance condition and is the principal feature used in the design of YIG filters and other resonance devices.

If the direct field is perpendicular to the direction of propagation of the alternating radio magnetic field, it is sometimes more convenient to take the former along the y-coordinate instead of the z-coordinate. The entries of the permeability tensor are then mapped into

$$[\mu_r] = \begin{bmatrix} \mu & 0 & -j\kappa \\ 0 & 1 & 0 \\ j\kappa & 0 & \mu \end{bmatrix} \tag{2.8a}$$

If the direct magnetic field is applied along the x-coordinate, then

$$[\mu_r] = \begin{bmatrix} 1 & 0 & 0 \\ 0 & \mu & -j\kappa \\ 0 & j\kappa & \mu \end{bmatrix} \tag{2.8b}$$

The permeability tensor with the direct magnetic field intensity at an arbitrary angle in the x-z plane is

$$[\mu_r] = \begin{bmatrix} \mu\cos^2\theta + \sin^2\theta & j\kappa\cos\theta & (1-\mu)\sin\theta\cos\theta \\ -j\kappa\cos\theta & \mu & j\kappa\sin\theta \\ (1-\mu)\sin\theta\cos\theta & -j\kappa\sin\theta & \mu\sin^2\theta + \cos^2\theta \end{bmatrix} \tag{2.9}$$

The diagonal and off-diagonal elements of the permeability tensor are often specified in terms of a normalized direct magnetization and internal direct magnetic field by

$$\mu = 1 - \frac{p\sigma}{1-\sigma^2} \tag{2.10a}$$

$$\kappa = \frac{-p}{1-\sigma^2} \tag{2.10b}$$

where

$$p = \frac{\gamma M_0}{\omega} \tag{2.11a}$$

$$\sigma = \frac{\gamma(H_0 - N_z M_0)}{\omega} \tag{2.11b}$$

It is sometimes desirable to be able to calculate p and σ in terms of μ and κ by reorganizing Eqs. (2.10a) and (2.10b). The required results are

$$\sigma = \frac{(\mu - 1)}{\kappa} \tag{2.12a}$$

$$p = \frac{\kappa(\mu-1)^2}{\kappa^2 - 1} \tag{2.12b}$$

If the independent variables are taken as p and μ, then rewriting Eq. (2.10a) gives σ as

$$\sigma = \frac{p + \sqrt{p^2 + 4(\mu-1)^2}}{2(\mu-1)} \tag{2.13}$$

and κ is fixed by Eq. (2.10b).

TABLE 2.1 Exact Equations

p, σ	$\mu = 1 - p\sigma/(1-\sigma^2),\ \kappa = -p/(1-\sigma^2)$
κ, μ	$p = (\mu - 1)^2/\kappa - \kappa,\ \sigma = (\mu - 1)/\kappa$
p, κ	$\mu = 1 + \sqrt{\kappa(p+\kappa)},\ \sigma = \sqrt{p/\kappa + 1}$
p, μ	$\kappa - \frac{1}{2}[-p + \sqrt{p^2 + 4(\mu-1)^2}],\ \sigma = \left[\frac{p + \sqrt{p^2 + 4(\mu-1)^2}}{-p + \sqrt{p^2 + 4(\mu-1)^2}}\right]^{1/2}$
σ, κ	$p = \kappa(\sigma^2 - 1),\ \mu = 1 + \sigma\kappa$
σ, μ	$p = (\mu - 1)(\sigma - 1/\sigma),\ \kappa = (\mu - 1)/\sigma$

If p and κ are fixed, then σ is given by combining Eqs. (2.12a) and (2.12b). The result is

$$\sigma = \sqrt{1 + p/\kappa} \tag{2.14}$$

There are six possible ways to relate the variables under consideration. Table 2.1 summarizes the situation.

2.3 DAMPING

To stabilize the motion of the magnetization vector at the resonance, a damping term must be introduced into the equation of motion. A form of phenomenological damping that is often used for engineering purposes is due to Gilbert. It amounts to adding an imaginary frequency term ($j\omega\alpha$) to the resonant frequency (ω_i). The damping term can therefore always be introduced by replacing ω_i by $\omega_i + j\omega\alpha$ in the loss-free components.

The entries of the susceptibility tensor based on the small signal solution of the equation of motion are then given by

$$\chi_{xx} = \chi_{yy} = \frac{-\omega_m(-\omega_i + j\omega\alpha)}{-\omega^2 + (\omega_i + j\omega\alpha)^2} \tag{2.15}$$

$$-\chi_{yx} = \chi_{xy} = \frac{-j\omega_m\omega}{-\omega^2 + (\omega_i + j\omega\alpha)^2} \tag{2.16}$$

The real and imaginary parts of the entries of the susceptibility tensor are defined by

$$\chi_{xx} = \chi'_{xx} + j\chi''_{xx} \tag{2.17}$$

$$\chi_{xy} = j(\chi'_{xy} + j\chi''_{xy}) \tag{2.18}$$

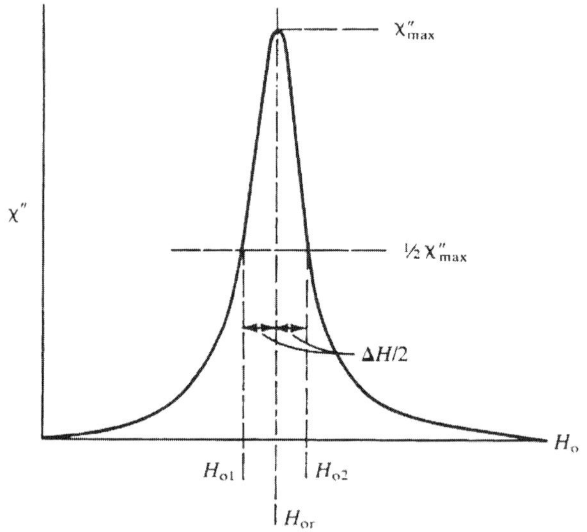

FIGURE 2.2 Linewidth of ferrites.

The phenomenological damping factor α is defined as the difference between the values of the magnetic field intensity (ΔH) at a constant frequency where the imaginary part, χ''_{xx}, of the diagonal component χ_{xx} of the susceptibility tensor attains a value that is half of its value at resonance (Fig. 2.2). It is related to the

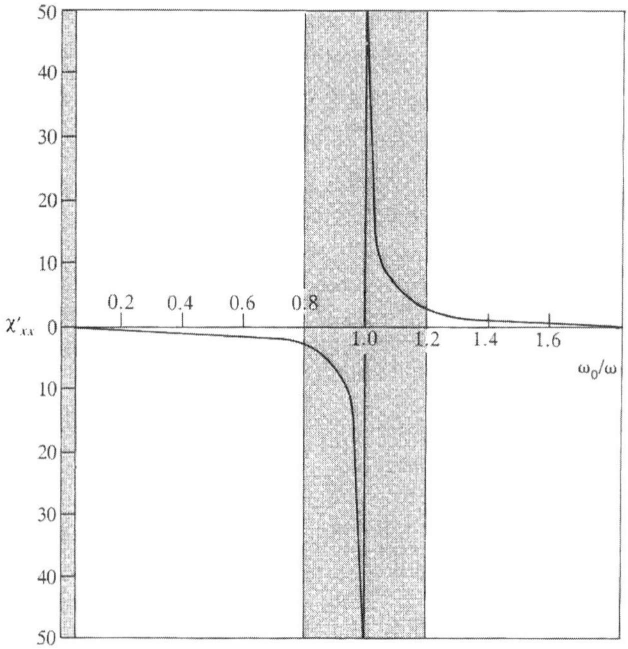

FIGURE 2.3 Real part of χ_{xx} component, for $\omega_m/\omega = 1$ and $\alpha = 0.01$.

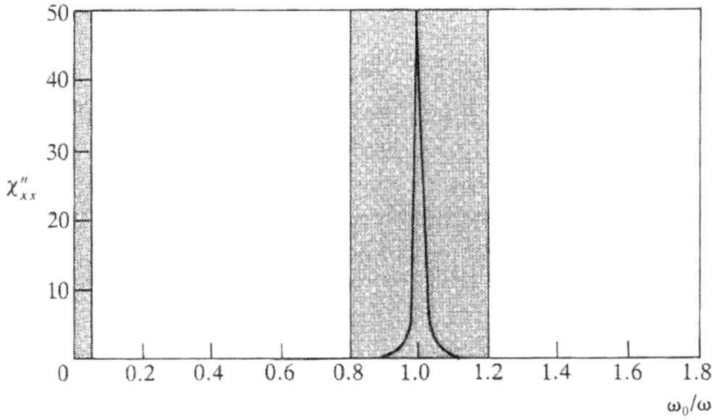

FIGURE 2.4 Imaginary part of χ_{xx} component, for $\omega_m/\omega = 1$ and $\alpha = 0.01$.

linewidth (ΔH) by

$$\omega\alpha = \gamma \Delta H/2 \tag{2.19}$$

The linewidth quoted by the material suppliers therefore provides a useful description of magnetic losses in ferrites or garnets at resonance. However, it does not adequately describe dissipation outside the skirts of the main resonance.

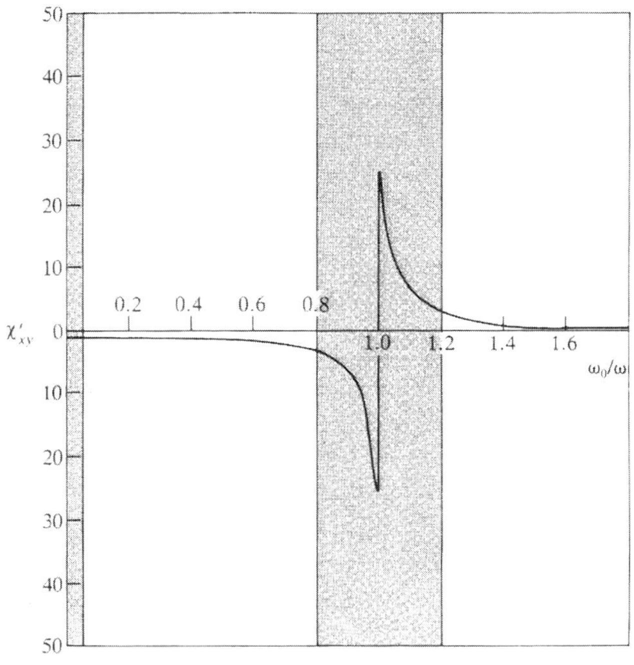

FIGURE 2.5 Real part of χ_{xy} component, for $\omega_m/\omega = 1$ and $\alpha = 0.01$.

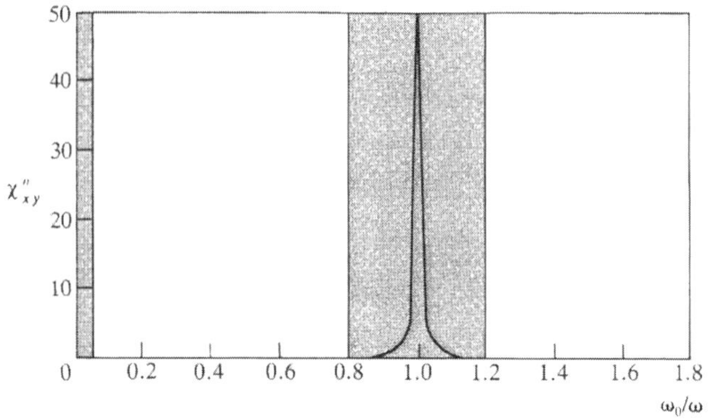

FIGURE 2.6 Imaginary part of χ_{xy} component, for $\omega_m/\omega = 1$ and $\alpha = 0.01$.

The relationships between the real and imaginary parts of the susceptibility tensor and ω_m/ω, ω_i/ω, and a are plotted in Figs. 2.3–2.6. The variation of κ with the magnetic parameters is the same as that of χ_{xy} in Figs. 2.3 and 2.4. The variation of μ with magnetic parameters may be determined from Figs. 2.5 and 2.6 by noting that $\chi_{xx} = \mu - 1$.

2.4 SCALAR PERMEABILITIES

A unique property of a gyromagnetic medium is that, whereas it is in general characterized by a tensor permeability, it displays a scalar permeability under one of three possible situations. It has one value if the alternating magnetic field intensity rotates in the same sense as the electron spin, another value if it rotates in the opposite sense, and a value of unity if it is aligned with the axis of the electron spin. One important application of this phenomenon is the nonreciprocal phase shifter. If the frequency of the radiofrequency (rf) magnetic field coincides with the natural precession frequency, the amplitude of the precession becomes particularly large in the situation for which the electron spin and the alternating magnetic field rotate in the same direction and the energy absorbed from the alternating magnetic field displays a maximum.

The eigenvalue equation to be solved is

$$\mu \mathbf{H} = [\mu]\mathbf{H} \tag{2.20}$$

μ is an eigenvalue, $\mathbf{H}$ is an eigenvector, and $[\mu]$ is the permeability tensor. The magnetic fields are proportional to the eigenvectors. The three scalar quantities correspond to the eigenvalues of the permeability tensor. If the direct magnetic field is orientated along the positive z-axis, then

$$\begin{vmatrix} \mu - \mu_i & j\kappa & 0 \\ -j\kappa & \mu - \mu_i & 0 \\ 0 & 0 & 1 - \mu_i \end{vmatrix} = 0 \tag{2.21}$$

The roots of this characteristic equation are

$$\mu_1 = \mu_+ = \mu - \kappa \tag{2.22}$$

$$\mu_2 = \mu_- = \mu + \kappa \tag{2.23}$$

$$\mu_3 = \mu_z = 1 \tag{2.24}$$

The counterrotating solutions are illustrated in Fig. 2.7. The natures of the eigenvalues and eigenvectors connected with the other two orientations of the direct magnetic field are deduced in a similar manner.

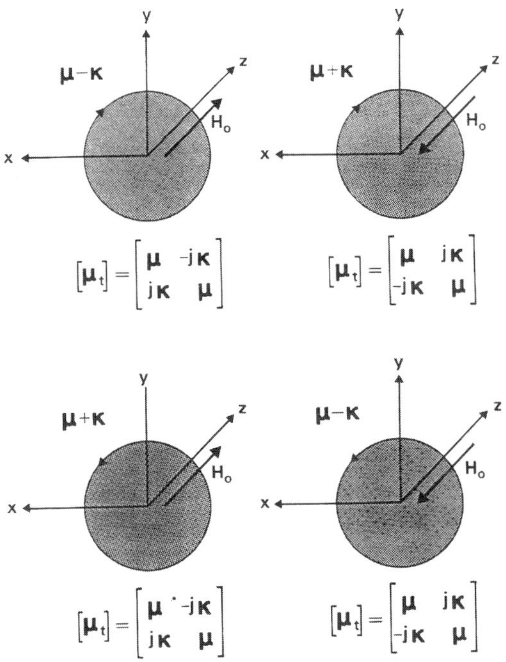

FIGURE 2.7 Eigensolutions of uniform mode in a magnetic insulator.

2.5 EFFECTIVE PERMEABILITY AND GYROTROPY

Two quantities that appear in the description of a microwave device based on the gyromagnetic effect are the gyrotropy (κ/μ) and the effective permeability (μ_{eff}). These are defined in terms of the direct normalized magnetic variables by

$$\frac{\kappa}{\mu} = \frac{p}{\sigma(p + \sigma) - 1} \tag{2.25}$$

$$\mu_{\text{eff}} = \frac{\mu^2 - \kappa^2}{\mu} = \frac{(p + \sigma^2)^2 - 1}{\sigma^2 + p\sigma - 1} \tag{2.26}$$

μ and κ are the diagonal and the off-diagonal elements of the permeability tensor of the magnetized magnetic insulator.

Figures 2.8 and 2.9 depict the gyrotropy and effective permeability in terms of p and σ. The condition for resonance of either quantity is

$$\sigma^2 + \sigma p - 1 = 0 \tag{2.27}$$

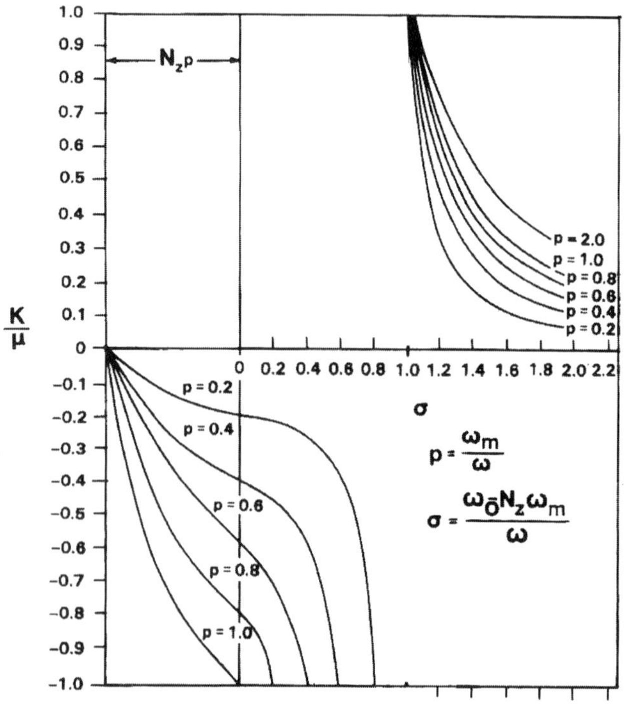

FIGURE 2.8 Gyrotropy versus internal direct magnetic field for parametric values of normalized magnetization.

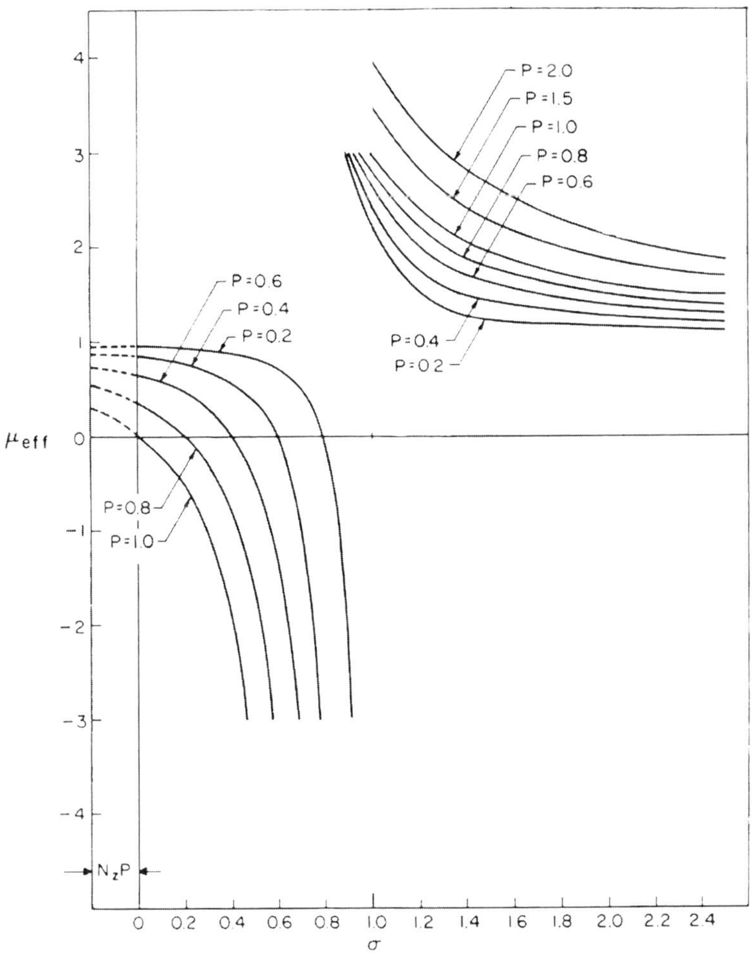

FIGURE 2.9 Effective permeability versus internal direct magnetic field for parametric values of normalized magnetization.

This condition differs from that met in connection with the entries of the permeability tensor μ and κ, which is given by

$$1 - \sigma^2 = 0 \tag{2.28}$$

If p is fixed by the material choice and κ/μ is fixed by the specification, then σ is set by

$$\sigma = \frac{-p + \sqrt{p^2 + 4[1 + p(\mu/\kappa)]}}{2} \tag{2.29}$$

and μ_{eff} is given in terms of p and σ by Eq. (2.26).

TABLE 2.2 Exact Equations

p, σ	$\dfrac{\kappa}{\mu} = \dfrac{\pm p}{\sigma(p+\sigma)-1}, \quad \mu_{\text{eff}} = \dfrac{(p+\sigma)^2-1}{\sigma(p+\sigma)-1}$
σ, μ_{eff}	$p = \dfrac{-\sigma(2-\mu_{\text{eff}}) \pm \sqrt{\sigma^2\mu_{\text{eff}}^2 + 4(1-\mu_{\text{eff}})}}{2}, \quad \dfrac{\kappa}{\mu} = \dfrac{\pm p}{\sigma(p+\sigma)-1}$
p, μ_{eff}	$\sigma = \dfrac{-p(2-\mu_{\text{eff}}) \pm \sqrt{p^2\mu_{\text{eff}}^2 + 4(\mu_{\text{eff}}-1)^2}}{2(1-\mu_{\text{eff}})}, \quad \dfrac{\kappa}{\mu} = \dfrac{\pm p}{\sigma(p+\sigma)-1}$
$\sigma, \kappa/\mu$	$p = \dfrac{\sigma^2-1}{[\pm(\mu/\kappa)-\sigma]}, \quad \mu_{\text{eff}} = \dfrac{(p+\sigma)^2-1}{\sigma(p+\sigma)-1}$
$p, \kappa/\mu$	$\sigma = \dfrac{-p \pm \sqrt{p^2+4[1+p(\mu/\kappa)]}}{2}, \quad \mu_{\text{eff}} = \dfrac{(p+\sigma)^2-1}{\sigma(p+\sigma)-1}$
$\kappa/\mu, \mu_{\text{eff}}$	$p = \left(1 - \dfrac{\mu_{\text{eff}}}{1-(\kappa/\mu)^2}\right)\left(\dfrac{1-\sigma^2}{\sigma}\right), \quad \dfrac{p(\mu/\kappa)}{-1+\sigma^2} = \dfrac{\mu_{\text{eff}}}{1-(\kappa/\mu)^2}$

There are again six different ways in which these quantities may be organized. These are indicated in Table 2.2.

The description of the elements of the permeability tensor given previously are only valid provided the material is saturated. This condition is met provided the internal direct magnetic field intensity H_i is equal to or greater than zero. At saturation

$$H_0 - N_z M_0 = 0 \tag{2.30}$$

N_z is a shape demagnetizing factor along the direction of the direct magnetic field intensity. Its nature is dealt with in some detail in Chapter 3.

2.6 KITTEL LINE

The relationship between the external and internal direct magnetic field intensity in an ellipsoidal magnetic insulator has been dealt with by introducing the concept of a shape demagnetizing factor along each coordinate of the magnetic insulator. The same technique has also been introduced in connection with the internal and external alternating magnetic fields.

$$h_x^i = h_x^e - N_x m_x \tag{2.31a}$$

$$h_y^i = h_y^e - N_y m_y \tag{2.31b}$$

$$h_z^i = h_z^e - N_z m_y \tag{2.31c}$$

The resonance condition of the electron spin is now obtained by retaining these additional terms in the expansion of the equation of motion of the magnetization vector. The result is

$$\omega_r^2 = \omega_x \omega_y \tag{2.32a}$$

$$\omega_x = \gamma[H_0 - (N_z - N_x)M_0] \tag{2.32b}$$

$$\omega_y = \gamma[H_0 - (N_z - N_y)M_0] \tag{2.32c}$$

The Kittel line produced by this arrangement now involves the shape factors N_x, N_y, and N_z, which are discussed in more detail in Chapter 3. These quantities are related, in a saturated material, by

$$N_x + N_y + N_z = 1 \tag{2.33}$$

The trajectory of the electron spin is now elliptically instead of circularly polarized, with an ellipticity specified by

$$e_r^2 = \omega_x / \omega_y \tag{2.34}$$

In the case of a disk or cylinder, the only geometry considered in the text,

$$N_x = N_y = N_t \tag{2.35a}$$

$$N_t \approx \frac{1}{2}(1 - N_z) \tag{2.35b}$$

The values of the demagnetizing factors introduced here may be experimentally deduced by making measurements of the Kittel line with different orientations of a particular geometry.

2.7 LOW-FIELD LOSSES IN UNSATURATED MAGNETIC INSULATOR

The general form of Kittel's resonance equation embodies a contribution to the natural frequency arising from shape demagnetizing fields within a single crystal domain even in the absence of a direct field. The origin of this absorption line may be appreciated by forming Kittel's resonance equation with $H_0 = 0$.

$$\omega_r = [(N_x - N_z)(N_y - N_z)]^{1/2} \omega_m \tag{2.36}$$

This relationship determines the lowest frequency at which the resonance condition may be satisfied in a saturated material. Any effective anisotropy fields must also, strictly speaking, be introduced into this condition.

In an unsaturated material, however, resonance absorption losses occur at frequencies up to at least twice the value predicted by Kittel's equation. The effect of so-called low-field loss on the shape of the resonance curve at microwave frequencies is illustrated in Fig. 2.10. The discrepancy between the resonant frequencies in a saturated and unsaturated material comes about because it is not appropriate to apply Kittel's formula to a solid unmagnetized ferrite containing randomly orientated

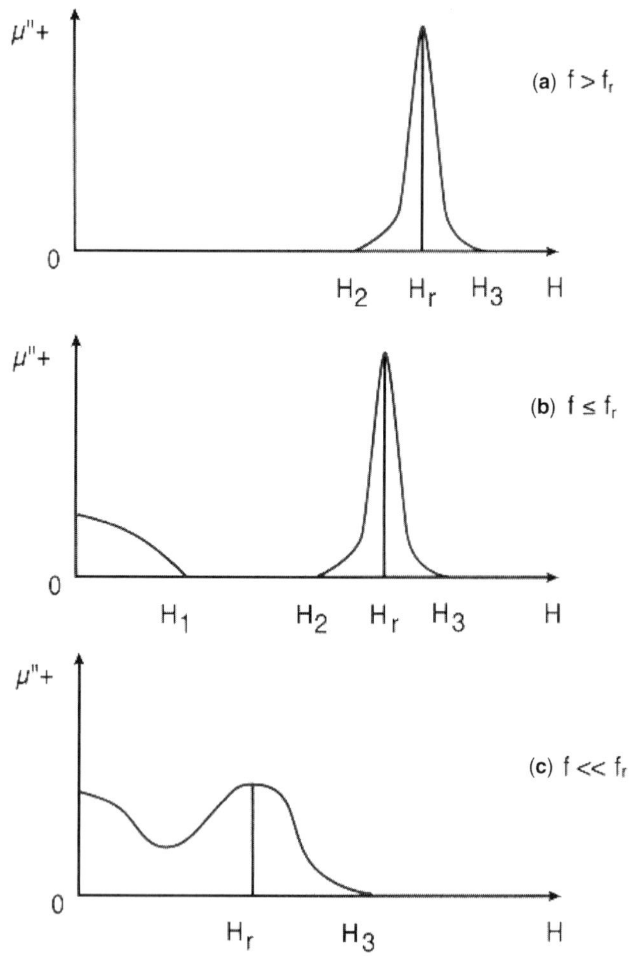

FIGURE 2.10 Effect of low-field losses on the shape of the resonance curve at microwave frequencies. (Adapted from B. Lax, Frequency and loss characteristics of microwave ferrite devices, *Proc. IRE*, Vol. 44, pp. 1368–1386, 1956.)

domains, although it may be applied to a single uniformly magnetized Weiss domain within the specimen. This essential restriction arises because the resonance condition was derived from the equation of motion of the macroscopic magnetization vector, where it is assumed that all spin dipoles are aligned and can be represented by a single magnetization vector. If this is not the case, additional demagnetizing factors must be introduced into Kittel's equation to account for the random orientation of the Weiss domains.

The contribution of the Weiss domains to the total demagnetizing field may be determined by considering a nonsaturated medium, which is still divided in such domains. A detailed investigation of this problem indicates that domain rotation resonance or so-called low-field loss can occur in the frequency interval between:

$$0 \leq \omega \leq \omega_{\mathrm{m}} \tag{2.37}$$

This type of loss may be avoided by either ensuring that the material is saturated or by reducing the magnetization of the material.

2.8 MAGNETIC BIAS POINTS ABOVE AND BELOW THE KITTEL LINE

Microwave ferrite devices can operate on either side of the Kittel line. Two separate demands, however, are placed on the direct magnetic field intensity. One is the so-called low-field loss and the other is the location of the Kittel absorption line. These constraints are sometimes difficult to reconcile at ultra-high frequencies (UHFs) below the Kittel line. The problem under consideration may be appreciated by restricting the discussion to the Kittel resonance line,

$$(\omega_{\mathrm{r}}/\gamma) = H_0 - N_z M_0 + N_{\mathrm{t}} M_0 \tag{2.38}$$

In order to avoid low-field loss, it is necessary for the material to be saturated. This condition is given by

$$H_l \geq H_0 - N_z M_0 \tag{2.39}$$

The direct magnetic field, however, cannot be increased at will without having the Kittel line intrude on the frequency response of the microwave circuit. It is also only possible to equally satisfy this condition in an ellipsoidal geometry such as a sphere or a cigar shaped one. For a nonellipsoidal geometry such as a disk, the internal field is nonuniform even in the presence of a uniform external one. This situation is sometimes catered for by introducing a radial dependent demagnetizing factor $N_z(r_i)$.

Taking the equal sign for the purpose of calculation gives

$$(\omega_r/\gamma) \geq N_t M_0 \tag{2.40}$$

For a sphere, $N_t = \frac{1}{3}$ and

$$(\omega_r/\gamma) \geq M_0/3 \tag{2.41}$$

For a disk,

$$(\omega_r/\gamma) \geq 0.025 M_0 \tag{2.42}$$

At 1 GHz, $\mu_0 M_0 = 0.1071$ T for the former case and 1.430 T for the latter case. The value of the magnetization produced in this way is an upper bound below the Kittel line and a lower one above it; a more complete description must cater for the effect of the finite linewidth of the Kittel line:

$$H_0 - n\,\Delta H/2 \quad \text{below resonance} \tag{2.43}$$

$$H_0 + n\,\Delta H/2 \quad \text{above resonance} \tag{2.44}$$

n is the integer that determines the number of half-linewidths above which the absorption of the Kittel line may be neglected.

2.9 SPINWAVE MANIFOLD

Under some circumstances, a magnetic insulator supports so-called spinwaves. The frequencies of these spinwaves are primarily determined by an effective exchange field. This field has its origin in the exchange energy, which is defined as that tending to orient parallel adjacent spins that are misaligned due to thermal agitation. A demagnetizing dipolar field is also introduced in connection with the exchange field in order to cater for demagnetizing effects. The nature of a typical spinwave is obtained by expanding the equation of motion of the magnetization vector under the influence of a direct magnetic field and the exchange and dipolar fields.

The frequency of a typical spinwave is

$$\omega_k^2 = (\omega_k^{12} \omega_k^{21}) \tag{2.45}$$

where

$$\omega_k^{12} = \gamma(H_0 - N_z M_0 + H_{ex}\, a^2 k^2) \tag{2.46}$$

$$\omega_k^{21} = \gamma(H_0 - N_z M_0 + H_{ex}\, a^2 k^2 + M_0 \sin^2\theta_k) \tag{2.47}$$

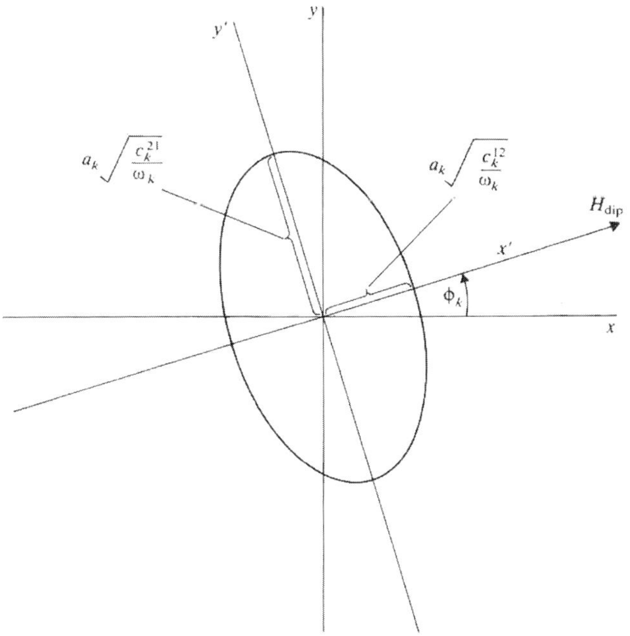

FIGURE 2.11 Primed coordinate system in transverse plane showing spinwave ellipse. (Adapted from J. Kemanis and S. Wang, Analysis of high power effects in ferrimagnetics from the point of view of energy transfer. Part 1. First order instabilities, *J. Appl. Phys.*, Vol. 35, pp. 1465–1470, 1964.)

The polarization of a typical spinwave is illustrated in Fig. 2.11. The spinwave is circularly polarised for a z-directed spinwave. The ellipticity is

$$e_k^2 = (\omega_k^{12}/\omega_k^{21}) \tag{2.48}$$

a is the distance between neighboring spins (m), k is the wavenumber of a typical spinwave (rad/m), H_{ex} is an equivalent exchange field (A/m) associated with the exchange energy, θ_k is the angle (rad) that a typical spinwave makes with the direct magnetic field, H_0 is the direct magnetic field intensity (A/m), M_0 is the direct magnetization (A/m), and N_z is a shape demagnetizing factor.

In order to proceed with a calculation of the spinwave manifold, it is necessary to have a complete understanding of the parameters entering into its description. The direct magnetic variables met in connection with the definition of the uniform line in a magnetic insulator are widely understood and need not be reiterated. The nature of the variables due to the exchange energy, however, do merit some attention. This is due to the fact that two different definitions of the exchange frequency have been introduced in the literature. In the Lax and Button formulation, used here, the exchange term in the description of the spinwave frequencies is written as the product of an exchange frequency ω_{ex} and the product a^2k^2:

$$\omega_{ex}a^2k^2 \tag{2.49}$$

ω_{ex} in the exchange frequency,

$$\omega_{ex} = \gamma H_{ex} \tag{2.50}$$

H_{ex} is the effective exchange field,

$$H_{ex} = \lambda M_0 \tag{2.51}$$

and λ is the molecular field coefficient.

For YIG material,

$$H_{ex} = 10^7\text{Oe} = (10^3/4\pi) \times 10^7\text{A/m} \tag{2.52}$$

and

$$\lambda = 5 \times 10^3 \tag{2.53}$$

The formulations employed by Gurewich and separately by Smit and Wijn differ in that the product a^2k^2 in the description of the spinwave frequencies is absorbed in both ω_{ex} and H_{ex}. These authors define the exchange frequency in the description of the spinwave frequencies as

$$\omega_{ex} = \gamma(\lambda M_0)a^2k^2 \tag{2.54}$$

The corresponding exchange field is

$$H_{ex} = (\lambda M_0)a^2k^2 \tag{2.55}$$

In practice, the two descriptions produce the same result.

A sketch of the spinwave manifold is indicated in Fig. 2.12. This is a plot of spinwave frequency as a function of the wavenumber. Kittel's equation for the uniform mode coincides with the middle formula on the axis of this diagram. The uniform mode can therefore be degenerate with a large number of spinwave modes, provided the condition $\omega_k = \omega_r$ is satisfied. In the cross-hatched area of this illustration, the normal modes are the so-called magnetostatic modes.

The existence of spinwaves has been demonstrated in a thin cobalt metallic film 29,440 Å (10^{-10} m) thick. Figure 2.13 indicates some lower order standing spinwaves in this arrangement. The experiment also enables the determination of the exchange field of the material.

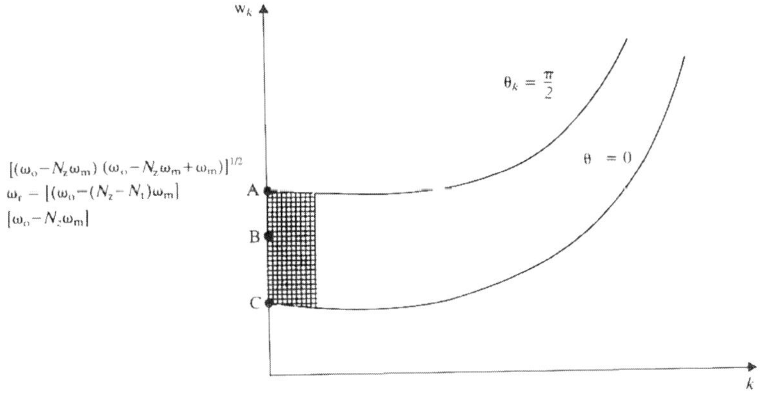

FIGURE 2.12 Spinwave manifold for isotropic ferrite ellipsoid. (Adapted from H. Suhl, The nonlinear behaviour of ferrites at high microwave signal levels, *Proc. IRE*, Vol. 41, pp. 1270–1284, 1953.)

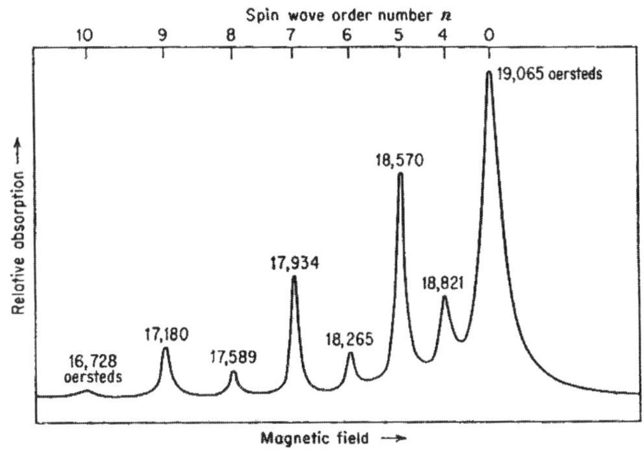

FIGURE 2.13 Standing spinwaves in cobalt metallic film 29,440 Å (10^{-10} m) thick. (Reprinted with permission from P. E. Tannenwald and R. Weber, *Phys. Rev.*, Vol. 121, p. 715, 1961.)

2.10 MAGNETIZATION VALUES OF VARIOUS FERRITES

The magnetization of ferrite materials has so far been discussed in terms of a normalized quantity p. In practice, the saturation magnetization ($\mu_0 M_0$) of the ferrites and garnets lies in the range 0.0150–0.5000 T for most commercial materials. For the garnet materials it is bracketed between 0.0150 and 0.1800 T. For magnesium manganese ferrites the magnetization is between 0.0800 and 0.3800 T; for nickel ferrites it varies between 0.1000 and 0.5000 T; and for lithium ferrites it lies between 0.0150 and 0.5000 T. Figure 2.14 is a table of the properties of yttrium gadolinium aluminum

CALCIUM VANADIUM IRON GARNETS

Type No.	$4\pi M_s$(G)	M_s(kA/m)	T_c(°C)	$\Delta H(O_E)$ (Maximum)	ΔH(kA/m) (Maximum)	ε'	$10^4 \tan \delta_\varepsilon$ (Maximum)
C7	1850	147.3	215	15	1.2	14.8	1
C6	1600	127.4	220	12	1.0	14.2	1
C5	1400	111.4	210	10	0.8	14.0	1
C4	1200	95.5	205	10	0.8	13.9	1
C3	1000	79.6	200	10	0.8	13.7	1
C2	800	63.7	190	10	0.8	13.5	1

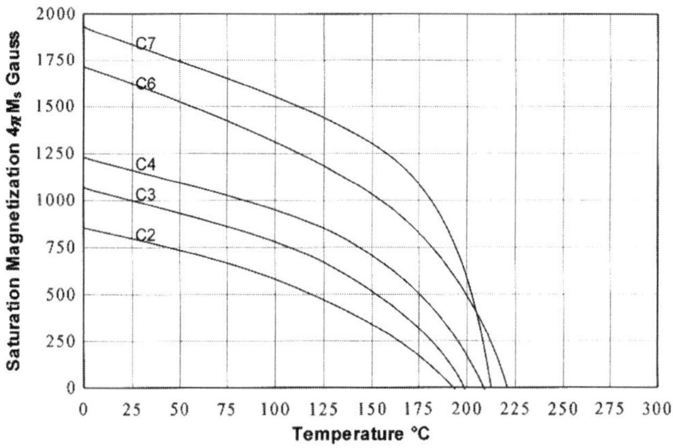

FIGURE 2.14 Properties of doped calcium vanadium iron garnets. (Courtesy of Hilteck Microwave Ltd.)

iron garnet with aluminum substitution for iron. Figure 2.15 depicts a similar table in the case of doped nickel chromium zinc iron spinels.

2.11 THE ORIGIN OF THE UNIFORM LINEWIDTH IN MAGNETIC INSULATORS

The origins of dissipation in a polycrystalline magnetic insulator involve a spin–lattice process and a spin–spin one. It is convenient, in order to separate the various processes, to replace the phenomenological linewidth met in connection with the Landau–Lifshitz (LL) damping term by a relaxation parameter (T). This quantity has the dimensions of time and is analogous to the time constant in a *RL* or *RC* circuit.

$$\frac{1}{\omega T} = \left(\frac{\gamma \Delta H}{2\omega}\right) \tag{2.56}$$

or

$$\frac{1}{T} = \left(\frac{\gamma \Delta H}{2}\right) \tag{2.57}$$

NICKEL CHROMIUM ZINC IRON SPINELS

Type No.	$4\pi M_s$(G)	M_s(kA/m)	T_c(°C)	$\Delta H(O_E)$ (Maximum)	ΔH(kA/m) (Maximum)	ε'	$10^4 \tan \delta_\varepsilon$ (Maximum)
N11	5000	398.0	375	180	14.3	13.5	20
N10	4600	366.2	450	200	15.9	13.3	20
N9	3850	306.5	500	210	16.7	13.1	20
N7	3000	238.8	590	250	19.9	13.0	20
N2	2350	187.1	525	300	23.9	12.4	20
N8	2000	159.2	500	300	23.9	12.0	20
N3	1900	151.2	500	300	23.9	12.0	20
N4	1775	141.3	473	300	29.3	11.6	20
N1	1500	119.4	490	680	54.2	11.5	20
N5	1450	115.4	450	350	27.8	11.2	20

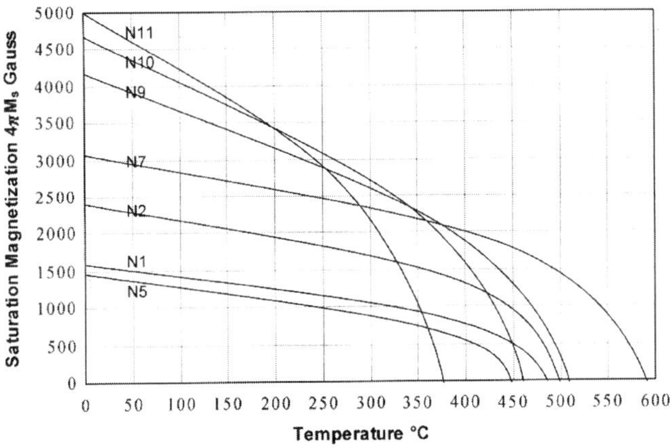

FIGURE 2.15 Properties of doped nickel chromium zinc iron spinels. (Courtesy of Hilteck Microwave Ltd.)

The spin–spin process involves the scattering of the uniform mode energy to spinwaves of the same frequency. The spin–lattice dissipation involves the direct scattering of the uniform mode energy to the lattice vibrations, where it is dissipated as heat. There is also a spin–lattice process from the spinwave reservoir to the lattice. Figure 2.16 is a schematic diagram of these processes. The relaxation directly to the lattice is often negligible compared with the transfer of energy to the lattice via the spinwaves. The total relaxation time is

$$\frac{1}{T} = \frac{1}{T1} + \frac{1}{T2} + \cdots \tag{2.58}$$

The aforementioned processes have been investigated extensively in the literature. The spin–spin process was experimentally clarified by demonstrating that the uniform line width of the polycrystalline material exhibited a peak when the

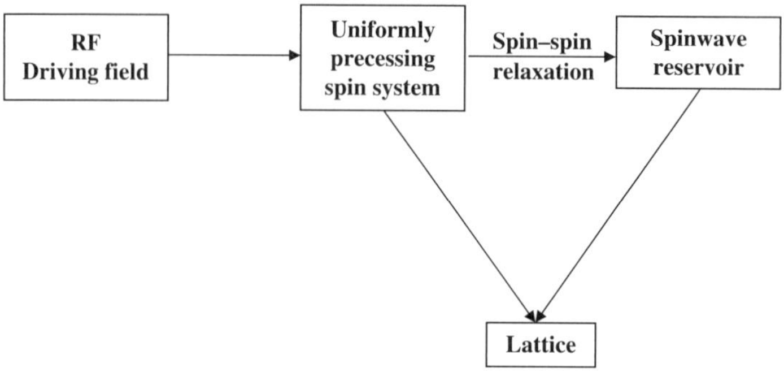

FIGURE 2.16 Relationships between relative density and uniform linewidth in various ferrite and garnet materials. (Reprinted with permission from A. G. Gureuch, *Ferrites at Microwave Frequencies*, Heywood and Co. Ltd., London, 1960.)

frequency of the pump is degenerate with the largest number of spinwaves. This condition is given experimentally and theoretically by

$$\omega = \tfrac{2}{3}\,\omega_{\mathrm{m}} \tag{2.59}$$

The pores of the polycrystalline material and the surface finish of its body provide two means whereby the energy from the uniform mode may be scattered to the spinwave reservoir and the lattice structure. Other contributions arise at large signal level in the presence of spinwave instabilities and must also be catered for.

In order to understand the connection between the uniform linewidth and the porosity, it is necessary to allow for internal demagnetizing fields at crystalline (grain) boundaries due to pores (cavities) between crystallites. These demagnetizing fields provide a means of coupling energy from the pump to spinwaves of the same frequency.

The contribution of the pores to the overall linewidth of the material is accurately given in terms of the porosity (p) by

$$(2\Delta H)_{\mathrm{pore}} = uM_0p \tag{2.60}$$

where p is defined by

$$p = v/V \tag{2.61}$$

u is a constant that is approximately equal to 1, v is the pore volume, and V is the total volume of the sample.

The porosity is related to the density (d) by

$$d = 1 - p \tag{2.62}$$

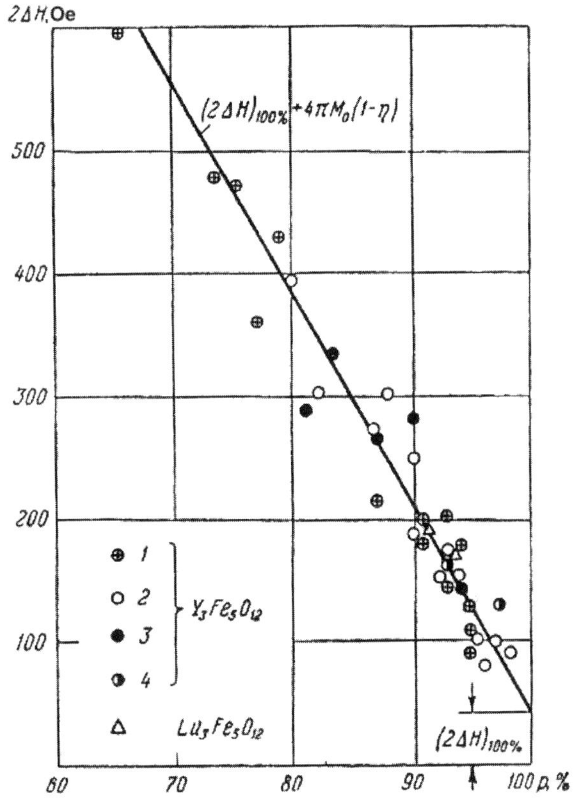

FIGURE 2.17 Relaxations in polycrystalline magnetic insulators. (Reproduced with permission from R. C. Fletcher, R. C. Lecraw, and E. G. Spencer, *Phys. Rev.*, Vol. 117, p. 955, 1960.)

The dependence of the resonance linewidth on the density of polycrystalline ferrites is indicated in Fig. 2.17. It shows good agreement between the variables under consideration.

The anisotropy energy in a single crystal is that which tends to align the magnetic domains along certain crystalline directions. One effect of the anisotropy in a single crystal is to shift the resonant frequency by an amount that depends on the angle between the direct field and the crystal axis. In a polycrystalline sample the angle varies through all values. The relaxation time involved in this process also manifests itself as a broadening of the uniform linewidth.

The LL relaxation model employed here assumes identical relaxation times along the direction of the direct magnetic field and in the plane transverse to it. The more general case is described by Bloch–Bloembergen (BB) but is outside the scope of this work.

CHAPTER THREE

Spatial Shape Demagnetizing Factors of Disk, Equilateral Triangle, and Irregular Hexagonal Magnetic Insulators

3.1 INTRODUCTION

The magnetic circuit of nonreciprocal ferrite devices is an important part in any design. The problem is fairly straightforward provided the direct magnetic field intensity in which the magnetic insulator or ferrite geometry is immersed may be assumed to be uniform and provided the geometry is an ellipsoid. The relationship between the external and internal direct magnetic field intensities is then given in terms of shape demagnetizing factors and the internal direct magnetic field intensity is also uniform. The simplest example of an ellipsoidal geometry is the sphere. The practical case for which the external direct magnetic field intensity is nonuniform and the geometry is not an ellipsoid is, of course, more complicated. An example of a nonellipsoidal geometry is a flat disk. The internal direct magnetic field intensity in this arrangement is nonuniform even when placed in a uniform field. This situation, in the case of a disk geometry, is sometimes dealt with by introducing a spatial dependent shape demagnetizing factor. The purpose of this chapter is to deal with the spatial demagnetizing factors of single and pairs of disk geometries in infinite space and to provide some data in finite space. One means of dealing with this problem is to evaluate the demagnetizing field necessary to saturate the magnetic insulator. This may be done by having recourse to one of a number of commercial finite element packages. The radial field distribution in a disk magnetic insulator can be made uniform by either

The Stripline Circulator: Theory and Practice. By J. Helszajn

shaping the external magnetic structure or by incorporating it in a semispherical pole piece geometry. Some data on the z-directed demagnetizing factor along the two principal axes of triangular and irregular hexagonal geometries are included for completeness. The chapter includes semiempirical descriptions of the entries of the permeability tensor in a partially magnetized magnetic insulator. It also includes some remarks about the magnetization in a partially magnetized magnetic insulator.

3.2 SHAPE DEMAGNETIZING FACTORS

The relationships between the magnetic field intensities inside and outside an ellipsoidal magnetic insulator, in a uniform field, may be dealt with exactly by introducing so-called shape demagnetizing factors. This is a classic approach and is reproduced here solely for the purpose of completeness. The internal field inside the geometry with the field along the z-coordinate is then defined in terms of that outside it by

$$H_{\mathrm{i}} = H_0 - N_z M \tag{3.1}$$

H_{i} is the magnetic field intensity inside the magnetic insulator (A/m), H_0 is that outside (A/m), M is the state of the magnetization (A/m), and N_z is a shape demagnetizing factor along the z-axis.

The three shape demagnetizing factors in an ellipsoidal geometry are connected by

$$N_x + N_y + N_z = 1 \tag{3.2}$$

There are several shapes that are of particular interest in the discussion of microwave ferrite devices. For a sphere,

$$N_x = N_y = N_z = \frac{1}{3} \tag{3.3}$$

The shape demagnetizing factors of a long cylinder of radius R and length L are approximately described by

$$N_z \approx 1 - [1 + (2R/L)^2]^{-1/2} \tag{3.4a}$$

$$N_x \approx N_y = \frac{1}{2}(1 - N_z) \tag{3.4b}$$

and for a thin disk,

$$N_z \approx 1 - (L/2R)[1 + (L/2R)^2]^{-1/2} \tag{3.4c}$$

$$N_x \approx N_y = \frac{1}{2}(1 - N_z) \tag{3.4d}$$

Figure 3.1 summarizes the demagnetizing factors for some simple ferrite configurations.

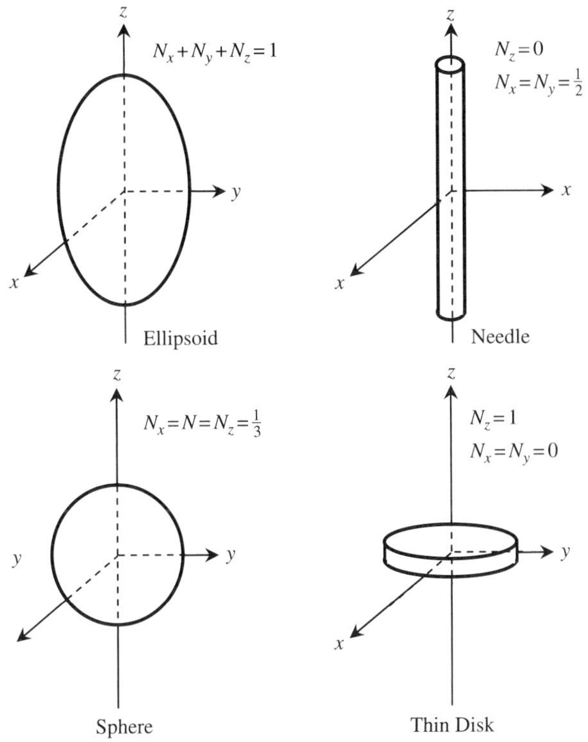

FIGURE 3.1 Shape demagnetizing factors of magnetic insulators.

Another possible relationship for N_z is

$$N_z = (1 + e^2)(e - \tan^{-1} e)/e^3 \tag{3.5}$$

where e is the approximate eccentricity of the ferrite disk:

$$e = [4(R/H)^2 - 1]^{1/2} \tag{3.6}$$

3.3 MAGNETIC FIELD INTENSITY AND FLUX DENSITY IN MAGNETIC INSULATOR

The normal magnetic flux densities are the same on either side of a boundary, but this is not generally so elsewhere except in the case of a thin plate in a uniform field. This property of the thin magnetic insulator sheet with the direct magnetic field intensity perpendicular to its surface may readily be demonstrated as a preamble to investigating some more specific cases. It starts by recalling the relationship, between the magnetic flux density, the partial magnetization, and the magnetic field intensity

(B_i, M, and H_i) inside a magnetic insulator.

$$B_i = \mu_0(H_i + M) \tag{3.7}$$

Outside the magnetic insulator the magnetic flux density is given in terms of the external direct magnetic field intensity (H_0) by

$$B_0 = \mu_0 H_0 \tag{3.8}$$

Writing H_i in terms of H_0 and the shape demagnetizing factor N_z in the first of these two conditions gives

$$B_i = \mu_0[H_0 - N_z M + M] \tag{3.9}$$

Taking N_z as unity throughout produces the required result without ado:

$$B_i = B_0 \tag{3.10}$$

This condition is a classic result in the literature.

3.4 THE SPATIAL DEMAGNETIZING FACTOR OF A FLAT DISK MAGNETIC INSULATOR

The shape demagnetizing factor perpendicular to a flat disk is often assumed to be independent of the radial coordinate of the geometry. This assumption is only correct in a ellipsoidal body. One consequence of this feature is that the magnetic field intensity inside a disk is nonuniform even in the presence of a uniform direct magnetic field. The exact internal field is

$$H_i(r) = H_0 - N_z(r)M \tag{3.11}$$

An accurate two part description for the radial variation of the demagnetizing field in a single flat disk in an infinite uniform field is available in the literature. In the central region it is described by

$$\frac{H_{\text{dem}}(r,0)}{M} = -1 + \frac{(L_f/2R_f)}{\left[1+(L_f/2R_f)^2\right]^{1/2}} + \left[\frac{3(L_f/2R_f)}{\left[1+(L_f/2R_f)^2\right]^{3/2}} - \frac{3(L_f/2R_f)^3}{\left[1+(L_f/2R_f)^2\right]^{5/2}}\right]\left(\frac{r}{2R_f}\right)^2 \tag{3.12}$$

At the edge of the disk it is specified by

$$\frac{H_{\mathrm{dem}}(r,0)}{M} = -\frac{1}{2} - \left(\frac{1}{\pi}\right)\tan^{-1}\left[\left(\frac{2R_{\mathrm{f}}}{L_{\mathrm{f}}}\right) - \left(\frac{2r}{L_{\mathrm{f}}}\right)\right] \tag{3.13}$$

r is the position within the magnetic insulator at which the demagnetizing field is measured, R_{f} is the radius of the geometry, and L_{f} is the thickness. Figure 3.2 illustrates the assembly under consideration.

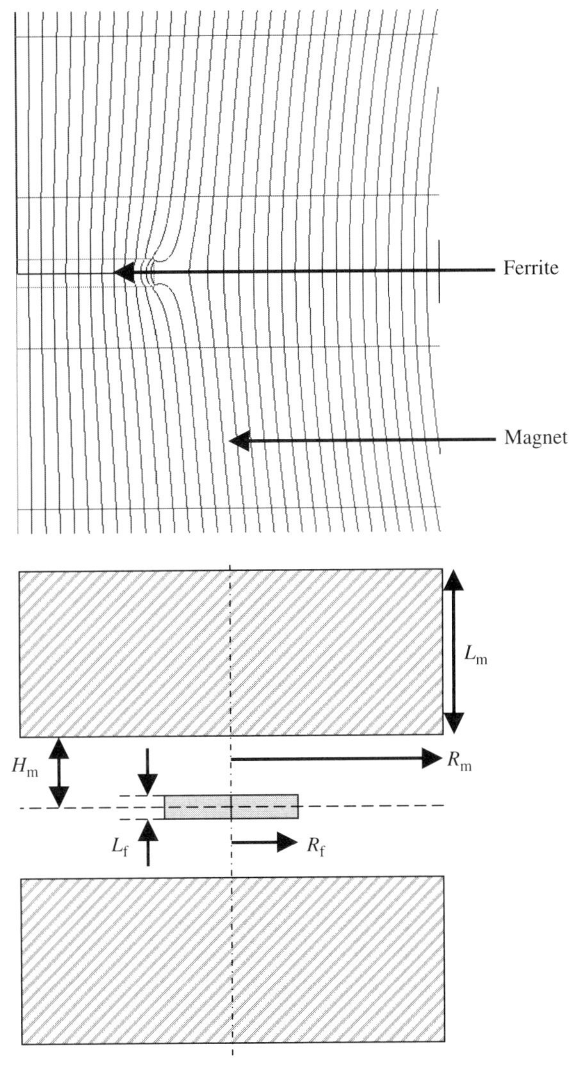

FIGURE 3.2 Magnetic flux density in flat disk magnetic insulator ($L_{\mathrm{f}}/2R_{\mathrm{f}} = 0.10$, $R_{\mathrm{m}}/R_{\mathrm{f}} = 10$).

The preceding two expressions may be smoothly connected. This is done for three different geometries in Fig. 3.3. This illustration suggests that the intercept between the two equations depends on the aspect ratio. One engineering approximation resides around $r/R_f = 0.825$.

The right-hand sides of the relationships in each region may be taken as one definition of the shape demagnetizing factor there.

$$\frac{H_{\text{dem}}(r,0)}{M} = -N_z(r,0) \tag{3.14}$$

At $r = 0$,

$$\frac{H_{\text{dem}}(r,0)}{M} = -1 + \left(\frac{L_f}{2R_f}\right)\left[1 + \left(\frac{L_f}{2R_f}\right)^2\right]^{-1/2} \tag{3.15}$$

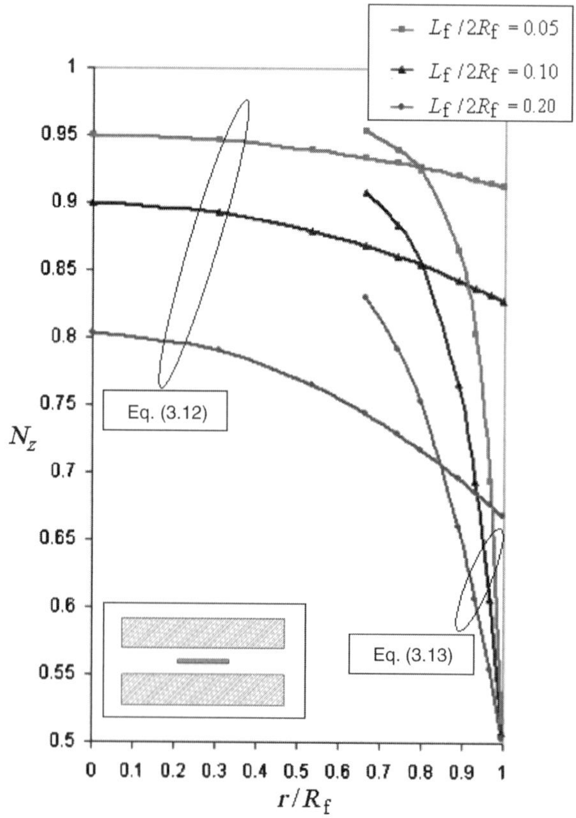

FIGURE 3.3 Closed form shape demagnetizing factor of flat disk magnetic insulator for parametric values of $L_f/2R_f$ ($R_m/R_f = 10$).

and at $r = R_f$,

$$\frac{H_{dem}(R_f, 0)}{M} = -\frac{1}{2} \tag{3.16}$$

One means to verify this two part relationship using a finite element (FE) solver is to saturate each region:

$$H_i(r, 0) = 0 \tag{3.17}$$

This gives

$$N_z(r, 0) = H_0(r, 0)/M_0 \tag{3.18}$$

This condition provides one means of evaluating the local shape demagnetizing factor of a disk. Figure 3.4 compares the demagnetizing factor of one geometry based on the approximate equations with that obtained using an FE solver. The

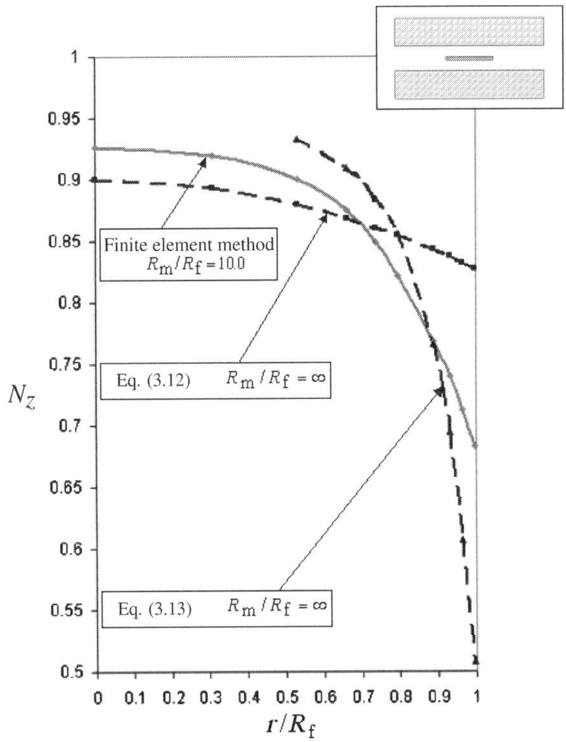

FIGURE 3.4 Comparison between FE and approximate closed form calculations of shape demagnetizing factor in flat disk magnetic insulator ($L_f/2R_f = 0.10$, $R_m/R_f = 10$).

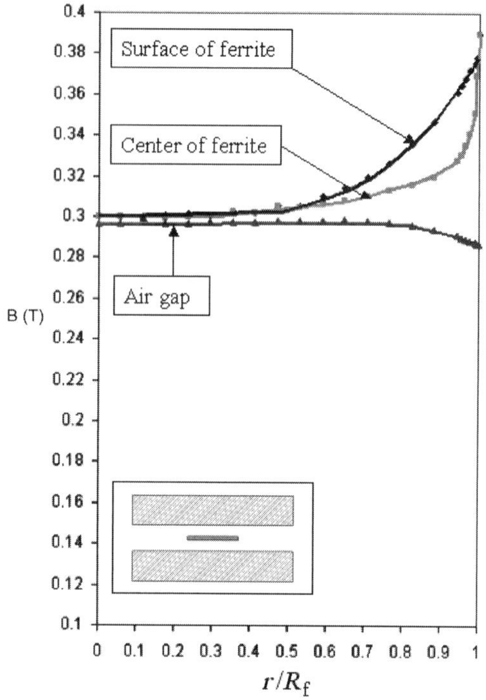

FIGURE 3.5 FE calculation of demagnetizing magnetic field intensity of a single disk magnetic insulator in uniform magnetic field ($L_f/2R_f = 0.10$, $R_m/R_f = 10$). Ferrite material: $B_r = 167.11$ kA/m, $H_c = 67.64$ A/m, $\mu_0 M_0 = 0.3000$ T. Magnet: $B_r = 692.32$ kA/m, $H_c = 636.62$ kA/m.

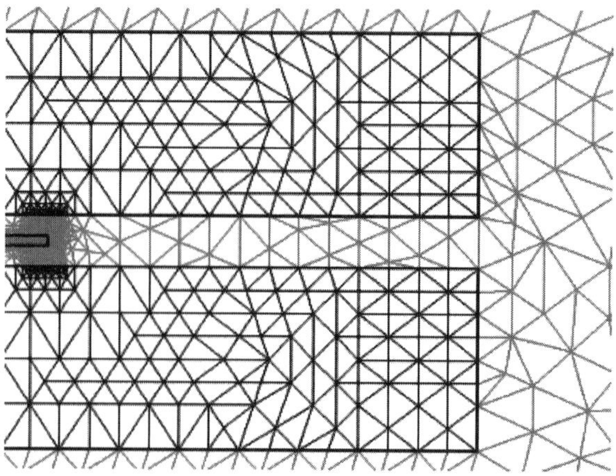

FIGURE 3.6 FE discretization of single disk magnetic insulator.

spatial flux densities of this arrangement in different regions are indicated in Fig. 3.5 with the insulator saturated on its axis. The FE discretization used to obtain these results is indicated in Fig. 3.6.

In practice, the magnets may be either larger than, equal to, or smaller than the magnetic insulator. The governing equations for the spatial shape demagnetizing factor of this arrangement is no longer applicable and must be obtained numerically. Figure 3.7 shows the effect of the size of the magnet on the internal magnetic flux density of a disk magnetic insulator for a given magnetization. It indicates that the flux there may be less than, equal to, or more than that at the axis. At the edge of the disk it becomes asymptotic to twice that on the axis of the geometry. This condition is in keeping with the approximate closed form formulation of the demagnetizing field.

One means of ensuring that a finite ferrite disk is uniformly magnetized is to place it between semispherical pole pieces with the same magnetization. Figure 3.8 indicates a solution for one geometry for which the pole piece has the same radius as that of the semispherical pole.

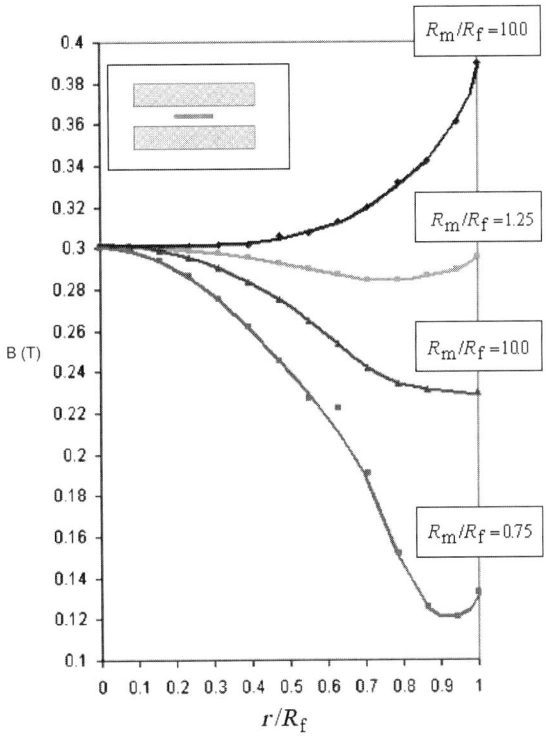

FIGURE 3.7 The effect of the size of a magnet on the internal flux density in a magnetic disk insulator. Ferrite material: $B_r = 167.11$ kA/m, $H_c = 67.64$ A/m, $\mu_0 M_0 = 0.3000$ T. Magnet: $B_r = 692.32$ kA/m, $H_c = 636.62$ kA/m.

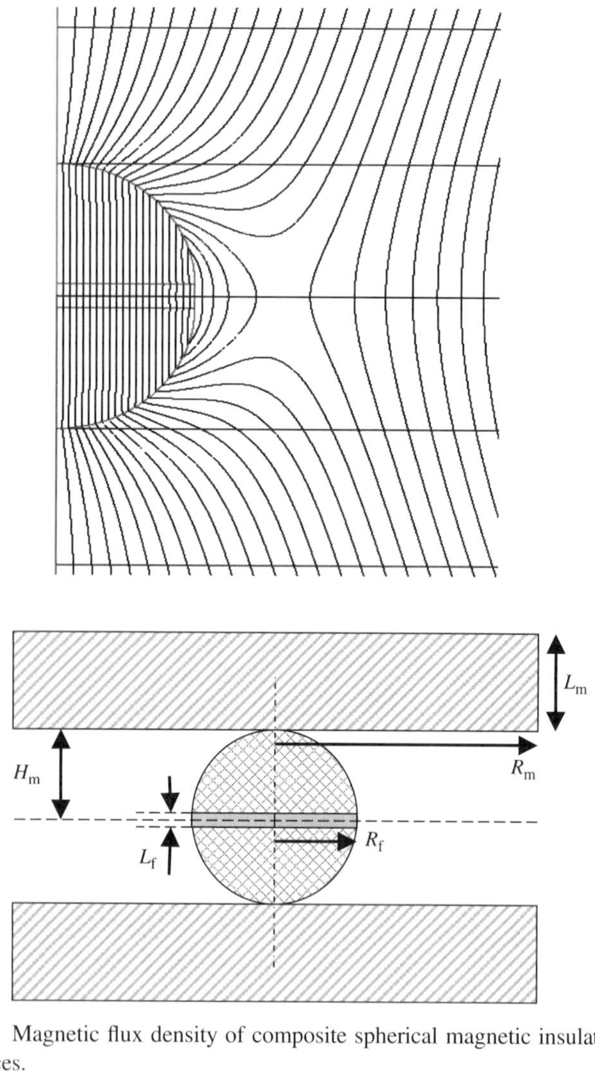

FIGURE 3.8 Magnetic flux density of composite spherical magnetic insulator with undersized pole pieces.

3.5 THE COUPLED DISK GEOMETRY

The case of a pair of disk magnetic insulators separated by a gap is also of some interest. The purpose of this section is to evaluate the effective demagnetizing field and shape factor on the axis of the arrangement.

The geometry under consideration is indicated in Fig. 3.9. The gap between the two disks is defined in terms of a filling factor k_f:

$$k_f = L_f/H_f$$

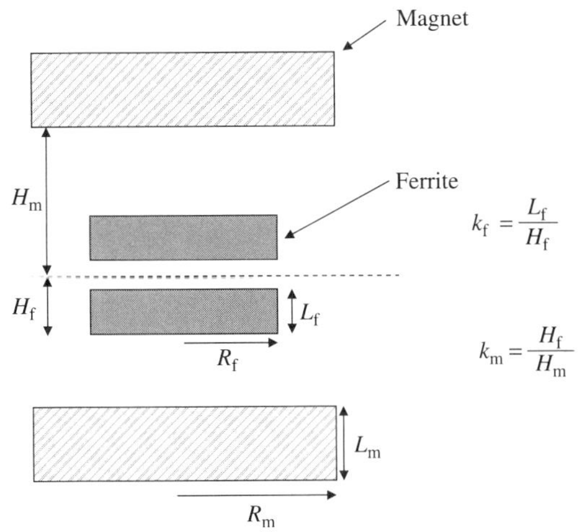

FIGURE 3.9 Schematic diagram of a pair of flat disk magnetic insulators.

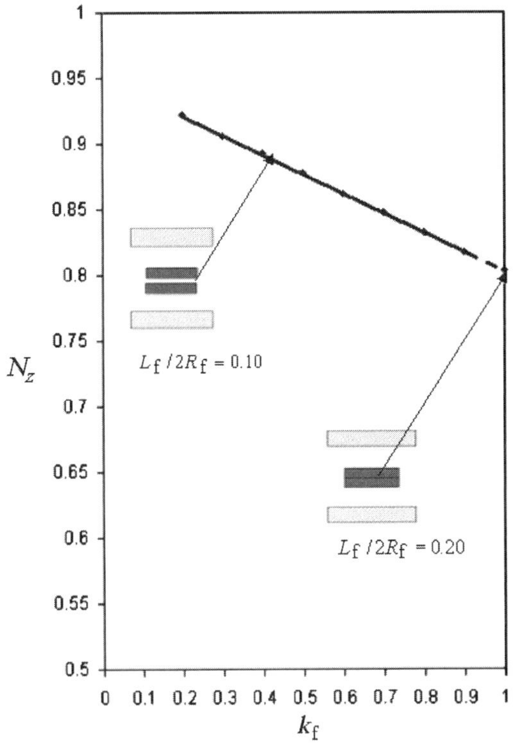

FIGURE 3.10 FE calculation of shape demagnetizing factor on the axis of a pair of disk magnetic insulators ($L_f/2R_f = 0.10$, $k_m = 0.1895$). Ferrite material: $B_r = 167.11$ kA/m, $H_c = 67.64$ A/m, $\mu_0 M_0 = 0.3000$ T. Magnet: $B_r = 692.32$ kA/m, $H_c = 636.62$ kA/m.

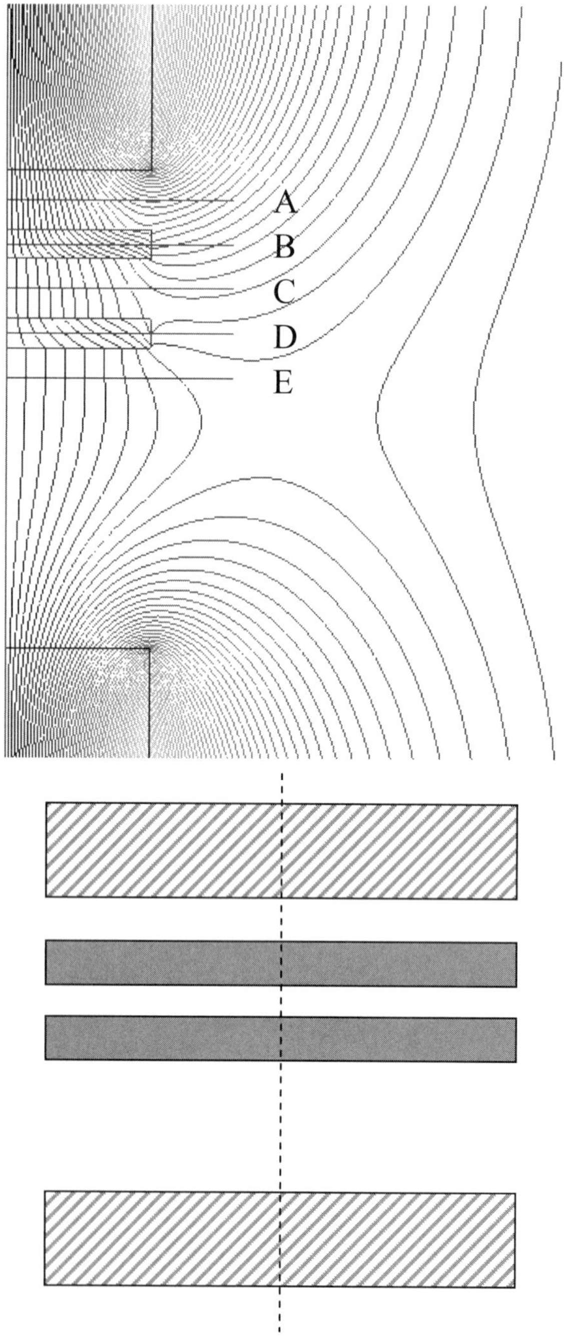

FIGURE 3.11 Magnetic flux density distribution in a pair of disk magnetic insulators in an asymmetric magnetic field. Ferrite material: $B_r = 167.11$ kA/m, $H_c = 67.64$ A/m, $\mu_0 M_0 = 0.3000$ T. Magnet: $B_r = 692.32$ kA/m, $H_c = 636.62$ kA/m.

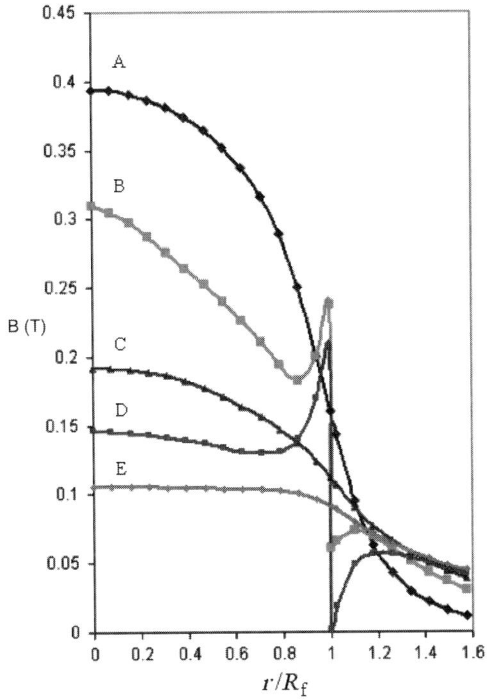

FIGURE 3.12 Flux densities in different regions in a pair of disk resonators in asymmetric magnetic field. Ferrite material: $B_r = 167.11$ kA/m, $H_c = 67.64$ A/m, $\mu_0 M_0 = 0.3000$ T. Magnet: $B_r = 692.32$ kA/m, $H_c = 636.62$ kA/m.

L_f is the length of a typical disk, H_f is the half-space defined by the outside face of a typical disk and half that of the gap. The case $k_f = 1$ reduces to a single disk of length $2L_f$; that for $k_f = 0$ corresponds to a pair of decoupled ones.

The gap or spacing between the magnets is specified by

$$k_m = H_f / H_m$$

The demagnetizing field is again obtained by locally saturating the material at the symmetry plane of a typical disk. The shape demagnetizing factor is obtained by having recourse to Eq. (3.12). Figure 3.10 shows the connection between the shape demagnetizing factor and the filling factor on the axis of the geometry. The demagnetizing factor is bounded by that of a single disk and that of a cylinder with half its aspect ratio.

The magnetic flux distribution of one arrangement with the symmetry plane of the ferrites different from that of the magnets is illustrated in Fig. 3.11. The flux densities inside the two ferrites are now different. This feature is indicated in more detail in Fig. 3.12.

3.6 THE IRREGULAR HEXAGONAL MAGNETIC INSULATOR

Another magnetic insulator of some interest is either a single or a pair of irregular hexagonal geometries. Each has two symmetry planes that have to be dealt with separately. The geometry under consideration is specified in Chapter 16. It is of note that these two coordinates are not perpendicular to each other. Figure 3.13 depicts the variation of the demagnetizing factor perpendicular to the plane of the plate along the inscribed radius of one equilateral geometry in infinite space. The corresponding results for which the magnetic fields coincide with the inscribed and circumscribed radii of the magnetic insulator are superimposed separately in this illustration. The variations of the shape demagnetizing factors along the circumscribed radius are illustrated separately in Fig. 3.14.

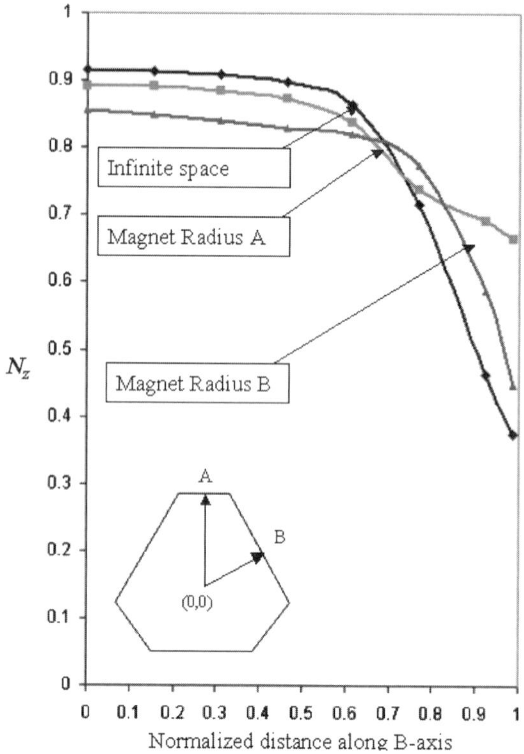

FIGURE 3.13 Shape demagnetizing factors of single irregular hexagonal magnetic insulator along B-dimension (A = 3.47 mm, B = 15.32, $\phi = 20°$). Ferrite material: $B_r = 167.11$ kA/m, $H_c = 67.64$ A/m, $\mu_0 M_0 = 0.3000$ T. Magnet: $B_r = 692.32$ kA/m, $H_c = 636.62$ kA/m. For $R_m/R_A = 10$, $R_m/R_A = 1.00$, and $R_m/R_B = 1.00$.

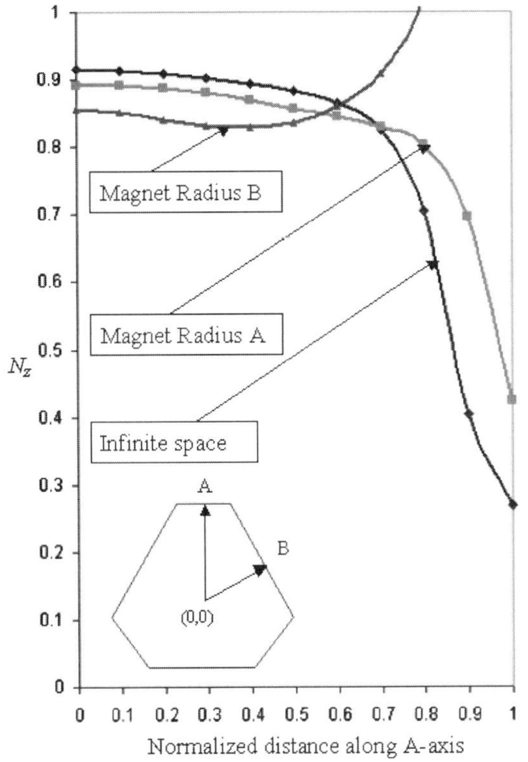

FIGURE 3.14 Shape demagnetizing factors of single irregular hexagonal magnetic insulator along A-dimension (A = 3.47 mm, B = 15.32, $\phi = 20^\circ$). Ferrite material: $B_r = 167.11$ kA/m, $H_c = 67.64$ A/m, $\mu_0 M_0 = 0.3000$ T. Magnet: $B_r = 692.32$ kA/m, $H_c = 636.62$ kA/m. For $R_m/R_A = 10$, $R_m/R_A = 1.00$, and $R_m/R_B = 1.00$.

3.7 REENTRANT MAGNETIC CIRCUITS

One practical application of the magnetostatic solver is in the adjustment of the direct magnetic circuits of UHF circulators. One common package is the reentrant geometry with a pair of ferrites between top and bottom magnets. The gyromagnetic resonator consists, in this instance, of either a disk or an irregular hexagonal structure. Another common geometry consists of side magnets instead of top and bottom magnets. In this instance the resonator must either have a triangular cross section or an irregular hexagonal one. In general, UHF circuits have to produce normalized magnetization values (B/M) of the order of 1.45 or less. Figure 3.15 illustrates the radial variation of B/M in the case of a pair of disk bodies assembled using top and bottom magnets. Figure 3.16 indicates one typical irregular hexagonal ferrite geometry with three side magnets. The adjustment of B/M is in this case achieved by varying the thickness of the pole pieces.

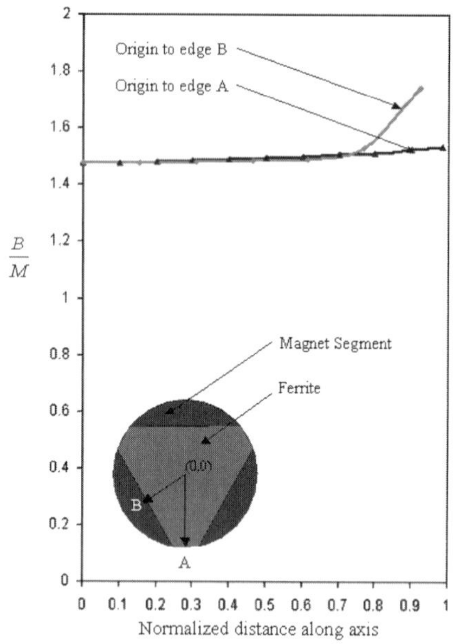

FIGURE 3.15 Normalized magnetic field intensity in a pair of disk magnetic insulators with top and bottom magnet configuration. Ferrite material: $B_r = 101.62$ kA/m, $H_c = 35.81$ A/m, $\mu_0 M_0 = 0.1780$ T. Magnet: $B_r = 318.31$ kA/m, $H_c = 266.58$ kA/m.

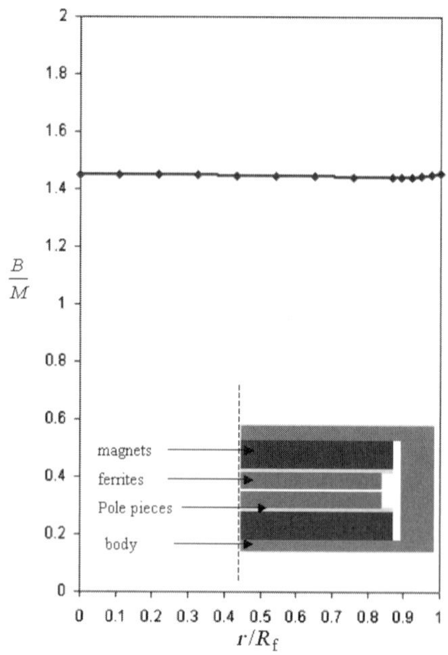

FIGURE 3.16 Normalized magnetic field intensity in a pair of irregular hexagonal magnetic insulators with side magnet configuration. Ferrite material: $B_r = 101.62$ kA/m, $H_c = 35.81$ A/m, $\mu_0 M_0 = 0.1780$ T. Magnet: $B_r = 855.5$ kA/m, $H_c = 795.77$ kA/m.

3.8 TRANSVERSE DEMAGNETIZING FACTOR OF MAGNETIC INSULATOR

The finite element method (FEM) may also be utilized to deduce the demagnetizing factor N_t of a disk magnetic insulator along its principal axis. This may be done by rotating the geometry 90° in order to place the magnetizing field in its plane instead of perpendicular to it.

3.9 TENSOR ELEMENTS IN PARTIALLY MAGNETIZED MAGNETIC INSULATOR

The usual descriptions of the elements of the permeability tensor introduced in Chapter 2 are only valid provided the material is saturated. This latter condition coincides with that for which the internal magnetic field intensity H_i is equal to or greater than zero. At saturation M is equal to M_0 and

$$H_0 - N_z M_0 = 0 \tag{3.19}$$

The right-hand side in the definition of saturation in a magnetic insulator here and elsewhere is usually taken to be zero. In reality it is a small positive number equal to the magnetic field intensity necessary to saturate the hysteresis loop of the material. This field is usually very small compared to the magnetization of the material and is frequently neglected compared to it.

The diagonal and off-diagonal elements of the permeability tensor in a magnetic insulator biased below saturation are described, on the basis of experiment, by partially magnetized quantities,

$$\kappa_p = \kappa(M_p/M_0) \tag{3.20}$$

$$\mu_p = \mu_d + (1 - \mu_d)(M_p/M_0)^{3/2} \tag{3.21}$$

where

$$\kappa = \gamma M_0/\omega \tag{3.22}$$

μ_d is the relative demagnetized permeability in the magnetic insulator,

$$\mu_d = \frac{1}{3} + \frac{2}{3}[1 - (\gamma M_0/\omega)^2]^{1/2} \tag{3.23}$$

M_p is the partial magnetization (A/m), κ_p is the partial off-diagonal entry of the permeability tensor, μ_p is the diagonal element, γ is the gyromagnetic ratio, and κ is the off-diagonal element at saturation.

At saturation

$$\kappa_p = \kappa \tag{3.24}$$

$$\mu_p = 1 \tag{3.25}$$

$$\mu_d = 1 \tag{3.26}$$

In a partially magnetized magnetic insulator, the real part of the longitudinal element μ_z is not equal to unity and must also be modified. One empirical representation is

$$\mu_z = \mu_d^{(1-M_p/M_0)^{5/2}} \tag{3.27}$$

The partial and saturated elements of the permeability tensor are separately related by

$$\mu^2 - \kappa^2 = \mu_p^2 - \kappa_p^2 \tag{3.28}$$

3.10 PARTIAL MAGNETIZATION IN A MAGNETIC INSULATOR

The tie between M_p/M_0 and B_0 may be obtained by having recourse to the connection between the magnetic flux densities inside and outside the geometry in conjunction with the magnetic intensities in the same two regions. The situation for which the direct magnetic field intensity is positive has to be dealt with separately from that for which it is negative. The result in the former case is obtained by having recourse to Eq. (3.9)

$$\frac{M_p}{M_0} = \frac{B_i - B_0}{\mu_0 M_0 (1 - N_z)} \tag{3.29}$$

This ratio may be evaluated by first determining N_z as a preamble to constructing the tie between B_i and B_0 using a suitable FE solver. The demagnetizing factor is determined separately by the flux density B_0 at magnetic saturation and the saturation magnetization $\mu_0 M_0$ of the magnetic insulator.

The partial magnetization with the polarization of the direct magnetization reversed is the same except for a change of sign.

The material is saturated provided

$$M_p/M_0 = 1$$

This condition coincides with

$$B_i - B_0 = \mu_0 M_0 (1 - N_z) \tag{3.30}$$

These two relationships are compatible with those already met in connection with the description of the internal magnetic flux density in a saturated material.

$$B_i = \mu_0 M_0 \tag{3.31}$$

$$B_0 = N_z \mu_0 M_0 \tag{3.32}$$

The first condition is in keeping with the definition of saturation in a magnetic insulator and the second one provides a definition of the demagnetizing factor of the geometry.

The partial magnetization M_p is zero provided

$$B_i - B_0 = 0$$

This is satisfied when the direct magnetic flux densities outside and inside the magnetic insulator are identically zero:

$$B_i = B_0 = 0$$

The robustness of this result may be further verified by utilizing it to evaluate M_p/M_0 on the axis of a gyromagnetic flat disk at magnetic saturation. The geometry of the disk is fixed by the radii of the magnet and ferrite bodies R_m/R_f (= 10), by the aspect ratio of the magnetic insulator $L_f/2R_f$ (= 0.10), and by its saturation magnetization

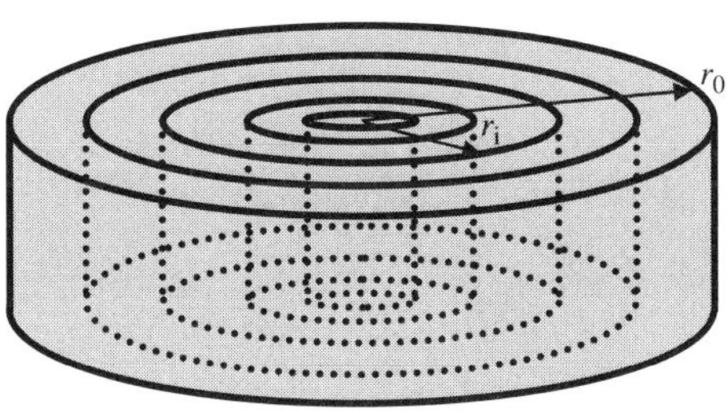

FIGURE 3.17 Disk magnetic insulator partitioned into concentric annular rings.

$\mu_0 M_0$ (0.3000 T). B_i, B_0, and N_z are deduced by having recourse to an FE process as

$$B_i = 0.3000 \text{ T}$$
$$B_0 = 0.2960 \text{ T}$$
$$N_z = 0.985$$

The magnetization obtained in this way is $M_p/M_0 = 1.0$.

The data in this chapter suggests that the demagnetizing factor in a nonellipsoidal geometry such as a disk in a uniform field may have significant radial variation. One way to cater for this feature is to partition the disk into a number of coarse or fine concentric annular rings with local values of demagnetizing factors and magnetization states as a preamble to tackling the microwave problem. Figure 3.17 shows one possibility.

CHAPTER FOUR

Scattering Matrix of *m*-Port Junction

4.1 INTRODUCTION

The definitions of the two-port gyrator circuit and the nonreciprocal three-port junction circulator were first articulated by Carlin on the basis of the unitary condition met in connection with the scattering matrices of these networks. The entries of this matrix are a set of quantities that relate incident and reflected waves at its ports or terminal planes, which describe the performance of a network under any specified terminating conditions. The coefficients along its main diagonal are reflection coefficients while those along its off-diagonal are transmission ones. The matrix is modified if one or more of the terminal planes are moved. A scattering matrix exists for every linear, passive, time invariant network. Figure 4.1 depicts a generalized junction enclosed by a surface S, which cuts the various transmission lines perpendicular to the axes and provides a definition for the ports or terminal planes of the structure. This text, however, is primarily concerned with symmetric m-port junctions. It is possible to deduce important general properties of junctions containing a number of ports by invoking such properties as symmetry, reciprocity, and energy conservation. An important property of the scattering matrix is that the permissible relationships between the entries of a lossless network are readily established by having recourse to the unitary condition. In particular, it shows that only a matched three-port lossless junction uniquely defines the properties of the ideal circulator, which is the topic of this text. The adjustment of such junctions, however, is best dealt with in terms of the eigenvalue problem in Chapter 5. In the presence of dissipation, all the existing relationships between the entries of the scattering matrix based on the unitary condition become invalid. In order to cater for this effect, it is usual to

The Stripline Circulator: Theory and Practice. By J. Helszajn

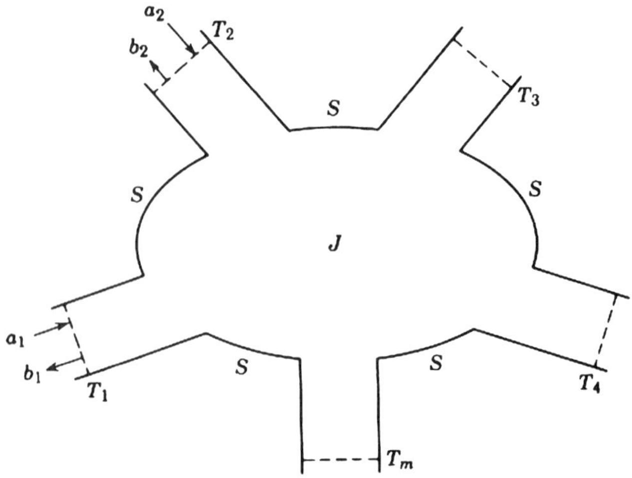

FIGURE 4.1 Schematic diagram of m-port junction showing incident and reflected waves.

introduce a so-called dissipation matrix. A description of this matrix is included for completeness.

4.2 THE SCATTERING MATRIX

The properties of the scattering matrix of an m-port network developed in this chapter are essentially restricted to circuits with the symmetries outlined in Figs. 4.2 and 4.3.

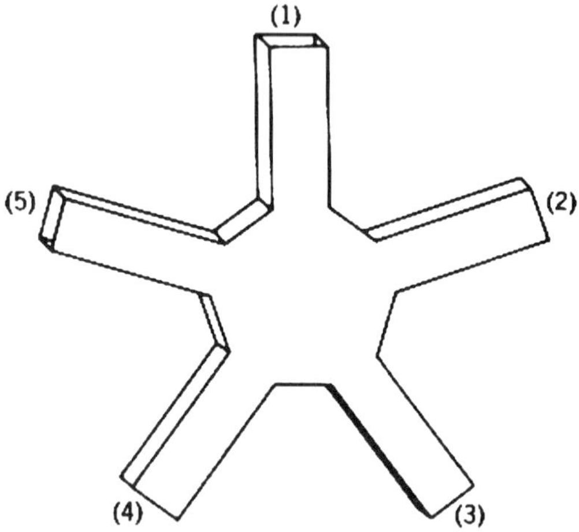

FIGURE 4.2 Schematic diagram of five-port symmetrical junction.

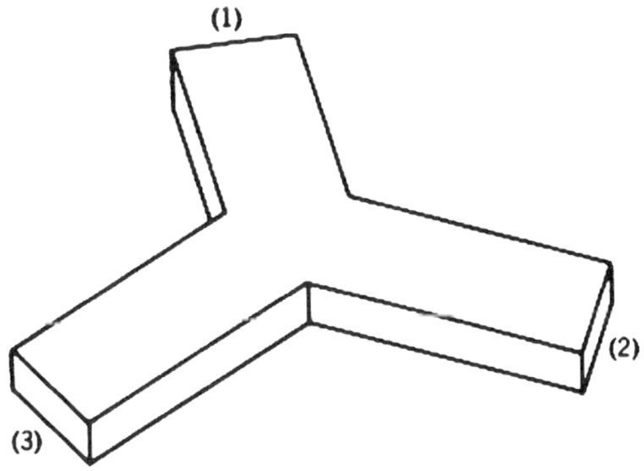

FIGURE 4.3 Schematic diagram of three-port symmetrical junction.

Its entries relate suitably chosen incident and reflected waves at the terminals of the network.

$$\bar{\mathbf{b}} = \bar{\mathbf{S}}\bar{\mathbf{a}} \tag{4.1}$$

$\bar{\mathbf{a}}$ and $\bar{\mathbf{b}}$ are incident and reflected vectors, which for a three-port network is given by

$$\bar{\mathbf{a}} = \begin{bmatrix} a_1 \\ a_2 \\ a_3 \end{bmatrix} \tag{4.2}$$

$$\bar{\mathbf{b}} = \begin{bmatrix} b_1 \\ b_2 \\ b_3 \end{bmatrix} \tag{4.3}$$

The corresponding scattering matrix $\bar{\mathbf{S}}$ is described by

$$\bar{\mathbf{S}} = \begin{bmatrix} S_{11} & S_{12} & S_{13} \\ S_{21} & S_{22} & S_{23} \\ S_{31} & S_{32} & S_{33} \end{bmatrix} \tag{4.4}$$

The elements along the main diagonal of this matrix are the reflection coefficients at the ports of the network; those along the off-diagonal represent the transmission coefficients between the same ports. Using this nomenclature, the relationships between the incoming and outgoing waves for a three-port network are therefore described by

$$b_1 = a_1S_{11} + a_2S_{12} + a_3S_{13} \tag{4.5a}$$

$$b_2 = a_1S_{21} + a_2S_{22} + a_3S_{23} \tag{4.5b}$$

$$b_3 = a_1S_{31} + a_2S_{32} + a_3S_{33} \tag{4.5c}$$

A schematic diagram of this relationship is depicted in Fig. 4.4.

A typical diagonal element of the scattering parameters of the three-port is defined in terms of the incident and reflected waves by the preceding equations as

$$S_{11} = \left.\frac{b_1}{a_1}\right|_{a_2=a_3=0} \tag{4.6a}$$

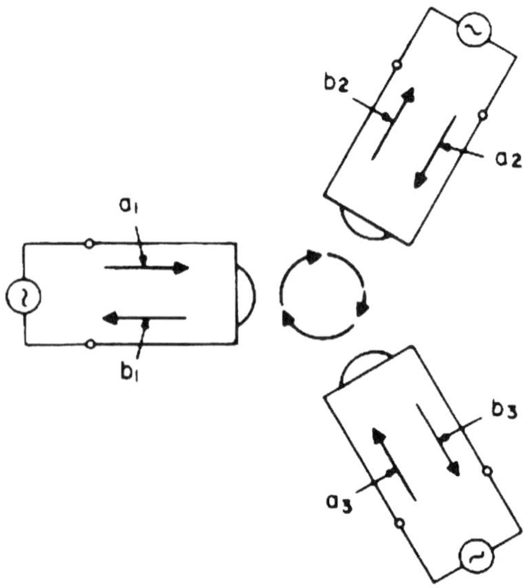

$$\begin{bmatrix} b_1 \\ b_2 \\ b_3 \end{bmatrix} = \begin{bmatrix} S_{11} & S_{21} & S_{31} \\ S_{31} & S_{11} & S_{21} \\ S_{21} & S_{31} & S_{11} \end{bmatrix} \begin{bmatrix} a_1 \\ a_2 \\ a_3 \end{bmatrix}$$

FIGURE 4.4 Definition of incident and reflected waves in three-port junction.

Typical off-diagonal elements are defined by

$$S_{13} = \left.\frac{b_1}{a_3}\right|_{a_1 = a_2 = 0} \tag{4.6b}$$

$$S_{31} = \left.\frac{b_3}{a_1}\right|_{a_2 = a_3 = 0} \tag{4.6c}$$

The a_i and b_i are normalized in such a way that $\frac{1}{2}a_i a_i^*$ is the available power at port i and $\frac{1}{2}b_i b_i^*$ is the emergent power at the same port.

$$a_i = \frac{1}{2}\left(V_i/\sqrt{Z_0} + \sqrt{Z_0}I_i\right), \quad i = 1, 2, 3 \tag{4.7}$$

$$b_i = \frac{1}{2}\left(V_i/\sqrt{Z_0} + \sqrt{Z_0}I_i\right), \quad i = 1, 2, 3 \tag{4.8}$$

Adopting these definitions indicates that the a's and b's at any port are linear combinations of the voltage and current variables at the same port. Such linear combinations are in fact also met in the description of a uniform transmission line for which the solution to the transmission line equations is given in terms of forward and backward traveling waves $A \exp(-\gamma z)$ and $B \exp(\gamma z)$ by

$$V = A \exp(-\gamma z) + B \exp(\gamma z) \tag{4.9}$$

$$Z_0 I = A \exp(-\gamma z) - B \exp(\gamma z) \tag{4.10}$$

A scrutiny of Eqs. (4.6b) and (4.6c) indicates that S_{13} connects port 3 to port 1, whereas S_{31} connects port 1 to port 3.

4.3 CIRCULATOR DEFINITION BY MEANS OF CYCLIC SUBSTITUTION

In the theory of finite groups, an operation, called cyclic substitution, is defined. This operation is usually illustrated as being performed on a sequence of letters or numbers. For example, the operation of cyclic substitution (*abcd*) means that *a* is replaced by *b*, *b* by *c*, *c* by *d*, and *d* by *a*. If this operation is performed on the sequence *bdac*, the result is

$$(abcd) \rightarrow bdac \rightarrow cabd$$

In the scattering relationship between incident and reflected waves the scattering matrix may be considered as indicating an operation performed on the incident waves, the result of which yields the reflected waves.

If the operator $\bar{\mathbf{S}}$ corresponds to a cyclic substitution, then the device having such a scattering matrix corresponds to the intuitive notion of the circulator property. Consequently, a circulator may be defined as follows: *A device, whose scattering matrix operates on the incident waves so as to produce the same result as the operation of a cyclic substitution on the incident voltages, is called a circulator.*

For a given m-port circulator, different systems of port numbering will lead to $(m-1)$ distinct scattering matrices. In the case of a structurally symmetrical circulator, the number of possibilities may be reduced, however, by specifying a standard numbering system. For an m-port junction having the symmetry in Figs. 4.2 and 4.3, it is always possible to number the ports to represent the cyclic substitution $(1, 2, 3, \ldots, m)$.

4.4 THE UNITARY CONDITION

The entries of the scattering matrix have the nature of reflection and transmission parameters so that the amplitude of any of its entries is bounded by zero and unity. The permissible relationships between these entries will now be deduced. The derivation of the required result starts by recognizing that the power dissipated in an m-port network is the difference between the incident power at all ports and the reflected power at the same ports.

$$P_{\text{diss}} = \frac{1}{2}\left(\sum_{i}^{m} a_i a_i^* - \sum_{i}^{m} b_i b_i^*\right) \tag{4.11}$$

It may readily be demonstrated that

$$\sum_{i}^{m} a_i a_i^* = (\bar{\mathbf{a}})^{\mathrm{T}}(\bar{\mathbf{a}}^*) \tag{4.12}$$

$$\sum_{i}^{m} b_i b_i^* = (\bar{\mathbf{a}})^{\mathrm{T}}(\bar{\mathbf{S}})^{\mathrm{T}}(\bar{\mathbf{S}}^*)(\bar{\mathbf{a}}^*) \tag{4.13}$$

The power dissipated in the circuit may therefore be expressed in matrix form as

$$P_{\text{diss}} = \tfrac{1}{2}(\bar{\mathbf{a}})^{\mathrm{T}}[\bar{\mathbf{I}} - (\bar{\mathbf{S}})^{\mathrm{T}}(\bar{\mathbf{S}}^*)](\bar{\mathbf{a}}^*) \tag{4.14}$$

$\bar{\mathbf{I}}$ is the unit matrix, $(\bar{\mathbf{S}})^{\mathrm{T}}$ is the transpose of $\bar{\mathbf{S}}$, and $(\bar{\mathbf{a}}^*)$ is the conjugate of $\bar{\mathbf{a}}$.

The quantity defined by the preceding equation is a Hermitian form in that the matrix inside the inner brackets

$$\overline{\mathbf{Q}} = \overline{\mathbf{I}} - (\overline{\mathbf{S}})^{\mathrm{T}}(\overline{\mathbf{S}}^{*}) \tag{4.15}$$

is its own conjugate transpose.

$$[\overline{\mathbf{I}} - (\overline{\mathbf{S}})^{\mathrm{T}}(\overline{\mathbf{S}}^{*})]^{*\mathrm{T}} = \overline{\mathbf{I}} - (\overline{\mathbf{S}}) \tag{4.16}$$

All energy functions are in fact either Hermitian or quadratic forms.

Since the power dissipated in a network is always positive, the Hermitian form in Eq. (4.14) is positive semidefinite.

$$\tfrac{1}{2}(\overline{\mathbf{a}})^{\mathrm{T}}[\overline{\mathbf{I}} - (\overline{\mathbf{S}})^{\mathrm{T}}(\overline{\mathbf{S}}^{*})](\overline{\mathbf{a}}^{*}) \geq 0 \tag{4.17}$$

For a reactance function the Hermitian form in the preceding equation is satisfied with the equal sign. The condition for a lossless junction therefore becomes

$$\overline{\mathbf{I}} - (\overline{\mathbf{S}})^{\mathrm{T}}(\overline{\mathbf{S}}^{*}) = 0 \tag{4.18}$$

This last statement indicates that the scattering matrix of a dissipationless network is unitary. It is widely used to establish the permissible relationships between the entries of the matrix in a reactance network.

4.5 NETWORK DEFINITION OF JUNCTION CIRCULATOR

Many of the properties of junction circulators are best discussed in terms of the scattering matrix of the device introduced in this chapter. The required nomenclature is defined in Fig. 4.3. It will now be demonstrated that a matched nonreciprocal symmetric three-port junction is necessarily a circulator. The scattering matrix of such a junction is given by

$$\overline{\mathbf{S}} = \begin{bmatrix} 0 & S_{21} & S_{31} \\ S_{31} & 0 & S_{21} \\ S_{21} & S_{31} & 0 \end{bmatrix} \tag{4.19}$$

Evaluating the unitary condition under this assumption indicates that

$$|S_{21}|^2 + |S_{31}|^2 = 1 \tag{4.20}$$

and

$$S_{21}S_{31}^* = 0 \tag{4.21}$$

These two equations are consistent provided

$$|S_{21}| = 1 \tag{4.22}$$

and

$$|S_{31}| = 0 \tag{4.23}$$

or

$$|S_{21}| = 0 \tag{4.24}$$

$$|S_{31}| = 1 \tag{4.25}$$

The above relationships indicate that if $S_{11} = 0$ (matched condition), then either $|S_{21}| = 1$ and $|S_{31}| = 0$ or $|S_{21}| = 0$ and $|S_{31}| = 1$. Either of these two conditions coincide with the definition of a perfect circulator.

The unitary condition may also be employed to demonstrate that it is impossible to match a lossless reciprocal three-port junction with threefold symmetry. If the junction is reciprocal it has the property that its scattering matrix is symmetric:

$$S_{ij} = S_{ji} \tag{4.26}$$

If it is matched its diagonal elements must be identically zero:

$$S_{11} = S_{22} = S_{33} = 0 \tag{4.27}$$

The scattering matrix for a matched reciprocal three-port junction is therefore

$$\overline{\mathbf{S}} = \begin{bmatrix} 0 & S_{21} & S_{21} \\ S_{21} & 0 & S_{21} \\ S_{21} & S_{21} & 0 \end{bmatrix} \tag{4.28}$$

Evaluating the unitary condition for this scattering matrix indicates that

$$2S_{21}S_{21}^* = 1 \tag{4.29}$$

and

$$S_{21}S_{21}^* = 0 \tag{4.30}$$

These two equations cannot be satisfied at the same time, so it is apparent that it is impossible to match such a three-port junction. It may be demonstrated separately that the best possible match for a reciprocal three-port symmetrical junction coincides with the condition $S_{11} = \frac{1}{3}$.

4.6 SEMI-IDEAL CIRCULATOR

It is also possible to obtain a relationship between the scattering coefficients of a lossless but nearly matched circulator. The scattering matrix for this network is

$$\mathbf{S} = \begin{bmatrix} S_{11} & S_{21} & S_{31} \\ S_{31} & S_{11} & S_{21} \\ S_{21} & S_{31} & S_{11} \end{bmatrix} \tag{4.31}$$

Applying the unitary condition gives

$$|S_{11}|^2 + |S_{21}|^2 + |S_{31}|^2 = 1 \tag{4.32}$$

and

$$S_{21}S_{31}^* + S_{21}S_{11}^* + S_{31}S_{11}^* = 0 \tag{4.33}$$

One solution with S_{21} close to unity and S_{11} and S_{31} small is

$$|S_{11}| \approx |S_{31}| \tag{4.34}$$

The transmission loss is specified by

$$|S_{21}| \approx 1 - 2|S_{11}|^2 \tag{4.35}$$

Thus minimum insertion loss corresponds to both maximum isolation and minimum VSWR looking into any of the three ports.

By virtue of Eq. (4.33), $|S_{21}|$ can be expressed in terms of $|S_{11}|$ and $|S_{31}|$. The three quantities in this equation can be interpreted as three vectors that span a triangle. Then inequality relations such as

$$|S_{21}|\,|S_{31}| \leq |S_{11}|\,(|S_{21}| + |S_{31}|) \tag{4.36}$$

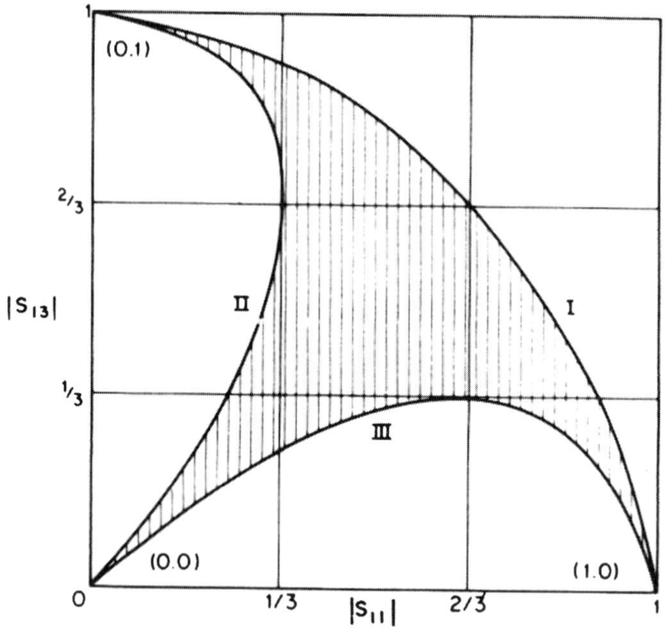

FIGURE 4.5 The close region of possible lossless and cyclic symmetrical three-ports. (Reproduced with permission from S. Hagelin, A flow graph analysis of 3- and 4 port junction circulators, *IEEE Trans. Microwave Theory Tech.* Vol. MTT-14(5), pp. 243–249, 1996.)

are valid. From these considerations Butterweck deduced that in the $|S_{11}|$, $|S_{21}|$ diagram the possible three ports are restricted to a region that is bounded by three ellipses:

$$|S_{11}|^2 + |S_{11}||S_{31}| + |S_{31}|^2 - |S_{11}| - |S_{31}| = 0 \tag{4.37a}$$

$$|S_{11}|^2 - |S_{11}||S_{31}| + |S_{31}|^2 + |S_{11}| - |S_{31}| = 0 \tag{4.37b}$$

$$|S_{11}|^2 - |S_{11}||S_{31}| + |S_{31}|^2 - |S_{11}| + |S_{31}| = 0 \tag{4.37c}$$

In Fig. 4.5 this region is given by the shaded area. The origin of the diagram $|S_{11}| = 0$, $|S_{31}| = 0$ represents an ideal clockwise circulator.

4.7 DISSIPATION MATRIX

In the presence of dissipation, the Hermitian form in Eq. (4.14), must be positive real,

$$\overline{\mathbf{Q}} = \bar{\mathbf{I}} - (\overline{\mathbf{S}}^*)^{\mathrm{T}}(\overline{\mathbf{S}}) \tag{4.38}$$

where $\bar{\mathbf{I}}$ is a unit matrix, $\bar{\mathbf{S}}$ is the scattering matrix, and $\bar{\mathbf{Q}}$ is a dissipation matrix, which is defined by

$$\overline{\mathbf{Q}} = \begin{bmatrix} q_{11} & -q_{21} & -q_{31} \\ -q_{31} & q_{11} & -q_{21} \\ -q_{21} & -q_{31} & q_{11} \end{bmatrix} \tag{4.39}$$

In the case of a symmetrical three-port junction, one has

$$q_{11} = -|S_{11}|^2 - |S_{21}|^2 - |S_{31}|^2 \tag{4.40a}$$

$$q_{21} = S_{11}S_{21}^* + S_{21}S_{31}^* + S_{31}S_{11}^* \tag{4.40b}$$

$$q_{31} = q_{21}^* \tag{4.40c}$$

The necessary and sufficient condition for the dissipation matrix to be positive real is that the principal minors of the dissipation matrix be nonnegative.

$$q_{11} \geq 0 \tag{4.41a}$$

$$\begin{vmatrix} q_{11} & -q_{21} \\ -q_{21}^* & q_{11} \end{vmatrix} \geq 0 \tag{4.41b}$$

$$\begin{vmatrix} q_{11} & -q_{21} & -q_{21}^* \\ -q_{21}^* & q_{11} & -q_{21} \\ -q_{21} & -q_{21}^* & q_{11} \end{vmatrix} \geq 0 \tag{4.41c}$$

4.8 TERMINAL PLANES OF JUNCTIONS

A classic network problem is the mapping of the entries of the scattering matrix under a shift of planes of the network. The required transformation is

$$\overline{\mathbf{S}}^1 = \overline{\mathbf{P}}\,\overline{\mathbf{S}}\,\overline{\mathbf{P}} \tag{4.42}$$

$\overline{\mathbf{S}}$ is the scattering matrix at the original set of ports, $\overline{\mathbf{S}}^1$ is that at the new set of ports, and $\bar{\mathbf{P}}$ is a diagonal matrix given by

$$\overline{\mathbf{P}} = \begin{bmatrix} \exp(-j\theta_1) & 0 & 0 & 0 & 0 & 0 \\ 0 & \exp(-j\theta_2) & 0 & 0 & 0 & 0 \\ 0 & 0 & \exp(-j\theta_3) & 0 & 0 & 0 \\ 0 & 0 & 0 & \exp(-j\theta_4) & 0 & 0 \\ 0 & 0 & 0 & 0 & \exp(-j\theta_5) & 0 \\ 0 & 0 & 0 & 0 & 0 & \exp(-j\theta_6) \end{bmatrix} \tag{4.43}$$

where θ_i is the electrical length at port i; θ_i is positive toward the generator and negative away from it. A scrutiny of the linear transformation of the scattering matrix indicates that

$$S_{ij} = S_{ij} \exp[-j(\theta_i - \theta_j)] \tag{4.44}$$

4.9 INSERTION PHASE SHIFT

A desirable quantity in the description of a three-port junction circulator is its phase shift between ports 1 and 2. The derivation of this quantity starts with a statement of S_{21}:

$$S_{21} = \frac{s_0 + \alpha s_+ + \alpha^2 s_-}{3} \tag{4.45}$$

where

$$s_0 = -1 \tag{4.46a}$$

$$s_+ = \exp[-j2(\phi_1 - \Delta\phi_1 + \pi/2)] \tag{4.46b}$$

$$s_- = \exp[-j2(\phi_1 + \Delta\phi_1 + \pi/2)] \tag{4.46c}$$

and

$$\alpha = \exp(j120) \tag{4.47a}$$

$$\alpha^2 = \exp(j240) \tag{4.47b}$$

The quantities s_0, s_+, and s_- are the so-called eigenvalues of the problem region and represent the reflection coefficients of the eigen-networks of the junction. These are defined in some detail in Chapter 5.

Combining the preceding equations readily gives

$$S_{21} = \frac{-1 + (\cos 2\Delta\phi_1 + \sqrt{3}\sin 2\Delta\phi_1)\exp(-j2\phi_1)}{3} \tag{4.48}$$

In the vicinity of the circulation condition

$$2\Delta\phi_1 = \pi/3$$

In the case of a clockwise circulator

$$S_{21} = \frac{-1 + 2\exp(-j2\phi_1)}{3} \tag{4.49}$$

At the midband frequency

$$\phi_1 = \pi/2$$

and

$$S_{21} = -1 \tag{4.50}$$

Writing S_{21} as

$$S_{21} = |S_{21}| \exp(-j2\phi_{21})$$

gives

$$|S_{21}| = \frac{\left[(-1 + 2\cos 2\phi_1)^2 + (2\sin 2\phi_1)^2\right]^{1/2}}{3} \tag{4.51}$$

and

$$\tan 2\phi_{21} = \frac{-2\sin 2\phi_1}{(-1 + 2\cos 2\phi_1)} \tag{4.52}$$

In the vicinity of $\phi_1 = \pi/2$

$$\tan 2\phi_{21} \approx \tfrac{2}{3} \sin 2\phi_1 \tag{4.53}$$

4.10 SPECIFICATION OF THREE-PORT CIRCULATORS WITH NONIDEAL LOADS

An ideal circulator is usually specified by its return loss (dB) at any port, its insertion loss (dB) between any two ports in the direction of circulation, and its isolation (dB) between any two ports in the other direction. The equipment maker, however, is usually more interested in the overall system specification rather than that of the circulator between ideal terminations. The VSWR at port 1 of a three-port circulator is in practice not only dependent on its specification but also on the load condition at port 2 and the specification of the termination at port 3. While an exact formulation of the problem is in principle available, it is too complicated for engineering practice. Upper and lower bounds to the problem in question may be derived by disregarding any reflection at port 1 in constructing the incident wave at port 2 and by neglecting any secondary reflections at port 1. Practical components are usually described in terms of a VSWR specification and this is the notation adopted here.

The required derivation rests on the rules governing the combination of two discrete VSWR S_1 and S_2 associated with two neighboring discontinuities. The worst case effective VSWR at port 1 is

$$S_{\mathrm{eff}} = \frac{(S_C S_A)(S_C S_L) + 1}{S_A + S_L} \tag{4.54}$$

S_C is the VSWR of the circulator at any port, S_A is that of the antenna at port 2, and S_L is that of the load at port 3. A schematic diagram of the arrangement considered here is indicated in Fig. 4.6.

This equation may also be readily solved for either S_A, S_C, or S_L in terms of S_{eff}.

$$S_A = \frac{-1 + S_{\mathrm{eff}} S_L}{-S_{\mathrm{eff}} + S_C^2 S_L} \tag{4.55a}$$

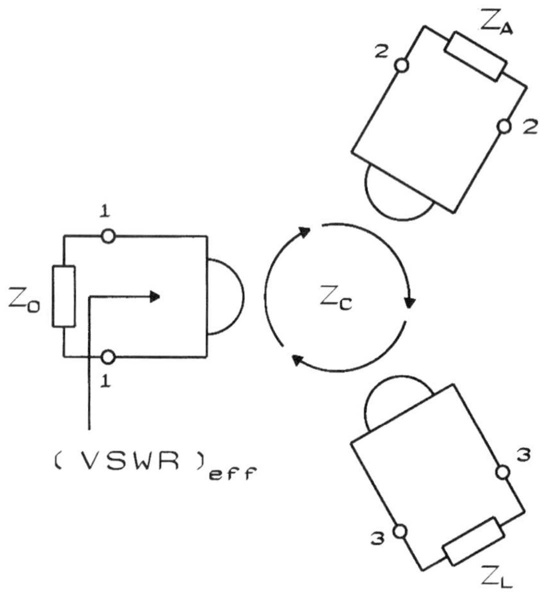

FIGURE 4.6 Schematic diagram of three-port junction circulator with nonideal loads.

$$S_{\mathrm{L}} = \frac{-1 + S_{\mathrm{A}} S_{\mathrm{eff}}}{-S_{\mathrm{eff}} + S_{\mathrm{C}}^2 S_{\mathrm{A}}} \tag{4.55b}$$

and

$$S_{\mathrm{C}} = \left(\frac{S_{\mathrm{eff}}(S_{\mathrm{A}} + S_{\mathrm{L}})}{S_{\mathrm{A}} S_{\mathrm{L}}} - 1 \right)^{1/2} \tag{4.55c}$$

The derivation of the best possible situation at port 1 is

$$S_{\mathrm{eff}} = \frac{S_{\mathrm{C}}^2 + S_{\mathrm{A}} S_{\mathrm{L}}}{S_{\mathrm{A}} + S_{\mathrm{L}}}, \quad S_{\mathrm{refl}} \geq S_{\mathrm{C}} \tag{4.56}$$

and

$$S_{\mathrm{A}} = \frac{S_{\mathrm{C}}^2 - S_{\mathrm{L}} S_{\mathrm{eff}}}{S_{\mathrm{eff}} - S_{\mathrm{L}}}, \quad S_{\mathrm{refl}} \geq S_{\mathrm{C}} \tag{4.57a}$$

$$S_{\mathrm{L}} = \frac{S_{\mathrm{A}} S_{\mathrm{eff}} - S_{\mathrm{C}}^2}{S_{\mathrm{A}} - S_{\mathrm{eff}}}, \quad S_{\mathrm{refl}} \geq S_{\mathrm{C}} \tag{4.57b}$$

$$S_{\mathrm{C}} = [S_{\mathrm{eff}}(S_{\mathrm{A}} + S_{\mathrm{L}}) - S_{\mathrm{A}} S_{\mathrm{L}}]^{1/2}, \quad S_{\mathrm{refl}} \geq S_{\mathrm{C}} \tag{4.57c}$$

Figure 4.7 indicates the relationship between the best and worst bounds on the VSWR at port 1 and the antenna load at port 2 for different circulator specifications and one typical load condition. The lower bounds in these curves are left blank between the origin and $S_{\mathrm{A}} \leq S_{\mathrm{C}}$.

The robustness of the approximations utilized in this work has been verified by introducing either the condition $S_{\mathrm{A}} = S_{\mathrm{C}}$ or $S_{\mathrm{L}} = S_{\mathrm{C}}$ in the lower bound relationship. A scrutiny of each case one at a time indicates that S_{eff} equals S_{C} in keeping with the exact result.

$$S_{\mathrm{eff}} = S_{\mathrm{C}} \tag{4.58}$$

The lower bound obtained here is in keeping with the definition of the complex gyrator circuit of this sort of junction when either port 2 or port 3 is terminated in its complex conjugate load.

Taking the particular case for which $S_{\mathrm{A}} = S_{\mathrm{L}} = S_{\mathrm{C}} = 1.22$ indicates that S_{eff} is bracketed between 1.22 and 1.32.

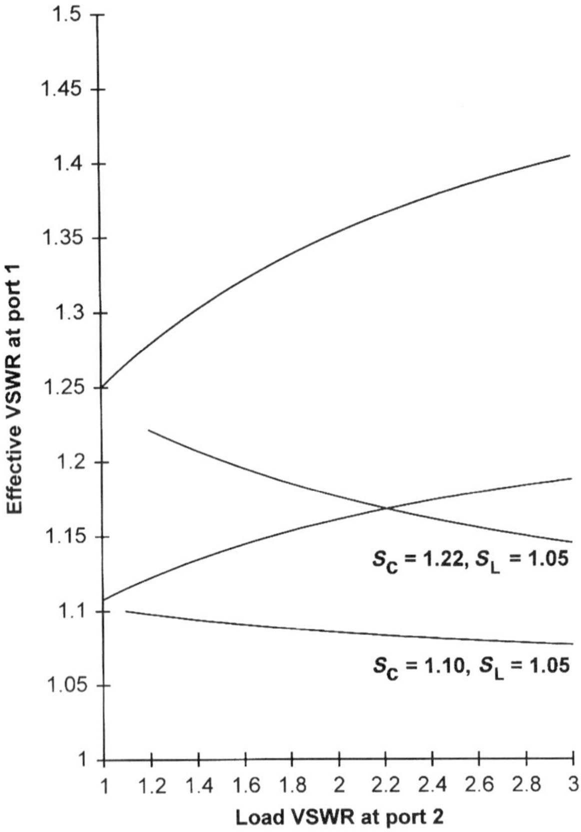

FIGURE 4.7 Upper and lower bounds on effective VSWR of load at port 1 versus VSWR at port 2 for different circulator and termination specifications. [Reproduced with permission from J. Helszajn, Standing-wave solution of 3-port junction circulator with 1-port terminated in a variable short circuit, *Proc. IEE*, Vol. 126, Part H (Microwaves, Optics and Acoustics), Mar. 1979.]

4.11 SWEPT FREQUENCY DESCRIPTION OF SCATTERING PARAMETERS

The scattering parameters of a junction circulator may be extracted from a knowledge of the frequency behavior of various one-port immittance statements. These types of measurements may also be separately generalized by either forming the input impedance or reflection coefficient at one port of the device with the other two terminated in various load conditions. The required relationship is given by

$$\rho_{in} = S_{11} + \frac{(\rho_2 + \rho_3)A + \rho_2\rho_3(B - 2S_{11}A)}{1 - (\rho_2 + \rho_3)S_{11} + \rho_2\rho_3(S_{11}^2 - A)} \tag{4.59}$$

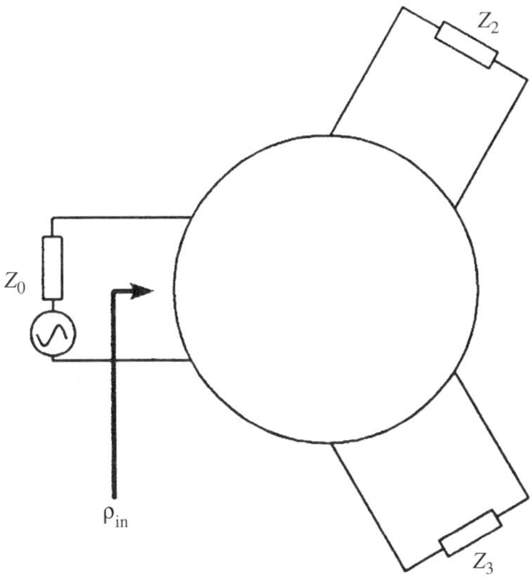

FIGURE 4.8 One-port experimental arrangement for extraction of scattering parameters.

where

$$A = S_{12}S_{13} \tag{4.60a}$$

$$B = S_{12}^3 + S_{13}^3 \tag{4.60b}$$

Each scattering parameter has both amplitude and phase and ρ_2 and ρ_3 are the reflection coefficients at ports 2 and 3:

$$\rho_i = \frac{Z_i - Z_0}{Z_i + Z_0}, \quad i = 2, 3 \tag{4.61}$$

The unknown amplitude and phase constants of S_{11}, S_{21}, and S_{31} may be deduced from measurements of ρ_{in} at port 1 by selecting six different load conditions at ports 2 and 3 and thereafter constructing suitable linear equations from which the unknown quantities may be calculated. If, for instance, $\rho_2 = \rho_3 = 0$ in the description of ρ_{in}, then the reflection coefficient is given for completeness as

$$\rho_{in} = S_{11} \tag{4.62}$$

as is readily understood. Figure 4.8 indicates the experimental arrangement.

CHAPTER FIVE

Eigenvalue Adjustment of Three-Port Circulator

5.1 INTRODUCTION

The unitary condition introduced in Chapter 4 may be used to define an ideal circulator but does not give any insight into its physical adjustment. In order to do so, it is necessary to establish the eigenvalues s_i and eigenvectors U_i associated with its scattering matrix description. A study of this problem indicates that the entries of the scattering matrix may be reduced to linear combinations of the so-called eigenvalues of the junction. A typical eigenvalue of a symmetrical m-port junction is a one-port reflection coefficient or immittance at any port produced by the application of a typical eigenvector at its ports. The eigenvectors are unique ways of exciting the junction and are determined by its symmetry only. The one-port circuits formed in this way are known as the eigen-networks of the problem region. The adjustment of the ideal three-port circulator requires that its reflection eigenvalues lie equally spaced on a unit circle. This condition is both necessary and sufficient and can be achieved by having recourse to only two independent variables. There are altogether four possible ideal eigenvalue diagrams of degree-1, each of which is associated with a unique complex gyrator circuit. One eigenvalue is usually associated with a so-called in-phase eigenvector and the other two by counterrotating ones. In a reciprocal three-port junction the latter two eigenvalues are degenerate, whereas in a nonreciprocal junction these are split by the gyrotropy of the gyromagnetic resonator.

The Stripline Circulator: Theory and Practice. By J. Helszajn

5.2 SCATTERING MATRIX, EIGENVALUES, AND EIGENVECTORS

The importance of the eigenvalue problem in network theory, to be described now, resides in the fact that the entries of a symmetrical or Hermitian square matrix may always be expressed in terms of the roots (eigenvalues) of its characteristic equation. One such matrix is the scattering one met in connection with the description of incident and reflected waves in an m-port circuit.

$$\overline{\mathbf{b}} = \overline{\mathbf{S}\mathbf{a}} \tag{5.1}$$

The formulation of the eigenvalue problem starts by examining the solutions of this standard equation in the case for which the m-dependent variables are related by a scalar quantity to the m-independent variables

$$\overline{\mathbf{a}} = \overline{\mathbf{U}}_i \tag{5.2}$$

$$\overline{\mathbf{b}} = s_i \overline{\mathbf{U}}_i \tag{5.3}$$

s_i is known as an eigenvalue and $\overline{\mathbf{U}}_i$ as an eigenvector.

The eigenvalue equation is now obtained by substituting the two preceding conditions into the original equation. The result is

$$\overline{\mathbf{S}\mathbf{U}}_i = s_i \overline{\mathbf{U}}_i \tag{5.4}$$

This equation has a nonvanishing value for $\overline{\mathbf{U}}_i$ provided

$$\det\left|\overline{\mathbf{S}} - s_i \overline{\mathbf{I}}\right| = 0 \tag{5.5}$$

where $\overline{\mathbf{I}}$ is a unit vector.

The determinant defined by this equation is a polynomial of degree m and its m roots are the m eigenvalues of the scattering matrix, some of which may be equal (degenerate). The corresponding m eigenvectors may be evaluated by substituting each eigenvalue one at a time into the eigenvalue equation. In the special case of the symmetrical m-port network considered here, the m-eigenvalues are one-port reflection coefficients at any port of the junction. For a lossless junction, these lie in the complex plane with unit amplitude. These eigenvalues can be calculated once the coefficients of the scattering matrix are specified. The m eigenvectors are the m possible voltage settings at the ports of the junction and are fixed by its symmetry only. The one-port circuits associated with the reflection eigenvalues are known as the eigen-networks of the device. The eigenvectors are completely determined by the symmetry of the junction so that a symmetrical perturbation of the junction alters the phase angles of the eigenvalues but leaves the eigenvectors unchanged. A schematic diagram of the eigenvalues equation is depicted in Fig. 5.1.

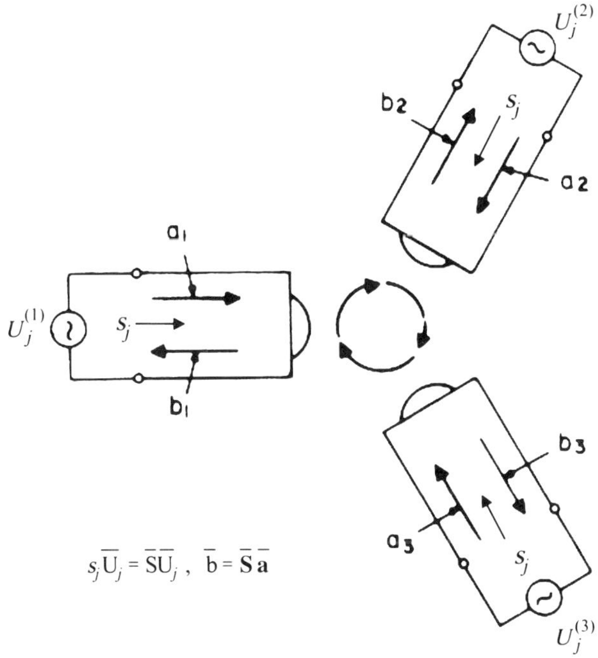

FIGURE 5.1 Schematic diagram of three-port junction indicating definition of eigenvalue equation.

5.3 EIGENVALUE ADJUSTMENT OF THREE-PORT CIRCULATOR

The eigenvalue arrangement of a reciprocal three-port junction for which S_{11} is a minimum coincides with the first of the two circulation conditions of one possible ideal three-port circulator. It will be derived first.

For a reciprocal three-port junction with threefold symmetry, the scattering matrix is

$$\bar{\mathbf{S}} = \begin{bmatrix} S_{11} & S_{21} & S_{21} \\ S_{21} & S_{11} & S_{21} \\ S_{21} & S_{21} & S_{11} \end{bmatrix} \tag{5.6}$$

The characteristic equation associated with this matrix is

$$(S_{11} - s_i)^3 - 3(S_{11} - s_i)S_{21}^2 + 2S_{21}^3 = 0 \tag{5.7}$$

Its three eigenvalues or roots are

$$s_0 = S_{11} + 2S_{21} \tag{5.8}$$

$$s_+ = s_- = S_{11} - S_{21} \tag{5.9}$$

This result indicates that two of the three eigenvalues of a reciprocal three-port junction are degenerate.

Taking linear combinations of these two equations indicates that the entries of the scattering matrix can also be written in terms of the eigenvalues as

$$S_{11} = \frac{s_0 + 2s_+}{3} \tag{5.10}$$

$$S_{21} = \frac{s_0 - s_+}{3} \tag{5.11}$$

It is observed from the first of these two relationships that the reflection coefficient S_{11} in such a junction is a minimum equal to $\left|\frac{1}{3}\right|$ provided

$$s_+ = -s_0 \tag{5.12}$$

The second of these equations suggests that the condition for which the reflection coefficient S_{11} is a minimum coincides with that for which the transmission one S_{21} is a maximum. The eigenvalue diagram for this situation is illustrated in Fig. 5.2.

One possible solution for the one-port reflection coefficients that meets the last two conditions is

$$s_+ = \exp\left[-j2\left(\phi_1 + \frac{\pi}{2}\right)\right] \tag{5.13a}$$

$$s_- = \exp\left[-j2\left(\phi_1 + \frac{\pi}{2}\right)\right] \tag{5.13b}$$

$$s_0 = \exp[-j2(\phi_0)] \tag{5.13c}$$

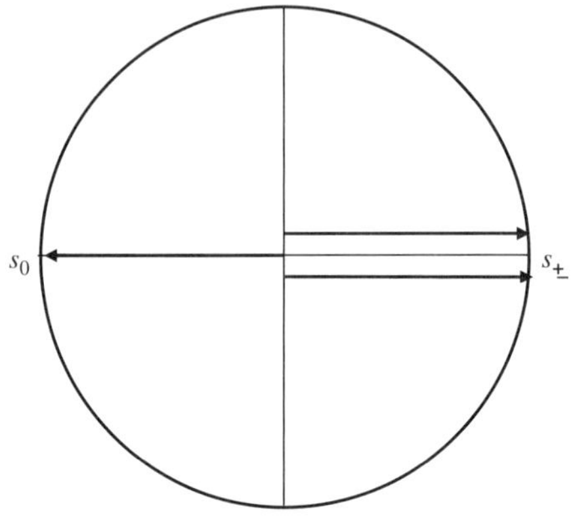

FIGURE 5.2 Eigenvalue diagram of reciprocal three-port junction for maximum power transfer.

provided that

$$\phi_0 = \phi_1 = \frac{\pi}{2} \tag{5.13d}$$

The reflection coefficients obtained in this way are associated with the one-port short- and open-circuited transmission lines illustrated in Fig. 5.3. The choice of these solutions will be discussed separately in connection with the description of the eigenvectors of the junction.

The scattering matrix for the arrangement for which S_{11} is a minimum is now obtained by combining Eq. (5.10) and Eq. (5.11) with Eqs. (5.13a)–(5.13c). The required result is

$$\overline{\mathbf{S}} = \begin{bmatrix} \frac{1}{3} & -\frac{2}{3} & -\frac{2}{3} \\ -\frac{2}{3} & \frac{1}{3} & -\frac{2}{3} \\ -\frac{2}{3} & -\frac{2}{3} & \frac{1}{3} \end{bmatrix} \tag{5.14}$$

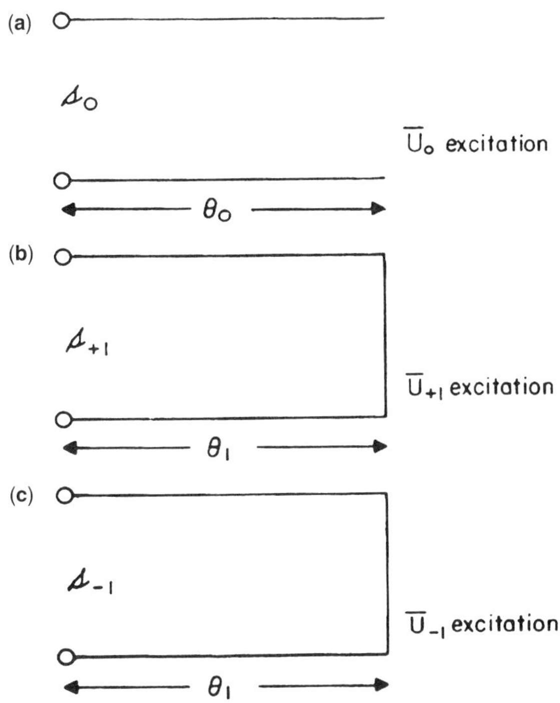

FIGURE 5.3 Eigen-networks of reciprocal three-port junction.

The derivation of the eigenvalue diagram of an ideal circulator likewise proceeds with the definition of its scattering matrix

$$\overline{\mathbf{S}} = \begin{bmatrix} 0 & S_{21} & 0 \\ 0 & 0 & S_{21} \\ S_{21} & 0 & 0 \end{bmatrix} \tag{5.15}$$

where S_{21}has unit amplitude.

The characteristic equation associated with this matrix is given without ado by

$$-s_i^3 + S_{21}^3 = 0 \tag{5.16}$$

One result for a device for which $S_{21} = -1$ is

$$s_+ = \exp\left[-j2\left(\phi_1 + \phi_+ + \frac{\pi}{2}\right)\right] \tag{5.17a}$$

$$s_{-1} = \exp\left[-j2\left(\phi_1 + \phi_- + \frac{\pi}{2}\right)\right] \tag{5.17b}$$

$$s_0 = \exp[-j(2\phi_0)] \tag{5.17c}$$

provided that

$$\phi_1 = \phi_0 = \frac{\pi}{2} \tag{5.18a}$$

$$\phi_+ = -\phi_- = -\frac{\pi}{6} \tag{5.18b}$$

The three reflection coefficients of an ideal circulator therefore lie equally spaced on a unit circle in the manner depicted in Fig. 5.4. One possible set of one-port eigen-networks that is compatible with this solution is indicated in Fig. 5.5.

In what follows it will be demonstrated that the eigenvectors $\overline{\mathbf{U}}_\pm$ corresponding to the eigenvalues s_+ and s_- produce counterrotating magnetic fields on the axis of the junction. Such magnetic fields establish different scalar permeabilities in a gyromagnetic medium so that a practical means of removing the degeneracy between the reflection eigenvalues exists.

5.4 EIGENVECTORS

The three eigenvectors of the problem region may be established by solving the eigenvalue equations one at a time. The equation connecting the eigenvector $\overline{\mathbf{U}}_0$

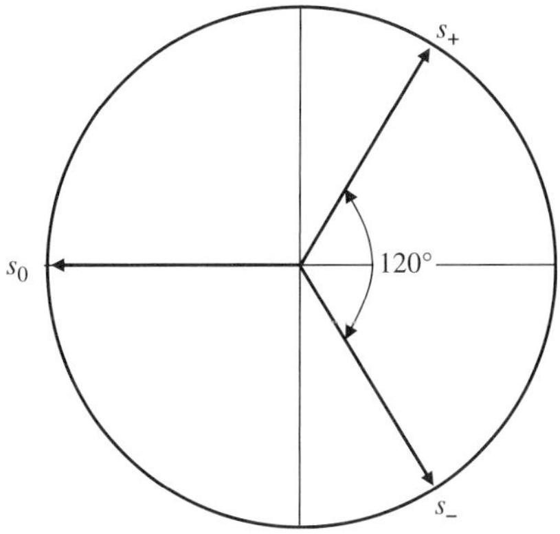

FIGURE 5.4 Eigenvalue diagram of ideal three-port junction circulator.

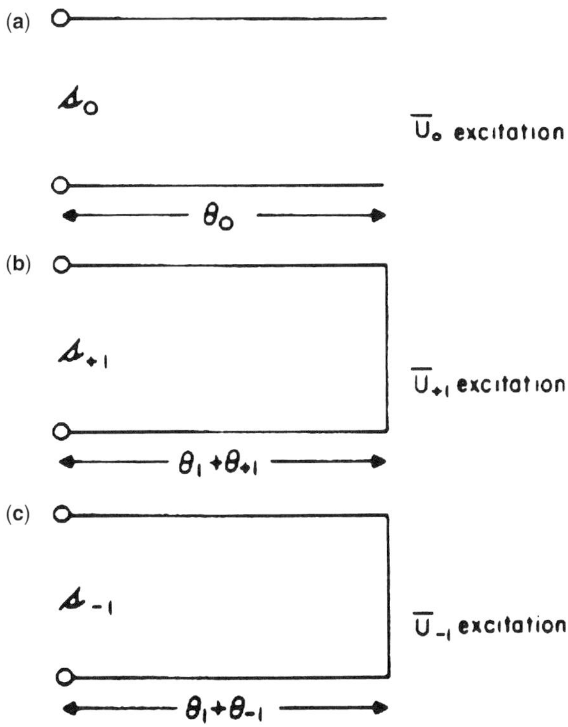

FIGURE 5.5 Eigen-networks of ideal three-port junction circulator.

and the eigenvalue s_0 of an ideal three-port circulator is

$$\begin{bmatrix} 0 & -1 & 0 \\ 0 & 0 & -1 \\ -1 & 0 & 0 \end{bmatrix} \begin{bmatrix} U_0^{(1)} \\ U_0^{(2)} \\ U_0^{(3)} \end{bmatrix} = s_0 \begin{bmatrix} U_0^{(1)} \\ U_0^{(2)} \\ U_0^{(3)} \end{bmatrix} \tag{5.19}$$

where

$$s_0 = \exp(-j\pi) \tag{5.20}$$

Taking $U_0^{(1)}$ as 1 readily gives one solution as

$$U_0^{(1)} = 1 \tag{5.21a}$$

$$U_0^{(2)} = 1 \tag{5.21b}$$

$$U_0^{(3)} = 1 \tag{5.21c}$$

The eigenvalue equation for the eigenvector $\overline{\mathbf{U}}_+$ and the eigenvalue s_+ is

$$\begin{bmatrix} 0 & -1 & 0 \\ 0 & 0 & -1 \\ -1 & 0 & 0 \end{bmatrix} \begin{bmatrix} U_+^{(1)} \\ U_+^{(2)} \\ U_+^{(3)} \end{bmatrix} = s_+ \begin{bmatrix} U_+^{(1)} \\ U_+^{(2)} \\ U_+^{(3)} \end{bmatrix} \tag{5.22}$$

where

$$s_+ = \exp(-j\pi/3) \tag{5.23}$$

Taking $U_+^{(1)}$ as 1 indicates that the solution, in this instance, is given by

$$U_+^{(1)} = 1 \tag{5.24a}$$

$$U_+^{(2)} = \alpha \tag{5.24b}$$

$$U_+^{(3)} = \alpha^2 \tag{5.24c}$$

where

$$\alpha = \exp(j120) \tag{5.25a}$$

$$\alpha^2 = \exp(j240) \tag{5.25b}$$

The solution of the eigenvalue equation for the eigenvector $\overline{\mathbf{U}}_-$ of the junction is given without ado by

$$U_-^{(1)} = 1 \tag{5.26a}$$

$$U_-^{(2)} = \alpha^2 \tag{5.26b}$$

$$U_-^{(3)} = \alpha \tag{5.26c}$$

The eigenvectors are normalized in such a way that

$$(\overline{\mathbf{U}}_\pm)^T \overline{\mathbf{U}}_\pm^* = 1 \tag{5.27}$$

One solution is therefore described by

$$\overline{\mathbf{U}}_0 = \frac{1}{\sqrt{3}}\begin{bmatrix} 1 \\ 1 \\ 1 \end{bmatrix}, \qquad \overline{\mathbf{U}}_+ = \frac{1}{\sqrt{3}}\begin{bmatrix} 1 \\ \alpha \\ \alpha^2 \end{bmatrix}, \qquad \overline{\mathbf{U}}_- = \frac{1}{\sqrt{3}}\begin{bmatrix} 1 \\ \alpha^2 \\ \alpha \end{bmatrix} \tag{5.28}$$

The three eigensolutions of the problem are illustrated in Fig. 5.6. Application of each eigenvector one at a time reveals the variables s_0, s_+, and s_- at any port.

Scrutiny of the magnetic field pattern in the transverse plane one at a time for the $\mathbf{U}_\pm$ eigenvectors indicates that each is circularly polarized on the axis of the junction with a different hand. A suitably magnetized magnetic insulator displays different scalar permeabilities $\mu \mp \kappa$, for such rotating fields so that such a magnetized junction may be employed to split the degeneracy between the $s_\pm$ reflection coefficients. This arrangement therefore provides a practical means for the construction of the ideal circulator.

A property of a symmetric junction that will now be verified is that a single signal at any port establishes all the eigenvectors of the problem region with equal amplitudes. This statement may be demonstrated without difficulty by constructing a linear combination of the eigenvectors associated with the problem under consideration.

$$\begin{bmatrix} 1 \\ 0 \\ 0 \end{bmatrix} = \frac{1}{3}\begin{bmatrix} 1 \\ 1 \\ 1 \end{bmatrix} + \frac{1}{3}\begin{bmatrix} 1 \\ \alpha \\ \alpha^2 \end{bmatrix} + \frac{1}{3}\begin{bmatrix} 1 \\ \alpha^2 \\ \alpha \end{bmatrix} \tag{5.29}$$

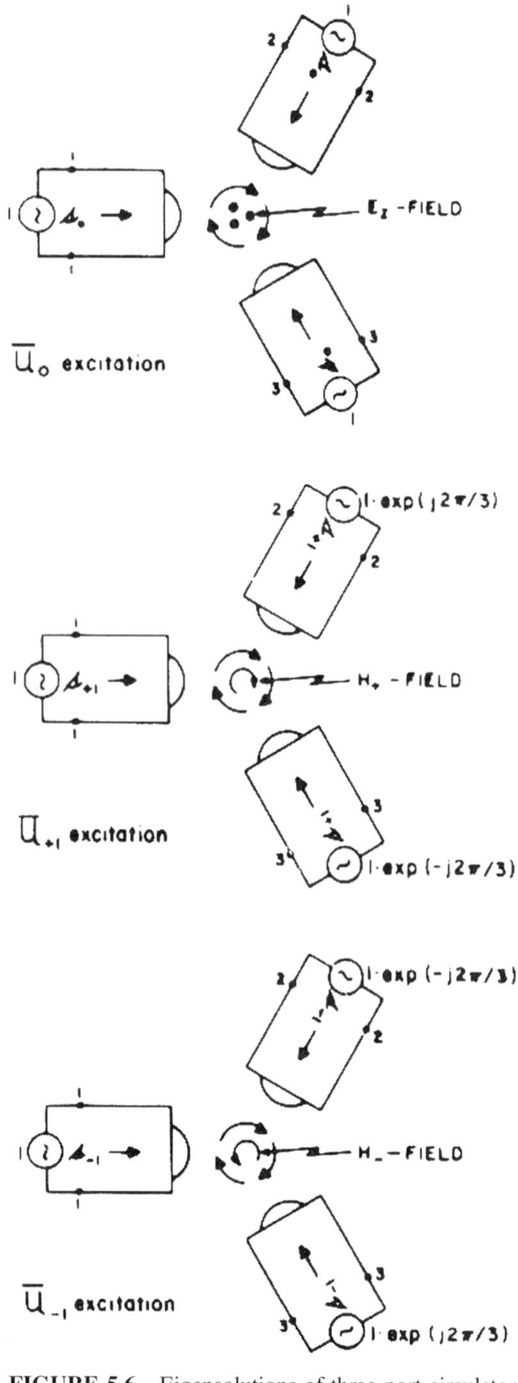

FIGURE 5.6 Eigensolutions of three-port circulator.

The equality is met provided the following identity is satisfied:

$$1 + \alpha + \alpha^2 = 0 \tag{5.30}$$

The eigenvectors are fixed by the junction symmetry; a symmetric perturbation of the junction alters the phases of the eigenvalues but leaves the eigenvectors unchanged.

5.5 SCATTERING MATRIX OF THREE-PORT JUNCTION CIRCULATOR

One technique that is often employed in the analysis of a symmetrical m-port network is to express the entries of its matrix description in terms of the solutions of more simple (usually one-port) circuits that reflect the symmetry of the problem region. One unique decomposition may be achieved by adopting the eigenvectors of the network for this purpose. The one-port circuits established in this way are known as the eigen-networks of the problem region. The eigenvalues of the scattering matrix are then the reflection coefficients associated with these eigen-networks. Furthermore, the entries of the scattering matrix are now reduced to simple linear combinations of these same eigenvalues. This procedure will now be demonstrated in connection with the description of the three-port junction circulator by considering each eigenvector one at a time as a preamble to laying out a more universal mathematical technique.

The required relationship between the scattering parameters and the eigenvalues of the problem region starts by constructing the incident and reflected waves for the first triplet of voltage settings or eigenvector.

$$\begin{bmatrix} b_1 \\ b_2 \\ b_3 \end{bmatrix} = \begin{bmatrix} S_{11} & S_{21} & S_{31} \\ S_{31} & S_{11} & S_{21} \\ S_{21} & S_{31} & S_{11} \end{bmatrix} \begin{bmatrix} \frac{1}{3} \\ \frac{1}{3} \\ \frac{1}{3} \end{bmatrix} \tag{5.31}$$

This gives

$$\frac{b_1}{\left(\frac{1}{3}\right)} = S_{11} + S_{21} + S_{31} \tag{5.32a}$$

$$\frac{b_2}{\left(\frac{1}{3}\right)} = S_{31} + S_{11} + S_{21} \tag{5.32b}$$

$$\frac{b_3}{\left(\frac{1}{3}\right)} = S_{21} + S_{31} + S_{11} \tag{5.32c}$$

The reflection coefficient at each port is therefore identical and in this instance is denoted by ρ_0:

$$\frac{b_1}{\left(\frac{1}{3}\right)} = \frac{b_2}{\left(\frac{1}{3}\right)} = \frac{b_3}{\left(\frac{1}{3}\right)} = \rho_0 \tag{5.33}$$

where

$$\rho_0 = S_{11} + S_{21} + S_{31} \tag{5.34}$$

The factor $\frac{1}{3}$ in the definition of the incident waves stems from the fact that each port is simultaneously excited.

Taking the second triplet of voltage settings or eigenvector gives

$$\begin{bmatrix} b_1 \\ b_2 \\ b_3 \end{bmatrix} = \begin{bmatrix} S_{11} & S_{21} & S_{31} \\ S_{31} & S_{11} & S_{21} \\ S_{21} & S_{31} & S_{11} \end{bmatrix} \begin{bmatrix} 1/3 \\ \alpha/3 \\ \alpha^2/3 \end{bmatrix} \tag{5.35}$$

or

$$\frac{b_1}{(1/3)} = S_{11} + \alpha S_{21} + \alpha^2 S_{31} \tag{5.36a}$$

$$\frac{b_2}{(\alpha/3)} = S_{11} + \alpha S_{21} + \alpha^2 S_{31} \tag{5.36b}$$

$$\frac{b_3}{(\alpha^2/3)} = S_{11} + \alpha S_{21} + \alpha^2 S_{31} \tag{5.36c}$$

provided $\alpha^3 = 1$.

The reflection coefficient at each port is again the same. Denoting it by ρ_+ gives

$$\rho_+ = S_{11} + \alpha S_{21} + \alpha^2 S_{31} \tag{5.37}$$

Likewise

$$\rho_- = S_{11} + \alpha^2 S_{21} + \alpha S_{31} \tag{5.38}$$

The required result is now established by taking suitable combinations of ρ_0, ρ_+, and ρ_-. Taking S_{11} by way of an example indicates that

$$S_{11} = \frac{\rho_0 + \rho_+ + \rho_-}{3} \tag{5.39a}$$

as asserted.

The other two relationships are obtained without ado:

$$S_{21} = \frac{\rho_0 + \alpha\rho_+ + \alpha^2\rho_-}{3} \tag{5.39b}$$

$$S_{31} = \frac{\rho_0 + \alpha^2\rho_+ + \alpha\rho_-}{3} \tag{5.39c}$$

The entries of the scattering matrix are therefore simple linear combinations of the one-port variables or eigenvalues revealed by the different eigenvectors of the problem region as asserted.

If the network is reciprocal, then

$$\rho_+ = \rho_- \tag{5.40}$$

and

$$S_{11} = \frac{\rho_0 + 2\rho_+}{3} \tag{5.41a}$$

$$S_{21} = S_{31} = \frac{\rho_0 - \rho_+}{3} \tag{5.41b}$$

in keeping with Eqs. (5.10) and (5.11).

5.6 DIAGONALIZATION

If the eigenvectors and eigenvalues are known, it is possible to expand the coefficients of the scattering matrix using a simple mathematical technique. This may be done by using the following similarity transformation:

$$\bar{\mathbf{S}} = \bar{\mathbf{U}}\,\bar{\boldsymbol{\lambda}}\,\bar{\mathbf{U}}^{-1} \tag{5.42}$$

The columns of the matrix $\bar{\mathbf{U}}$ are constructed in terms of the eigenvectors $\bar{\mathbf{U}}_i$ of the problem region, and $\bar{\mathbf{U}}^{-1}$ is the inverse of $\bar{\mathbf{U}}$. The matrix $\bar{\boldsymbol{\lambda}}$ is a diagonal matrix with the eigenvalues of $\bar{\mathbf{S}}$ along its main diagonal. If the eigenvectors of $\bar{\mathbf{S}}$ are linearly independent and $\bar{\mathbf{S}}$ is a real symmetric or Hermitian matrix, it may be demonstrated that

$$\bar{\mathbf{U}}^{-1} = (\bar{\mathbf{U}}^*)^{\mathrm{T}} \tag{5.43}$$

where $(\bar{\mathbf{U}}^*)^{\mathrm{T}}$ is the transpose of the complex conjugate of $\bar{\mathbf{U}}$. The relationship between the eigenvalue and the coefficients of the scattering matrix is then obtained by multiplying out the similarity identity.

The diagonalization procedure will now be developed in the case of an ideal circulator circuit by constructing the square matrices $\overline{\mathbf{U}}$ and $(\overline{\mathbf{U}}^*)^{\mathrm{T}}$ in terms of the eigenvectors of the problem region.

$$\overline{\mathbf{U}} = \frac{1}{\sqrt{3}}\begin{bmatrix} 1 & 1 & 1 \\ 1 & \alpha & \alpha^2 \\ 1 & \alpha^2 & \alpha \end{bmatrix} \tag{5.44}$$

$$\left(\overline{\mathbf{U}}^*\right)^T = \frac{1}{\sqrt{3}}\begin{bmatrix} 1 & 1 & 1 \\ 1 & \alpha^* & (\alpha^*)^2 \\ 1 & (\alpha^*)^2 & \alpha^* \end{bmatrix} \tag{5.45}$$

In forming the matrix $\overline{\mathbf{U}}$ in terms of the m eigenvectors of the problem region, care has been taken to ensure that the resultant matrix is symmetrical.

The diagonal matrix $\overline{\boldsymbol{\lambda}}$ is constructed using the eigenvalues of $\overline{\mathbf{S}}$:

$$\overline{\boldsymbol{\lambda}} = \begin{bmatrix} s_0 & 0 & 0 \\ 0 & s_+ & 0 \\ 0 & 0 & s_- \end{bmatrix} \tag{5.46}$$

Diagonalizing the matrix $\overline{\mathbf{S}}$ gives

$$\begin{bmatrix} S_{11} & S_{21} & S_{31} \\ S_{31} & S_{11} & S_{21} \\ S_{21} & S_{31} & S_{11} \end{bmatrix} = \frac{1}{3}\begin{bmatrix} 1 & 1 & 1 \\ 1 & \alpha & \alpha^2 \\ 1 & \alpha^2 & \alpha \end{bmatrix}\begin{bmatrix} s_0 & 0 & 0 \\ 0 & s_+ & 0 \\ 0 & 0 & s_- \end{bmatrix} \times \begin{bmatrix} 1 & 1 & 1 \\ 1 & \alpha^* & (\alpha^*)^2 \\ 1 & (\alpha^*)^2 & \alpha^* \end{bmatrix} \tag{5.47}$$

The required result is in accord with that described by Eqs. (5.39), (5.40), and (5.41).

The eigenvalues appearing in the scattering description of the junction correspond to one-port reflection variables. Such variables are displayed at any port provided the generator settings coincide with the corresponding eigenvectors.

This result again indicates that the reflection eigenvalues of an ideal three-port junction circulator are displaced by 120° on a unit circle.

5.7 DISSIPATION EIGENVALUES

One way to cater for dissipation in an m-port symmetrical junction is to introduce a dissipation matrix. If the scattering and dissipation matrices have common eigenvectors, the respective eigenvalues are related by the following theorem. *If*

$$\overline{\mathbf{Q}}\overline{\mathbf{U}}_i = q_i\overline{\mathbf{U}}_i$$

then

$$f(\overline{\mathbf{Q}})\overline{\mathbf{U}}_i = f(q_i)\overline{\mathbf{U}}_i$$

where $\overline{\mathbf{U}}_i$ is an eigenvector.

This theorem may be used to deduce the eigenvalues of the dissipation matrix in terms of those of the scattering matrix.

Making use of the relationship between the two matrices in Chapter 4 gives, in the case of the three-port junction,

$$q_0 = 1 - s_0 s_0^* \tag{5.48a}$$

$$q_{+1} = 1 - s_{+1} s_{+1}^* \tag{5.48b}$$

$$q_{-1} = 1 - s_{-1} s_{-1}^* \tag{5.48c}$$

The last three equations may be used to construct the scattering matrix in terms of the eigenvalues of the dissipation matrix. In a lossless junction, the amplitudes of the scattering matrix eigenvalues are unity, while if the junction is lossy the amplitudes will depart from unity.

The above discussion indicates that the eigenvalues of the dissipation matrix represent the dissipation associated with each possible way of exciting the junction. These eigenvalues are real quantities that become zero when those of the scattering matrix become unity.

The entries of the dissipation matrix are

$$q_{11} = \frac{q_0 + q_+ + q_{-1}}{3} \tag{5.49a}$$

$$q_{12} = \frac{q_0 + \alpha q_{+1} + \alpha^2 q_{-1}}{3} \tag{5.49b}$$

$$q_{13} = \frac{q_0 + \alpha^2 q_{+1} + \alpha q_{-1}}{3} \tag{5.49c}$$

Here, q_{11} represents the total dissipation of the junction, and q_{12} is a complex quantity that determines the allowable relationships between the scattering parameters.

In practice, the entries of the dissipation and scattering matrices may be evaluated directly by relating their eigenvalues to the loaded and unloaded Q-factors of the junction eigen-networks.

5.8 EVALUATION OF DEGENERATE COUNTERROTATING EIGENVALUE

One possible way in which the degenerate eigenvalue s_1 of a reciprocal three-port junction can be experimentally deduced without the need to fabricate its eigen-network consists of placing a sliding short-circuit at port 2 and a matched load at port 3. The variable short-circuit is varied until there is total reflection at port 1. The reflection coefficient at port 1 is then the eigenvalue s_1. This technique is especially appropriate in the case of a waveguide junction for which the reference terminals are usually ill defined. The experimental arrangement is shown in Fig. 5.7.

The derivation of this technique starts with the relationships between the incoming and outgoing waves at the three ports of the junction.

$$\begin{bmatrix} b_1 \\ b_2 \\ b_3 \end{bmatrix} = \begin{bmatrix} S_{11} & S_{21} & S_{21} \\ S_{21} & S_{11} & S_{21} \\ S_{21} & S_{21} & S_{11} \end{bmatrix} \begin{bmatrix} a_1 \\ a_2 \\ a_3 \end{bmatrix} \tag{5.50}$$

where for a reciprocal three-port junction

$$S_{11} = \frac{s_0 + 2s_1}{3} \tag{5.51a}$$

$$S_{21} = \frac{s_0 - s_1}{3} \tag{5.51b}$$

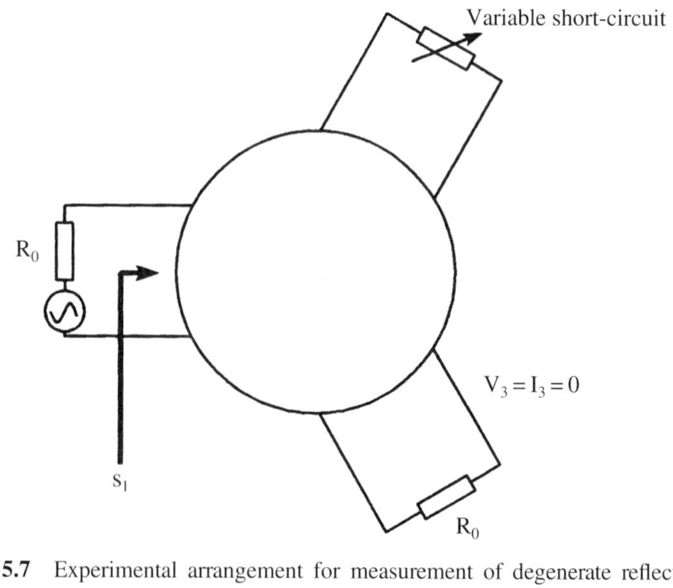

FIGURE 5.7 Experimental arrangement for measurement of degenerate reflection eigenvalue s_1.

In the arrangement considered here a variable short-circuit at port 2 is located at a suitable plane in order to decouple port 3 from the other two. The conditions to be satisfied are

$$b_2 = a_2 \exp\,[-j2(\phi + \pi/2)] \tag{5.52}$$

and

$$a_3 = b_3 = 0 \tag{5.53}$$

ϕ is the electrical length between the position of the short-circuit and the reference plane of b_2/a_2. Introducing the second condition into Eq. (5.50) gives

$$b_1 = S_{11}a_1 + S_{21}a_2 \tag{5.54a}$$

$$b_2 = S_{21}a_1 + S_{11}a_2 \tag{5.54b}$$

$$0 = S_{21}a_1 + S_{21}a_2 \tag{5.54c}$$

Scrutiny of the third of these three equations indicates that the waves a_1 and a_2 at ports 1 and 2 are out-of-phase there.

$$a_1/a_2 = -1 \tag{5.55}$$

Examination of the first and second equations separately indicates that the network between ports 1 and 2 formed in this way is symmetrical. Substituting the above condition into Eqs. (5.54a) and (5.54b) indicates that

$$b_1/a_1 = S_{11} - S_{21} = s_1 \tag{5.56}$$

and

$$b_1/a_1 = b_2/a_2 \tag{5.57}$$

respectively. The reflection coefficient at port 1 is then the degenerate eigenvalue s_1 of a junction:

$$s_1 = \exp\,[-j2(\phi + \pi/2)] \tag{5.58}$$

This measurement therefore provides a means of measuring s_1 with reference to the short-circuit position. The planes at port 1 at which the standing wave is zero are known as the characteristic planes of the junction. The characteristic plane is defined by the short-circuit positions given by $\phi = 0$, π, 2π, and so on.

It is of note that the condition $a = b_3 = 0$ is compatible with that met in the development of the degenerate counterrotating eigen-network; namely, $V_3 = I_3 = 0$.

5.9 EVALUATION OF IN-PHASE EIGENVALUE

The phase angle of the in-phase eigen-network of a demagnetized junction may be measured directly using the eigenvalue approach by applying equal in-phase signals at the three ports of the junction. It may also be deduced by making use of the relationship between the scattering variable S_{11} and the in-phase and degenerate counterrotating eigenvalues s_0 and s_1.

In a lossless device for which the in-phase eigen-network may be synthesized by an open-circuited network and the degenerate counterrotating ones by short-circuited ones,

$$s_0 = 1\exp(-j2\phi_0) \tag{5.59}$$

$$s_1 = 1\exp[-j2(\phi_1 + \pi/2)] \tag{5.60}$$

and

$$S_{11} = |S_{11}|\exp(-j2\phi_{11}) \tag{5.61}$$

where

$$|S_{11}| = \frac{(\text{VSWR}) - 1}{(\text{VSWR}) + 1} \tag{5.62}$$

The reflection phase angle ϕ_0 may be evaluated using Eq. (5.51a) by forming one of four possible relationships between the independent variables $|S_{11}|$, ϕ_{11}, and ϕ_1.

$$2\phi_0 = \cos^{-1}(3|S_{11}|\cos 2\phi_{11} + 2\cos 2\phi_1) \tag{5.63a}$$

$$2\phi_0 = 2\phi_1 + \pi + \cos^{-1}\left(\frac{9|S_{11}|^2 - 5}{4}\right) \tag{5.63b}$$

$$2\phi_0 = 2\phi_{11} + \cos^{-1}\left(\frac{3|S_{11}|^2 - 1}{2|S_{11}|}\right) \tag{5.63c}$$

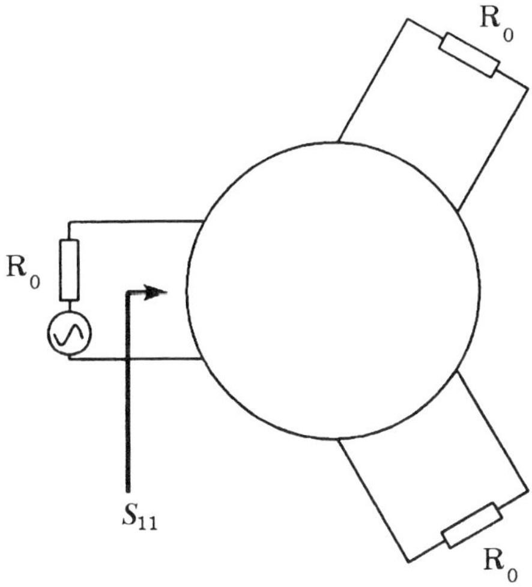

FIGURE 5.8 Experimental arrangement for measurement of S_{11}.

The first identity is constructed by taking a linear combination of s_0 and s_0^*, and the other two are obtained by forming the products of $S_{11}S_{11}^*$ and $s_1s_1^*$. The first relationship for ϕ_0 requires a knowledge of all three independent variables, whereas the second two need only a measurement of $|S_{11}|$ and a knowledge of ϕ_1 or ϕ_{11}.

The fourth relationship between ϕ_0, ϕ_1, and ϕ_{11} may also be formed by eliminating $|S_{11}|$ by Eqs. (5.63a) and (5.63b). The result at $2\phi_1 = \pi$ involves ϕ_0 and ϕ_{11} only:

$$\cos^2 2\phi_0 + (4 - 4\cos^2 2\phi_{11})\cos 2\phi_0 + (4 - 5\cos^2 2\phi_{11}) = 0 \tag{5.64}$$

S_{11} may be evaluated by terminating the two output ports by matched loads in the manner indicated in Fig. 5.8, and s_1 may be determined by decoupling port 3 from port 1 by placing a variable short-circuit at port 2 as already described in connection with Fig. 5.9. The angles of the scattering variables may be located at the reference plane of the junction ($d_{s/c}$) by replacing the resonator by a metal plug. The result for ϕ_{11} is

$$2\phi_{11} = (4\pi/\lambda_g)(d_{s/c} - d_{min}) \tag{5.65}$$

where d_{min} is the position of a minimum in the VSWR along the line.

A similar relationship applies to $2\phi_1 + \pi$.

$$2\phi_1 + \pi = (4\pi/\lambda_g)(d_{s/c} - d_{min}) \tag{5.66}$$

5.10 SPLIT FREQUENCIES OF GYROMAGNETIC RESONATORS

Useful information about junction circulators may also be deduced from a knowledge of the frequency variation of the reflection coefficient at port 1 with ports 2 and 3 terminated in matched load. This method is particularly attractive with swept frequency

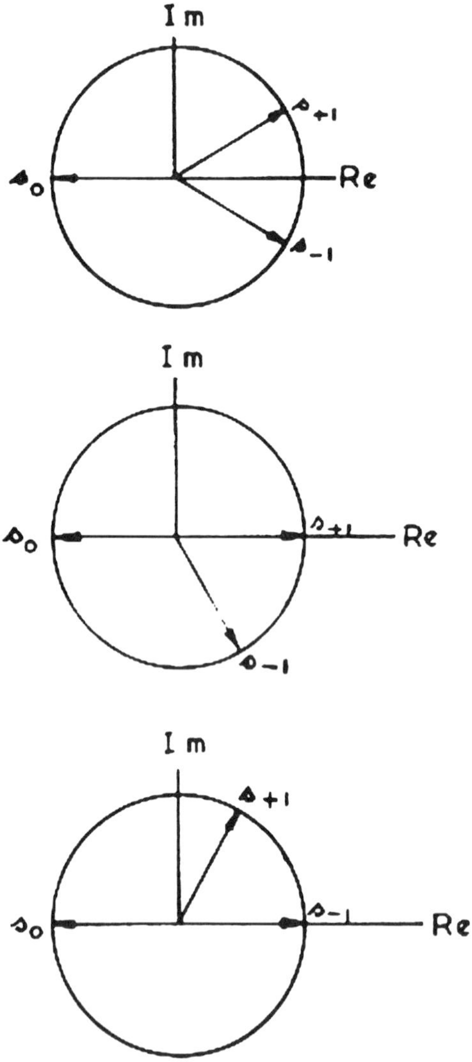

FIGURE 5.9 Eigenvalue diagrams of magnetized circulator. (Reproduced with permission from J. Helszajn, Microwave measurement techniques for junction circulators, *Trans. IEEE Microwave Theory Tech.*, Vol. MTT-21, pp. 347–351, 1973.)

instrumentation. The two split frequencies of the device coincide with the frequencies for which

$$|S_{11}| = \frac{1}{3} \tag{5.67}$$

which corresponds with a VSWR of 2:1 or a return loss of 9.5 dB. The derivation of this condition begins by making use of the standard relationship between the scattering parameters and its eigenvalues:

$$S_{11} = \frac{s_0 + s_+ + s_-}{3} \tag{5.68}$$

It continues by replacing s_0 by

$$s_0 = -1 \tag{5.69}$$

and by recognizing that the required result coincided with the conditions for which either s_+ or s_- is in antiphase with s_0.

The first condition gives

$$|S_{11}| = \left|\frac{s_+}{3}\right| \tag{5.70}$$

and the second condition gives

$$|S_{11}| = \left|\frac{s_-}{3}\right| \tag{5.71}$$

The required result is obtained by recalling that

$$|s_+| = |s_-| = 1 \tag{5.72}$$

Figure 5.9 depicts the required eigenvalue diagrams.

5.11 PHASE ANGLE OF IN-PHASE EIGENVALUE

A scrutiny of the relationship between the reflection coefficient at port 1 of the terminated circulator and the eigenvalues of the junction indicates that the amplitude of the former takes the value of $\frac{1}{3}$ whenever any two of the three eigenvalues are in antiphase. The angle of the reflection coefficient is then that of the remaining eigenvalue. The in-phase reflection angle at discrete frequencies is obtained by setting the arithmetic mean of the counterrotating eigenvalues to zero.

$$s_+ + s_- = 0 \tag{5.73}$$

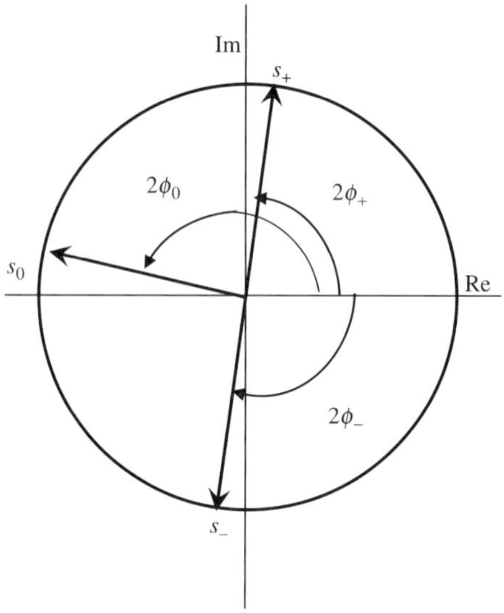

FIGURE 5.10 Extraction of in-phase reflection angle ($\phi_{11} = \phi_0$, $|S_{11}| = \frac{1}{3}$, $g > \sqrt{3}$).

Introducing this condition into the relationship between the reflection coefficient at port 1 of the junction and the eigenvalues of the problem region gives the required result.

$$|S_{11}| = \frac{1}{3} \tag{5.74}$$

$$\phi_{11} = 2\phi_0 \tag{5.75}$$

In obtaining this result S_{11} is defined as

$$S_{11} = |S_{11}| \exp(-j\phi_{11}) \tag{5.76}$$

The angles ϕ_{11} of the reflection coefficient S_{11} and that of the in-phase eigen-network $2\phi_0$ are therefore equal in this situation. This condition is satisfied in the vicinity of the midband frequency at a value of gyrotropy for which the return loss has a single value of 9.5 dB and at two discrete frequencies above this value of gyrotropy. The immittance of the terminated circulator at each of these points is essentially real. The eigenvalue diagram at which the former condition applies is depicted in Fig. 5.10.

5.12 ADJUSTMENT OF TRIPLE POLE CIRCULATOR

The usual eigenvalue adjustment of the classic circulator involves the extraction of a pair of degenerate counterrotating impedance poles and one impedance zero. This may be understood by recalling that the reflection eigenvalue of a typical eigen-network has a value of $+1$ at a typical impedance pole and a value of -1 at a zero. It is also understood that this sort of adjustment is associated with a degree-1 frequency response. It is, however, not unique. Another realization is based on the extraction of an impedance pole from each eigen-network. Its adjustment is tied to a degree-2 frequency response. The various steps involved in this

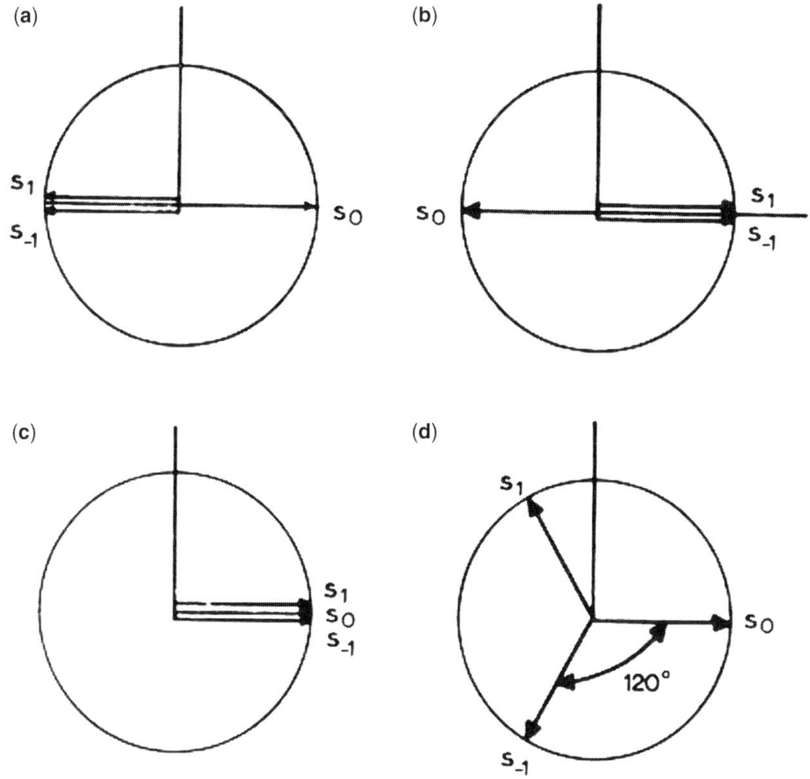

FIGURE 5.11 (a) Initial location of eigenvalues for wideband junction. (b) Location of eigenvalues after first circulation for wideband junction. (c) Location of eigenvalues after second circulation adjustment for wideband junction. (d) Location of eigenvalues after third circulation adjustment for wideband junction. (Reproduced with permission from J. Helszajn, Wideband circulator adjustment using $n = \pm 1$ and $n = 0$ electromagnetic-field patterns, *Electron. Lett.*, Vol. 6, pp. 729–731, Nov. 12, 1970.)

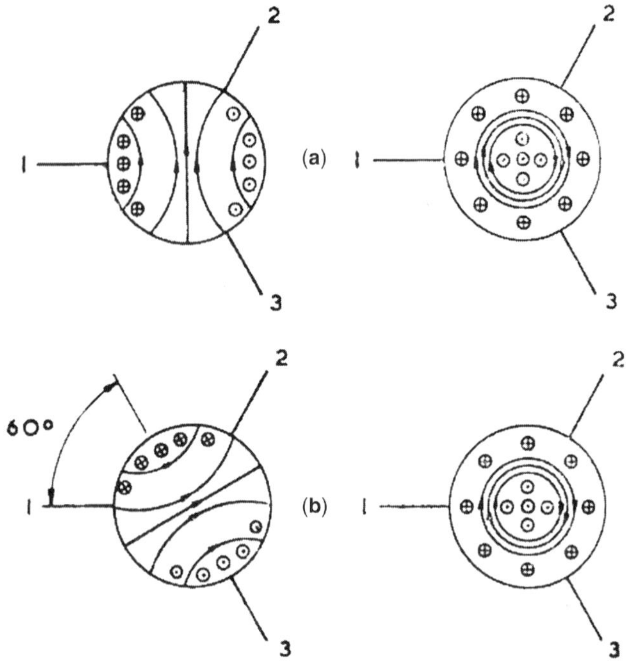

FIGURE 5.12 Phenomenological adjustment of triple pole circulator using the $TM_{0,1,0}$ and $TM_{\pm 1,1,0}$ modes in a disk resonator. (Reproduced with permission from J. Helszajn, Three resonant mode adjustment of the waveguide circulator, *Radio Electron. Engineer*, Vol. 42, No. 4, pp. 1–4, Apr. 1972.)

solution are illustrated in Fig. 5.11. It involves access to three instead of two independent variables. A phenomenological adjustment of this sort of circulator in terms of the $TM_{0,1,0}$ and $TM_{\pm 1,1,0}$ in a simple disk gyromagnetic resonator with the top and bottom electric walls and a magnetic node wall is indicated in Fig. 5.12.

CHAPTER SIX

Impedance Matrix of Junction Circulator

6.1 INTRODUCTION

There are in general six different ways in which the boundary conditions of a circulator can be introduced. Its scattering matrix and its immittance matrices provide three possible descriptions and the corresponding eigenvalues provide three others. The actual choice in any particular situation is usually determined by the physical problem. It is always possible to construct the scattering matrix of a junction but its impedance or admittance matrix need not be realizable. It has already been seen that a knowledge of the scattering matrix of the junction together with the unitary condition is sufficient for its definition and that one of its corresponding eigenvalues is all that is required for its adjustment. One-port equivalent circuits, however, separately require a definition of one of its immittance matrices. There are altogether four degree-1 circuits and four degree-2. This chapter includes the derivation of each commonly encountered degree-1 and degree-2 circuits. It also develops the quality factor of the basic circulation arrangement in terms of the gyrotropy of the gyromagnetic resonator as well as other fundamental conditions met elsewhere in the text in connection with the operation of the stripline circulator.

An impedance matrix can be constructed after each symmetrical adjustment of the impedance eigenvalues of the junction. This means that it is possible, in principle, to form $(m - 1)$ equivalent networks, each of which corresponds to the $(m - 1)$ eigenvalue adjustments of the matrix.

The Stripline Circulator: Theory and Practice. By J. Helszajn

6.2 IMPEDANCE MATRIX OF JUNCTION CIRCULATOR

The derivation of equivalent circuits requires a knowledge of an immittance rather than the scattering description of the network. The impedance matrix of a three-port symmetrical but nonreciprocal network is

$$\overline{\mathbf{Z}} = \begin{bmatrix} Z_{11} & Z_{12} & Z_{13} \\ Z_{13} & Z_{11} & Z_{12} \\ Z_{12} & Z_{13} & Z_{11} \end{bmatrix} \tag{6.1}$$

This matrix can be diagonalized to reveal the eigenvalues of the junction in a similar way to that met in connection with the scattering one. This may be done by noting that the eigenvectors of $\overline{\mathbf{S}}$ are also those of $\overline{\mathbf{Z}}$ since the two commute. Hence

$$\overline{\mathbf{Z}} = \overline{\mathbf{U}}\bar{\mathbf{z}}\overline{\mathbf{U}}^{-1} \tag{6.2}$$

where $\bar{\mathbf{z}}$ is a diagonal matrix with the eigenvalues of the impedance matrix $\overline{\mathbf{Z}}$. It is here assumed that the matrix $\overline{\mathbf{Z}}$ exists. This statement requires that the matrix $\overline{\mathbf{I}} - \overline{\mathbf{S}}$ is nonsingular. The voltages and currents on this sort of circuit are indicated in Fig. 6.1.

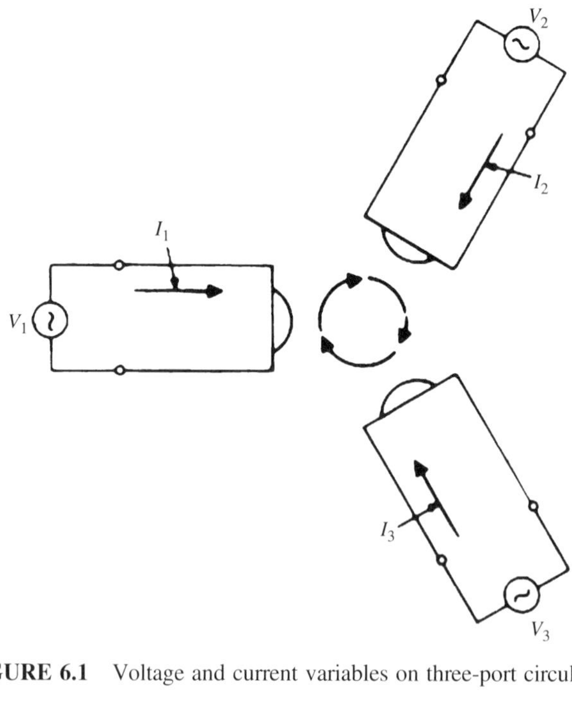

FIGURE 6.1 Voltage and current variables on three-port circulator.

Following the procedure outlined in Chapter 5 to derive the relationships between the scattering coefficients and its eigenvalues, one obtains in the case of the $\overline{\mathbf{Z}}$ matrix

$$Z_{11} = Z_{22} = Z_{22} = \frac{Z^0 + Z^+ + Z^-}{3} \tag{6.3a}$$

$$Z_{12} = Z_{23} = Z_{31} = \frac{Z^0 + \alpha Z^+ + \alpha^2 Z^-}{3} \tag{6.3b}$$

$$Z_{13} = Z_{21} = Z_{32} = \frac{Z^0 + \alpha^2 Z^+ + \alpha Z^-}{3} \tag{6.3c}$$

The magnetic field in the transverse plane for the in-phase set of the generator settings ($\overline{\mathbf{U}}_0$) is zero on the axis of the junction while the electric field along the axis is a maximum there. A possible one-port equivalent circuit in this instance is an open-circuited one. The solution for the counterrotating generator settings ($\overline{\mathbf{U}}_+$ and $\overline{\mathbf{U}}_-$) produces counterrotating circularly polarized magnetic fields in the transverse plane on the axis of the junction and produces an electric field that is zero there. One-port short-circuited transmission lines are therefore possibilities in this instance. The eigensolutions of a three-port junction and the three eigen-networks of the circuit are identical to those met in connection with the scattering matrix of the junction, with the scattering variables replaced by immittance ones.

The open-circuit parameters of the single four-port junction are likewise related to the impedance eigenvalues of the network by

$$Z_{11} = Z_{22} = Z_{33} = Z_{44} = \frac{Z^0 + Z^{+1} + Z^{-1} + Z^{+2}}{4} \tag{6.4a}$$

$$Z_{12} = Z_{23} = Z_{34} = Z_{41} = \frac{Z^0 + jZ^{+1} - jZ^{-1} - Z^{+2}}{4} \tag{6.4b}$$

$$Z_{13} = Z_{24} = Z_{31} = Z_{42} = \frac{Z^0 - Z^{+1} - Z^{-1} + Z^{+2}}{4} \tag{6.4c}$$

$$Z_{14} = Z_{43} = Z_{32} = Z_{21} = \frac{Z^0 - jZ^{+1} + jZ^{-1} - Z^{+2}}{4} \tag{6.4d}$$

6.3 EIGENVALUES OF IMMITTANCE MATRICES

A typical immittance matrix of the junction may be constructed in terms of the scattering matrix, without the need to invert matrices by having recourse to the connections between the various eigenvalues. This may be done by making use of the following theorem, which has already been introduced in Chapter 5. *If*

$$\overline{\mathbf{S}}\overline{\mathbf{U}}_i = s_i\overline{\mathbf{U}}_i \tag{6.5}$$

then

$$f(\overline{\mathbf{S}})\,\overline{\mathbf{U}}_i = f(s_i)\overline{\mathbf{U}}_i \tag{6.6}$$

This theorem will first be applied to deduce the relationship between the eigenvalues of the $\overline{\mathbf{S}}$ and $\overline{\mathbf{Z}}$ matrices. The bilinear transformation between the two matrices is

$$\overline{\mathbf{S}} = \left(\overline{\mathbf{Z}} - Z_0\overline{\mathbf{I}}\right)\left(\overline{\mathbf{Z}} + Z_0\overline{\mathbf{I}}\right)^{-1} \tag{6.7}$$

Z_0 is the characteristic impedance of the ports.

Making use of the above theorem immediately produces the classic relationship between s_i and Z_i:

$$s_i = \frac{Z_i - Z_0}{Z_i + Z_0} \tag{6.8}$$

Writing Z_i in terms of s_i gives

$$\frac{Z_i}{Z_0} = \frac{1 + s_i}{1 - s_i} \tag{6.9}$$

Z_i is an absolute eigenvalue that satisfies the eigenvalue equation

$$\overline{\mathbf{Z}}\,\overline{\mathbf{U}}_i = Z_i\overline{\mathbf{U}}_i \tag{6.10}$$

The derivation of the relationship between the eigenvalues of the scattering and admittance matrices starts again with the bilinear relationship between the two:

$$\overline{\mathbf{S}} = \left(Y_0\overline{\mathbf{I}} - \overline{\mathbf{Y}}\right)\left(Y_0\overline{\mathbf{I}} + \overline{\mathbf{Y}}\right)^{-1} \tag{6.11}$$

The connection between the eigenvalue in this instance is given without ado by

$$s_i = \frac{Y_0 - Y_i}{Y_0 + Y_i} \tag{6.12}$$

The corresponding linear transformation between Y_i and s_i is given by

$$\frac{Y_i}{Y_0} = \frac{1 - s_i}{1 + s_i} \tag{6.13}$$

Y_i is a normalized eigenvalue that satisfies the eigenvalues equation defined by

$$\overline{\mathbf{Y}}\,\overline{\mathbf{U}}_i = Y_i \overline{\mathbf{U}}_i \tag{6.14}$$

6.4 COMPLEX GYRATOR IMMITTANCE OF THREE-PORT CIRCULATOR

The complex gyrator immittance of the three-port junction circulator is a fundamental quantity in its description. It is defined by

$$Z_{\text{in}} = V_1/I_1 \tag{6.15}$$

with

$$V_3 = I_3 = 0 \tag{6.16}$$

The origin of this definition may be understood by writing down the voltage–current relationships of the network:

$$\begin{bmatrix} V_1 \\ V_2 \\ 0 \end{bmatrix} = \begin{bmatrix} Z_{11} & Z_{21} & Z_{31} \\ Z_{31} & Z_{11} & Z_{21} \\ Z_{21} & Z_{31} & Z_{11} \end{bmatrix} \begin{bmatrix} I_1 \\ I_2 \\ 0 \end{bmatrix} \tag{6.17}$$

The required result is

$$Z_{\text{in}} = Z_{11} - Z_{21}^2/Z_{31} \tag{6.18}$$

The condition at port 2 is then given by

$$Z_{\text{out}} = V_2/(-I_2) = Z_{\text{in}}^* \tag{6.19}$$

This relationship indicates that terminating each port of the junction by Z_{in}^* in a cyclic manner is sufficient to match the device. Figure 6.2 illustrates the schematic diagram of this arrangement.

If the frequency variation of the in-phase impedance eigenvalue Z^0 may be neglected compared to those of the degenerate or split ones, then it is possible to

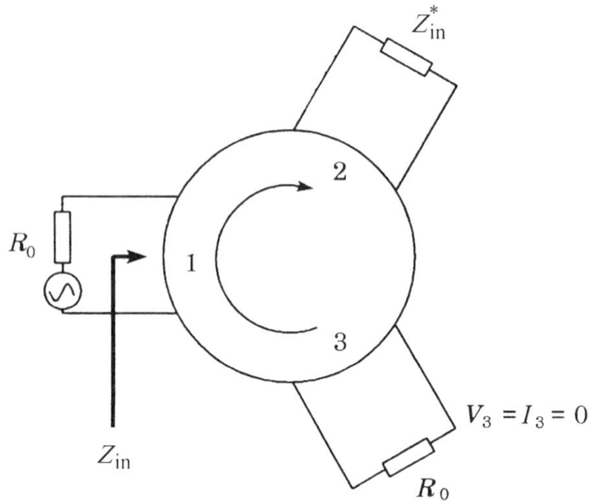

FIGURE 6.2 Definition of complex gyrator circuit. (Reproduced with permission from J. Simon, Broadband strip-transmission line Y junction circulators, *IEEE Trans. Microwave Theory Tech.*, Vol. MTT-13, pp. 335–345, May 1965.)

deduce an especially simple model for this class of device (see Fig. 6.3). The required realization starts by assuming that

$$Z^0 = 0 \tag{6.20}$$

It continues by developing the open-circuit parameters under this condition:

$$Z_{11} \approx \frac{Z^+ + Z^-}{3} \tag{6.21a}$$

$$Z_{21} \approx -\left(\frac{Z^+ + Z^-}{6}\right) + j\sqrt{3}\left(\frac{Z^+ - Z^-}{6}\right) \tag{6.21b}$$

$$Z_{31} \approx -\left(\frac{Z^+ + Z^-}{6}\right) - j\sqrt{3}\left(\frac{Z^+ - Z^-}{6}\right) \tag{6.21c}$$

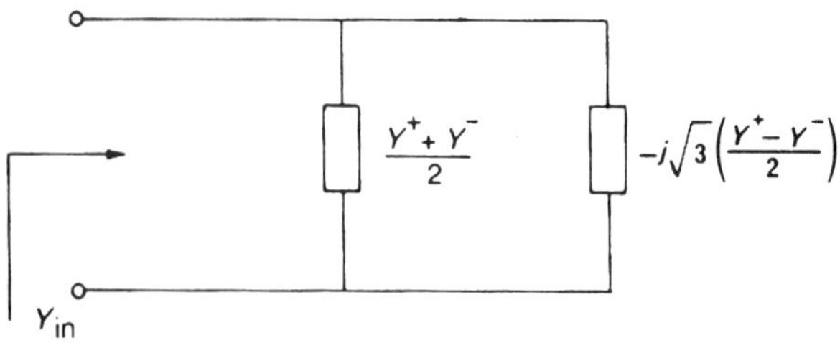

FIGURE 6.3 Schematic diagram of degree-1 complex gyrator circuit.

The immittance eigenvalues here and elsewhere in the text are pure imaginary numbers. Forming Y_{in} instead of Z_{in} gives

$$Y_{in} = \frac{1}{Z_{in}} = \frac{Z_{31}}{Z_{11}Z_{31} - Z_{21}^2} \tag{6.22}$$

Noting that

$$Z_{11}Z_{31} - Z_{21}^2 = -Z^+Z^-/3$$

gives

$$Y_{in} = \frac{(Z^+ + Z^-) + j\sqrt{3}(Z^+ - Z^-)}{2Z^+Z^-}$$

or

$$Y_{in} = \left(\frac{Y^+ + Y^-}{2}\right) - j\sqrt{3}\left(\frac{Y^+ - Y^-}{2}\right) \tag{6.23}$$

The imaginary and real parts of Y_{in} are therefore related to the sum and difference of the split admittance eigenvalues in a surprisingly simple way:

$$jB_{in} = \left(\frac{Y^+ + Y^-}{2}\right) \tag{6.24a}$$

$$G_{in} = -j\sqrt{3}\left(\frac{Y^+ - Y^-}{2}\right) \tag{6.24b}$$

Figure 6.4 gives an experimental Smith chart representation of the complex gyrator circuit of one arrangement for parametric values of H_0/M_0. Figures 6.5 and 6.6 indicate the same data in Cartesian form. The solid lines in these illustrations indicate the best fit on the experimental data. The normalized susceptance slope parameter, b', of the complex gyrator circuit, in the vicinity of the midband frequency ($\omega_0 = 2\pi f_0$), is separately obtained from Fig. 6.5 by forming

$$b' = \left.\frac{\omega}{2}\frac{\partial b}{\partial \omega}\right|_{\omega=\omega_0} \tag{6.25}$$

The quality factor of the complex gyrator circuit may be calculated in terms of b' and g without ado.

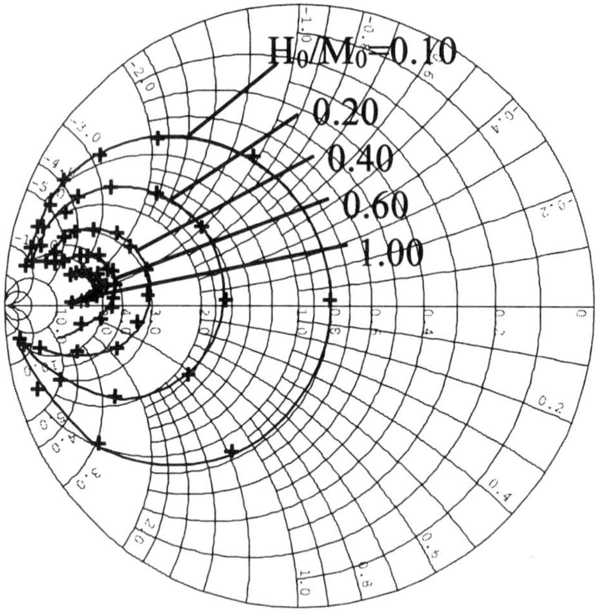

FIGURE 6.4 Smith chart representation of complex gyrator circuit of 2.0 GHz circulator for different magnetizing fields ($\psi = 0.22$ rad, $2R = 25.4$ mm).

One equivalent circuit of the three-port junction circulator is therefore a simple one-port LCR network. This classic result is illustrated in Fig. 6.3. Furthermore, a knowledge of Y^+ and Y^- is sufficient to describe this class of device.

6.5 SYNTHESIS OF JUNCTION CIRCULATORS USING RESONANT IN-PHASE EIGEN-NETWORK

While the in-phase eigen-network of a three-port junction circulator may often be idealized by a frequency-independent short-circuit boundary condition at its input terminals, it may also be adjusted to exhibit a series resonance there. The complex gyrator impedance in this instance is approximately given by

$$Z_{in} \approx \frac{8Z^0 - (Z^+ + Z^-)}{6} + j\frac{(Z^+ - Z^-)}{2\sqrt{3}} \tag{6.26}$$

The in-phase eigenvalue Z^0 has been idealized by a short-circuit boundary condition in forming the real part of the gyrator immittance but has been retained in describing its imaginary part. This impedance is readily realized in the form indicated in Fig. 6.7

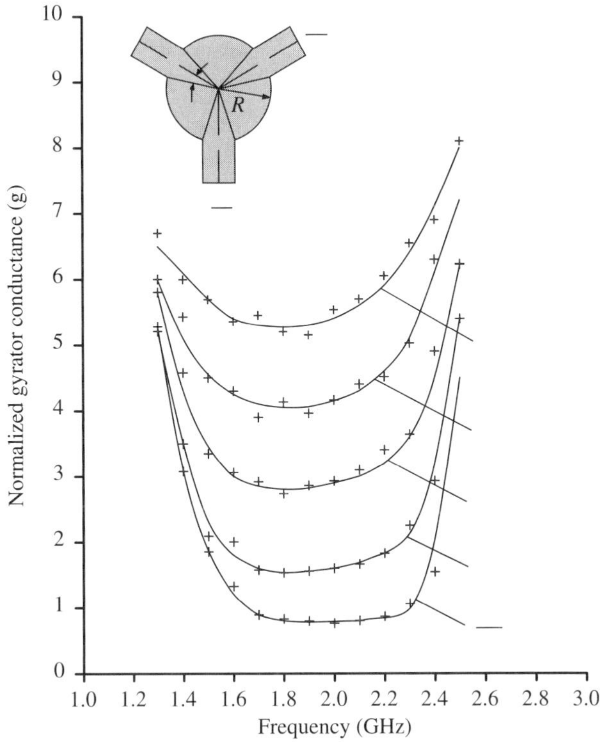

FIGURE 6.5 Cartesian representation of normalized gyrator conductance of circulator for different magnetizing fields using complex gyrator method ($\psi = 0.222$ rad, $2R = 25.4$ mm).

by expanding Z_{in} as

$$Z_{\text{in}} = Z_1 + 1/Y_1 \tag{6.27a}$$

where

$$Z_1 \approx 4Z^0/3 \tag{6.27b}$$

$$Y_1 \approx \frac{(Y^+ + Y^-)}{2} - j\sqrt{3}\frac{(Y^+ - Y^-)}{2} \tag{6.27c}$$

The nature of the second term in this expansion is obtained provided the following assumption holds:

$$\frac{(Y^+ - Y^-)}{4Y^+Y^-} \approx -1 \tag{6.28}$$

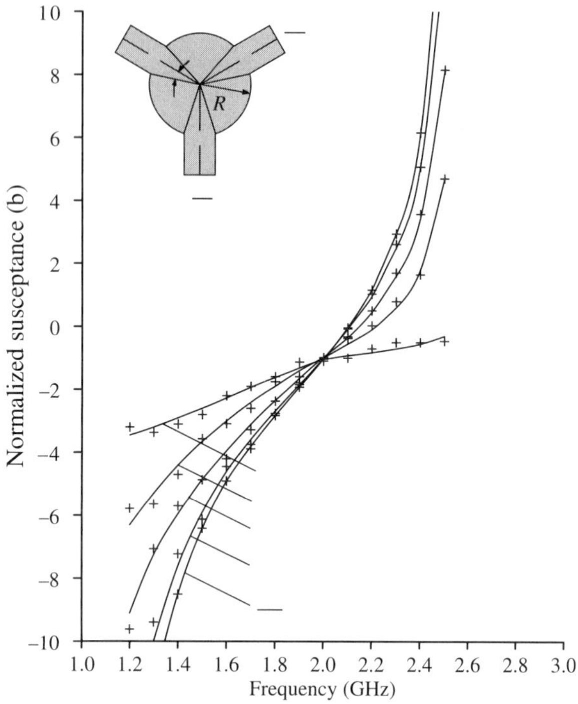

FIGURE 6.6 Cartesian representation of normalized gyrator susceptance of circulator for different magnetizing fields using complex gyrator method ($\psi = 0.22$ rad, $2R = 25.4$ mm).

The latter condition is met in the vicinity of the midband frequency, which is defined by

$$Y^+ + Y^- = 0$$

The equivalent circuit in Fig. 6.7 reduces to the usual approximation by omitting the Z^0 term in the derivation. This circuit has the nature of a bandpass filter, which may be adjusted to display a reflection or transmission characteristic akin to that of a quarter-wave coupled junction with its in-phase eigen-network idealized by a short-circuit boundary condition.

6.6 EQUIVALENT CIRCUIT OF THREE-PORT CIRCULATOR

An equivalent three-port circuit based on ideal gyrators can be synthesized from the impedance matrix by having recourse to simple matrix addition. Making use of the bilinear mapping between the reflection eigenvalues of an ideal circulator and

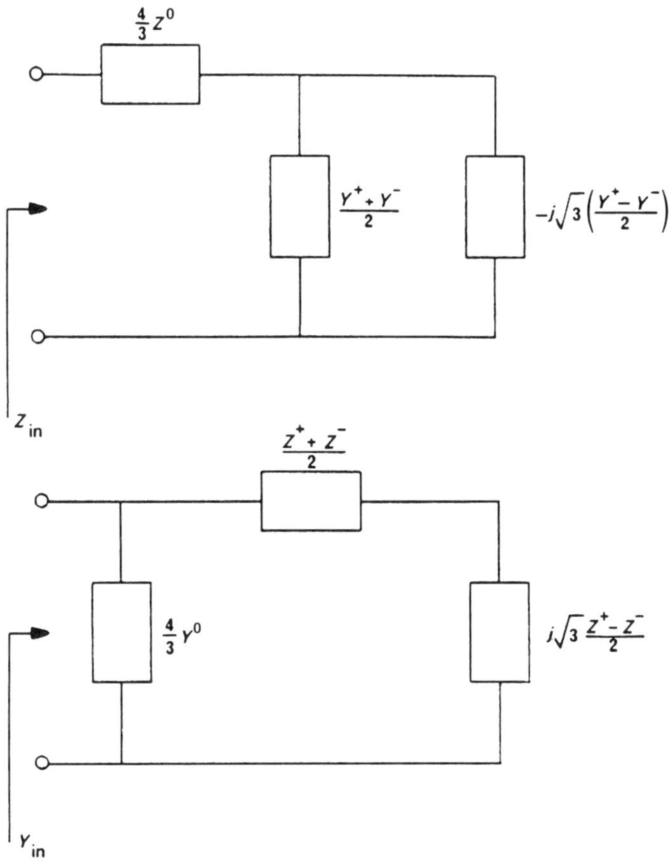

FIGURE 6.7 Schematic diagrams of degree-2 complex gyrator circuits.

the impedance eigenvalues gives

$$Z^0 = 0 \tag{6.29a}$$

$$Z^+ - j\sqrt{3}Z_0 \tag{6.29b}$$

$$Z^- = j\sqrt{3}Z_0 \tag{6.29c}$$

The open-circuit parameters of an ideal circulator are therefore

$$Z_{11} = 0 \tag{6.30a}$$

$$Z_{12} = R_0 \tag{6.30b}$$

$$Z_{13} = -R_0 \tag{6.30c}$$

The corresponding impedance matrix of an ideal three-port circulator is

$$\overline{\mathbf{Z}} = \begin{bmatrix} 0 & R_0 & -R_0 \\ -R_0 & 0 & R_0 \\ R_0 & -R_0 & 0 \end{bmatrix} \tag{6.31}$$

Making use of the rule of matrix addition indicates that

$$\begin{aligned} \overline{\mathbf{Z}} = \begin{bmatrix} 0 & R_0 & -R_0 \\ -R_0 & 0 & R_0 \\ R_0 & -R_0 & 0 \end{bmatrix} &= \begin{bmatrix} 0 & R_0 & 0 \\ -R_0 & 0 & 0 \\ 0 & 0 & 0 \end{bmatrix} \\ &+ \begin{bmatrix} 0 & 0 & -R_0 \\ 0 & 0 & 0 \\ R_0 & 0 & 0 \end{bmatrix} + \begin{bmatrix} 0 & 0 & 0 \\ 0 & 0 & R_0 \\ 0 & -R_0 & 0 \end{bmatrix} \end{aligned} \tag{6.32}$$

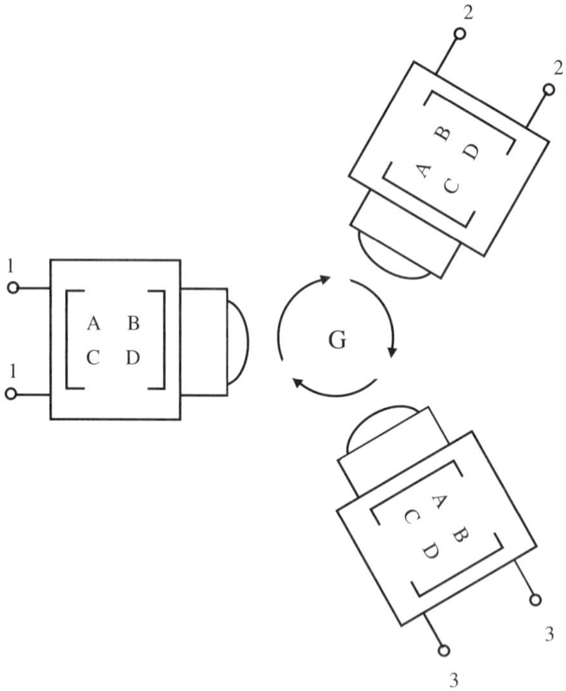

FIGURE 6.8 Topology of degree-2 three-port junction circulator in terms of two-port gyrator circuits.

The first impedance matrix represents an ideal gyrator circuit connected between ports 1 and 2 of the junction. The other two matrices represent similar gyrator circuits between ports 2 and 3 and ports 3 and 1. The equivalent circuit is shown in Fig. 6.8.

6.7 QUALITY FACTOR OF JUNCTION CIRCULATOR

The real part of the complex gyrator admittance of a three-port junction circulator is sometimes written in terms of the susceptance slope parameter of the degenerate counterrotating eigen-networks and the resonant frequencies of the split eigen-networks. The relationship obtained here appears naturally in both Bosma's and Fay and Comstock's developments. It is separately employed to define the quality factor of the complex gyrator circuit solely in terms of the split frequencies of the gyromagnetic resonator. This approach is of special interest in the network problem of the device. The derivation of the required formulation begins by writing

$$Y_+ \approx -j(Y_1 + \Delta Y_1)\cot(\theta_1 - \Delta\theta_1) \tag{6.33a}$$

$$Y_- \approx -j(Y_1 - \Delta Y_1)\cot(\theta_1 + \Delta\theta_1) \tag{6.33b}$$

$\theta_1 \mp \Delta\theta_1$ are the electrical lengths of the split eigen-networks:

$$\theta_1 \mp \Delta\theta_1 = (\pi/2)(1 + \delta_\pm) \tag{6.34}$$

The normalized frequency variables $\delta_\pm$ are separately expanded about the frequencies $\omega_\pm$ of each eigen-network:

$$\delta_\pm = \frac{\omega - \omega_\pm}{\omega_\pm} \tag{6.35}$$

The admittance $\pm\Delta Y_1$ represents the perturbations of the split wave or characteristic admittances of the substrate.

At $\theta_1 = \pi/2$,

$$Y_+ = j(Y_1 + \Delta Y_1)\tan(\pi\delta_+/2) \tag{6.36a}$$

$$Y_- = j(Y_1 - \Delta Y_1)\tan(\pi\delta_-/2) \tag{6.36b}$$

Introducing these relationships into the normalized real part of the complex gyrator admittance readily produces the desired result:

$$G \approx \sqrt{3}B'\left(\frac{\omega_+ - \omega_-}{\omega_0}\right) \tag{6.37}$$

provided

$$\tan(\pi\delta_{\pm}/2) \approx (\pi\delta_{\pm}/2)$$

B' is the absolute value of the susceptance slope parameter of the degenerate counter-rotating eigen-network.

$$B' = \pi Y_1/4 \tag{6.38}$$

Equation (6.37) is also sometimes written

$$\frac{1}{Q_L} = \frac{G}{B'} = \sqrt{3}\left(\frac{\omega_+ - \omega_-}{\omega_0}\right) \tag{6.39}$$

This relationship permits any one of the above variables to be deduced from a knowledge of the other two. It is a recurring identity in the description of the conventional three-port circulator.

The imaginary part B of the complex gyrator circuit may be evaluated separately from B' without difficulty. The result, in the vicinity of the midband frequency, is

$$B = 2\delta\,(4B'/\pi) \tag{6.40}$$

where

$$\delta = (\omega - \omega_0)/\omega_0 \tag{6.41}$$

6.8 DEGENERATE COUNTERROTATING EIGEN-NETWORK (s_1)

While the degenerate eigenvalue may be represented by an equivalent transmission line, a knowledge of its exact topology is desirable for numerical purposes. One possible way to do so will now be derived. The derivation of this eigen-network starts by placing a magnetic wall at port 3 of the junction and constructing the relationship between the other two ports in the manner indicated in Fig. 6.9. This gives

$$I_3 = 0 \tag{6.42}$$

and

$$\begin{bmatrix} V_1 \\ V_2 \\ V_3 \end{bmatrix} = \begin{bmatrix} Z_{11} & Z_{12} & Z_{12} \\ Z_{12} & Z_{11} & Z_{12} \\ Z_{12} & Z_{12} & Z_{11} \end{bmatrix} \begin{bmatrix} I_1 \\ I_2 \\ 0 \end{bmatrix} \tag{6.43}$$

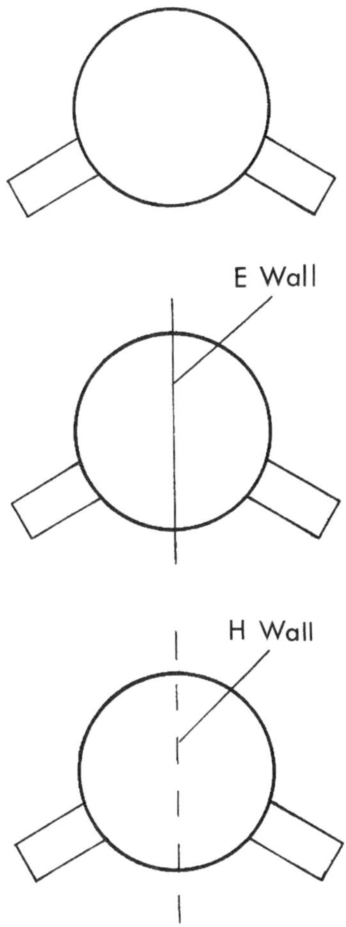

FIGURE 6.9 Construction of degenerate eigen-network.

Z_{11} and Z_{12} are linear combinations of the eigenvalues of the three-port circuit in the usual way.

$$Z_{11} = (Z^0 + 2Z^1)/3 \tag{6.44a}$$

$$Z_{12} = (Z^0 - Z^1)/3 \tag{6.44b}$$

Z^0 is the in-phase eigenvalue and Z^1 is the degenerate counterrotating one.

The connection between ports 1 and 2 of the reduced symmetrical network in Fig. 6.9 is then given by

$$\begin{bmatrix} V_1 \\ V_2 \end{bmatrix} = \begin{bmatrix} Z_{11} & Z_{12} \\ Z_{12} & Z_{11} \end{bmatrix} \begin{bmatrix} I_1 \\ I_2 \end{bmatrix} \tag{6.45}$$

The in-phase and out-of-phase eigenvalues $Z_{o/c}$ and $Z_{s/c}$ of the new circuit are now deduced by taking the sum and difference of the open-circuit parameters:

$$Z_{o/c} = Z_{11} + Z_{12} \tag{6.46a}$$

$$Z_{s/c} = Z_{11} - Z_{12} \tag{6.46b}$$

Evaluating these quantities in terms of the eigenvalues of the original circuit indicates that

$$Z_1' = (Z_0 - Z_1)/3 \tag{6.47a}$$

$$Z_2' = Z_1 \tag{6.47b}$$

This result indicates that the out-of-phase eigen-network of the reduced two-port circuit coincides with that of the degenerate counterrotating one of the original three-port circuit.

It is also recognized that a knowledge of Z_1 and the in-phase eigen-network of the reduced two-port circuit may be employed separately to evaluate the in-phase eigen-network of the original three-port junction.

It is of separate note that for the out-of-phase eigen-network

$$I_1 = -I_2$$

and the voltage at port 3 is

$$V_3 = 0$$

in addition to the condition $I_3 = 0$ at the same port. If this is the case, then

$$V_1/I_1 = V_2/I_2 = Z_{11} - Z_{12} \tag{6.48}$$

in keeping with the eigenvalue problem.

The experimental arrangement in question is shown in Fig. 6.10.

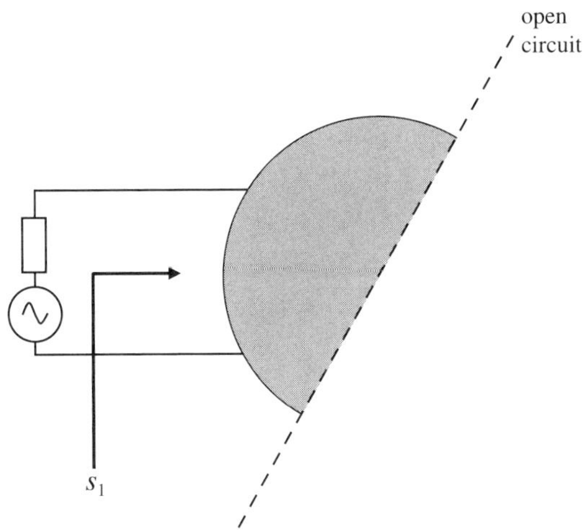

FIGURE 6.10 Experimental arrangement for measurement of degenerate reflection eigenvalue s_1.

6.9 IN-PHASE EIGEN-NETWORK

The one-port in-phase eigen-network is also a classic circuit, which may be fabricated separately. Figure 6.11 indicates this circuit in the case of a junction using a planar disk resonator. It is obtained by partitioning the circuit with appropriate magnetic walls. This one-port circuit accurately displays the admittance and reflection eigenvalues of the circuit provided its boundaries are idealized by perfect magnetic walls. It therefore represents the simplest test fixture for the characterization of this eigen-network. A shortcoming of this eigen-network, however, is that the open walls exposed by this boundary condition are ill defined in practical circuits.

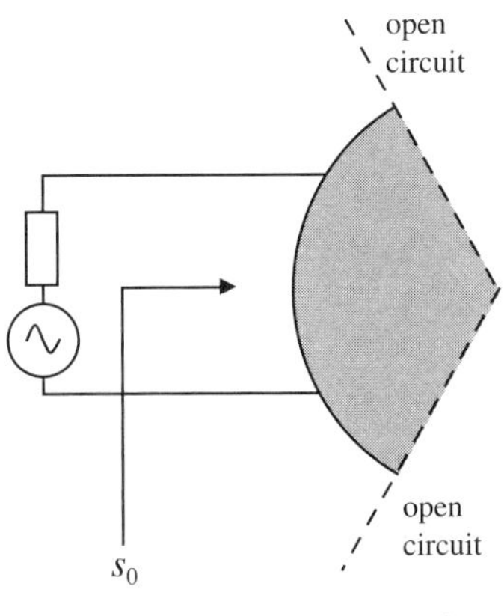

FIGURE 6.11 Topology of in-phase eigen-network.

The phase angle of the in-phase eigen-network of a demagnetized junction may be measured directly using the eigenvalue approach by applying equal in-phase signals at the three ports of the junction or by making use of the relationship between the scattering variable S_{11} and the in-phase and degenerate counterrotating eigenvalues s_0 and s_1:

$$S_{11} = (s_0 + 2s_1)/3 \tag{6.49}$$

One possible way in which the eigenvalue s_1 of a reciprocal three-port junction can be experimentally determined without the need to fabricate its eigen-network consists of placing a sliding short-circuit at port 2 and a matched load at port 3. The variable short-circuit is varied until there is total reflection at port 1. The reflection coefficient at port 1 is then the eigenvalue s_1. This technique is especially appropriate in the case of a waveguide junction for which the reference terminals are usually ill defined.

6.10 SPLIT EIGEN-NETWORKS OF JUNCTION CIRCULATOR

An especially simple method to obtain the split admittance eigen-networks of a circulator for which the in-phase eigen-network may be idealized by a short-circuit boundary condition at the terminals of the junction can be deduced from a knowledge of its complex gyrator circuit. This quantity is defined by Eqs. (6.24) and (6.25) by

$$Y_{in} = G + jB \tag{6.50}$$

The split admittance eigenvalues may now be evaluated by taking linear combinations of the real and imaginary parts of the complex gyrator admittance. The result is

$$Y_+ = j(B + G/\sqrt{3}) \tag{6.51a}$$

$$Y_- = j(B - G/\sqrt{3}) \tag{6.51b}$$

The susceptance of the degenerate counterrotating eigenvalues is negative at Y_+ and positive at Y_-. B is a pure real number and Y_1 is a pure imaginary number.

$$jB = Y_1 \tag{6.52}$$

The split frequencies may also be determined from this sort of data by recognizing that these coincide with the frequencies at which the split admittance eigenvalues are zero:

$$Y_\pm = 0 \tag{6.53}$$

The two frequencies therefore coincide with the conditions

$$G = -B/\sqrt{3}, \qquad \omega = \omega_- \tag{6.54a}$$

$$G = +B/\sqrt{3}, \qquad \omega = \omega_+ \tag{6.54b}$$

One experimental arrangement, in keeping with the definition of the complex gyrator circuit, is obtained by decoupling port 3 from port 1 by adjusting a triple stub tuner at port 2. It is illustrated in Fig. 6.2.

The immitances of the complex gyrator circuit at the split frequencies are given without ado by

$$Y_{\text{in}} = \frac{Y_+}{2} - j\frac{\sqrt{3}Y_+}{2}, \qquad \omega = \omega_- \tag{6.55a}$$

$$Y_{\text{in}} = \frac{Y_-}{2} + j\frac{\sqrt{3}Y_-}{2}, \qquad \omega = \omega_+ \tag{6.55b}$$

6.11 GYRATOR CONDUCTANCE OF CIRCULATOR

A relationship between the reflection coefficient at port 1 and the normalized gyrator conductance g with ports 2 and 3 terminated in 50 Ω loads also exists. Its derivation starts with the relationships in Chapter 5 between the one-port reflection coefficient and the reflection angles associated with each possible eigen-network.

At midband

$$\phi_0 = \phi_1 = \pi/2 \tag{6.56}$$

and for symmetrical splitting

$$\phi_- = -\phi_+ \tag{6.57}$$

One relationship is now constructed by evaluating S_{11} in terms of its reflection eigenvalues and making use of the bilinear transformation between the scattering and immittance eigenvalues:

$$S_{11} = \frac{1 - 3\tan^2(\phi_+)}{3 + 3\tan^2(\phi_+)} \tag{6.58}$$

A second relationship is obtained by forming the gyrator conductance in terms of the admittance eigenvalues:

$$g = G/Y_0 = \sqrt{3}\tan(\phi_+) \tag{6.59}$$

It is now possible to eliminate ϕ_+ between S_{11} and g. The result in terms of the VSWR is

$$(\text{VSWR}) = \frac{|3 + g^2| + |1 - g^2|}{|3 + g^2| - |1 - g^2|} \tag{6.60}$$

For g larger than unity

$$g = \sqrt{2(\text{VSWR}) - 1} \tag{6.61}$$

For g smaller than unity

$$g = \sqrt{\frac{2 - (\text{VSWR})}{(\text{VSWR})}} \tag{6.62}$$

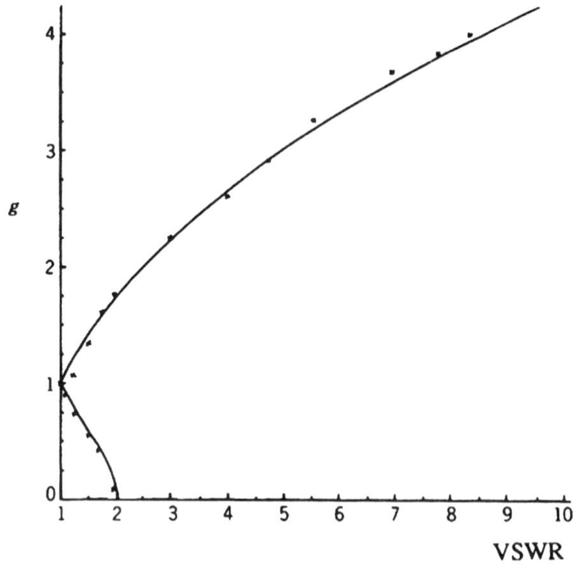

FIGURE 6.12 Relationship between VSWR and gyrator conductance of junction circulator. (Reproduced with permission from J. Helszajn, Microwave measurement techniques for junction circulators, *IEEE Trans. Microwave Theory Tech.*, Vol. MTT-21, pp. 347–351, 1973.)

The connection between (VSWR) and g is indicated in Fig. 6.12. Whether g is larger or smaller than unity is determined by whether there is a voltage maximum or minimum at the load. The condition on this curve for which g equals zero and the VSWR equals 2 coincides with that of a demagnetized junction.

6.12 THE GYRATOR NETWORK

The basic two-port circuit met in connection with nonreciprocal networks is the gyrator one introduced in Chapter 4.

It will now be demonstrated that both the impedance and admittance matrices exist for such a network. The impedance eigenvalues are

$$\frac{Z_+}{R_0} = \frac{1+s_+}{1-s_+} = \frac{1+j}{1-j} = j \tag{6.63a}$$

$$\frac{Z_-}{R_0} = \frac{1+s_-}{1-s_-} = \frac{1-j}{1+j} = -j \tag{6.63b}$$

The corresponding admittance eigenvalues are the reciprocal of the impedance eigenvalues.

$$Y_+/Y_0 = R_0/Z_+ = -j \tag{6.64a}$$

$$Y_-/Y_0 = R_0/Z_- = j \tag{6.64b}$$

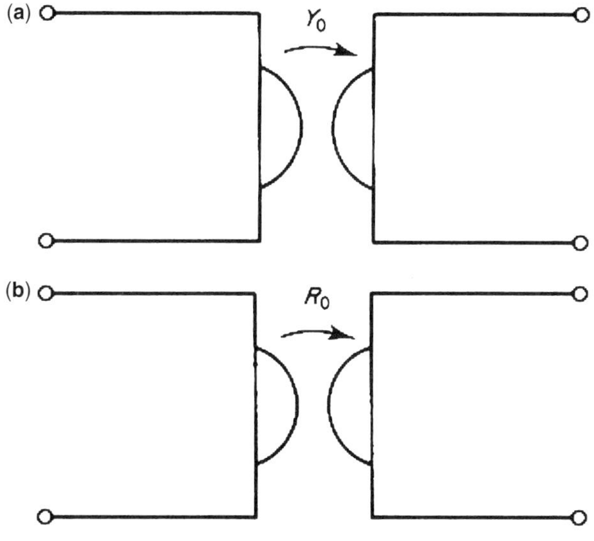

FIGURE 6.13 Schematic diagram of two-port gyrator network using (a) impedance gyrator and (b) admittance gyrator.

The admittance matrix is now diagonalized using Eq. (6.2)

$$\overline{\mathbf{Y}} = \frac{1}{2}\begin{bmatrix} (Y_+ + Y_-) & j(Y_+ - Y_-) \\ -j(Y_+ - Y_-) & (Y_+ + Y_-) \end{bmatrix} \tag{6.65}$$

The result, in terms of the original variables, is

$$\overline{\mathbf{Y}} = \begin{bmatrix} 0 & Y_0 \\ -Y_0 & 0 \end{bmatrix} \tag{6.66}$$

The $\overline{\mathbf{Y}}$ matrix schematic of the two-port gyrator is illustrated in Fig. 6.13a. The result for the $\overline{\mathbf{Z}}$ matrix is

$$\overline{\mathbf{Z}} = \begin{bmatrix} 0 & -R_0 \\ R_0 & 0 \end{bmatrix} \tag{6.67}$$

The $\overline{\mathbf{Z}}$ matrix schematic of the two-port gyrator is depicted in Fig. 6.13b.

CHAPTER SEVEN

The One-Port Topology of the Degree-1 and Degree-2 Terminated Circulator

7.1 INTRODUCTION

The ideal symmetrical three-port circulator is a matched lossless nonreciprocal junction with one decoupled port and perfect transmission between the other two. The possibility of transforming a general nonreciprocal three-port junction into an ideal circulator by the addition of external two-port circuits at each port has also been demonstrated separately. A possible model of a nonideal circulator in terms of an ideal three-port circulator with two-port reactive elements connected at each port has also been proposed. The immittances at port 1 that enter into the descriptions of the terminals of a three-port junction circulator are either its gyrator immittance, already dealt with in Chapter 6, or that with ideal resistive loads at ports 2 and 3. The notion of the gyrator immittance is appropriate in the synthesis of degree-2 or degree-3 circulators while that of the terminated one is appropriate in the optimization problem. The topology of the terminated circulator is the topic of this chapter. Its purpose is to synthesize its one-port topology in the admittance plane in the case for which ports 2 and 3 are terminated in real resistive loads and in the case for which the ports are loaded by broadband matching networks. This is done in terms of the eigenvalues of the problem region under the assumption that the in-phase eigen-network may be idealized by a short-circuited boundary condition and an ideal two-port gyrator circuit. The real part circuit, also dealt with in Chapter 6, and those at the split frequencies of the counterrotating eigenvalues are also deduced as a preamble to realizing the general circuit. The one-port circuit obtained

The Stripline Circulator: Theory and Practice. By J. Helszajn

in this way is compatible with existing experimental procedures. This equivalent circuit is sometimes referred to as a degree-1 circuit in that its frequency response corresponds to that of a Chebyshev specification with the same degree. The chapter also includes the equivalent circuit of a degree-2 arrangement. Its frequency response coincides with an equal ripple degree-2 Chebyshev specification.

7.2 INPUT IMMITTANCE OF TERMINATED CIRCULATOR IN TERMS OF OPEN-CIRCUIT PARAMETERS

The input impedance at port 1 of a symmetrical three-port circulator with ports 2 and 3 terminated in matched loads is of interest in its characterization. The schematic diagram of the arrangement in question is shown in Fig. 7.1.

The derivation of the required result starts with a statement of the voltage–current relationship at the ports of the junction:

$$\begin{bmatrix} V_1 \\ V_2 \\ V_3 \end{bmatrix} = \begin{bmatrix} Z_{11} & Z_{12} & Z_{13} \\ Z_{13} & Z_{11} & Z_{12} \\ Z_{12} & Z_{13} & Z_{11} \end{bmatrix} \begin{bmatrix} I_1 \\ I_2 \\ I_3 \end{bmatrix} \tag{7.1}$$

It continues by terminating ports 2 and 3 in ideal real loads. These two conditions are satisfied provided

$$V_2 = -I_2 Z_0 \tag{7.2a}$$

$$V_3 = -I_3 Z_0 \tag{7.2b}$$

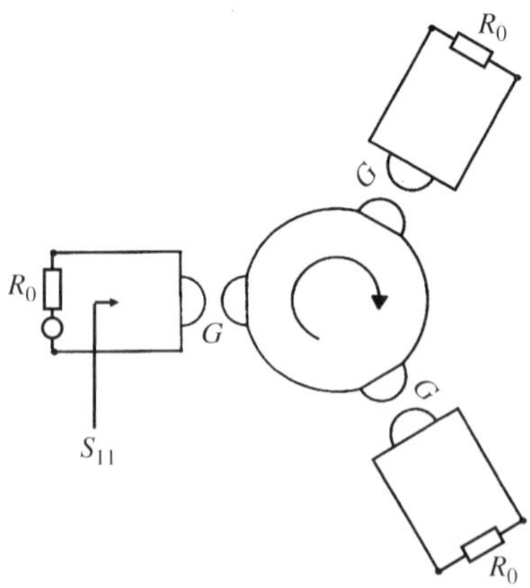

FIGURE 7.1 Schematic diagram of three-port circulator.

Combining the SE equations gives

$$(V_1/I_1) = Z_{11} + Z_{12}(I_2/I_1) + Z_{13}(I_3/I_1) \tag{7.3a}$$

$$(-I_2/I_1)\,Z_0 = Z_{13} + Z_{11}(I_2/I_1) + Z_{12}(I_3/I_1) \tag{7.3b}$$

$$(-I_3/I_1)\,Z_0 = Z_{12} + Z_{13}(I_2/I_1) + Z_{11}(I_3/I_1) \tag{7.3c}$$

The impedance Z_{in} at port 1 is now defined as

$$Z_{in} = V_1/I_1 \tag{7.4}$$

The required result is readily deduced by rearranging the preceding equations:

$$Z_{in} = Z_{11} + \frac{Z_{12}^3 - Z_{12}^{*3} + 2Z_{12}Z_{12}^*(Z_{11} + Z_0)}{(Z_{11} + Z_0)^2 + Z_{12}Z_{12}^*} \tag{7.5}$$

In obtaining this result use has also been made of the fact that

$$Z_{13} = -Z_{12}^* \tag{7.6}$$

This important identity may be verified by having recourse to Eqs. (7.7b) and (7.7c) in the next section.

7.3 INPUT IMPEDANCE OF TERMINATED CIRCULATOR IN TERMS OF EIGENVALUES

It is preferable for the purpose of synthesis to express the input immittance of the terminated circulator in terms of its eigenvalues rather than by its open-circuit parameters. Its derivation begins by expressing the open-circuit parameters of the junction in terms of its in-phase (Z_0) and counterrotating ($Z_\pm$) eigenvalues and it continues by assuming that the in-phase eigen-network may be realized by a frequency independent short-circuit boundary condition at the terminals of the junction. The open-circuit parameters of the circulator then reduce to those met in connection with the development of the complex gyrator circuit of the junction in Chapter 6.

$$Z_{11} = \frac{Z_+ + Z_-}{3} \tag{7.7a}$$

$$Z_{12} = \frac{-(Z_+ + Z_-)}{6} + j\frac{\sqrt{3}(Z_+ - Z_-)}{6} \tag{7.7b}$$

$$Z_{13} = \frac{-(Z_+ + Z_-)}{6} - j\frac{\sqrt{3}(Z_+ - Z_-)}{6} \tag{7.7c}$$

Introducing these relationships into the one-port impedance function of the terminated circulator at port 1 gives

$$Z_{in} = \frac{Z_0[(Z_+ + Z_-)\,Z_0 + 2Z_+Z_-]}{2(Z_+ + Z_-)\,Z_0 + Z_+Z_- + 3Z_0^2} \tag{7.8}$$

The phenomenological eigen-networks in question are indicated in Chapter 5.

The first element in the realization of the immittance of the circuit is assumed, in this work, to be a shunt element. Forming Y_{in} in terms of Z_{in} gives

$$Y_{in} = 1/Z_{in} \tag{7.9}$$

The required result is given in terms of the counterrotating admittance eigenvalues by

$$Y_{in} = \frac{Y_0^2 + 2Y_0(Y_+ + Y_-) + 3Y_+Y_-}{2Y_0 + (Y_+ + Y_-)} \tag{7.10}$$

The same result may also be readily deduced by having recourse to the reflection coefficient S_{11} and the bilinear transformation between reflection and impedance. This gives

$$3S_{11} = -1 + \left(\frac{Y_0 - Y_+}{Y_0 + Y_+}\right) + \left(\frac{Y_0 - Y_-}{Y_0 + Y_-}\right) \tag{7.11}$$

and

$$Y_{in} = \frac{1 - S_{11}}{1 + S_{11}} \tag{7.12}$$

S_{11} is also given in terms of the original variables by

$$3S_{11} = \frac{Y_0^2 - (Y_+ + Y_-)Y_0 - 3Y_+Y_-}{Y_0^2 + (Y_+ + Y_-)Y_0 + Y_+Y_-} \tag{7.13}$$

The corresponding admittance is in keeping with Eq. (7.10).

A property of the terminated circulator is that the reflection coefficient at the split frequencies of the counterrotating eigen-networks coincides with a 9.5 dB point in the return loss. This property is displayed by S_{11} without ado in keeping with the result in Chapter 5.

$$|S_{11}| = \frac{1}{3}, \quad Y_+ = 0 \tag{7.14a}$$

$$|S_{11}| = \frac{1}{3}, \quad Y_- = 0 \tag{7.14b}$$

It is appropriate to note that $Y_\pm$ throughout the text are pure imaginary quantities.

7.4 GYRATOR CIRCUIT

Before attempting to extract an ideal gyrator circuit in the synthesis of a one-port immittance function with an admittance Y_{in}, it is perhaps worthwhile to recall its definition,

$$Y = \begin{bmatrix} 0 & G_0 \\ -G_0 & 0 \end{bmatrix} \tag{7.15}$$

The mapping between the input and output terminals of an ideal gyrator circuit starts with the voltage–current relationship at its port.

$$\begin{bmatrix} I_1 \\ I_2 \end{bmatrix} = \begin{bmatrix} 0 & G_0 \\ -G_0 & 0 \end{bmatrix} \begin{bmatrix} V_1 \\ V_2 \end{bmatrix} \tag{7.16}$$

It continues by expressing the output voltage in terms of the load condition:

$$I_2 = -V_2 Y_{\text{out}} \tag{7.17}$$

The relation between the input and output terminal is

$$Y_{\text{in}} = I_1/V_1 = G_0^2/Y_{\text{out}} \tag{7.18}$$

The topology of the gyrator circuit is indicated in Fig. 7.2.

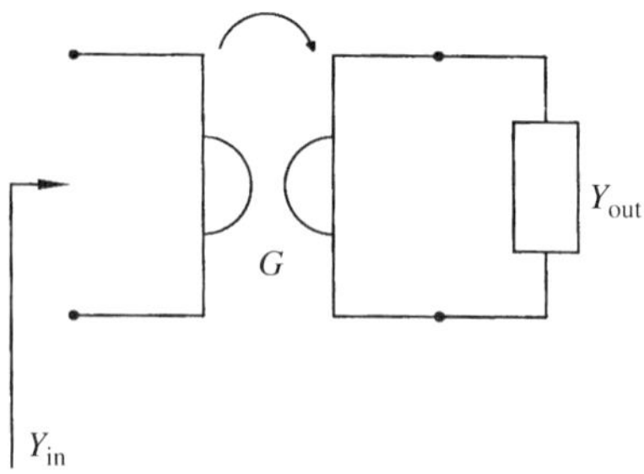

FIGURE 7.2 Topology of ideal two-port gyrator.

The gyrator conductance of the ideal two-port gyrator is given in terms of its eigenvalues by

$$G_0 = (-j/2)(Y_+ - Y_-) \tag{7.19}$$

where

$$Y_+ = -jG_0 \tag{7.20a}$$

$$Y_- = jG_0 \tag{7.20b}$$

As will be seen in a later section, the gyrator conductance met in the synthesis of a three-port junction is defined by

$$G_0 = \left(-j\sqrt{3}/2\right)(Y_+ - Y_-) \tag{7.21}$$

If the impedance eigenvalues rather than the admittance eigenvalues are the independent variables, then

$$G_0 = \left(\frac{j\sqrt{3}}{2}\right)\left(\frac{Z_- - Z_+}{Z_+ Z_-}\right) \tag{7.22}$$

7.5 REAL PART CONDITION

In order to facilitate the synthesis procedure it is helpful to first establish the topology of the one-port immittance function at the midband frequency. This condition is met provided

$$Y_+ + Y_- = 0 \tag{7.23}$$

The corresponding admittance at port 1 is then described by

$$Y_{in} = \frac{Y_0^2 + 3Y_+Y_-}{2Y_0} \tag{7.24}$$

In the case of an ideal circulator

$$Y_{\pm} = \mp jY_0/\sqrt{3} \tag{7.25}$$

and

$$Y_{in} = Y_0 \tag{7.26}$$

A scrutiny of this one-port immittance function indicates that it does not at first sight embody a gyrator element. In order to reveal such a network it is necessary to have recourse to the following identity:

$$(Y_+ - Y_-)^2 = (Y_+^2 + Y_-^2) - 2Y_+Y_-$$

or

$$\frac{1}{2}\left[\frac{(Y_+^2 + Y_-^2)}{Y_+Y_-} - \frac{(Y_+ + Y_-)^2}{Y_+Y_-}\right] = 1$$

In the present situation

$$Y_+ = -jG$$
$$Y_- = jG$$

or

$$Y_+Y_- = G^2$$
$$Y_+^2 = -G^2$$
$$Y_-^2 = -G^2$$

where G is an arbitrary real constant.

Combining the preceding equations gives

$$(Y_+^2 + Y_-^2)/Y_+Y_- = -2 \tag{7.27}$$

and

$$Y_+Y_- = \left[-j\frac{(Y_+ - Y_-)}{Y_+Y_-}\right]^2 \tag{7.28}$$

The one-port immittance of the terminated circulator may therefore be written in terms of the gyrator conductance of the three-port circulator defined by Eq. (7.26):

$$Y_{\text{in}} = Y_0/2 + G^2/2Y_0 \tag{7.29}$$

The gyrator conductance of the three-port circulator is defined in Eq. (7.24).

The required circuit is indicated in Fig. 7.3. The junction is demagnetized when the gyrator conductance is zero. The input admittance is then

$$Y_{\text{in}} = Y_0/2 \tag{7.30}$$

This admittance may also be written, with $\theta = 90^\circ$, as

$$Y_{\text{in}} = Y_0^2/2Y_0 \tag{7.31}$$

This result is in keeping with the equivalent circuit of the H-plane reciprocal junction in Fig. 7.4.

For the normalized input conductance g_{in} larger than unity,

$$g_{\text{in}} = G_{\text{in}}/Y_0 = (VSWR) \tag{7.32}$$

The corresponding gyrator conductance is

$$g = \sqrt{2(VSWR) - 1} \tag{7.33}$$

For g_{in} smaller than unity,

$$g_{\text{in}} = \frac{G_{\text{in}}}{Y_0} = \frac{1}{(VSWR)} \tag{7.34}$$

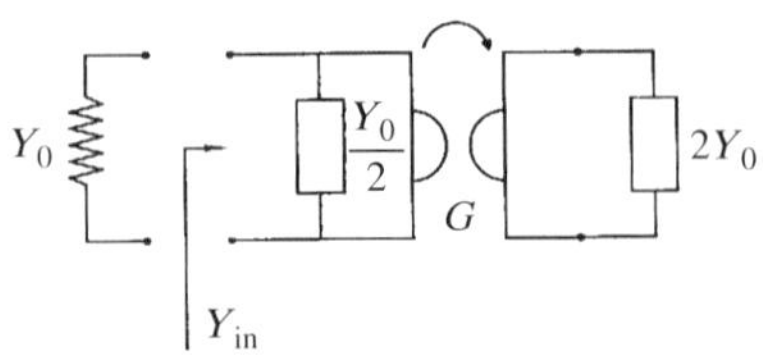

FIGURE 7.3 Real part realization of terminated circulator.

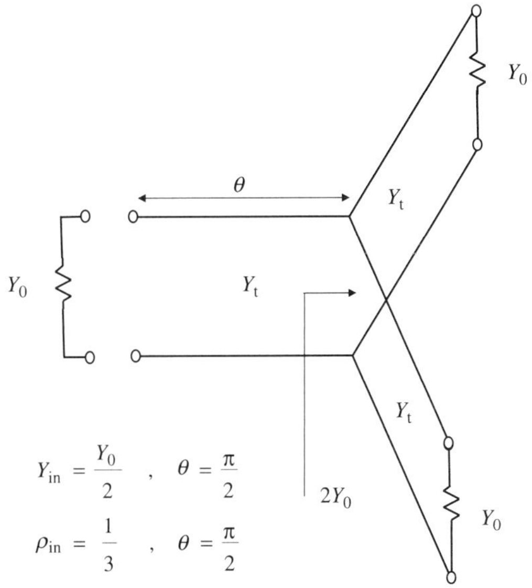

FIGURE 7.4 Equivalent circuit of *H*-plane three-port reciprocal junction.

In this instance, the gyrator conductance is given by

$$g = \sqrt{\frac{2 - (VSWR)}{(VSWR)}} \tag{7.35}$$

This classic result provides one means of experimentally evaluating the gyrator conductance of this class of circulator and is in keeping with the result in Chapter 6. While the topology at port 1 of the terminated circulator deduced here is a proper representation of its input admittance, it does not specifically embody the separate contributions of ports 2 and 3. One way of doing so is to terminate ports 2 and 3 by different load conditions as a preamble to synthesis. The general form of the reflection coefficient form which this quantity may be realized can readily be established.

7.6 SYNTHESIS OF COMPLEX ADMITTANCE OF TERMINATED CIRCULATOR

The synthesis of the one-port complex immittance of the terminated circulator is the subject of this section. It begins by extracting an admittance, $Y_0/2$, from Y_{in}. The remainder admittance is

$$Y_{in} = \frac{3(Y_+ + Y_-)Y_0 + 6Y_+Y_-}{2Y_0 + (Y_+ + Y_-)}$$

The synthesis procedure continues by removing an admittance, $3(Y_+ + Y_-)/4$, from Y_{in}. The extraction of this term produces a remainder admittance given by

$$Y_{in} = \frac{G^2}{2Y_0 + (Y_+ + Y_-)}$$

This admittance is readily realized by an ideal two-port gyrator circuit terminated by a parallel combination of $2Y_0$ and $(Y_+ + Y_-)$. The required topology is depicted in Fig. 7.5. The shunt elements in front of the gyrator circuit coincide with the one-port immittance of the isotropic junction. The effects of the gyrotropy are represented by the two-port gyrator circuit and its complex load.

The real and imaginary parts of the input admittance of the terminated circulator may readily be constructed by decomposing Y_{in} in a second Foster form:

$$Y_{in} = \frac{Y_0}{2} + \frac{3(Y_+ + Y_-)}{4} + \frac{G^2}{2Y_0 + (Y_+ + Y_-)} \tag{7.36}$$

The real and imaginary parts of this quantity are

$$G_{in} = \frac{Y_0}{2} + \frac{2G^2 Y_0}{4Y_0^2 + (Y_+ + Y_-)^2} \tag{7.37}$$

$$B_{in} = \frac{3(Y_+ + Y_-)}{4} - \frac{G^2(Y_+ + Y_-)^2}{4Y_0^2 + (Y_+ + Y_-)^2} \tag{7.38}$$

Specializing this relationship to a degree-1 circulator in the vicinity of its midband frequency indicates that

$$G_{in} = Y_0 \tag{7.39}$$

$$B_{in} = (Y_+ + Y_-)/2 \tag{7.40}$$

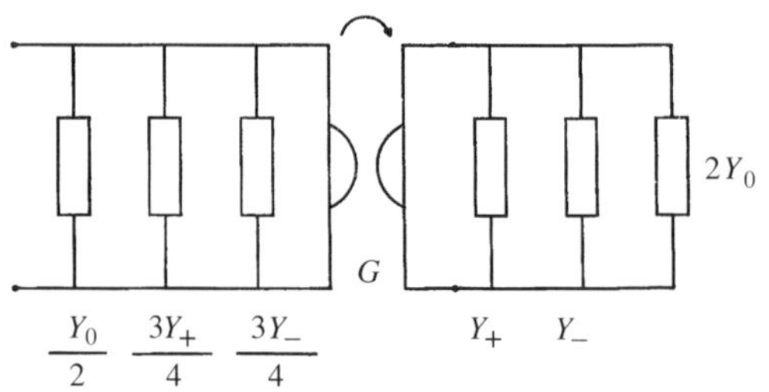

FIGURE 7.5 Complex admittance realization of terminated circulator.

provided

$$G = Y_0 \tag{7.41}$$

$$(Y_+ + Y_-) << 2Y_0 \tag{7.42}$$

This suggests that the susceptance, in the vicinity of the midband frequency of the degree-1 terminated circulator, coincides with that of the complex gyrator circuit of the three-port circulator. Elsewhere, the susceptance slope parameters of the two arrangements bear no obvious relationship.

7.7 TOPOLOGIES OF TERMINATED CIRCULATOR AT THE SPLIT FREQUENCIES

The equivalent circuit of the terminated circulator at the split frequencies is also of some interest. It provides one means of experimentally evaluating the susceptances of the split counterrotating eigen-networks of the topology. The two circuits are indicated in Fig. 7.6. These are obtained by inspection from Fig 7.5. The realization of a typical result starts by putting $Y_- = 0$ in Y_{in} in Eq. (7.10). This gives

$$Y_{in} = \frac{Y_0^2 + 2Y_0Y_+}{2Y_0 + Y_+} \tag{7.43}$$

It continues by extracting an admittance $Y_0/2$ from Y_{in}. The remainder admittance is

$$Y_{in} = \frac{3Y_0Y_+}{4Y_0 + 2Y_+} \tag{7.44}$$

The synthesis procedure proceeds by forming Z_{in} from Y_{in}:

$$Z_{in} = \frac{4Y_0 + 2Y_+}{3Y_0Y_+} \tag{7.45}$$

The required decomposition is

$$Z_{in} = \frac{2Z_0}{3} + \frac{4Z_+}{3} \tag{7.46}$$

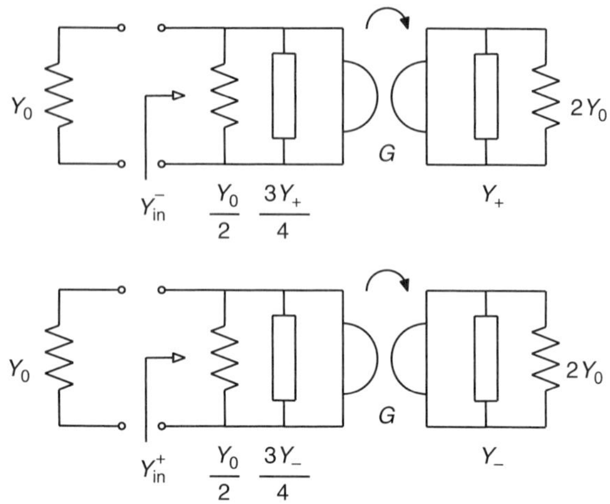

FIGURE 7.6 Topologies of terminated circulator at split frequencies in terms of two-port gyrator circuits.

The result can also be deduced from the general solution in Fig. 7.7 by recognizing that the gyrator conductance is now defined by

$$G = (j\sqrt{3}/2)(Y_+) \tag{7.47}$$

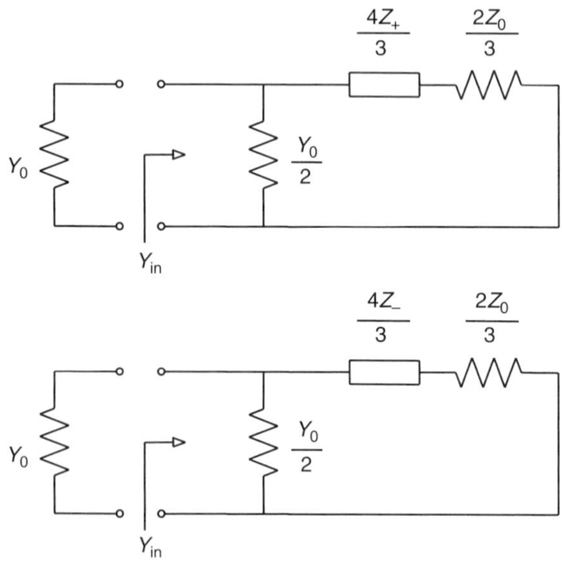

FIGURE 7.7 Topologies of terminated circulator at the split frequencies.

The reflection coefficient of either arrangement is in keeping with the conditions in Eq. (7.14). The solution at $\omega = \omega_+$ follows without ado.

7.8 DEGREE-2 TOPOLOGY

One way to broadband a directly coupled (degree-1) circulator is to introduce a matching circuit at each port. The most common arrangement consists of a single quarter-wave long impedance transformer. This structure is referred to as a degree-2 topology in that its frequency response is associated with a degree-2 Chebyshev polynomial. The extension of the degree-1 problem to that of the degree-2 one is straightforward. Figure 7.8 indicates the configuration in question. The derivation of the required result begins by replacing the load conditions at ports 2 and 3 by

$$V_2/-I_2 = V_3/-I_3 = Z_L \tag{7.48}$$

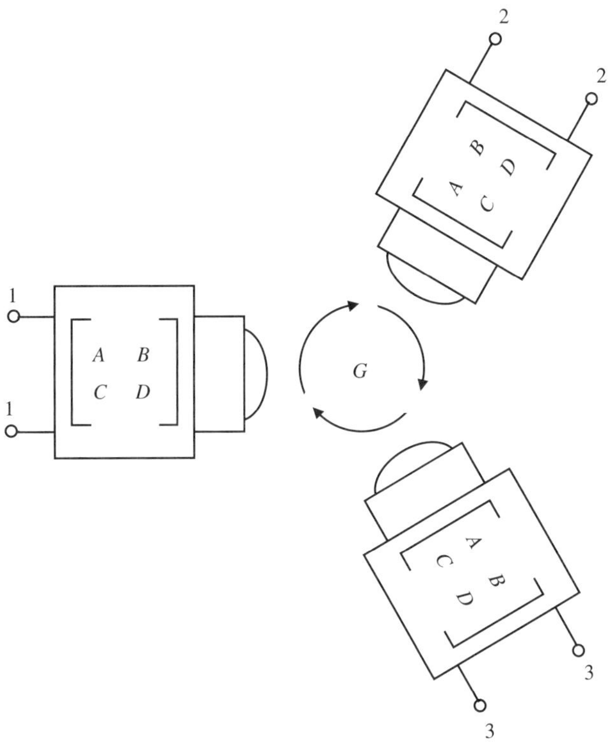

FIGURE 7.8 Schematic diagram of a degree-2 three-port circulator.

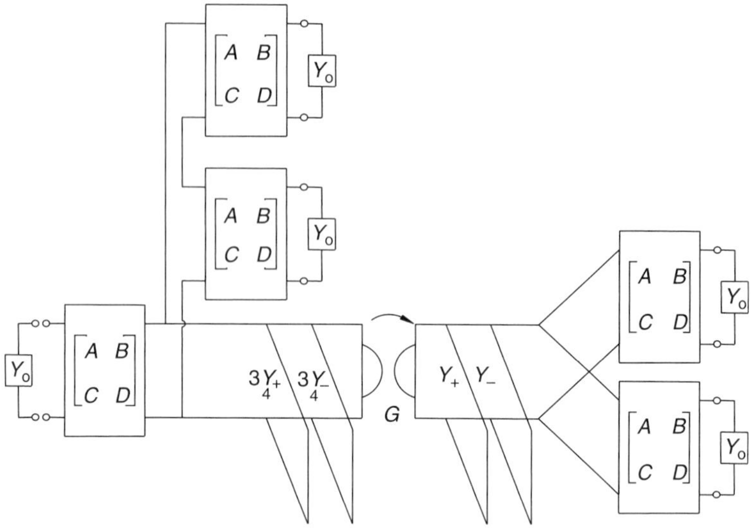

FIGURE 7.9 Equivalent circuit of a degree-2 circulator.

Z_L is the input impedance at the resonator terminals of a unit element (UE), terminated in the characteristic impedance of the arrangement.

$$Z_L = \frac{D + jCZ_0}{AZ_0 + jB} \tag{7.49}$$

The *ABCD* parameters coincide with a UE of impedance Z_0 and electrical length θ. The one-port equivalent circuit at the resonator terminals is then realized by replacing Y_0 in the degree-1 circuit by Y_L. The arrangement of the circulator at the terminals of the overall network is obtained separately by mapping the resonator terminals through the two-port circuit defined by the *ABCD* parameters of the matching circuit. Figure 7.9 depicts the one-port circuit obtained in this way.

CHAPTER EIGHT

Cutoff Space of Cloverleaf Resonators with Magnetic Walls

8.1 INTRODUCTION

An important planar resonator with a magnetic side wall and top and bottom electric walls with the symmetry of the three-port junction circulator is the cloverleaf geometry. In practice, some other possible structures of the three-port circulator, to be dealt with separately, include disk, triangular, wye, ring, and irregular hexagonal geometries. The purpose of this chapter is to employ the finite element method to analyze the cloverleaf gyromagnetic resonator with either threefold or fourfold symmetry. It involves the construction of an energy functional, which when extremized produces the required eigenvalues and eigenvectors of the problem region. These topologies are of value in the design of three- and four-port junction circulators and other circuits. It is completely fixed by two radii. The chapter includes the descriptions of the cutoff space and the field patterns of this type of isotropic circuit with threefold and fourfold symmetries. This is done for the first four modes of each configuration. The mode notation employed in the description of this geometry is that met in connection with a simple circular geometry. The chapter also provides calculations on the split cutoff space of gyromagnetic cloverleaf resonators with threefold symmetry.

The Stripline Circulator: Theory and Practice. By J. Helszajn

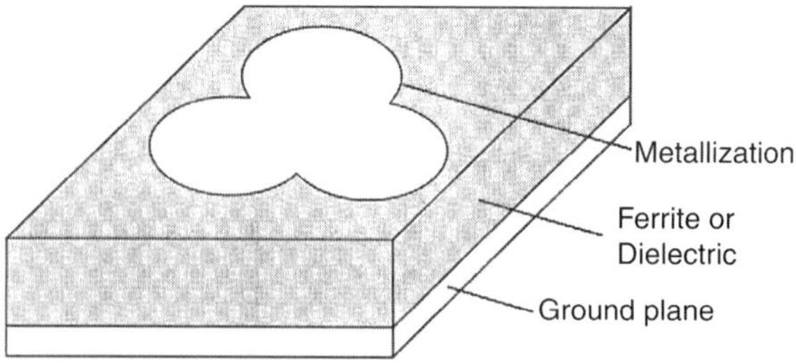

FIGURE 8.1 Topology of cloverleaf planar resonator with threefold symmetry.

8.2 CLOVERLEAF RESONATOR

The schematic diagram of the cloverleaf circuit with threefold symmetry discussed in this chapter is illustrated in Fig. 8.1. It may be described by two radii. Figure 8.2 depicts the coordinate system employed here as well as three typical configurations. The minimum outside radius is related to the inside one by

$$R_o(\min) = (\sqrt{3}/2)R_i \tag{8.1}$$

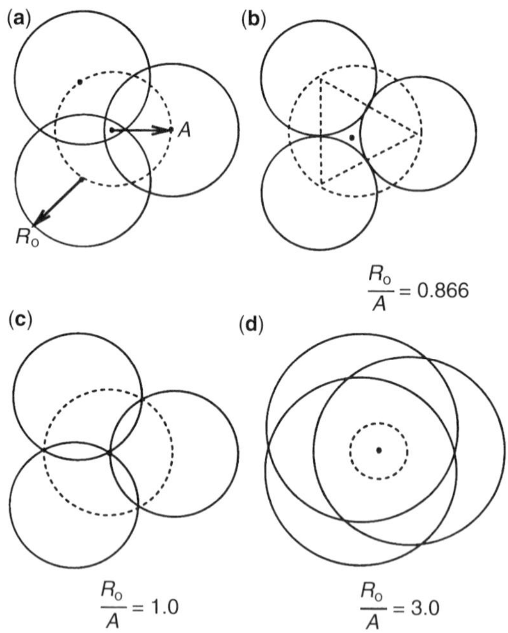

FIGURE 8.2 (a) Coordinate system used in description of cloverleaf resonator with threefold symmetry. Topology of cloverleaf resonator with (b) minimum surface area $R_o/A = 0.86$, (c) $R_o/A = 1.0$, and (d) $R_o/A = 3.0$.

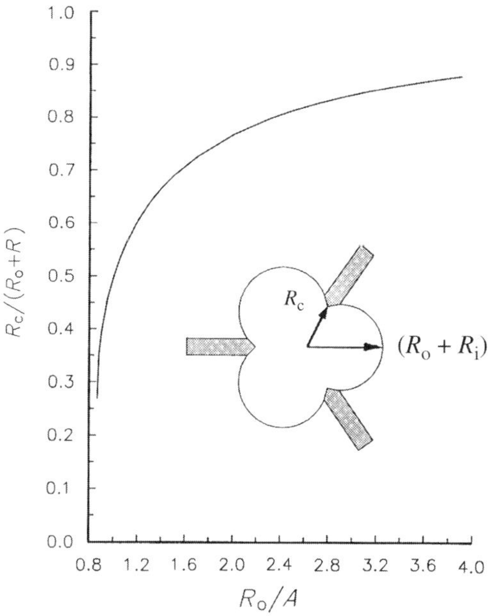

FIGURE 8.3 Relationship between aspect ratio of cloverleaf resonator with threefold symmetry and the ratio of the possible coupling port radii.

One feature of the cloverleaf geometry is the possibility of coupling to it either at the intersection of the lobes employed to define its geometry or at its extremities. The former possibility affords some degree of miniaturization in the layout of microwave junction circulators. Figure 8.3 indicates the relationship between the aspect ratio of the resonator and the ratio of its radii. The choice of any coupling port is related of course to the external Q-factor of the circuit. This problem, however, is outside the scope of this work.

8.3 FINITE ELEMENT METHOD

The solution of the cloverleaf circuit with top and bottom electric walls and a magnetic side wall investigated in this chapter is not compatible with a closed form formulation. It provides a good example of the use of the finite element method.

The mathematical formulation of the finite element method relies on the construction of an energy functional, which is then discretized and extremized to form a variational solution of the problem. One suitable functional for the isotropic problem region, which has a magnetic wall boundary condition included as a natural term, is

$$F(E_z) = \iint_S \left[-|\nabla_t E_z|^2 + k_e^2 |E_z|^2 \right] ds \tag{8.2}$$

S is the surface of the planar circuit. ∇_t is the transverse differential operator in Cartesian coordinates. The wavenumber (k_e) is defined in the usual way by

$$k_e^2 = k_0^2 \varepsilon_f \tag{8.3}$$

where

$$k_0 = 2\pi/\lambda_0 \tag{8.4}$$

and ε_f is the relative dielectric constant of the dielectric substrate.

The finite element method commences by subdividing the surface of the planar circuit into a number of elementary triangles, m. It continues by approximating the true field solution, E_z, within each triangular finite element by a trial function expansion of the form

$$E_z'(x, y) = \sum_{k=1}^{n} u_k \alpha_k(x, y), \quad k = 1, 2, \ldots, n \tag{8.5}$$

where α_k are a suitable set of real basis functions that embody the spatial variation of the problem and u_k are complex coefficients that represent the unknown fields at the nodes of the finite element mesh.

The number of nodes, n, within each triangle is defined by the degree, q, of the approximation problem as

$$n = (q + 1)(q + 2)/2 \tag{8.6}$$

The eigenvalue problem is now established by extremizing the energy functional using the Rayleigh–Ritz procedure. The required eigenvalue equation is

$$\{[\mathbf{D}]\}\,\overline{\mathbf{U}} = k_e^2 [\mathbf{B}]\,\overline{\mathbf{U}} \tag{8.7}$$

The elements appearing in the matrices [**B**] and [**D**] are defined by

$$B_{ij} = \iint_S (\alpha_i \cdot \alpha_j) ds \tag{8.8}$$

$$D_{ij} = \iint_S \left((\nabla_t \alpha_i) \cdot (\nabla_t \alpha_j) \right) ds \tag{8.9}$$

where

$$i = 1, 2, 3, \ldots, p \quad \text{and} \quad j = 1, 2, 3, \ldots, p$$

Once the basis functions have been selected the general matrix eigenvalue problem may be solved for the p eigenvalues k_e^2 and p eigenvectors $\overline{\mathbf{U}}$. The eigenvalues are the normalized cutoff frequencies of the resonator and the eigenvectors are discrete values of the approximated field at the finite element nodes. $\overline{\mathbf{U}}$ is a column matrix whose dimension is equal to the number of nodes in the finite element mesh.

8.4 CUTOFF SPACE OF ISOTROPIC CLOVERLEAF RESONATOR WITH THREEFOLD SYMMETRY

The relationship between the details of the circuit and the cutoff wavenumbers for the dominant mode and a number of higher order modes for a resonator with magnetic walls is illustrated in Fig. 8.4. The number of discrete triangles in the finite element mesh is designated by the integer m, the number of nodes within each triangle by the integer n, the degree of the approximation problem is denoted by the integer q, and the number of nodes within the final mesh by p. This kind of problem produces p eigenvalues and p eigenvectors. The finite element mesh utilized in the solution of this circuit is illustrated in Fig. 8.5. The independent mesh variables are given for this arrangement by

$$q = 2 \quad \text{and} \quad m = 57$$

and the dependent one by

$$n = 6$$

The number of nodes before assembly is

$$m \times n = 342$$

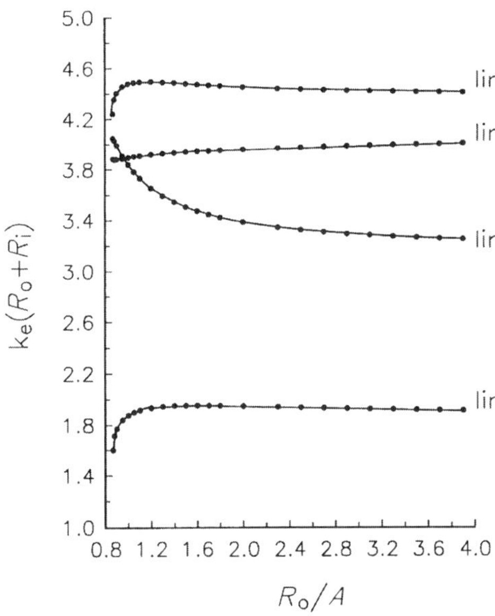

FIGURE 8.4 Cutoff space of cloverleaf planar resonator with threefold symmetry with magnetic side walls.

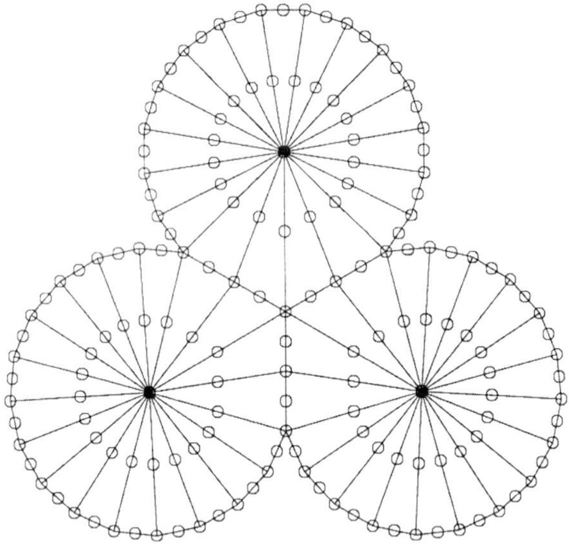

FIGURE 8.5 Finite element mesh for cloverleaf planar resonator with threefold symmetry.

and, after merging coincident nodes, the number reduces to

$$p = 160$$

The cutoff numbers of the first four modes in this type of resonator for the condition in Eq. (8.9) are given by

$$k_e(R_o + R_i) = 1.59, \; R_o = \left(\sqrt{3}/2\right) R_i$$

$$k_e(R_o + R_i) = 3.88, \; R_o = \left(\sqrt{3}/2\right) R_i$$

$$k_e(R_o + R_i) = 4.05, \; R_o = \left(\sqrt{3}/2\right) R_i$$

$$k_e(R_o + R_i) = 4.24, \; R_o = \left(\sqrt{3}/2\right) R_i$$

One possible notation for this type of circuit may be established by recognizing that as R_o/R_i increases, its cutoff space is asymptotic to that of a circular disk resonator. If this convention is adopted, then the modes of the cloverleaf resonator may be referred to as limit TM_{mno} modes of the disk resonator. In this nomenclature m refers to the number of half cycles of the magnetic field along the azimuthal direction, n refers to the number of half cycles along the radius of the circuit, and o indicates that there is no variation of the fields along the axis of the resonator.

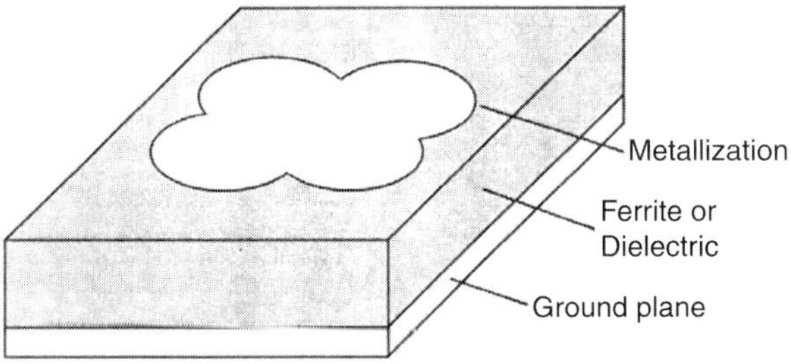

FIGURE 8.6 Topology of cloverleaf planar resonator with fourfold symmetry.

8.5 CUTOFF SPACE OF ISOTROPIC CLOVERLEAF RESONATOR WITH FOURFOLD SYMMETRY

A cloverleaf resonator with fourfold symmetry with magnetic and electric side walls may also be visualized without difficulty. Figure 8.6 illustrates its topology. Figure 8.7 depicts its construction and three typical geometries. This geometry is defined by variables R_o and A. Its minimum outside radius is

$$R_o(\min) = A/2 \tag{8.10}$$

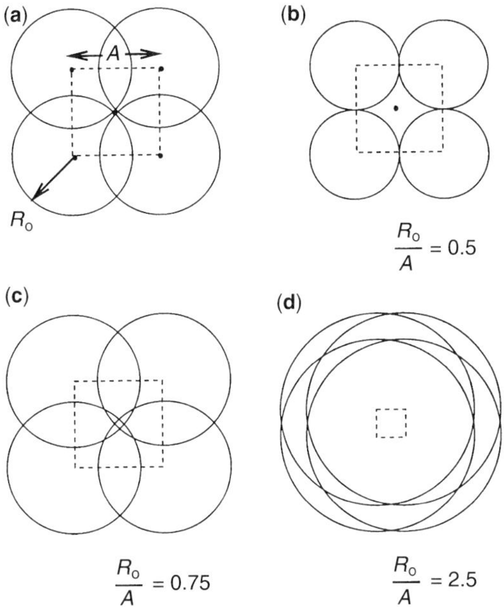

FIGURE 8.7 (a) Coordinate system used in description of cloverleaf resonator with fourfold symmetry. Topology of cloverleaf resonator with (b) minimum surface area $R_o/A = 0.5$, (c) $R_o/A = 0.757$, and $R_o/A = 2.5$.

The finite element mesh used in the analysis of this resonator is indicated in Fig. 8.8. This mesh is described by

$$q = 2$$
$$m = 76$$
$$n = 6$$
$$m \times n = 456$$
$$p = 213$$

The cutoff numbers of this type of circuit with top and bottom electric walls bounded by a perfect magnetic wall are indicated in Fig. 8.9. The mode nomenclature in this instance may also be deduced by referring to a simple disk resonator. A notable feature of this circuit is that as the ratio R_o/A decreases from the disk resonator limit, the degenerate $n = 2$ modes are split by the geometry.

The cutoff numbers of the first four modes in this resonator, for the condition in Eq. (8.10), are given by

$$k_e\left(R_o + \frac{A}{\sqrt{2}}\right) = 1.76, \; R_o = \frac{A}{2}$$

$$k_e\left(R_o + \frac{A}{\sqrt{2}}\right) = 2.18, \; R_o = \frac{A}{2}$$

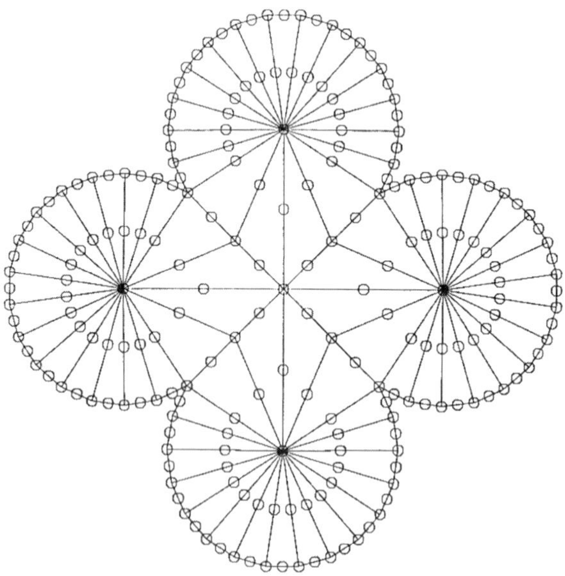

FIGURE 8.8 Finite element mesh for cloverleaf planar resonator with fourfold symmetry.

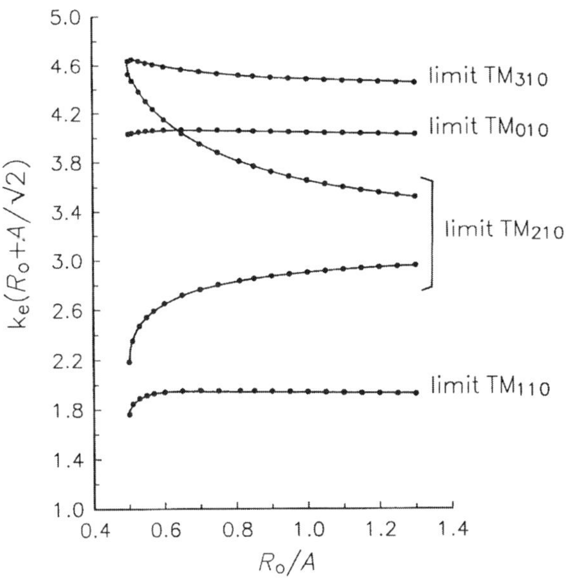

FIGURE 8.9 Cutoff space of cloverleaf planar resonator with fourfold symmetry with magnetic side wall.

$$k_e\left(R_o + \frac{A}{\sqrt{2}}\right) = 4.03, \ R_o = \frac{A}{2}$$

$$k_e\left(R_o + \frac{A}{\sqrt{2}}\right) = 4.52, \ R_o = \frac{A}{2}$$

8.6 FIELD PATTERNS IN CLOVERLEAF RESONATORS

The equipotential lines of the first four modes in an isotropic cloverleaf resonator with threefold symmetry for three different aspect ratios are indicated in Figs. 8.10–8.13. The symmetric ($n = 0$ limit) mode, illustrated in Fig. 8.10, has the property that it has a magnetic wall on both the axis and on the open side wall of the planar resonator. The fundamental ($n = 1$ limit) mode is indicated in Fig. 8.11. Its field pattern, unlike the symmetric mode, displays an electric wall on the axis of the circuit. As the ratio R_o/R_t approaches its lower bound, its field pattern reduces to that of the fundamental mode in a symmetric wye resonator with threefold symmetry. The first two higher order ($n = 2$ and $n = 3$ limit) modes encountered in this type of resonator are shown separately in Figs. 8.12 and 8.13. The equipotential lines of both these modes also exhibit an electric wall along the axis of the resonator.

FIGURE 8.10 Equipotential lines of cloverleaf resonator with threefold symmetry, for $n = 0$ limit disk ($R_o/A = 0.86$, 1.0, 5.0). (Reproduced with permission from J. Helszajn and D. J. Lynch, Cut-off space of cloverleaf resonators with electric and magnetic walls, *IEEE Trans. Microwave Theory Tech.*, Vol. MTT-40, No. 8, pp. 1620–1629, 1992.)

The electric field distributions for the first four modes in a cloverleaf resonator with fourfold symmetry are depicted in Figs. 8.14–8.17. This is, again, done for three typical geometries. The symmetric $n = 0$ limit mode

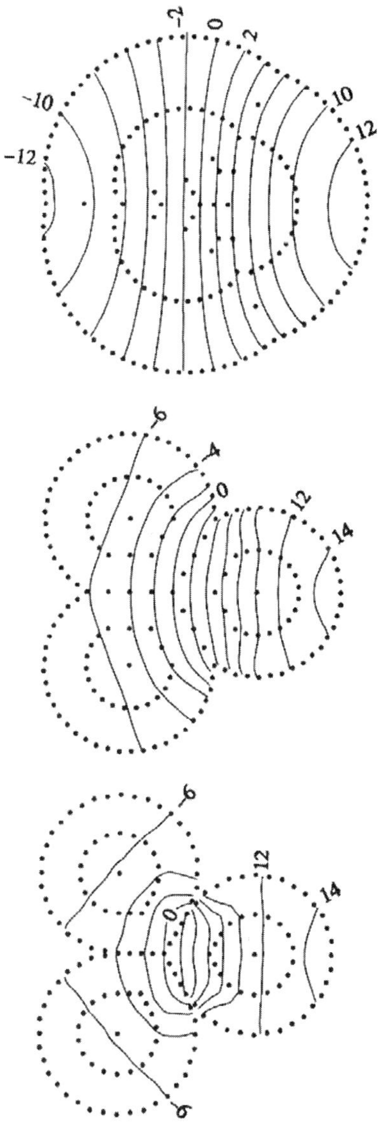

FIGURE 8.11 Equipotential lines of cloverleaf resonator with threefold symmetry, for $n = 1$ limit disk mode $R_o/A = 0.86$, 1.0, 5.0). (Reproduced with permission from J. Helszajn and D. J. Lynch, Cut-off space of cloverleaf resonators with electric and magnetic walls, *IEEE Trans. Microwave Theory Tech.*, Vol. MTT-40, No. 8, pp. 1620–1629, 1992.)

in this sort of circuit displays once more a magnetic wall at the center of the resonator, whereas the $n = 1$ and $n = 2$ limit modes have an electric wall there.

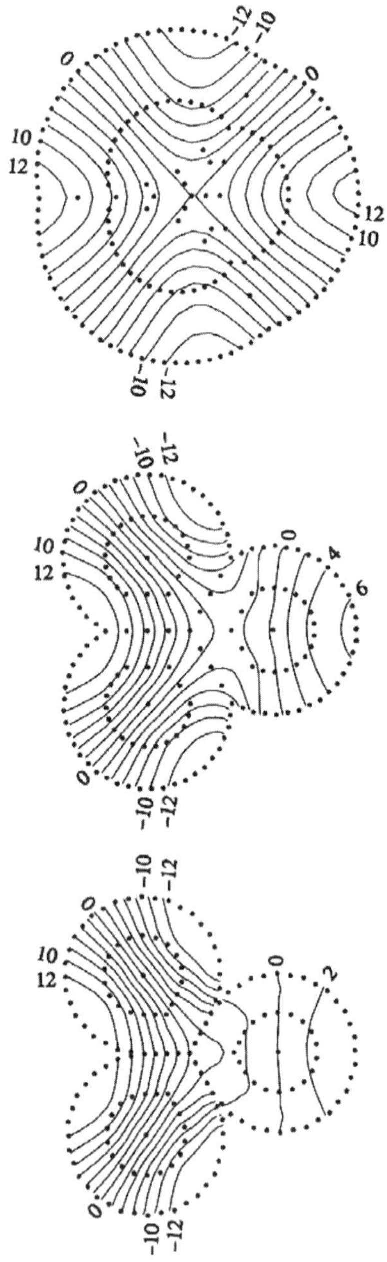

FIGURE 8.12 Equipotential lines of cloverleaf resonator with threefold symmetry, for $n = 2$ limit disk mode ($R_o/A = 0.86, 1.0, 5.0$). (Reproduced with permission from J. Helszajn and D. J. Lynch, Cut-off space of cloverleaf resonators with electric and magnetic walls, *IEEE Trans. Microwave Theory Tech.*, Vol. MTT-40, No. 8, pp. 1620–1629, 1992.)

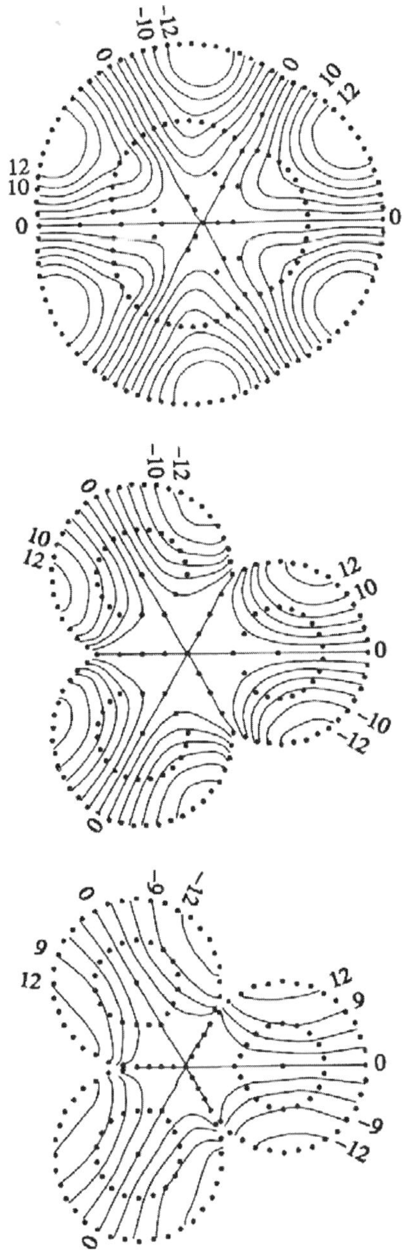

FIGURE 8.13 Equipotential lines of cloverleaf resonator with threefold symmetry, for $n = 3$ limit disk mode ($R_o/A = 0.86$, 1.0, 5.0). (Reproduced with permission from J. Helszajn and D. J. Lynch, Cut-off space of cloverleaf resonators with electric and magnetic walls, *IEEE Trans. Microwave Theory Tech.*, Vol. MTT-40, No. 8, pp. 1620–1629, 1992.)

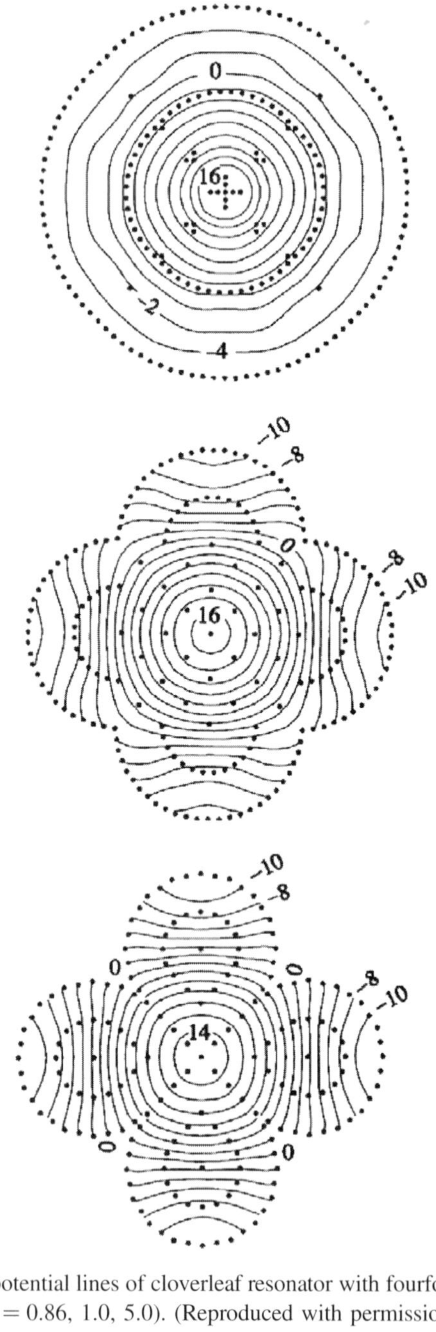

FIGURE 8.14 Equipotential lines of cloverleaf resonator with fourfold symmetry, for $n = 0$ limit disk mode ($R_o/A = 0.86$, 1.0, 5.0). (Reproduced with permission from J. Helszajn and D. J. Lynch, Cut-off space of cloverleaf resonators with electric and magnetic walls, *IEEE Trans. Microwave Theory Tech.*, Vol. MTT-40, No. 8, pp. 1620–1629, 1992.)

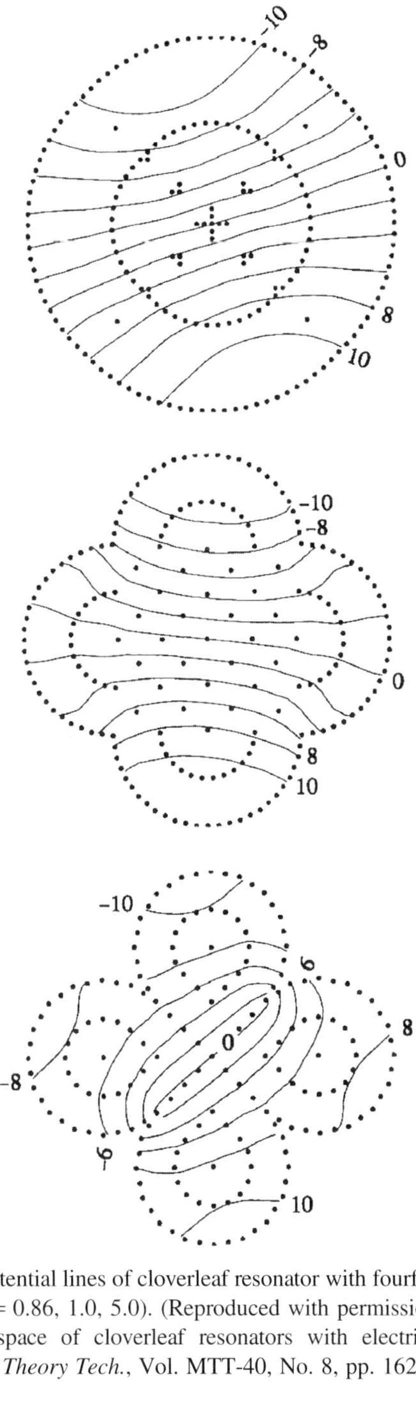

FIGURE 8.15 Equipotential lines of cloverleaf resonator with fourfold symmetry, for $n = 1$ limit disk mode ($R_o/A = 0.86$, 1.0, 5.0). (Reproduced with permission from J. Helszajn and D. J. Lynch, Cut-off space of cloverleaf resonators with electric and magnetic walls, *IEEE Trans. Microwave Theory Tech.*, Vol. MTT-40, No. 8, pp. 1620–1629, 1992.)

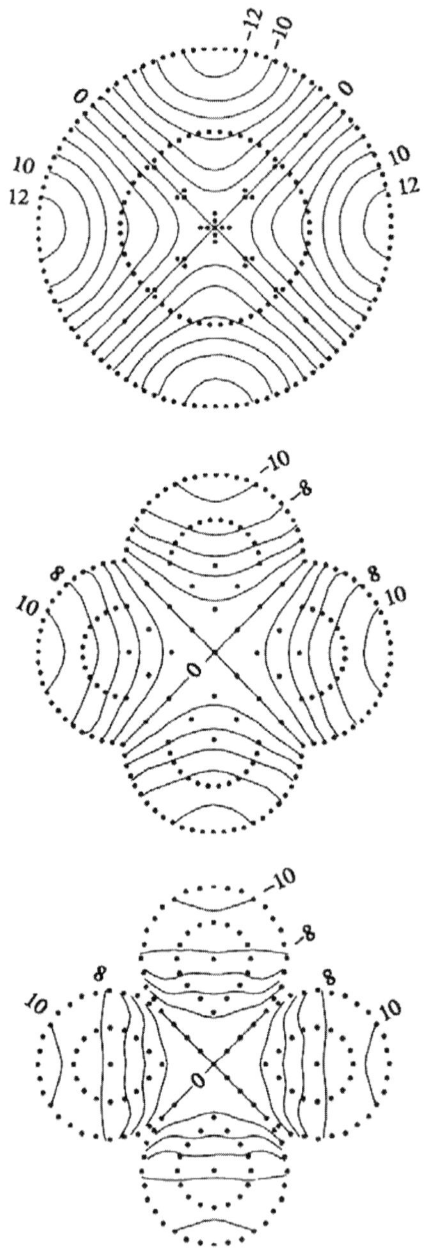

FIGURE 8.16 Equipotential lines of cloverleaf resonator with fourfold symmetry, for $n = 2$ limit disk (lower) mode ($R_o/A = 0.86, 1.0, 5.0$). (Reproduced with permission from J. Helszajn and D. J. Lynch, Cut-off space of cloverleaf resonators with electric and magnetic walls, *IEEE Trans. Microwave Theory Tech.*, Vol. MTT-40, No. 8, pp. 1620–1629, 1992.)

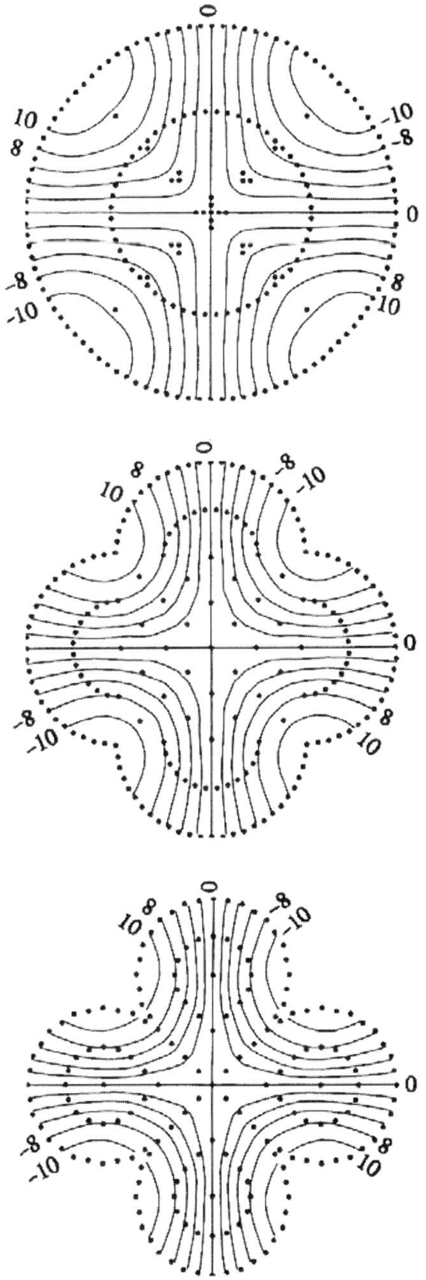

FIGURE 8.17 Equipotential lines of cloverleaf resonator with fourfold symmetry, for $n = 2$ limit disk (upper) mode ($R_o/A = 0.86, 1.0, 5.0$). (Reproduced with permission from J. Helszajn and D. J. Lynch, Cut-off space of cloverleaf resonators with electric and magnetic walls, *IEEE Trans. Microwave Theory Tech.*, Vol. MTT-40, No. 8, pp. 1620–1629, 1992.)

8.7 GYROMAGNETIC CLOVERLEAF RESONATOR

The functional met in connection with the gyromagnetic circuit is a standard result in the literature. It differs from that of the isotropic problem region in that it involves an additional contour integral term:

$$F(E_z) = \iint_S \left[|\nabla_t E_z|^2 - k_e^2 |E_z|^2 \right] ds - j\frac{\kappa}{\mu} \int_{\xi} E_z \frac{\partial E_z}{\partial t} dt \tag{8.11}$$

This functional has the magnetic wall boundary condition included as a natural term. s is the surface of the planar circuit, ξ represents the periphery, and t is the boundary tangent defined in a counterclockwise direction. μ and κ are the diagonal and off-diagonal elements of the tensor permeability. When the gyrotropy κ/μ is set to zero the functional reduces to that of an isotropic resonator. ∇_t is the transverse differential operator in Cartesian coordinates. The wavenumber (k_e) is defined in the usual way by

$$k_e^2 = k_0^2 \mu_{eff} \varepsilon_f \tag{8.12}$$

where

$$\mu_{eff} = \frac{\mu^2 - \kappa^2}{\mu} \tag{8.13}$$

ε_f is the relative dielectric constant of the ferrite or dielectric substrate.

The discretization and extremization of this functional proceed in a like manner to that of the isotropic problem region. The details are dealt with in Chapter 15.

8.8 SPLIT CUTOFF SPACE OF GYROMAGNETIC CLOVERLEAF RESONATOR WITH THREEFOLD SYMMETRY

The split mode chart of a gyromagnetic cloverleaf resonator is also of some interest. Figure 8.18 depicts the relationship between the splitting of the dominant pair of degenerate modes and the aspect ratio of the resonator for six different values of gyrotropy. It is obtained by retaining the gyrotropy term in the functional described in Eq. (8.11).

Scrutiny of this result indicates that the gyromagnetic effect in this type of resonator depends on the aspect ratio and the gyrotropy of the geometry and that there are a number of different combinations of these variables for a given gyromagnetic effect. Another feature of this result is that the splitting between the dominant pair of degenerate modes in a weakly magnetized cloverleaf resonator is proportional, in the usual way, to the gyrotropy of the material. The angle between the degenerate

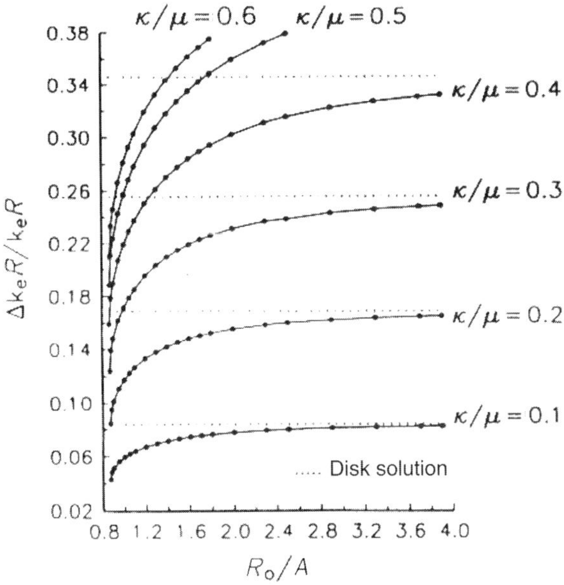

FIGURE 8.18 Split radial wavenumbers of weakly magnetized gyromagnetic cloverleaf resonator. (Reproduced with permission from J. Helszajn and D. J. Lynch, Cut-off space of cloverleaf resonators with electric and magnetic walls, *IEEE Trans. Microwave Theory Tech.*, Vol. MTT-40, No. 8, pp. 1620–1629, 1992.)

split radial wavenumbers of the resonator approaches that of a simple disk in an orderly fashion.

$$\frac{(k_e R)^+ - (k_e R)^-}{(k_e R)} = \frac{2n}{(k_e R)^2 - n^2}\left(\frac{\kappa}{\mu}\right) \tag{8.14}$$

The quality factor of a junction circulator using such a planar resonator is related to the gyrotropy by

$$\frac{1}{Q_L} = \sqrt{3}\left(\frac{(k_e R)^+ - (k_e R)^-}{(k_e R)}\right) \tag{8.15}$$

Values of Q_L between 2 and $2\frac{1}{2}$ are suitable for the design of quarter-wave coupled devices with modest specifications. This suggests that the normalized wavenumbers must be bounded by

$$0.231 < \left(\frac{(k_e R)^+ - (k_e R)^-}{(k_e R)}\right) < 0.288$$

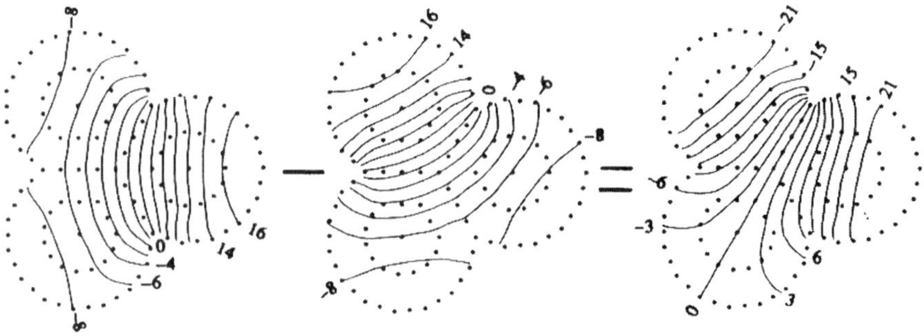

FIGURE 8.19 Standing wave circulation solution of circulator using dominant mode in cloverleaf resonator with threefold symmetry ($R_o/A = 1.0$). (Reproduced with permission from J. Helszajn and D. J. Lynch, Cut-off space of cloverleaf resonators with electric and magnetic walls, *IEEE Trans. Microwave Theory Tech.*, Vol. MTT-40, No. 8, pp. 1620–1629, 1992.)

One attractive feature of the cloverleaf resonator is the possibility of coupling to it at the terminals defined by the intersection of the lobes used in its construction. Adopting a value of $(R_o/R_i) = 1.0$, for instance, indicates that the interval defined by the normalized split cutoff numbers can be accommodated with

$$0.40 \leq (\kappa/\mu) \leq 0.60$$

Once the quality factor of this type of device is established, a knowledge of its susceptance slope parameter is sufficient for design. This quantity is defined by the resonator shape, its thickness, and the choice of coupling terminals. Its solution is not at this time available in the open literature.

8.9 STANDING WAVE SOLUTION OF CIRCULATORS USING CLOVERLEAF RESONATORS

The standing wave solution of a planar junction circulator using this type of resonator is also of some interest. It may be obtained by taking a suitable linear combination of the modes of the isotropic circuit. Figure 8.19 shows the construction of such a solution for the dominant mode in such a gyromagnetic cloverleaf resonator. An ideal circulator may be formed with this type of circuit with ports attached to either two possible triplets of terminals.

CHAPTER NINE

Standing Wave Solution of Wye Gyromagnetic Planar Resonator

9.1 INTRODUCTION

An important planar resonator that has the symmetry of the three-port junction circulator is the wye resonator with top and bottom electric walls and a magnetic side wall. This sort of resonator consists of a central circular region symmetrically loaded by three open-circuited transmission lines. One exact means of dealing with such a structure is to have recourse to various numerical techniques. The method utilized here is based on the decomposition of the problem region into a symmetrical three-port region bounded by three open-circuited lines. The resonant frequencies of the symmetrical and counterrotating families of modes of the overall circuit are then obtained by satisfying the boundary conditions between the impedance eigenvalues of the circular region and the impedance of a typical open-circuited stub. Some calculations based on a finite element solver are included for completeness. It provides one means of constructing standing wave solutions of the various modes of the geometry. A closed form description of a wye resonator consisting of an inner gyromagnetic region to which are connected three stubs is included separately. It provides one means by which the split frequencies of the resonator under the application of the gyrotropy may be approximately deduced. Although it is difficult to visualize rotation of the equipotential lines in a magnetized wye resonator, it is nevertheless possible to do so at one of two possible triplets of ports. This may be done by taking suitable linear combinations of those of the demagnetized geometry. The first two circulation modes of the geometry are dealt with.

The Stripline Circulator: Theory and Practice. By J. Helszajn

9.2 CUTOFF SPACE OF WYE PLANAR RESONATOR

The equipotential lines and cutoff numbers of the first three modes of a wye resonator have been computed by having recourse to a finite element program. The schematic diagram of the resonator under consideration is depicted in Fig. 9.1. The coupling angle of a typical stub is, in this geometry, restricted to $\pi/3$ rad. It is defined by its width (W) and length (R). The geometry is subdivided into 13 triangular elements and a third order polynomial approximation is made to the EM fields in each triangle. The degree of the polynomial is fixed by the volume of labor involved in setting up the matrix problem. The number of triangles chosen is determined by the fact that the amount of computer time taken to solve the problem is not linearly dependent on the number of triangles. A typical segmentation of a wye resonator is illustrated in Fig. 9.2. The equipotential lines of the fundamental TM mode in this resonator are indicated in Fig. 9.3. Its cutoff number has also been computed. It is specified by

$$kR = 1.643 \tag{9.1}$$

where

$$k = \omega\sqrt{\mu_o \mu_r \varepsilon_f \varepsilon_r} \tag{9.2}$$

ε_r is the relative dielectric constant, μ_r the relative permeability of the ferrite material, and ω is the radian frequency (rad/s). This cutoff number applies for $W/R = 0.4$.

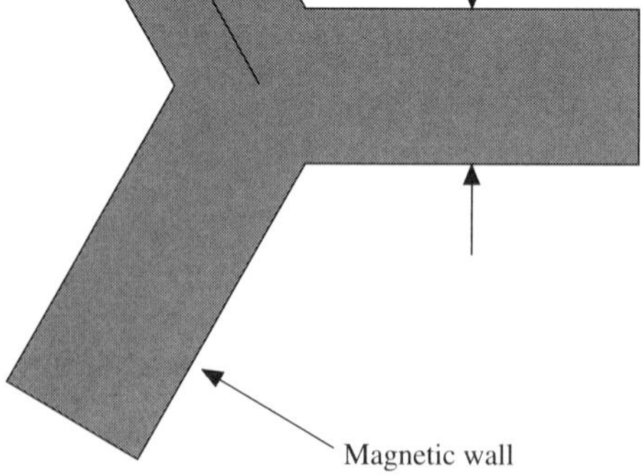

FIGURE 9.1 Schematic diagram of wye resonator.

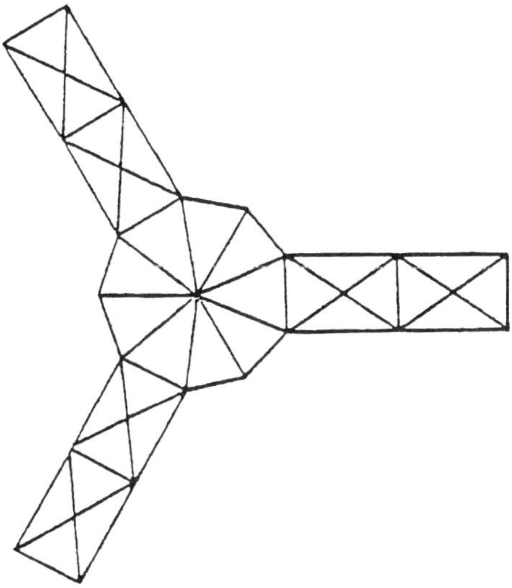

FIGURE 9.2 Segmentation of wye resonator into finite elements.

Figures 9.4a and 9.4b depict the first symmetric and the first higher order TM standing wave patterns in this type of geometry. The cutoff numbers of these two modes are

$$kR = 3.33 \quad \text{and} \quad kR = 4.91$$

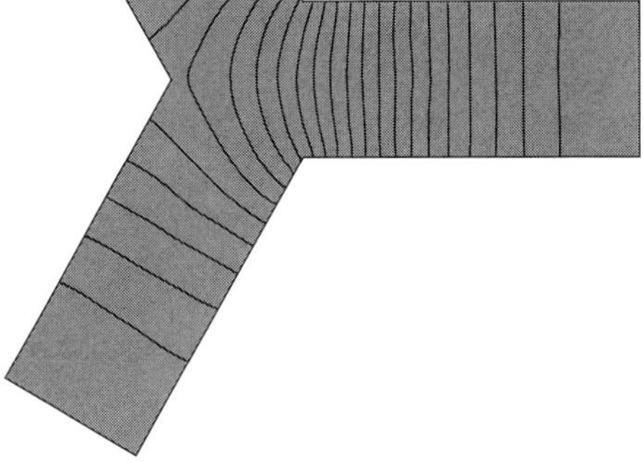

FIGURE 9.3 Equipotential lines of dominant mode in wye resonator.

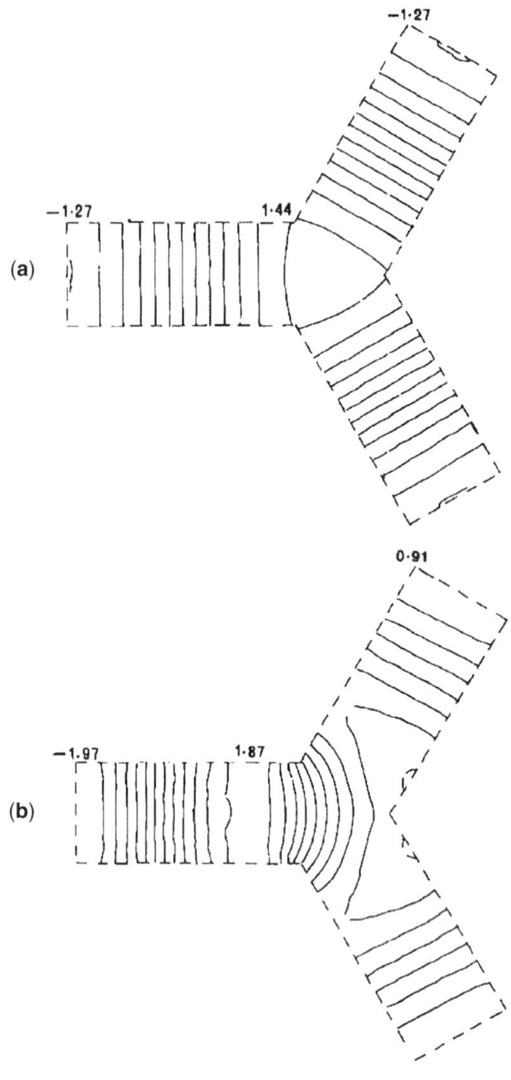

FIGURE 9.4 (a) First symmetric mode in wye resonator; (b) first higher order dominant mode in wye resonator. (Reproduced with permission from J. Helszajn, Standing wave solution of planar irregular hexagonal and wye resonators, *IEEE Trans. Microwave Theory Tech.*, Vol. MTT-29, No. 6, pp. 562–567, 1981.)

The equipotential lines of the symmetric mode in Fig. 9.4a has a maximum value both at the center and at the end of the stub and has a zero approximately midway along the stub at

$$kR \approx 1.67$$

The standing wave solution in Fig. 9.4b is also suitable for the construction of a three-port planar circulator.

The fields at the nodes have been normalized so that the field distribution of a typical mode satisfies the condition

$$\iint \phi_a^2 \, ds = 1 \tag{9.3}$$

The resonant modes produced by the program are orthogonal.

9.3 STANDING WAVE SOLUTION OF JUNCTION CIRCULATORS USING WYE RESONATORS

The equipotential lines of the dominant mode in the isotropic resonator have the symmetry encountered in the construction of planar circulators on magnetized substrates. It may therefore be utilized in the construction of a three-port planar junction circulator. Circulation solutions in magnetized wye resonators may be constructed by taking a linear combination of two standing wave patterns of the demagnetized wye resonator with one pattern rotated through 120°. This construction is depicted in Fig. 9.5 in the case of the dominant mode. It indicates that an ideal circulation condition can be realized by coupling to the wye resonator at one of two triplets of ports in the manner illustrated in Fig. 9.6. The arrangement in Fig. 9.6a produces a widely used commercial quarter-wave coupled three-port junction circulator whose outside radius is on the order of a quarter-wave at the operating frequency of the device. Figure 9.7 indicates the frequency response of one commercial quarter-wave coupled device. The standing wave solution for the first higher order circulation mode in a wye resonator is depicted in Fig. 9.8.

9.4 RESONANT FREQUENCIES OF UE LOADED DISK MAGNETIZED RESONATORS

One possible variation of the wye resonator is the disk-stub gyromagnetic arrangement consisting of a circular plate to which are connected three short UEs. The frequency may be deduced in this instance by visualizing it as a circular region loaded by three UEs or by a six-port arrangement with three of its ports closed by magnetic walls and the other three terminated by suitable stubs. The equivalence between the two models suggests that the first six impedance poles of the problem region are strictly speaking necessary in order to reproduce the boundary conditions of the resonator. The topology under consideration is indicated in Fig. 9.9. It is fixed by a coupling or shape angle (ψ) and the ratio of the radii R_i and R_o. Its degenerate or split resonance may be deduced

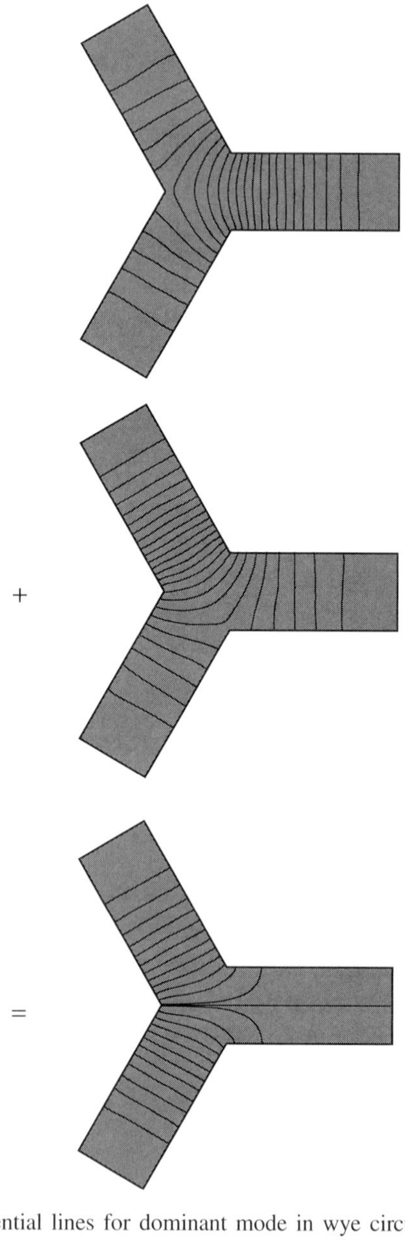

FIGURE 9.5 Equipotential lines for dominant mode in wye circulator. (Reproduced with permission from J. Helszajn, Standing wave solution of planar irregular hexagonal and wye resonators, *IEEE Trans. Microwave Theory Tech.*, Vol. MTT-29, No. 6, pp. 562–567, 1981.)

by resonating the stubs with the eigen-networks of the circular gyromagnetic region. The description of a typical impedance pole of an m-port symmetrical region is a standard result in the literature.

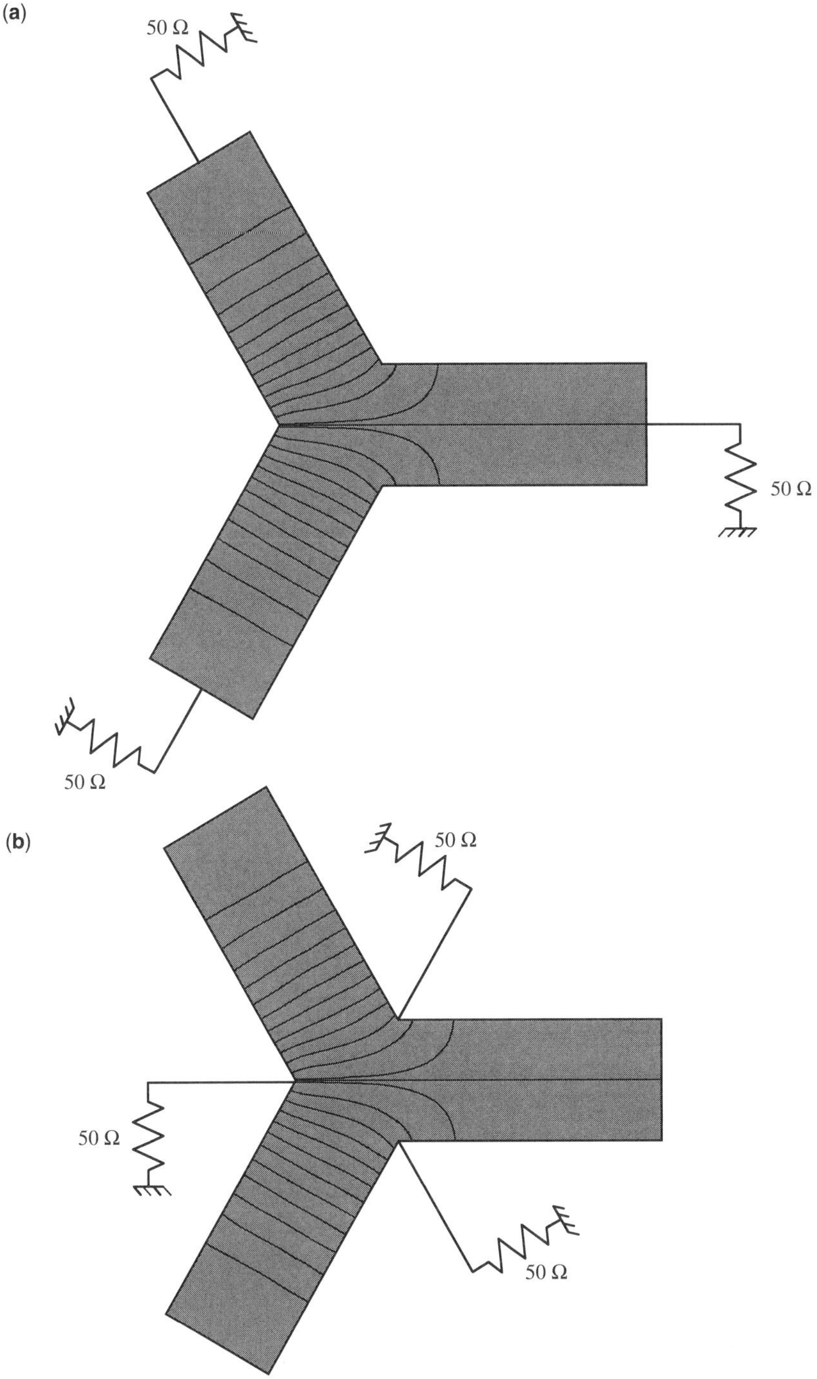

FIGURE 9.6 Circulation terminals of wye resonator: (a) “aa” terminals; (b) “bb” terminals.

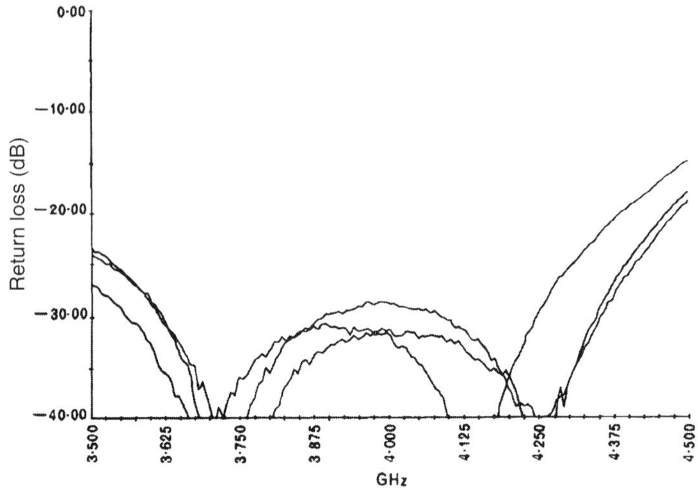

FIGURE 9.7 Frequency response of a quarter-wave coupled wye resonator. (Reproduced with permission from J. Helszajn, Standing wave solution of planar irregular hexagonal and wye resonators, *IEEE Trans. Microwave Theory Tech.*, Vol. MTT-29, No. 6, pp. 562–567, 1981.)

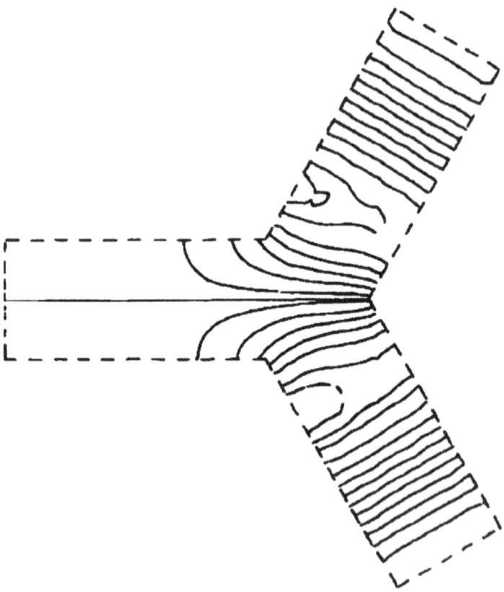

FIGURE 9.8 Equipotential lines for the first higher order circulation mode in wye resonator. (Reproduced with permission from J. Helszajn, Standing wave solution of planar irregular hexagonal and wye resonators, *IEEE Trans. Microwave Theory Tech.*, Vol. MTT-29, No. 6, pp. 562–567, 1981.)

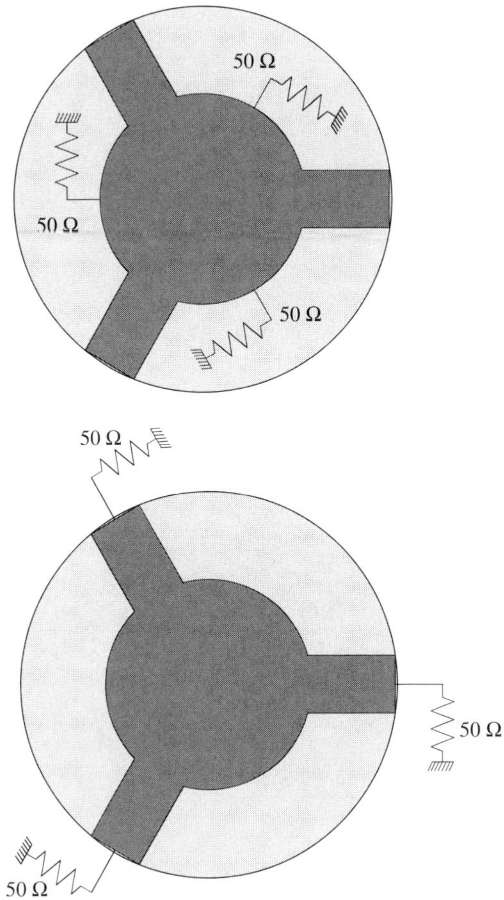

FIGURE 9.9 Schematic diagram of wye resonator using circular disk loaded with UEs.

If the loosely coupled junction is visualized as a symmetrical six-port network with alternate ports open-circuited, then the eigenvalues at the reactive walls of the three-port network are given in terms of the poles of the six-port one by

$$Z^0 \approx \frac{Z_0 + Z_3}{2} \tag{9.4a}$$

$$Z^+ \approx \frac{Z_{+1} + Z_{-2}}{2} \tag{9.4b}$$

$$Z^- \approx \frac{Z_{-1} + Z_{+2}}{2} \tag{9.4c}$$

If the eigenvalues are written in terms of the impedance poles of the three-port network, the result is

$$Z^0 \approx Z_0 + Z_3 \tag{9.5a}$$

$$Z^+ \approx Z_{+1} + Z_{-2} \tag{9.5b}$$

$$Z^- \approx Z_{-1} + Z_{+2} \tag{9.5c}$$

A typical pole of a three-port symmetrical isotropic junction is given by

$$Z_n = \frac{j3\psi\eta_r Z_r}{\pi}\left(\frac{\sin n\psi}{n\psi}\right)^2\left(\frac{J_n'(kR_i)}{J_n(kR_i)}\right)^{-1} \tag{9.6}$$

It is useful, for the purpose of computation, to write this quantity as

$$Z_n = \frac{j3\psi\eta_r Z_r}{\pi}\left(\frac{\sin n\psi}{n\psi}\right)^2\left(\frac{J_{n-1}(kR_i)}{J_n(kR_i)} - \frac{n}{kR_i}\right)^{-1} \tag{9.7}$$

The connection between the two descriptions may be derived by having recourse to the Bessel relationship:

$$xJ_n'(x) = xJ_{n-1}(x) - nJ_n(x) \tag{9.8}$$

The shape or coupling angle of a typical strip (ψ) is defined by

$$\sin\psi = (W/2R_i) \tag{9.9}$$

η_r is the constituent wave impedance of the substrate:

$$\eta_r = \sqrt{\mu_r/\varepsilon_f} \tag{9.10}$$

Z_r is the characteristic impedance of a typical stripline:

$$Z_r = 30\pi\ \ln\left(\frac{W + t + 2H}{W + t}\right) \tag{9.11}$$

The phase constant (k) has the meaning in Eq. (9.2). W, t, and H are the linear dimensions of the uniform striplines.

Useful polynomial approximations for the Bessel functions $J_0(x)$ and $J_1(x)$ for x between 0 and 3 are

$$J_0(x) = 1 - 2.2499997(x/3)^2 + 1.2656208(x/3)^4 - 0.3163866(x/3)^6 + 0.0444479(x/3)^8 - 0.0039444(x/3)^{10} + 0.0002100(x/3)^{12}$$

$$J_1(x) = x\left[0.50 - 0.56249985(x/3)^2 + 0.21093573(x/3)^4 - 0.03954289(x/3)^6 + 0.00443319(x/3)^8 - 0.00031761(x/3)^{10} + 0.00001109(x/3)^{12}\right]$$

and the recurrence formulas

$$J_{n+1}(x) = \frac{2n}{x} J_n(x) - J_{n-1}(x)$$

$$J_{-n}(x) = (-1)^n J_n(x)$$

are sufficient for computational purposes.

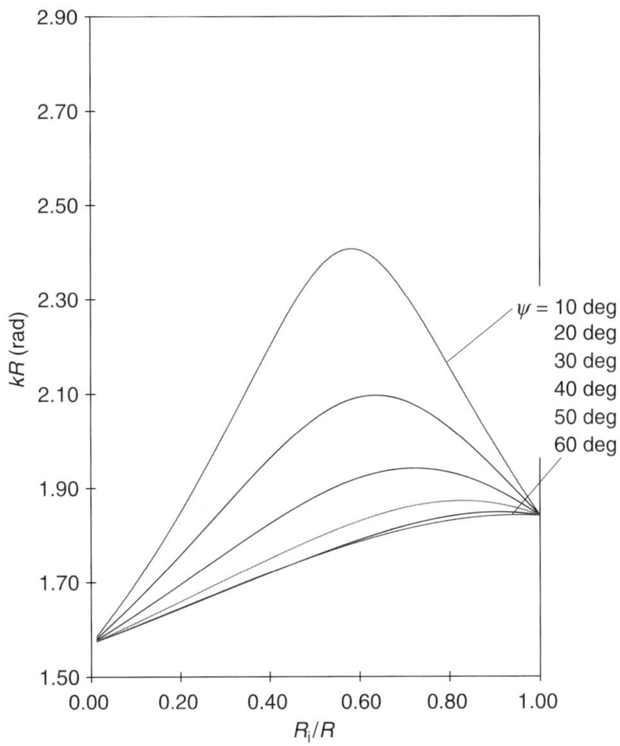

FIGURE 9.10 Cutoff wavenumber for the fundamental mode in demagnetized wye resonator.

The characteristic equation for the frequencies of the resonator is obtained by resonating the degenerate impedance eigenvalues using suitable open-circuited uniform transmission lines. The characteristic equation for the first two pair of degenerate resonances is now established by constructing a transverse resonance condition between the radial and uniform lines:

$$Z^{\pm} = j\eta_r Z_r \cot(kL)$$

L is the length of a typical open-circuited stub:

$$L = R_o - R_i \tag{9.12}$$

The required condition is

$$\left(\frac{3\psi}{\pi}\right)\left(\frac{\sin\psi}{\psi}\right)^2\left(\frac{J_0(kR_i)}{J_1(kR_i)} - \left(\frac{1}{kR_i}\right)\right)^{-1} + \left(\frac{3\psi}{\pi}\right)\left(\frac{\sin 2\psi}{2\psi}\right)^2\left(\frac{J_1(kR_i)}{J_2(kR_i)} - \left(\frac{2}{kR_i}\right)\right)^{-1} - \cot(kL) = 0 \tag{9.13}$$

for $n = 1$ and 2.

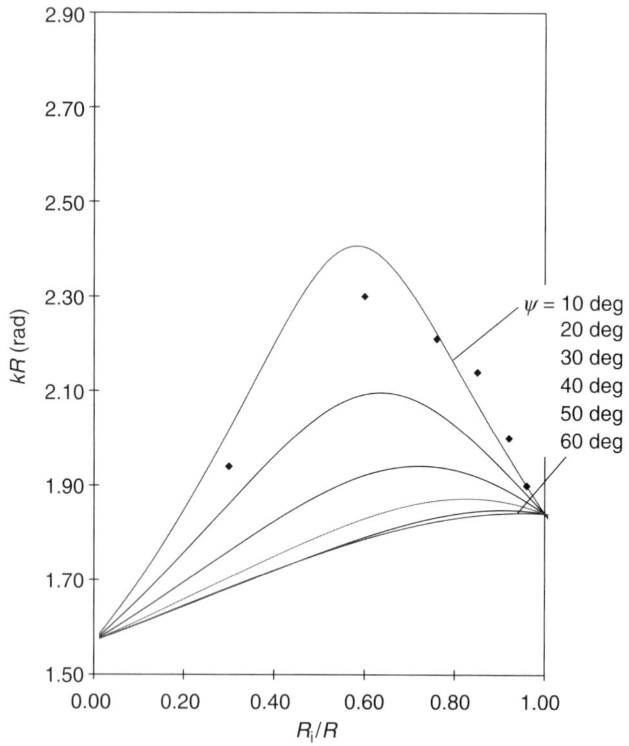

FIGURE 9.11 Comparison between closed form and FEM calculations of cutoff space of wye resonator.

The first two modes of the circular region have been retained in this formulation. The mode chart of the demagnetized wye resonator is depicted in Fig. 9.10 for $n = 1$.

The agreement between the calculations undertaken here and some FEM calculations is indicated separately in Fig. 9.11.

9.5 THE GYROMAGNETIC CUTOFF SPACE

The split cutoff space of a gyromagnetic resonator with m-ports is also readily established. The description of a typical pole of this sort of problem region is also a classic result in the literature.

$$Z_n \approx \frac{j3\psi\eta_r Z_r}{\pi}\left(\frac{\sin n\psi}{n\psi}\right)^2\left(\frac{J_{n-1}(kR_i)}{J_n(kR_i)} - n\left(\frac{1+\kappa/\mu}{kR_i}\right)\right)^{-1} \tag{9.14}$$

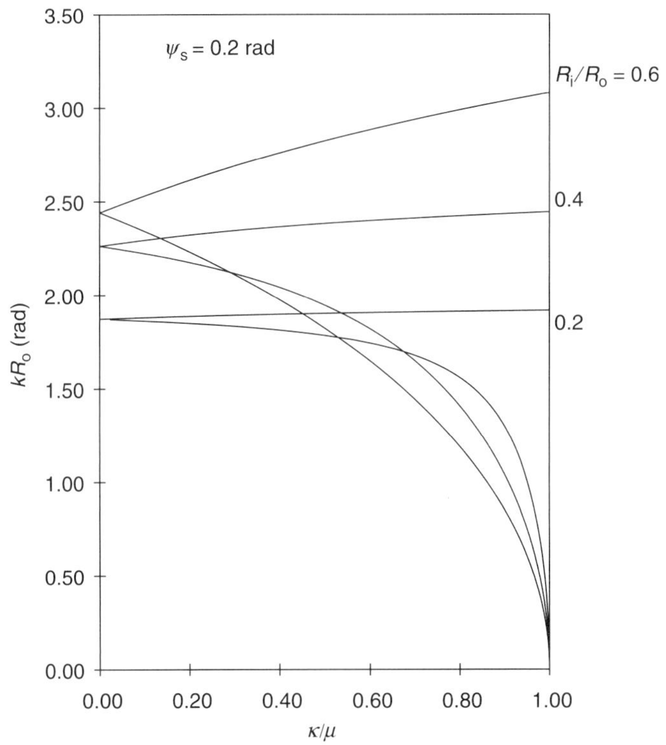

FIGURE 9.12 Split frequencies of planar gyromagnetic wye resonator in kR space using closed form formulation ($\psi = 0.20$ rad, $R_i/R_o = 0.20$, 0.40, and 0.60).

The corresponding cutoff numbers of the split gyromagnetic space are fixed by

$$\left(\frac{3\psi}{\pi}\right)\left(\frac{\sin\,\psi}{\psi}\right)^2\left(\frac{J_0(kR_{\rm i})}{J_1(kR_{\rm i})}-\left(\frac{1+\kappa/\mu}{kR_{\rm i}}\right)\right)^{-1}$$
$$+\left(\frac{3\psi}{\pi}\right)\left(\frac{\sin\,2\psi}{2\psi}\right)^2\left(\frac{J_{-3}(kR_{\rm i})}{J_{-2}(kR_{\rm i})}+2\left(\frac{1+\kappa/\mu}{kR_{\rm i}}\right)\right)^{-1}-\cot(kL)=0 \qquad (9.15)$$

for $n = +1$ and $n = -2$;

$$\left(\frac{3\psi}{\pi}\right)\left(\frac{\sin\,\psi}{\psi}\right)^2\left(\frac{J_{-2}(kR_{\rm i})}{J_{-1}(kR_{\rm i})}+\left(\frac{1+\kappa/\mu}{kR_{\rm i}}\right)\right)^{-1}$$
$$+\left(\frac{3\psi}{\pi}\right)\left(\frac{\sin\,2\psi}{2\psi}\right)^2\left(\frac{J_1(kR_{\rm i})}{J_2(kR_{\rm i})}-2\left(\frac{1+\kappa/\mu}{kR_{\rm i}}\right)\right)^{-1}-\cot(kL)=0 \qquad (9.16)$$

for $n = -1$ and $n = +2$.

μ and κ are the usual diagonal and off-diagonal entries of the permeability tensor. The ratio of κ and μ is known as the gyrotropy of the problem region.

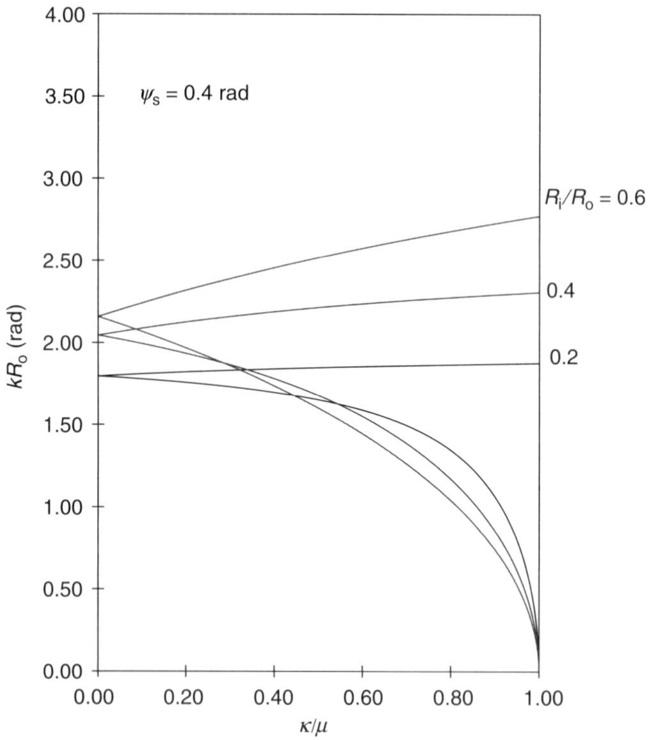

FIGURE 9.13 Split frequencies of planar wye resonator in kR space using closed form formulation ($\psi = 0.40$ rad, $R_i/R_o = 0.20$, 0.40, and 0.60).

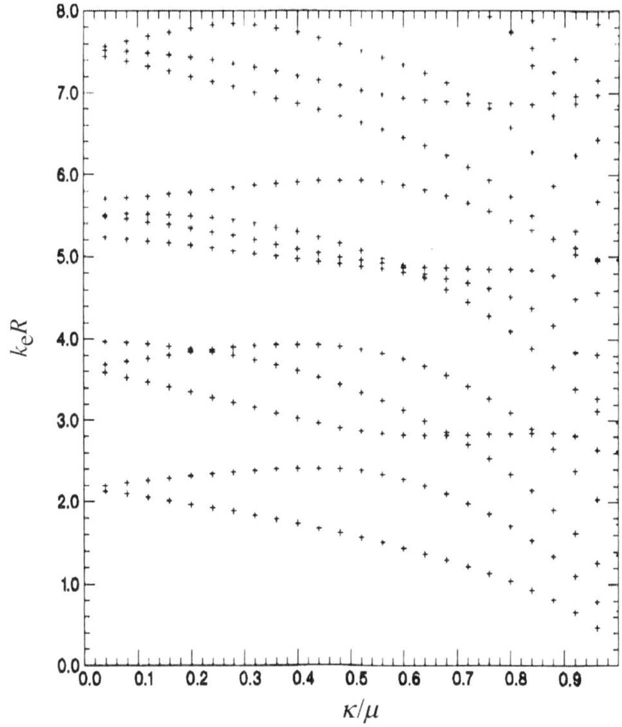

FIGURE 9.14 Split frequencies of planar wye resonator using finite elements.

Figure 9.12 indicates the split mode chart with $R_i/R_o = 0.20$, 0.40 and 0.60 and $\psi_s = 0.20$ rad obtained by disregarding the $n = \pm 2$ modes. Figure 9.13 depicts another result. The opening between the split branches is essentially independent of the shape angle but deteriorates rapidly below $R_i/R_o = 0.40$. A typical result based on a finite element solver is indicated in Fig. 9.14.

9.6 OPEN-CIRCUIT PARAMETERS OF CIRCULATORS USING WYE RESONATORS

In order to obtain the open-circuit parameters of a circulator using a wye resonator, it is necessary to construct its eigenvalues at its input terminals. The required mappings between the eigenvalues at the circular gyromagnetic region and those at the terminals of the UEs are given by

$$Z^0 = \eta_r Z_r \left(\frac{Z^0 + j\eta_r Z_r \tan kL}{\eta_r Z_r + jZ^0 \tan kL} \right) \tag{9.18a}$$

$$Z^{\pm} = \eta_r Z_r \left(\frac{Z^{\pm} + j\eta_r Z_r \tan kL}{\eta_r Z_r + jZ^{\pm} \tan kL} \right) \tag{9.18b}$$

The eigen-networks formed in this way are depicted in Fig. 9.15.

The relationships between the open-circuit parameters of the overall junction and the eigenvalues of the problem region are described in the usual way by

$$Z_{11} \approx \frac{Z^0 + Z^+ + Z^-}{3} \tag{9.19a}$$

$$Z_{21} \approx \frac{Z^0 + \alpha Z^+ + \alpha^2 Z^-}{3} \tag{9.19b}$$

$$Z_{31} \approx \frac{Z^0 + \alpha^2 Z^+ + \alpha Z^-}{3} \tag{9.19c}$$

where

$$\alpha = \exp(j120)$$

$$\alpha^2 = \exp(j240)$$

$$\alpha^3 = 1.0$$

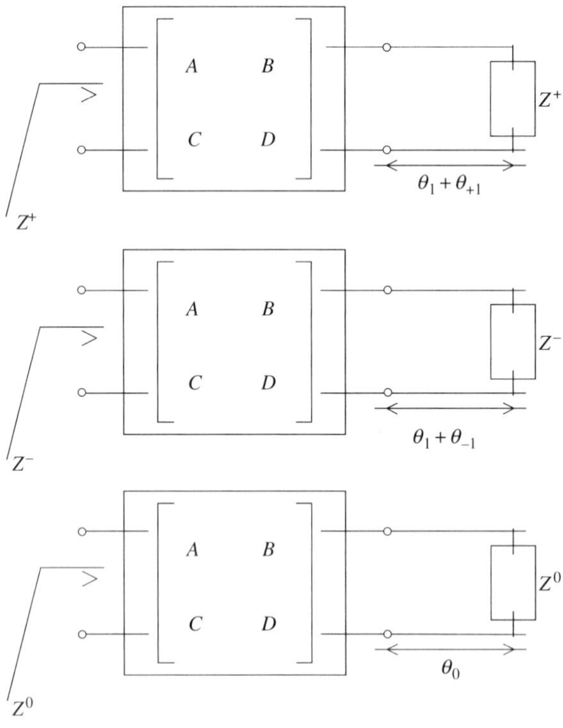

FIGURE 9.15 Eigen-networks of circulator using wye resonator.

For a weakly magnetized resonator with the in-phase eigen-network idealized by a short-circuit boundary condition,

$$Y_{in} = \left(\frac{Y^+ + Y^-}{2}\right) - j\sqrt{3}\left(\frac{Y^+ - Y^-}{2}\right) \tag{9.20}$$

9.7 THE SHORT UE

The nature of the stubs loading the circular region of the resonator may be understood by having recourse to the equivalent circuit of a typical UE. The equivalent circuit in question is depicted in Fig. 9.16. Its entries are given by

$$Z_1 = Z_3 = Z_0 \tanh(\gamma l/2) \tag{9.21a}$$

$$Z_2 = \frac{Z_0}{\sinh(\gamma l)} \tag{9.21b}$$

Neglecting dissipation and replacing a typical trigonometric function by its small angle approximation gives

$$\tanh(\gamma l) = j\beta l \tag{9.22a}$$

$$\sinh(\gamma l) = j\beta l \tag{9.22b}$$

This gives

$$Z_1 = Z_3 \approx jZ_0(\beta l/2) \tag{9.23}$$

$$Z_2 \approx -jZ_0/\beta l \tag{9.24}$$

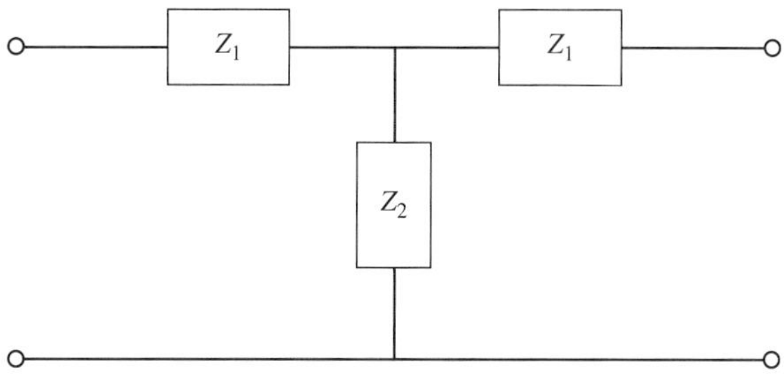

FIGURE 9.16 Equivalent circuit of uniform transmission line.

Z_0 and β are defined in terms of inductance and capacitance per unit length of the line by

$$Z_0 = \sqrt{(\mu_r L_0/\varepsilon_r C_0)} \tag{9.25}$$

$$\beta = \omega\sqrt{\mu_r \varepsilon_r L_0 C_0} \tag{9.26}$$

Combining the previous relationships gives

$$Z_1 = Z_3 = j\omega(\mu_r L_0 l)/2 \tag{9.27a}$$

$$Z_2 = -j/\omega(\varepsilon_r C_0 l) \tag{9.27b}$$

A scrutiny of these relationships indicates that a high impedance line UE may be represented by a series inductor equal to twice that of either series branch:

$$Z_1 = Z_3 \approx j\frac{\omega(\mu_r L_0 l)}{2} \tag{9.28a}$$

$$Z_2 \approx 0 \tag{9.28b}$$

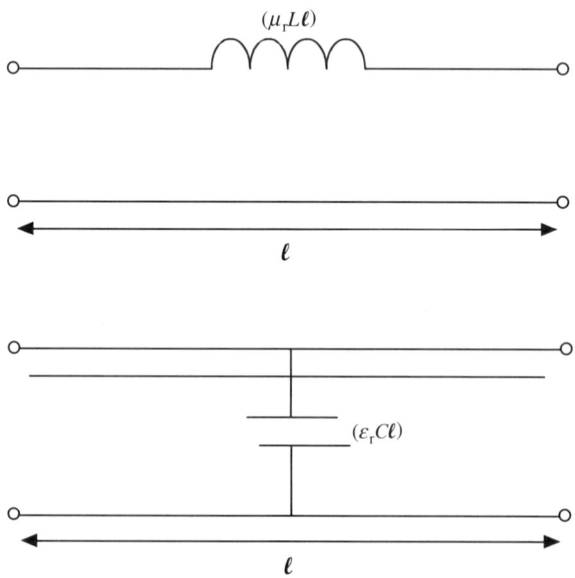

FIGURE 9.17 Equivalent circuits of high and low impedance uniform transmission lines.

Likewise, a low impedance line may be approximated by a shunt capacitance:

$$Z_1 = Z_3 = 0 \tag{9.29a}$$

$$Y_2 = 1/Z_2 = j\omega(\varepsilon_r C_0 l) \tag{9.29b}$$

The equivalent circuits under consideration are indicated in Fig. 9.17.

CHAPTER TEN

Planar Resonators with Triplets of Radial and Circumferential Magnetic Walls

10.1 INTRODUCTION

Planar resonators with threefold symmetry and with top and bottom electric walls and a magnetic side wall may be perturbed by suitable triplets of radial and circumferential magnetic walls. Figure 10.1 illustrates the two possibilities under discussion. The effect of a triplet of radial magnetic wall ridges on the cutoff space of a disk resonator may be understood by employing duality between a planar circuit with a magnetic wall boundary condition and a circular waveguide at cutoff with three metal ridges with an electric wall boundary condition. Its effect is to perturb its cutoff space except for its symmetric branches. This latter feature may be understood by recognizing that the current induced in the metallization by the location of a typical radial magnetic wall is, for this type of mode, parallel to such a wall and is therefore unaffected by it. The effect of a triplet of circumferential slots is discussed separately. A careful consideration of the effects of the two orthogonal families of walls suggests that these provide independent means of perturbing each type of eigen-network entering into the design of a three-port junction circulator. As the slots define magnetic walls, these are not expected to radiate. This is in keeping with practice. One means of investigating either geometry is, again, the finite element method. Planar resonators of triangular shape are also finding increasing use and this circuit has also been analyzed. This chapter also includes statements about the loaded Q-factor of circulators using a triplet of radial walls in circular and triangular plates. It therefore provides a complete description of circulators using weakly

The Stripline Circulator: Theory and Practice. By J. Helszajn

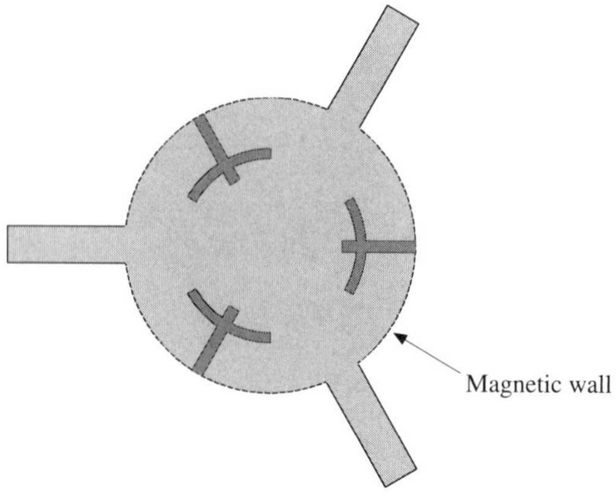

FIGURE 10.1 Planar resonator loaded with triplets of radial and circumferential walls.

magnetized planar resonators symmetrically loaded with magnetic wall ridges. Some experimental mode charts, obtained in this way at S-band on disk and triangular circuits, are superimposed on the theoretical results in Figs. 10.5 and 10.6. The comparison between theory and experiment disregards any perturbation in frequency of the resonant mode due to the effects of imperfect magnetic wall boundary conditions or that of coupling of the striplines to the junction.

The geometry of the disk circuit is fixed by the thickness t of the center conductor (0.40 mm), the width of the magnetic wall ridges (0.80 and 1.80 mm), the ground plane spacing $2H$ (4.00 mm), and the radius of the disk (10.70 mm). The length of the ridge was varied between 0 and 10 mm. The garnet material had a magnetization (μ_0 M_0) of 0.0550 T and a dielectric constant of 15.1. The triangular resonator has similar linear dimensions with the exceptions of the side of the triangle (A), which was 18.4 mm, and the material, which had a magnetization (μ_0 M_0) of 0.0600 T.

10.2 DOMINANT MODE CHARTS FOR PLANAR RESONATORS WITH RADIAL MAGNETIC WALL RIDGES

The cutoff spaces of planar circuits with top and bottom electric walls and magnetic side walls with magnetic wall ridges may be understood by using duality at cutoff between a planar circuit with magnetic walls and an equivalent waveguide with electric walls. The two circuits dealt with in this section are illustrated in Fig. 10.2. The dualities between the planar disk and triangular resonators and the corresponding waveguides are illustrated in Fig. 10.3.

The two circuits under consideration can be readily analyzed by having recourse to a standard finite element calculation. The finite element discretization of each geometry is depicted in Fig. 10.4. The inherent flexibility of the finite element method allows different resonator shapes and magnetic wall ridges to be characterized without difficulty. The resonator is here subdivided into 21 triangles and a fourth

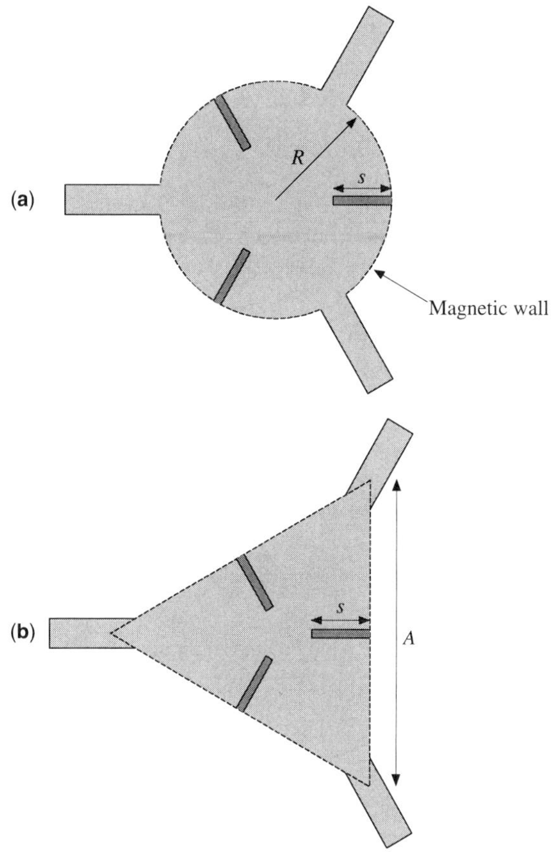

FIGURE 10.2 (a) Planar disk resonator with three symmetrical magnetic wall ridges. (b) Planar triangular resonator with three symmetrical magnetic wall ridges.

order polynomial approximation is made to the fields in each triangle. The mode charts obtained in this way, for two different magnetic wall ridge sizes, are illustrated in Figs. 10.5 and 10.6.

The cutoff number for the dominant mode of the circular configuration in the absence of electric or magnetic ridges is given by

$$kR = 1.841 \quad (1.917) \tag{10.1}$$

The cutoff numbers for the triangular configuration is

$$kA = \frac{4\pi}{3} \quad (4.227) \tag{10.2}$$

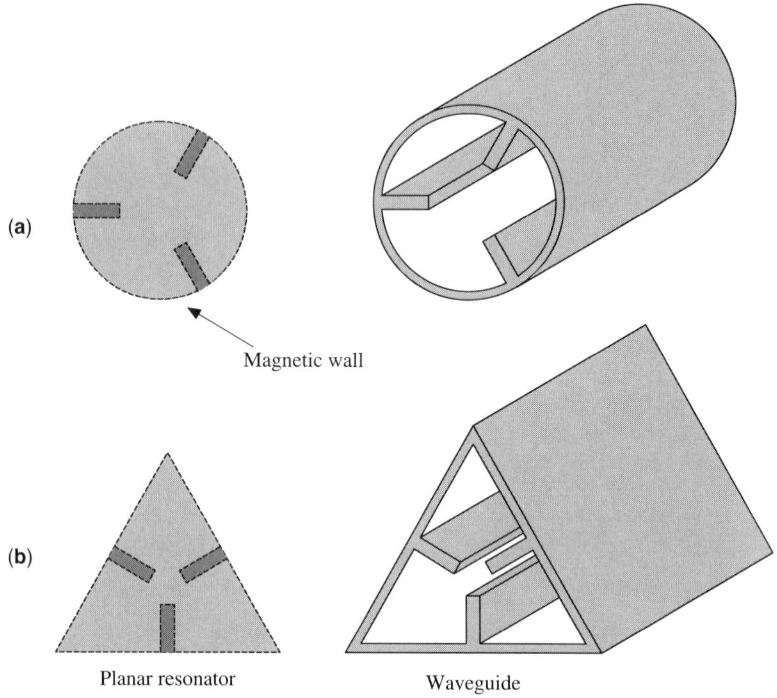

FIGURE 10.3 (a) Duality between planar disk resonator with magnetic wall ridges and circular waveguide with electric wall ridges. (b) Duality between planar triangular resonator with magnetic wall ridges and triangular waveguide with electric wall ridges.

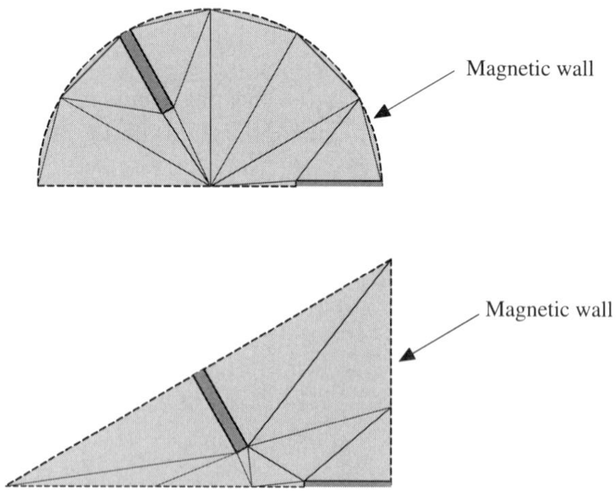

FIGURE 10.4 Discretization of planar resonators loaded by radial magnetic walls.

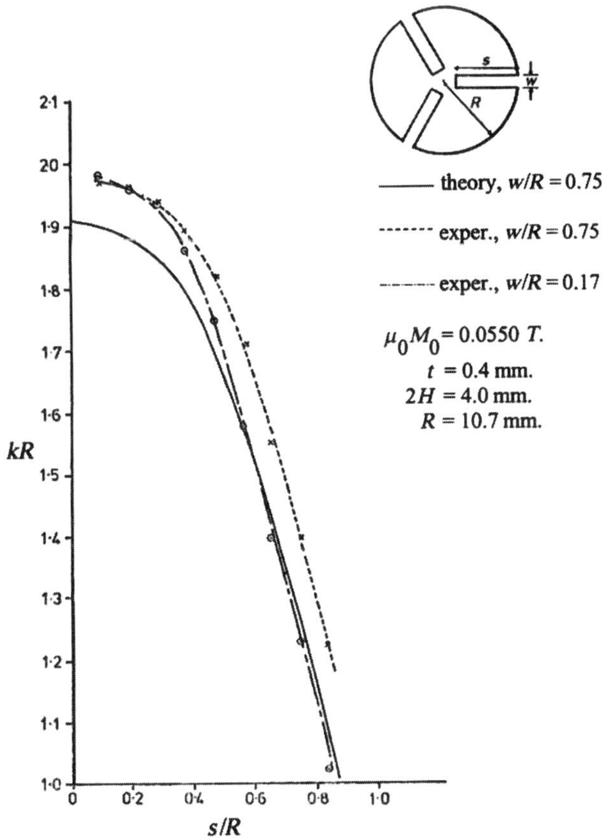

FIGURE 10.5 Theoretical and experimental mode chart for planar disk resonator with magnetic wall ridges. (Reproduced with permission from J. Helszajn, R. D. Baars, and W. T. Nisbet, Characteristics of circulators using planar triangular and disk resonators symmetrically loaded with magnetic walls, *IEEE Trans. Microwave Theory Tech.*, Vol. MTT-28, pp. 616–621, June 1980.)

R is the radius of the planar disk or circular waveguide (meters), A is the length of the side of the planar triangular resonator or waveguide (meters), and k is the radial propagation constant (rad/s):

$$k = \frac{2\pi}{\lambda_0} \sqrt{\varepsilon_f \mu_{eff}} \tag{10.3}$$

ε_f is the relative dielectric constant, μ_{eff} is the relative effective permeability of the substrate, and λ_0 is the free space wavelength (meters). The quantities in parentheses in Eqs. (10.1) and (10.2) refer to the computed values obtained by having recourse to a dedicated FE solver.

A feature of a triplet of radial magnetic walls is that the in-phase mode of the resonator is not perturbed by the introduction of such magnetic walls. This may also

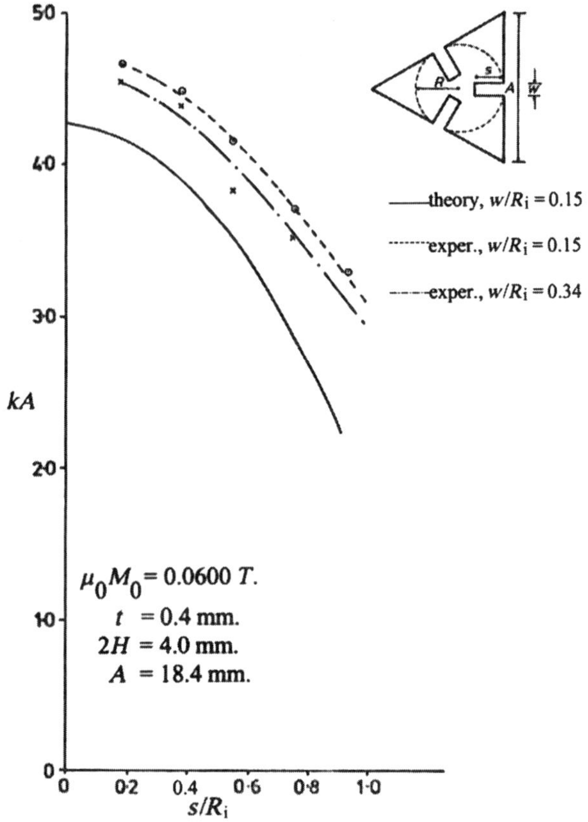

FIGURE 10.6 Theoretical and experimental mode chart for planar triangular resonator with magnetic wall ridges. (Reproduced with permission from J. Helszajn, R. D. Baars, and W. T. Nisbet, Characteristics of circulators using planar triangular and disk resonators symmetrically loaded with magnetic walls, *IEEE Trans. Microwave Theory Tech.*, Vol. MTT-28, pp. 616–621, June 1980.)

be understood with reference to the in-phase eigen-network, as shown in Fig. 10.7. It indicates that for narrow ridges the magnetic wall coincides with the magnetic wall boundary condition of the eigen-network.

10.3 HIGHER ORDER MODE CHARTS IN PLANAR RESONATORS WITH RADIAL MAGNETIC WALL RIDGES

The effects of three symmetrically located magnetic walls on the first two higher order modes in disk and triangular planar resonators with top and bottom electric walls have also been investigated. The cutoff numbers of the unperturbed modes

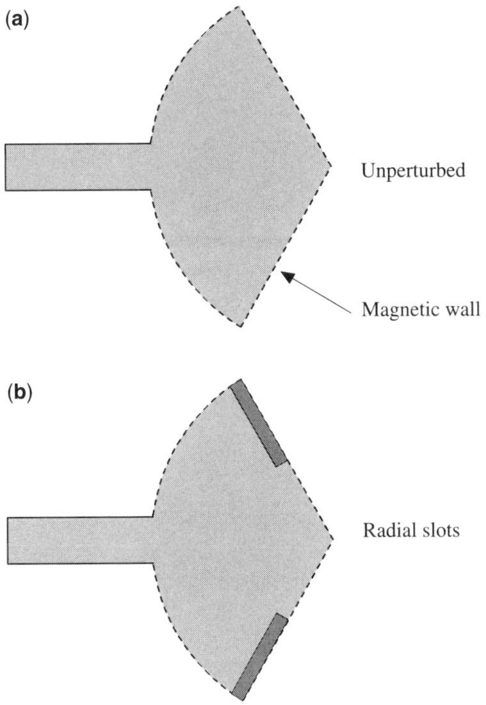

FIGURE 10.7 The in-phase eigen-network of a circulator using a planar disk resonator (a) unperturbed and (b) loaded with radial ridges. (Reproduced with permission from J. Helszajn, R. D. Baars, and W. T. Nisbet, Characteristics of circulators using planar triangular and disk resonators symmetrically loaded with magnetic walls, *IEEE Trans. Microwave Theory Tech.*, Vol. MTT-28, pp. 616–621, June 1980.)

are given by

$$kR = 3.054 \quad (3.165) \tag{10.4}$$

$$kR = 3.832 \quad (3.879) \tag{10.5}$$

The corresponding values for the triangular geometry are

$$kA = 4\pi\sqrt{3} \quad (7.328) \tag{10.6}$$

$$kA = \frac{8\pi}{3} \quad (8.448) \tag{10.7}$$

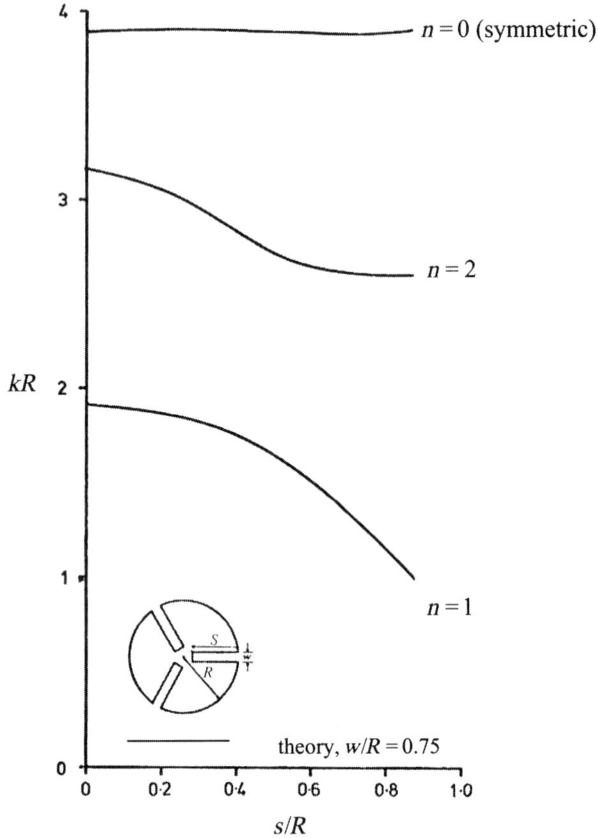

FIGURE 10.8 Theoretical mode spectrum for planar disk resonator with magnetic wall ridges (Reproduced with permission from J. Helszajn, R. D. Baars, and W. T. Nisbet, Characteristics of circulators using planar triangular and disk resonators symmetrically loaded with magnetic walls, *IEEE Trans. Microwave Theory Tech.*, Vol. MTT-28, pp. 616–621, June 1980.)

Figures 10.8 and 10.9 depict the mode spectrum of the first three modes versus the length of the magnetic walls for disk and triangular resonators, respectively.

A separate study of the field patterns in such planar resonators indicates that the first higher order mode in the case of a triangle is a symmetric one, whereas in the case of a planar disk resonator it is the third rather than the second mode that is the symmetric one. The magnetic walls have little or no effect on the symmetric modes of each geometry. This feature may be understood by recognizing that the magnetic walls are in each instance perpendicular to the magnetic field so that the currents in the center conductor are parallel to the magnetic walls. This result is in keeping with the in-phase eigen-network in Fig. 10.7.

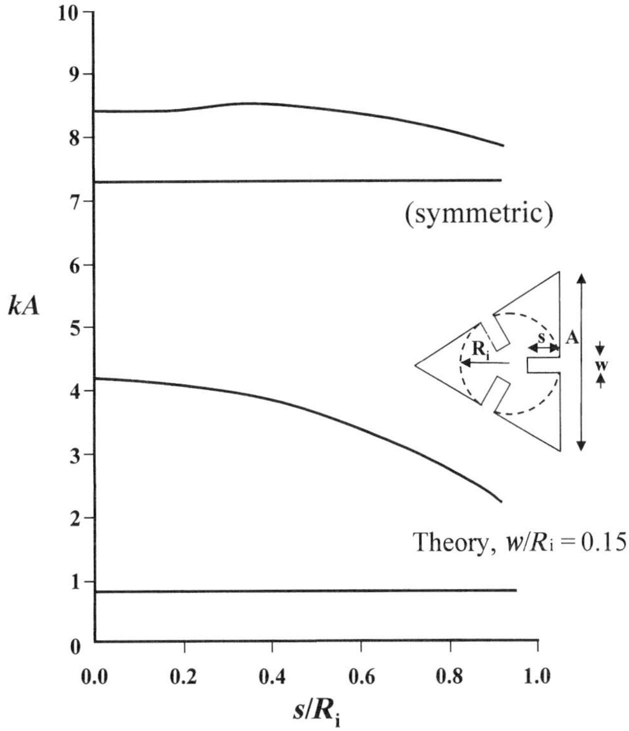

FIGURE 10.9 Theoretical mode spectrum for planar triangular resonator with magnetic wall ridges. (Reproduced with permission from J. Helszajn, R. D. Baars, and W. T. Nisbet, Characteristics of circulators using planar triangular and disk resonators symmetrically loaded with magnetic walls, *IEEE Trans. Microwave Theory Tech.*, Vol. MTT-28, pp. 616–621, June 1980.)

10.4 FAY AND COMSTOCK CIRCULATION SOLUTION

Two definitions of the quality factor of the complex gyrator circuit of a weakly magnetized junction have been introduced in the text. One is based on the gyrotropy of the magnetic insulator; the other is based on the physical details of the structure under consideration. One circulation solution due to Fay and Comstock is obtained by reconciling the two descriptions. The perturbation of the former description by the introduction of a triplet of magnetic walls is the topic of the next section. This shift in the quality factor may be inferred by having recourse to perturbation theory.

One of the two definitions of the loaded Q-factor (Q_L) of a resonator employed by Fay and Comstock is

$$Q_L = \frac{\omega_0 U_0}{P_{r1} + P_{r2} + P_{r3}} \tag{10.8}$$

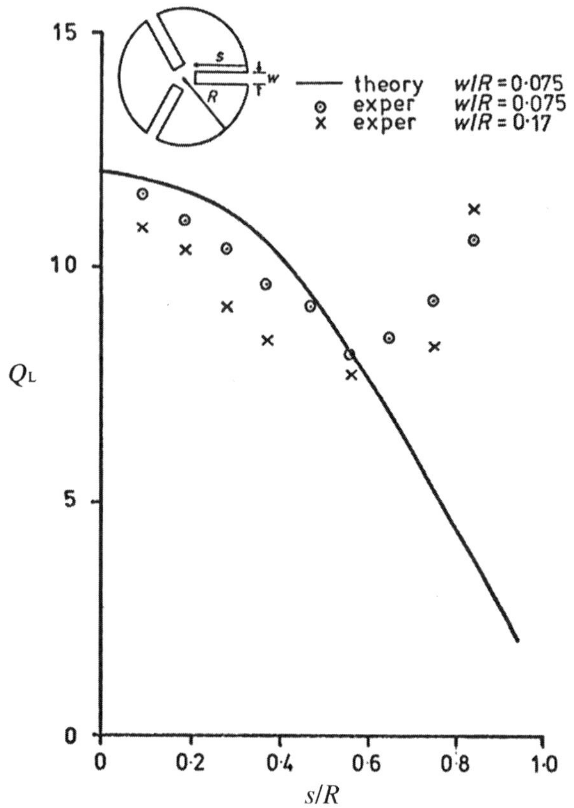

FIGURE 10.10 Theoretical and experimental loaded Q-factor of a circulator using planar disk resonators with magnetic wall ridges. (Reproduced with permission from J. Helszajn, R. D. Baars, and W. T. Nisbet, Characteristics of circulators using planar triangular and disk resonators symmetrically loaded with magnetic walls, *IEEE Trans. Microwave Theory Tech.*, Vol. MTT-28, pp. 616–621, June 1980.)

U_0 is the stored energy, ω_0 is the radian resonant frequency, and P_{ri} is the power dissipated in a typical termination. In the case of a circulator,

$$P_{r1} = P_{r3} = 0 \tag{10.9}$$

The quality factor obtained in this way in the case of a circulator using a disk resonator is given in Chapter 16 in the absence of magnetic wall ridges by

$$Q_L = \frac{1.48\varepsilon_0\varepsilon_r\omega_0R^2}{Y_0H} \tag{10.10}$$

Y_0 is the conductance at the output stripline and also that looking into the resonator circuit, and H is the thickness of each ferrite disk.

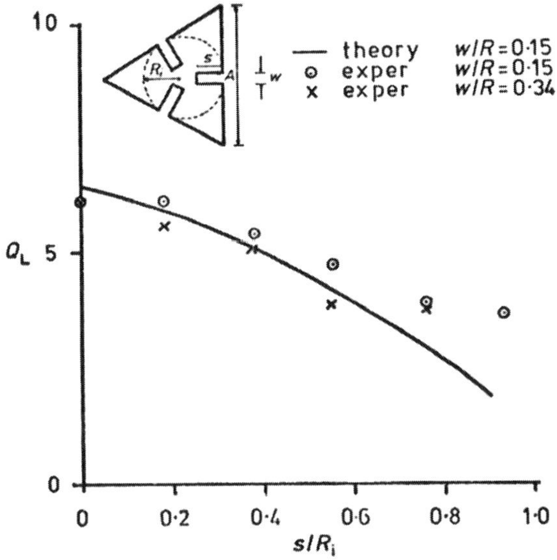

FIGURE 10.11 Theoretical and experimental loaded Q-factor of a circulator using a planar triangular resonator loaded with magnetic wall ridges. (Reproduced with permission from J. Helszajn, R. D. Baars, and W. T. Nisbet, Characteristics of circulators using planar triangular and disk resonators symmetrically loaded with magnetic walls, *IEEE Trans. Microwave Theory Tech.*, Vol. MTT-28, pp. 616–621, June 1980.)

Once H is set by Q_L, the width of the striplines is approximately defined by the output conductance in the usual way:

$$Y_0 = \left[30\pi \ln\left(\frac{W + t + 2H}{W + t}\right)\right]^{-1} \tag{10.11}$$

A statement of the quality factor in terms of the split frequency of the function on the gyrotropy of the resonator completes the Fay and Comstock solution. The corresponding quantities in the cases of apex and side coupled triangular resonators are also available in Chapter 16.

10.5 QUALITY FACTOR OF CIRCULATORS USING PLANAR RESONATORS WITH TRIPLETS OF RADIAL MAGNETIC WALLS

The relation between the shift in resonant frequency and the change in the stored energy in a planar resonator with top and bottom electric walls and a magnetic side wall due to the introduction of magnetic walls may be obtained by having recourse to perturbation theory.

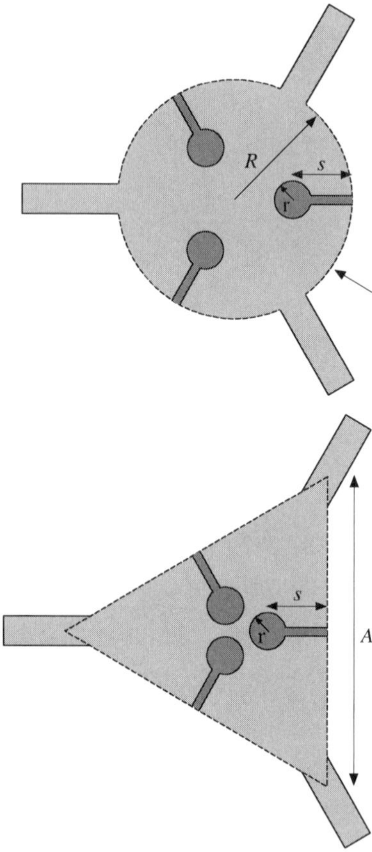

FIGURE 10.12 Planar disk and triangular resonators with three symmetrical hole-slot configurations. (Reproduced with permission from J. Helszajn, R. D. Baars, and W. T. Nisbet, Characteristics of circulators using planar triangular and disk resonators symmetrically loaded with magnetic walls, *IEEE Trans. Microwave Theory Tech.*, Vol. MTT-28, pp. 616–621, June 1980.)

The governing condition, assuming that the original and perturbed alternating fields are similar and that the introduction of magnetic wall ridges does not significantly perturb the electric field, is

$$\frac{\Delta\omega_0}{\omega_0} = -\frac{\Delta U_0}{U_0} \tag{10.12}$$

$\Delta\omega_0$ and ΔU_0 denote the shift in resonant frequency and the change in stored energy of the cavity, respectively. The sign in the preceding equation is compatible with a

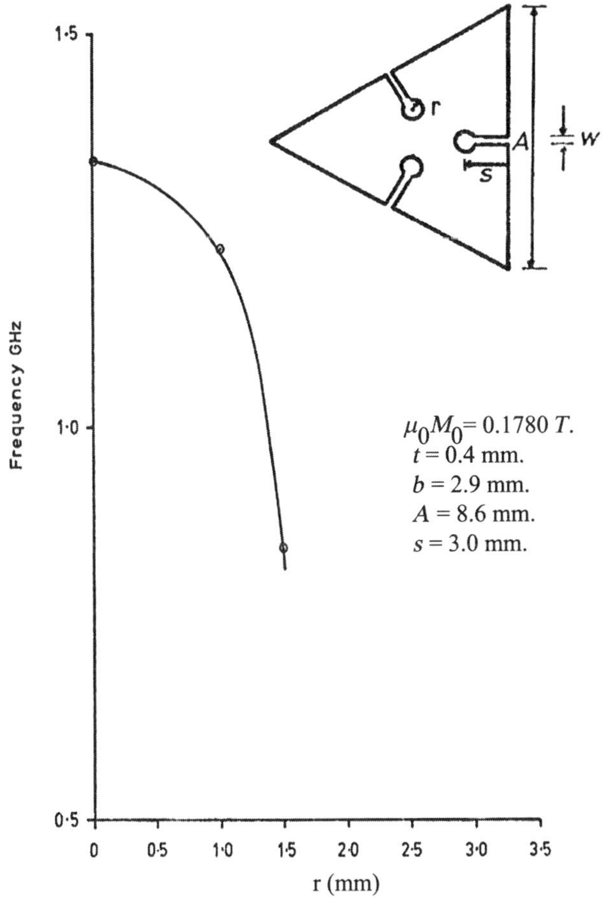

FIGURE 10.13 Experimental mode chart for a planar triangular resonator with a hole-slot tuning arrangement. (Reprpduced with permission from J. Helszajn, R. D. Baars, and W. T. Nisbet, Characteristics of circulators using planar triangular and disk resonators symmetrically loaded with magnetic walls, *IEEE Trans. Microwave Theory Tech.*, Vol. MTT-28, pp. 616–621, June 1980.)

change in magnetic wall boundary condition instead of an electric one in a circular cavity and is consistent with experiment.

The perturbation formulation introduced here is usually used to calculate the shift in resonant frequency of a resonator from a calculation of the energy stored and the change in the stored energy. The change in stored energy is here obtained in terms of the unperturbed stored energy and a calculation of frequency obtained from a finite element analysis. It is therefore expected that the perturbation formulation of the loaded Q-factor should be accurate over a larger interval than is normally found to be the case when applying the theory in the standard manner.

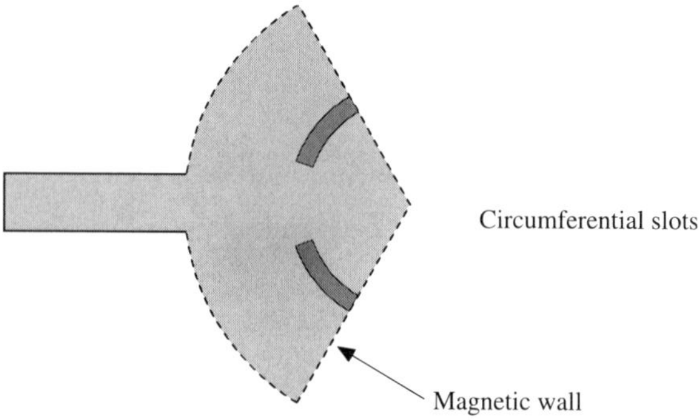

FIGURE 10.14 In-phase eigen-network loaded with circumferential slots.

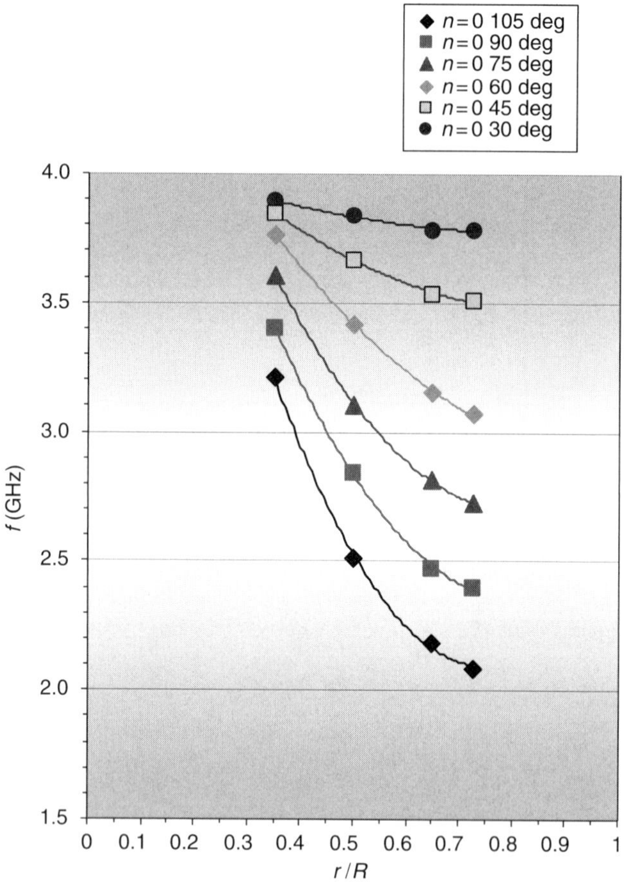

FIGURE 10.15 Perturbation of in-phase TM_{010} mode by circumferential slots. (Reproduced with permission from J. Helszajn, I. Macfarlane, M. McKay, A. Bunce, J. Sharp, and M. Hocine, Planar resonators with three-fold symmetry with triplets of circumferential walls, *IEE Proc – Microwaves, Antennas and Propagation*, Vol. 152, No. 5, pp. 285–291, October 2005.)

The derivation of this result starts by replacing Eq. (10.8) by

$$Q_L - \Delta Q_L = \frac{(\omega_0 - \Delta\omega_0)(U_0 - \Delta U_0)}{P_{r2}} \tag{10.13}$$

The required result is now obtained by combining the two preceding equations:

$$Q'_L = Q_L(1 - (\Delta\omega_0/\omega_0))^2 \tag{10.14}$$

Q'_L is the loaded Q-factor at the perturbed frequency $(\omega_0 - \Delta\omega_0)$ of the resonator, ω_0 is the unperturbed radian frequency of the resonator, Q_L is the loaded Q-factor of the unperturbed resonator, and $\Delta\omega_0$ is the shift in the resonant frequency of the resonator due to the introduction of the radial magnetic walls.

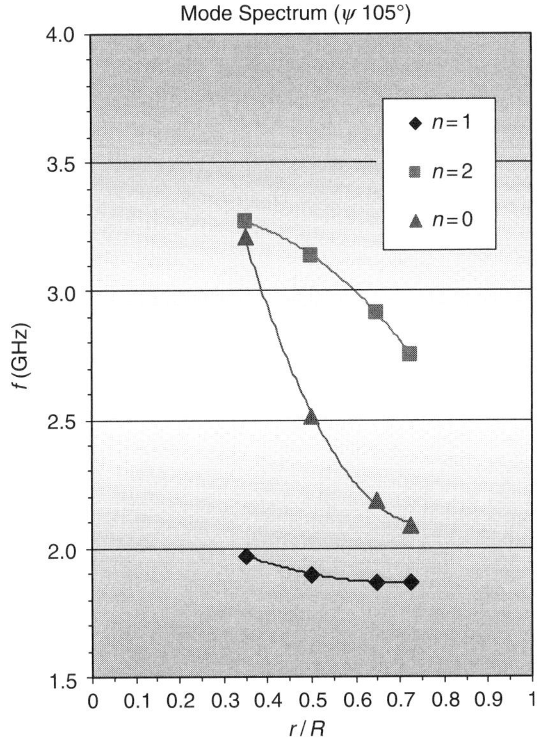

FIGURE 10.16 Mode chart of disk resonator loaded by circumferential magnetic walls ($\psi = 105°$). (Reproduced with permission from J. Helszajn, I. Macfarlane, M. McKay, A. Bunce, J. Sharp, and M. Hocine, Planar resonators with three-fold symmetry with triplets of circumferential walls, *IEE Proc. Microwaves, Antennas and Propagation*, Vol. 152, No. 5, pp. 285–291, October 2005.)

Figure 10.10 shows the correlation between measurement and theory in the case of a circulator using a planar triangular resonator loaded with magnetic ridges. Figure 10.11 depicts the same result for a circulator using a planar disk resonator. The agreement between theory and experiment is adequate for engineering purposes up to s/R (or s/R_i) equal to about 0.60.

10.6 EXPERIMENTAL MODE CHARTS OF CIRCULATORS USING PLANAR RESONATORS SLOT–HOLE

Figure 10.12 shows a hole-slot center conductor configuration having an extended tuning range. Its mode chart, obtained on an above-resonance UHF circulator, is depicted in Fig. 10.13. The dual waveguide geometry for this configuration is understood by inspection.

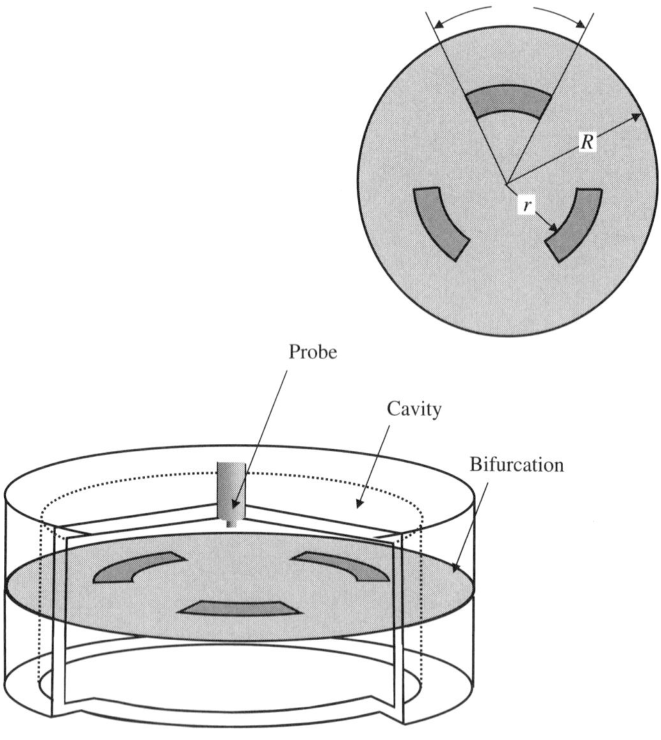

FIGURE 10.17 Impedance zero fixture. (Reproduced with permission from J. Helszajn, I. Macfarlane, M. McKay, A. Bunce, J. Sharp, and M. Hocine, Planar resonators with threefold symmetry with triplets of circumferential walls, *IEE Proc. Microwaves, Antennas and Propagation*, Vol. 152, No. 5, pp. 285–291, October 2005.)

10.7 CIRCUMFERENTIAL MAGNETIC WALL

One means of perturbing the degenerate counterrotating eigen-networks of a symmetrical three-port junction is to locate a triplet of radial magnetic walls into its central plate in the mannar indicated in Fig. 10.1. While this type of wall alters the cutoff number of the counterrotating eigen-networks, it leaves that of the in-phase one unperturbed. Another possible triplet of magnetic walls is circumferential ones around the resonator plate. Such slots do not affect the counterrotating modes provided these are located in the inner region of the plate, where the degenerate magnetic fields are essentially circularly polarized but have, however, a strong influence on the in-phase eigen-network. The in-phase eigen-network under consideration is indicated in Fig. 10.14. Some experimental data of the effect of a triplet of circumferential slots on the first symmetric mode for parametric values of θ are summarized in

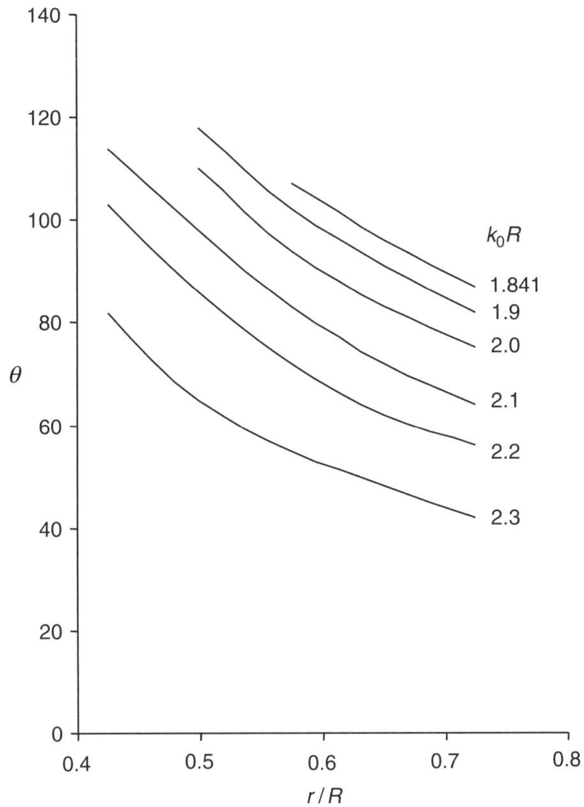

FIGURE 10.18 TM_{010} impedance zero in θ, kr plane. (Reproduced with permission from J. Helszajn, I. Macfarlane, M. McKay, A. Bunce, J. Sharp, and M. Hocine, Planar resonators with three-fold symmetry with triplets of circumferential walls, *IEE Proc. Microwaves, Antennas and Propagation*, Vol. 152, No. 5, pp. 285–291, October 2005.)

Fig. 10.15. Figure 10.16 separately summarizes the mode charts of the first three modes in such a resonator in the r/R, θ space.

10.8 IMPEDANCE ZERO

In order to adjust a three-port junction circulator it is necessary not only to realize a pair of degenerate counterrotating impedance poles at its terminals but also to place a virtual in-phase or symmetric impedance there. One way to establish this latter condition is again to introduce a triplet of circumferential slots in the resonator plate. The experimental fixture is a resonator plate in a bifurcated cavity with an electric rather than a magnetic side wall. The necessary arrangement is illustrated in Fig. 10.17. The TM_{010} solution in the θ, r/R plane is indicated in Fig. 10.18.

CHAPTER ELEVEN

Unloaded Quality Factors of Junction Circulators

11.1 INTRODUCTION

One model of a junction circulator is a gyromagnetic resonator at the junction of three transmission lines. A knowledge of its counterrotating unloaded quality factor and split frequencies is sufficient to specify both the gain bandwidth and the insertion loss of the device. One means of measuring these quantities consists of forming the immittance at port 1 by open-circuiting ports 2 and 3 of the circulator. The constituent resonator established in this way displays the impedance poles of the problem region without ado. These poles have, in the presence of dissipation, both real and imaginary parts. A characterization of the latter quantities and that of the loaded quality factor of the gyrator circuit are the main endeavors of the chapter. Both the unloaded split quality factors and the loaded quality factor vary with the gyrotropy so that each quantity has to be established separately in terms of the direct magnetization and direct magnetic field intensity. The chapter summarizes some experimental data on a stripline arrangement using a simple disk gyromagnetic resonator biased below the Kittel line. This is done for a number of coupling angles that the ports subtend at the terminals of the resonator.

11.2 EIGENVALUE DIAGRAMS OF SEMI-IDEAL CIRCULATION

The effects of dissipation on the performance of three-port circulators has been dealt with in a number of publications. The purpose of this section is to examine its insertion loss parameter in terms of the eigenvalues of its dissipation matrix.

The Stripline Circulator: Theory and Practice. By J. Helszajn

The insertion loss of symmetrical circulators is given in terms of the reflection eigenvalues of the junction by

$$S_{21} = \frac{s_0 + \alpha s_+ + \alpha^2 s_-}{3} \tag{11.1}$$

where

$$s_0 = (1 - q_0/2)\exp(-j2\phi_0) \tag{11.2a}$$

$$s_+ = (1 - q_+/2)\exp\left[-j2(\phi_+ + \pi/2)\right] \tag{11.2b}$$

$$s_- = (1 - q_-/2)\exp\left[-j2(\phi_- + \pi/2)\right] \tag{11.2c}$$

q_0 and $q_\pm$ are the in-phase and counterrotating eigenvalues of the dissipation matrix ($\overline{\mathbf{Q}}$). In this discussion the angles of the reflection eigenvalues are specialized to those of an ideal circulator.

$$\phi_0 = \pi/2 \tag{11.3a}$$

$$\phi_+ = \pi/2 - \pi/6 \tag{11.3b}$$

$$\phi_- = \pi/2 + \pi/6 \tag{11.3c}$$

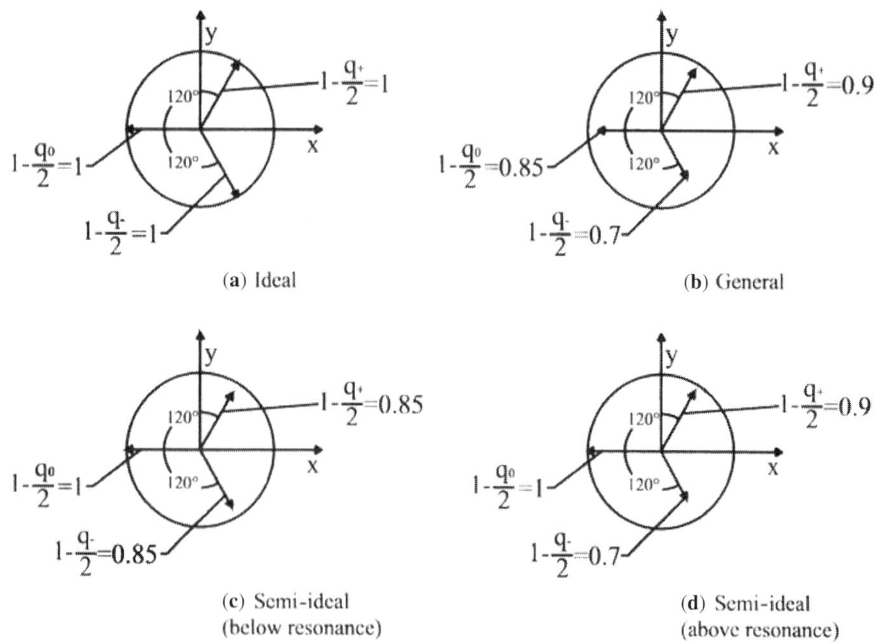

FIGURE 11.1 Eigenvalue diagrams of lossy circulator.

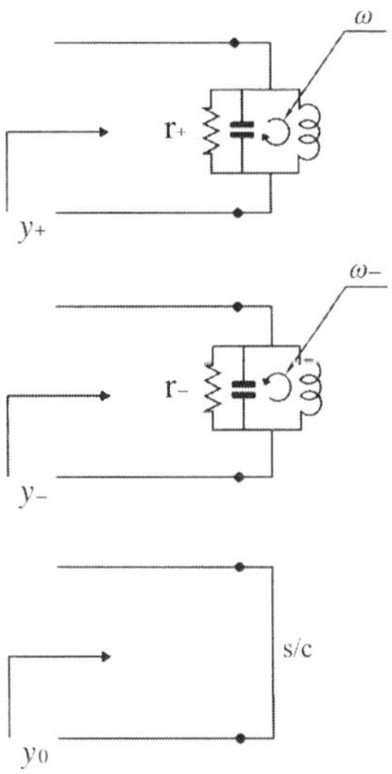

FIGURE 11.2 Lumped element eigen-networks of lossy three-port circulator.

It is furthermore assumed that the dissipation of the in-phase eigen-network may be neglected compared to those of the split counterrotating ones.

$$q_0 = 0 \tag{11.4}$$

The adjustment of the lossy three-port circulator is not unique. One possible adjustment optimizes the return loss but produces a finite isolation. Another idealizes the isolation but results in a finite return loss. Still another produces equal but finite values of return loss and isolation. Four possible eigenvalue diagrams are illustrated in Fig. 11.1. A lumped element representation of the eigen-networks of a three-port junction circulator in the presence of damping terms is indicated in Fig. 11.2.

11.3 CONSTITUENT RESONATOR

One means of investigating the gyromagnetic space of a three-port circulator is to form its constituent resonator. A feature of this circuit is that its impedance coincides with the one-port network defined by the main diagonal of the original impedance matrix of the three-port junction. It therefore displays all the poles of the problem region. It is obtained at one typical port by open-circuiting the other two.

$$Z_{11} = \frac{V_1}{I_1}\bigg|_{I_2 = I_3 = 0} \tag{11.5}$$

It is separately understood that Z_{11} may be expanded in terms of the in-phase Z^0 and counterrotating $Z^{\pm}$ impedance eigenvalues of the junction. This gives

$$Z_{11} = \frac{Z^0 + Z^+ + Z^-}{3} \tag{11.6}$$

Each eigenvalue can be expanded separately in terms of the in-phase and counterrotating impedance poles of the problem region in a first Foster form without ado.

$$Z^0 = Z_0 + Z_3 + \cdots \tag{11.7a}$$

$$Z^+ = Z_{+1} + Z_{-2} + \cdots \tag{11.7b}$$

$$Z^- = Z_{-1} + Z_{+2} + \cdots \tag{11.7c}$$

Figure 11.3 illustrates the experimental constituent arrangement.

A one-port first Foster expansion in terms of the poles of the problem region of Z_{11} is shown separately in Fig. 11.4. The characters of the split unloaded Q-factors of the dominant split modes of a gyromagnetic resonator may be revealed by introducing damping terms in the idealized first Foster form expansion of the constituent resonator. If the split frequencies of the dominant pair of counterrotating poles are sufficiently separated, then the unloaded Q-factor of each pole may be experimentally evaluated using standard one-port techniques.

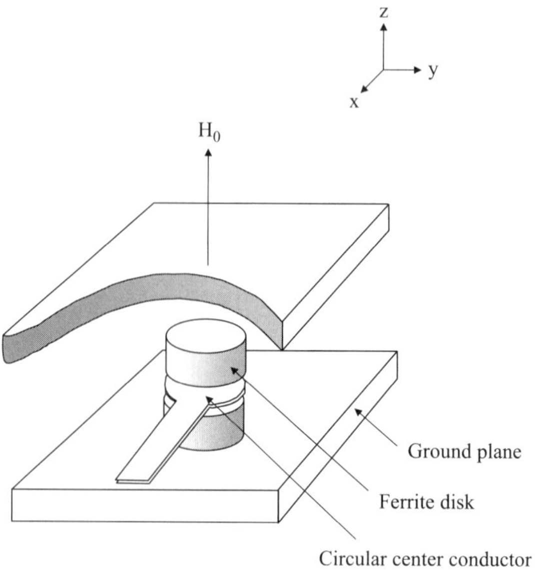

FIGURE 11.3 Definition of constituent resonator.

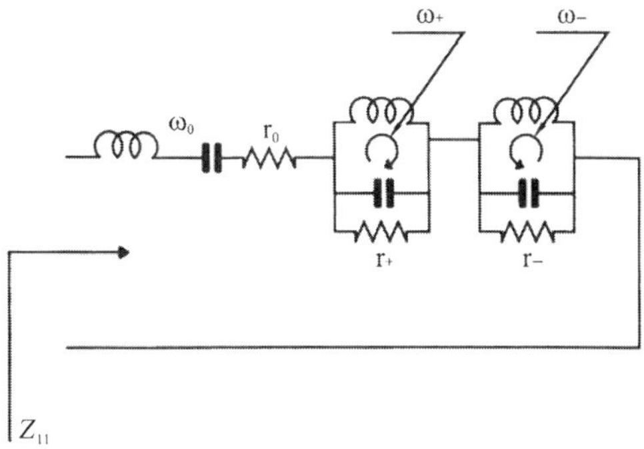

FIGURE 11.4 First Foster form of constituent resonator.

11.4 UNLOADED, EXTERNAL, AND LOADED *Q*-FACTORS

The insertion loss of a two-port resonant network is often expressed in terms of an unloaded, an external, and a loaded quality factor. Since an ideal circulator has some of the features of a transmission line cavity between two of its ports, it is appropriate to do so here also.

In defining the *Q*-factor, the power that is dissipated within the circuit is normally separately expressed from that within the external circuit. The total power dissipated within the circuit is described by an unloaded *Q*-factor, Q_U. This quantity is defined in terms of the resonant frequency of the junction, ω_0, and the stored and dissipated energies of the circuit:

$$Q_U = \frac{\omega_0 \text{ (energy stored in the circuit)}}{\text{power dissipated in circuit}} \tag{11.8}$$

If coupling ports, loops, or probes are introduced into the resonator, the power dissipated in each load may be expressed separately by a single external quality factor, $Q_{ex,n}$, defined by

$$Q_{ex,n} = \frac{\omega_0 \text{ (energy stored in the circuit)}}{\text{power reflected at port } n} \tag{11.9}$$

The total dissipated power is described by a loaded *Q*-factor, Q_L, given by

$$Q_L = \frac{\omega_0 \text{ (energy stored in the circuit)}}{\text{power dissipated} + \text{power reflected at all ports}} \tag{11.10}$$

It is readily demonstrated that

$$\frac{1}{Q_L} = \frac{1}{Q_U} + \frac{1}{Q_{ex,1}} + \frac{1}{Q_{ex,2}} + \cdots \tag{11.11}$$

Equations (11.8) and (11.9) are usually expressed in abbreviated form as

$$Q_U = \frac{\omega_0 U_0}{P_d} \tag{11.12}$$

$$Q_{ex,n} = \frac{\omega_0 U_0}{P_{r,n}} \tag{11.13}$$

U_0 is the energy stored in the circuit, P_d is the power dissipated in the circuit, and $P_{r,n}$ is the power dissipated in the external terminations.

11.5 MEASUREMENT OF UNLOADED SPLIT *Q*-FACTORS

The unloaded quality factor of a one-port resonator may be deduced experimentally in more than one way. One method involves a determination of whether the resonator is over- or undercoupled, a measurement of the midband VSWR and the frequencies at which the return loss is 3 dB below its midband value. The other method proceeds in a similar manner except that the loaded quality factor is obtained by a measurement of the phase of the reflection coefficient instead of its amplitude. The two methods produce identical results.

The first of the two methods makes use of the connection between its loaded (Q_L), external (Q_{ex}) and unloaded (Q_U) quality factors. The quantities appearing in this relationship may be evaluated experimentally by making use of some standard conditions. Before doing so, it is recalled that a cavity is either undercoupled, critically coupled, or overcoupled. If it is overcoupled then

$$\text{VSWR} = \frac{R}{n^2 Z_0} \tag{11.14a}$$

and

$$\left(\frac{1}{Q_L}\right) = \left(\frac{1}{Q_{ex}}\right)(1 + \text{VSWR}) \tag{11.14b}$$

$$\left(\frac{1}{Q_L}\right) = \left(\frac{1}{Q_U}\right)\left(\frac{1 + \text{VSWR}}{\text{VSWR}}\right) \tag{11.14c}$$

where

$$\left(\frac{1}{Q_L}\right) = \pm\left(\frac{\omega_0}{\omega_0 - \omega}\right) \tag{11.15}$$

The preceding equations are compatible provided

$$\mathrm{VSWR} = \left(\frac{Q_{ex}}{Q_U}\right) \tag{11.16}$$

VSWR is the voltage standing wave ratio at resonance, and ω is the radian frequency at which the power reflected is 3 dB down on its value at resonance. It is readily recognized that the two descriptions of VSWR in Eqs. (11.14a) and (11.16) are one and the same.

If the resonator is undercoupled then

$$\mathrm{VSWR} = \frac{n^2 Z_0}{R} \tag{11.17a}$$

and

$$\left(\frac{1}{Q_L}\right) = \left(\frac{1}{Q_{ex}}\right)\left(\frac{1 + \mathrm{VSWR}}{\mathrm{VSWR}}\right) \tag{11.17b}$$

$$\left(\frac{1}{Q_L}\right) = \left(\frac{1}{Q_U}\right)(1 + \mathrm{VSWR}) \tag{11.17c}$$

The latter two equations are again compatible provided

$$\mathrm{VSWR} = \left(\frac{Q_U}{Q_{ex}}\right) \tag{11.18}$$

The loaded quality factor entering into the experimental determination of the unloaded quality factor of the resonator is not to be confused with that of the complex gyrator circuit.

The quality factors of a one-port cavity may also be deduced from a Smith chart display of its frequency response. Figure 11.5 shows one typical construction. The

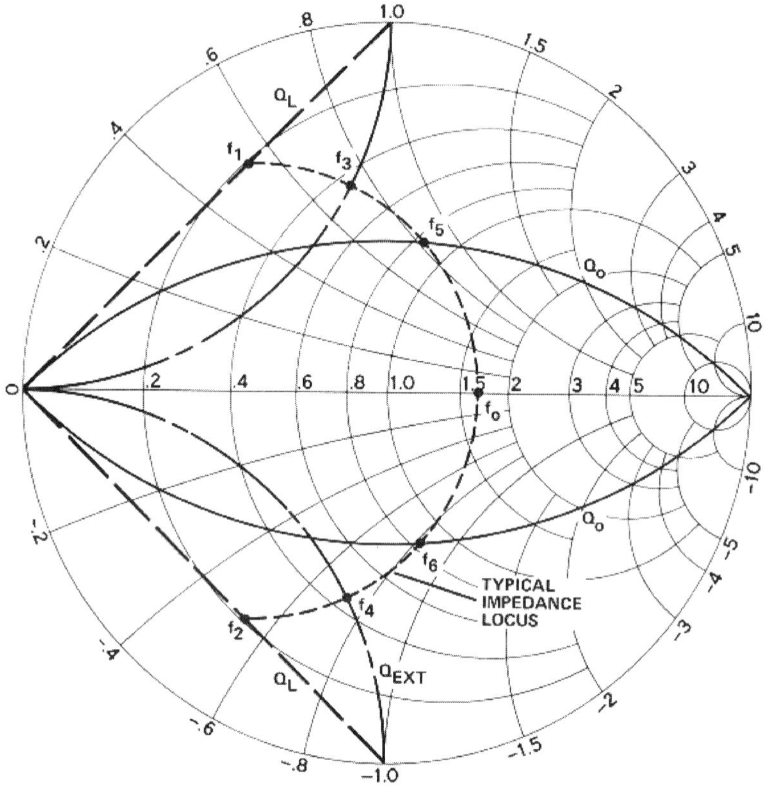

FIGURE 11.5 Smith chart display. Loci Q_o, Q_{ext} and Q_L of one-port cavity.

quality factors in question are defined in terms of the frequency markers by

$$Q_o = \frac{f_{res}}{(f_6 - f_5)} \tag{11.19a}$$

$$Q_{ext} = \frac{f_{res}}{(f_4 - f_3)} \tag{11.19b}$$

$$Q_L = \frac{f_{res}}{(f_2 - f_1)} \tag{11.19c}$$

11.6 EXPERIMENTAL DATA

Some experimental data on the split unloaded quality factors of the dominant pair of split modes of a simple gyromagnetic disk resonator are summarized in this section. This is done for a number of materials and for a number of different coupling angles that the strips subtend at a typical port. The experimental arrangement is the one-port circuit obtained by open-circuiting ports 2 and 3 of the circulator.

Figure 11.6 indicates the relationships between the direct magnetic flux density and the lower and upper split unloaded quality factors of the arrangement.

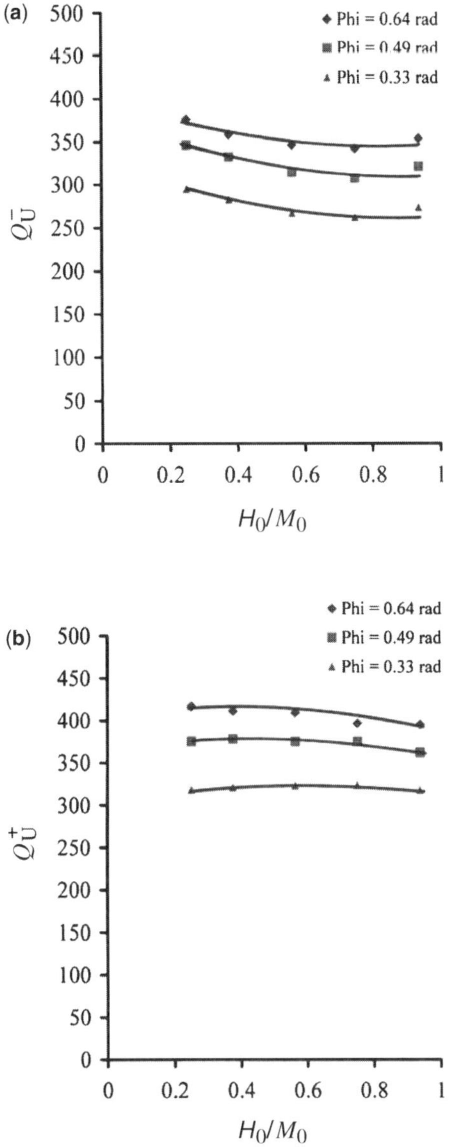

FIGURE 11.6 (a) Upper and (b) lower unloaded quality factors of gyromagnetic disk resonator for parametric values of coupling angles ($\mu_0 M_0 = 0.0400$ T). (Reproduced with permission from J. Helszajn, T. Vincent, and B. Pierce, Unloaded quality factors of junction circulators, *IEE Proc. Microwaves, Antennas and Propagation*, Vol. 152, No. 2, pp. 82–88, April 2005.)

This is done for three typical coupling angles, $\psi = 0.33$ rad, $\psi = 0.49$ rad, and $\psi = 0.64$ rad. The results apply to a gyromagnetic resonator biased below the main Kittel resonance. The ground plane spacing is adjusted in each instance to maintain the feed impedance at 50 Ω. The material employed in obtaining this result is a garnet one with a magnetization $\mu_0 M_0$ equal to 0.0400 T. The operating frequency is 2.0 GHz. A scrutiny of this data suggests that the split unloaded Q-factors show some dependence on the coupling angle. This feature is no doubt related to the ground plane spacing of the resonator. Since the main endeavor of this work is the effect of the magnetization on the split unloaded Q-factors of the circuit, the work on the other materials is restricted to the intermediate coupling angle. The aspect ratio of the resonator (R/L) is 3.0. Figure 11.7 depicts, for completeness sake, the relationship between the direct magnetic field intensity and the external quality factor for parametric values of coupling angle. Figure 11.8 illustrates a typical frequency response. Figure 11.9 depicts the upper and lower unloaded quality factors for materials with $\mu_0 M_0 = 0.0400$ T, 0.0550 T, 0.0680 T, and 0.0800 T for a constituent resonator with a coupling angle equal to 0.49 rad. The composition of the garnet is in every instance a gadolinium aluminum doped material. This material may be fired with values of saturation magnetization $(\mu_0 M_0)$ between 0.0400 T and 0.1400 T. The 3 dB linewidth $(\mu_0 \Delta H)$ is typically 0.0070 T; its spinwave linewidth $(\mu_0 \Delta H_k)$ is about 0.0004 T. The dielectric constant varies between 14 and 15 for the materials in question. One property of these results is that the unloaded quality factor of each lower branch shows a significant degradation as the direct magnetic field

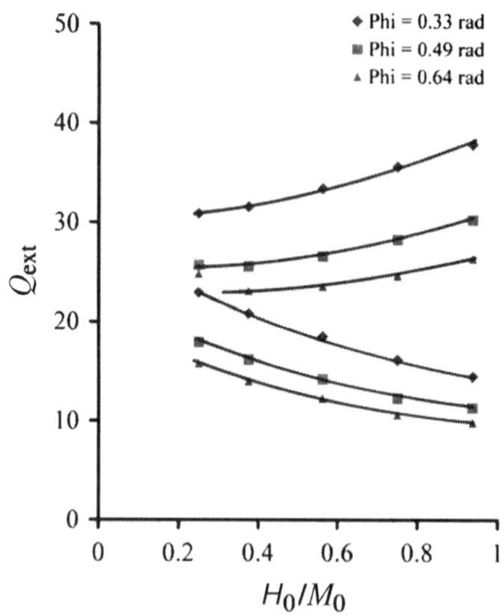

FIGURE 11.7 Split external quality factors of gyromagnetic resonator ($\mu_0 M_0 = 0.0400$ T). (Reproduced with permission from J. Helszajn, T. Vincent, and B. Pierce, Unloaded quality factors of junction circulators, *IEE Proc. Microwaves, Antennas and Propagation*, Vol. 152, No. 2, pp. 82–88, April 2005.)

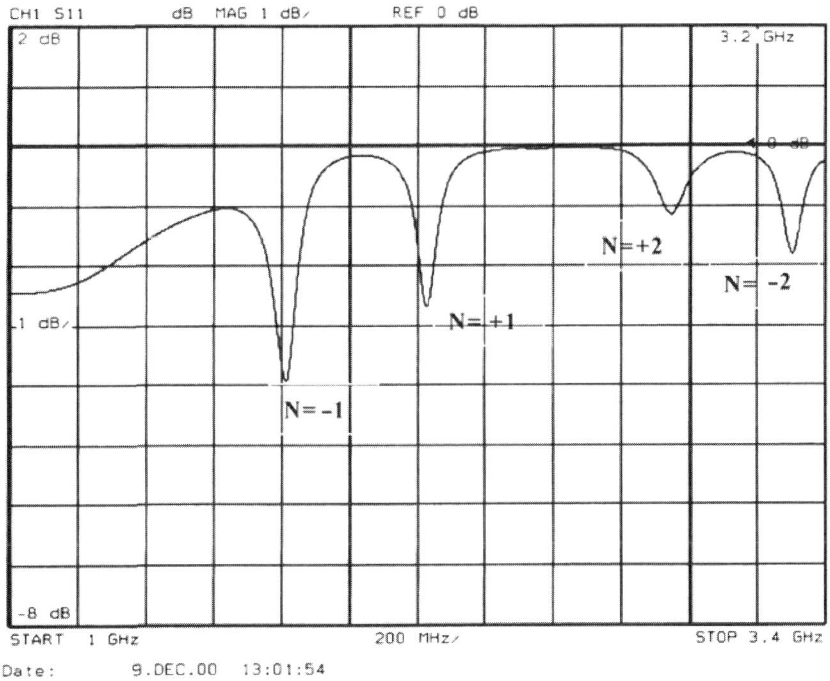

FIGURE 11.8 Frequency response of constituent resonator for a typical direct magnetic field.

increases. This may be understood by recognizing that, in keeping with the frequency response in Fig. 11.8, these branches encroach on the skirt of the uniform Kittel line. Another feature of these results is that the unloaded quality factors of the upper and lower branches deteriorate as the direct magnetization is increased. In obtaining this data all the measurements have been kept at a frequency of about 2.0 GHz by adjusting individually the radii of the different ferrites in order to cater for the permeability effects of the magnetization.

11.7 INSERTION LOSS OF JUNCTION CIRCULATORS

The insertion loss (L) between ports 1 and 2 of a junction circulator is defined in terms of the amplitude of the transmission parameter (S_{21}) by

$$L(\text{dB}) = -20\log_{10}\left|\frac{1}{S_{21}}\right| \tag{11.20}$$

The fundamental aspects of dissipation in a junction circulator are best discussed in terms of its dissipation eigenvalues. For engineering purposes, however, it is preferable to do so in terms of equivalent loaded and unloaded quality factors. The

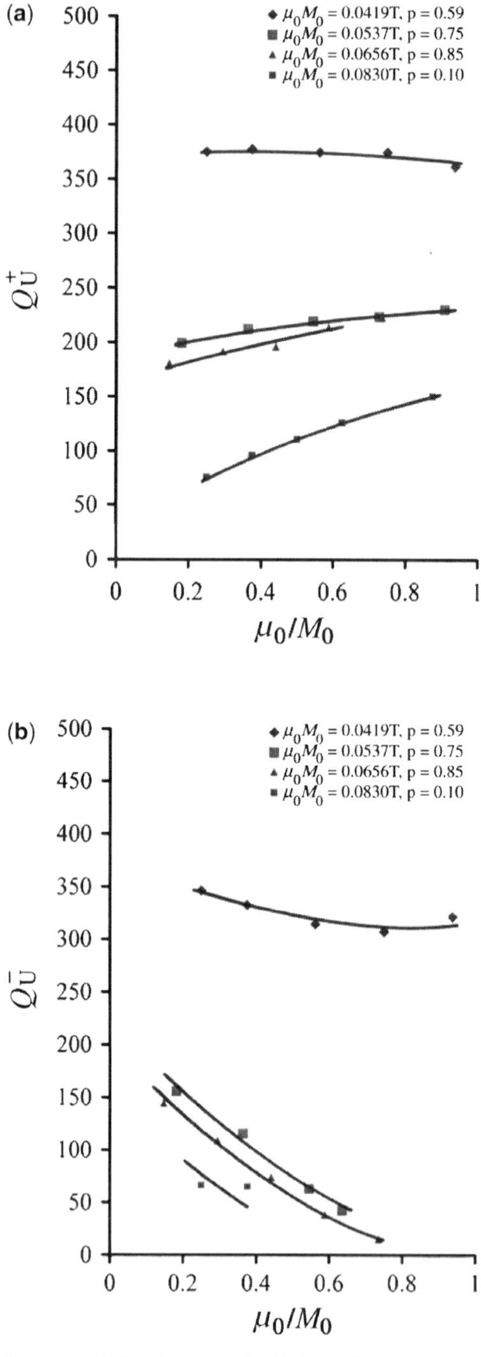

FIGURE 11.9 (a) Upper and (b) lower unloaded quality factors of gyromagnetic disk resonator ($\mu_0 M_0 = 0.0400$ T, 0.0550 T, 0.0680T , and 0.0800 T). (Reproduced with permission from J. Helszajn, T. Vincent, and B. Pierce, Unloaded quality factors of junction circulators, *IEE Proc. Microwaves, Antennas and Propagation*, Vol. 152, No. 2, pp. 82–88, April 2005.)

required result in terms of the magnetic parameters of the resonator is given by

$$L(\mathrm{dB}) \approx -20 \log_{10}\left|1 + \frac{Q_{\mathrm{L}}}{Q_{\mathrm{eff}}}\right| \tag{11.21}$$

Q_{L} is the quality factor of the complex gyrator circuit, and Q_{eff} is the effective unloaded quality factor associated with the gyrotropy of the gyromagnetic resonator.

The effective midband unloaded Q-factor (Q_{eff}) appearing in the insertion loss function is defined in terms of the split unloaded ones and the gyrotropy by

$$\frac{1}{Q_{\mathrm{eff}}} = \frac{1}{2}\left[\left(\frac{\mu - \kappa}{\mu}\right)\left(\frac{1}{Q_{\mathrm{U}}^{+}}\right) + \left(\frac{\mu + \kappa}{\mu}\right)\left(\frac{1}{Q_{\mathrm{U}}^{-}}\right)\right] \tag{11.22}$$

μ and κ are the diagonal and off-diagonal entries of the permeability tensor in Chapter 2.

The unloaded quality factors, in a gyromagnetic resonator biased below the Kittel line, are not in practice too unequal. This condition corresponds to the eigenvalue diagram in Fig. 11.1b. If this is the case, then

$$\frac{1}{Q_{\mathrm{eff}}} \approx \frac{1}{2}\left(\frac{1}{Q_{\mathrm{U}}^{+}} + \frac{1}{Q_{\mathrm{U}}^{-}}\right) \tag{11.23}$$

The situation in an above-resonance circulator is somewhat more difficult in that

$$Q_{\mathrm{U}}^{+} \neq Q_{\mathrm{U}}^{-} \tag{11.24}$$

In this sort of device Q_{eff} increases while Q_{L} decreases as the direct magnetic field is reduced. A compromise between bandwidth and insertion loss is therefore necessary in practice. This situation corresponds to the eigenvalue diagram in Fig. 11.1d.

The loaded quality factor of the complex gyrator circuit, Q_{L}, is given separately in the usual way in terms of the split frequencies of the gyromagnetic resonator and elsewhere in this text

$$\frac{1}{Q_{\mathrm{L}}} = \sqrt{3}\left(\frac{\omega_{+} - \omega_{-}}{\omega_0}\right) \tag{11.25}$$

The loaded quality factor and insertion loss of a typical commercial junction circulator are

$$Q_{\mathrm{L}} = 2.5 \quad \text{and} \quad L(\mathrm{dB}) = 0.25$$

The effective unloaded quality factor necessary to meet this specification is

$$Q_{\mathrm{eff}} = 87$$

If

$$Q_{\mathrm{L}} = 2.0$$

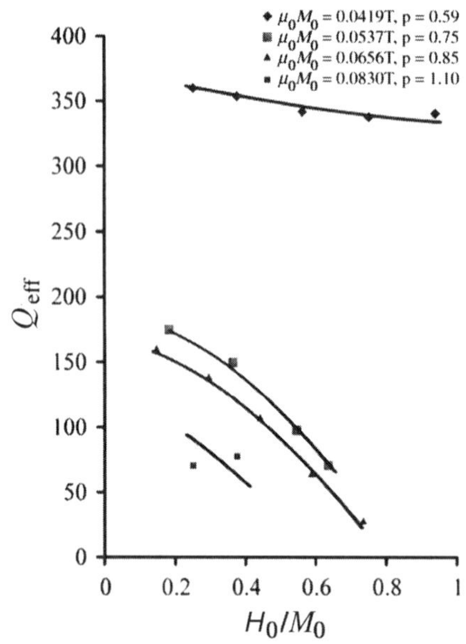

FIGURE 11.10 Effective unloaded quality factor versus magnetic field intensity. (Reproduced with permission from J. Helszajn, T. Vincent, and B. Pierce, Unloaded quality factors of junction circulators, *IEE Proc. Microwaves, Antennas and Propagation*, Vol. 152, No. 2, pp. 82–88, April 2005.)

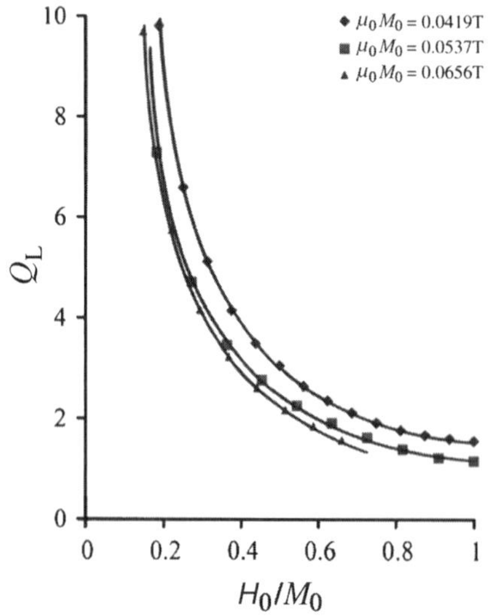

FIGURE 11.11 Loaded quality factor of complex gyrator circuit.

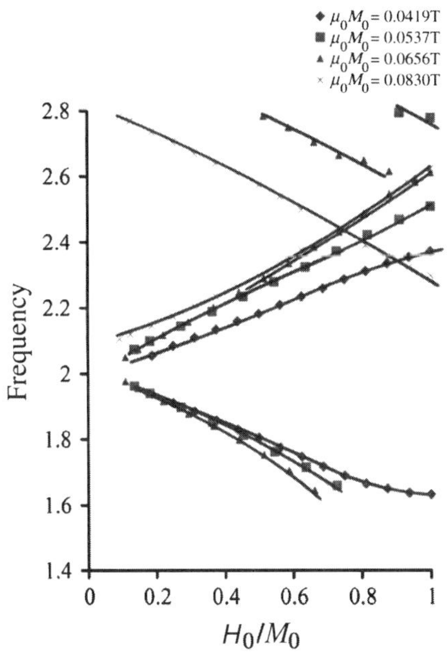

FIGURE 11.12 Split frequencies of gyromagnetic resonator versus magnetic field intensity.

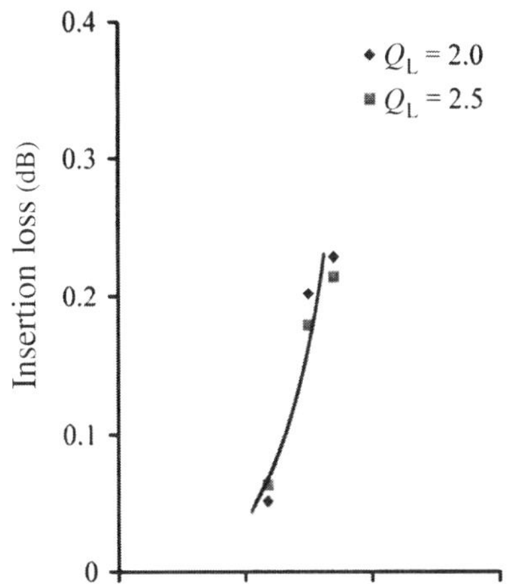

FIGURE 11.13 Relationship between insertion loss of junction circulator and normalized magnetization. (Reproduced with permission from J. Helszajn, T. Vincent, and B. Pierce, Unloaded quality factors of junction circulators, *IEE Proc. Microwaves, Antennas and Propagation*, Vol. 152, No. 2, pp. 82–88, April 2005.)

and

$$L(\text{dB}) = 0.50$$

then

$$Q_{\text{eff}} = 35$$

In order to construct the insertion loss of the circulator it is necessary to calculate the effective quality factor of the gyromagnetic resonator and the loaded quality factor of the complex gyrator circuit. Figure 11.10 depicts the relationship between the effective quality factor and direct magnetic field intensity for each material utilized in this work. Figure 11.11 indicates the loaded quality factor. This latter quantity is obtained in terms of the split frequencies of the resonator. Figure 11.12 gives a typical plot. The insertion loss of the overall circulator may be deduced once Q_{eff} and Q_{L} are established. Figure 11.13 shows the relationship between the insertion loss and the normalised magnetization of the gyromagnetic resonator for two typical values of Q_{L}.

The nature of the relationship between the insertion loss and the normalized magnetization is similar to that met in connection with the design of below-resonance ferrite phase shifters. It is apparent from this result that the use of materials with values of normalized magnetization above 0.60 is unsuitable for the design of low loss circulators.

11.8 INSERTION LOSS OF UHF CIRCULATORS

A simple technique whereby the intrustion of the Kittel line on the passband of a circulator may be established is a topic of this section. It consists of displaying the return loss of the circulator at port 1 with ports 2 and 3 open circuited. The displacement of the reflection coefficient from the rim of the charts is a qualitative measure of the dissipation of the junction. Its deviation from a constant displacement indicates the intrusion of the skirt of the Kittel line onto the passband.

The connection between the return loss of the constituent circuit and that of the circulator has not been formally derived at this time. One semiempirical relationship based on a sample of five above-resonance units at 2 GHz is

$$L(\text{dB}) = A(RL)\,\text{dB}$$

where

$$A = 1/4.6$$

taking RL as 1.0 dB by way of example gives L(dB) as 0.23 dB.

11.9 SCATTERING MATRIX OF SEMI-IDEAL CIRCULATORS

Semi-ideal circulators are those for which dissipation exists. In such circulators either

$$S_{11} = 0, \quad S_{21} \neq 1, \quad \text{and} \quad S_{31} \neq 0 \qquad (11.26\text{a})$$

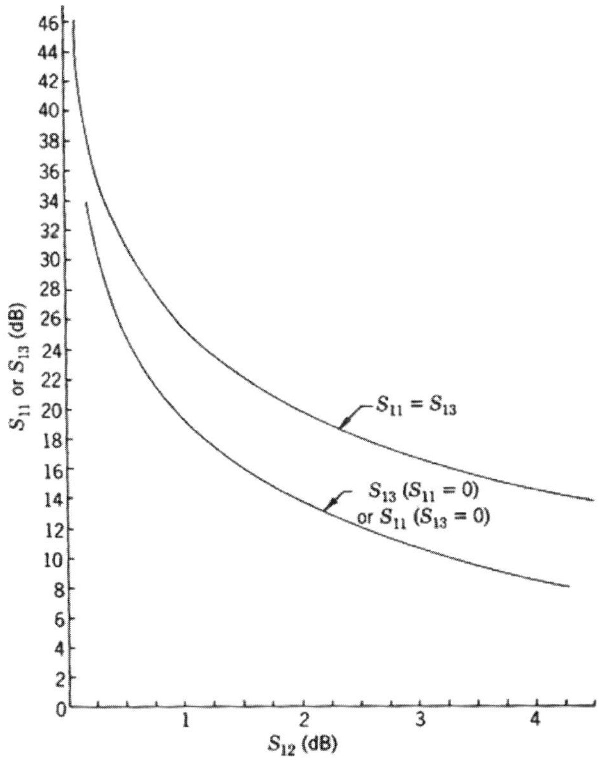

FIGURE 11.14 Relationships between scattering parameters of semi-ideal circulator. (Reproduced with permission from J., Helszajn, Dissipation and Scattering Matrices of Lossy Junctions, *IEEE Trans. Microwave Theory Tech.*, Vol. MTT-29, pp. 779–782, 1972.)

or

$$S_{11} \neq 0, \quad S_{21} \neq 1, \quad \text{and} \quad S_{31} = 0 \tag{11.26b}$$

The first situation is obtained when the angle between s_{+1} and s_{-1} is less than 120°. The second case is obtained when it is larger than 120°.

The scattering parameters are given in terms of the dissipation eigenvalues introduced in Chapter 4:

$$3S_{11} = -1 + \left(1 - \frac{q_{+1}}{2}\right)\exp(-j2\phi_+) + \left(1 - \frac{q_{-1}}{2}\right)\exp(j2\phi_+) \tag{11.27a}$$

$$3S_{21} = -1 + \alpha\left(1 - \frac{q_{+1}}{2}\right)\exp(-j2\phi_+) + \alpha^2\left(1 - \frac{q_{-1}}{2}\right)\exp(j2\phi_+) \tag{11.27b}$$

$$3S_{31} = -1 + \alpha^2\left(1 - \frac{q_{+1}}{2}\right)\exp(-j2\phi_+) + \alpha\left(1 - \frac{q_{-1}}{2}\right)\exp(j2\phi_+) \tag{11.27c}$$

where it has been assumed that the splitting is symmetrical:

$$\phi_{-1} = -\phi_{+1} \tag{11.28}$$

In semi-ideal circulators S_{11} and S_{31} are completely determined by S_{21} provided it is assumed that the amplitudes of s_{+1} and s_{-1} are equal. This means that the latter quantity can be obtained simply by measuring either S_{11} or S_{31}.

The first case to be considered is that in which the angle between s_{+1} and s_{-1} is such that $S_{11} = 0$. This condition is obtained by setting $S_{11} = 0$. The second case to be considered here is that for which the angle between s_{+1} and s_{-1} is such that $S_{31} = 0$. This condition is obtained by setting $S_{31} = 0$. The two possibilities are illustrated in Fig. 11.14.

CHAPTER TWELVE

The Lumped Element Circulator

12.1 INTRODUCTION

One common engineering preoccupation is that of miniaturization. The classic means of breaking the relationship between the wavelength and the size of a microwave circuit is to have recourse to lumped element technology. The purpose of this chapter is to deal with the theory of the classic lumped element circulator. Its adjustment is a straightforward eigenvalue problem so that it provides an excellent introduction to that of the more demanding distributed configuration. It consists of a ferrite disk with three coils wound on it so that the alternating magnetic fields of the coils are oriented at 120° with respect to each other. A direct magnetic field intensity is applied normal to the plane of the assembly. Both series and shunt junctions have been described in the literature but this chapter deals exclusively with the former configuration. One possible planar coil structure of the shunt geometry is an interwoven mesh arrangement of three short sections of short-circuited stripline at 120°, which are insulated from each other. If the striplines are electrically short the energy within the strips is essentially magnetic. The degenerate resonances of the junction are then established by connecting shunt capacitances at the three terminals, while the gyrator impedance is set by adjusting the direct magnetic field.

Circulators using hybrid technology comprised of lumped elements and distributed circuits have also evolved over time. One possibility is a distributed resonator matched by a printed lumped element network; it will be dealt with in some detail. Another is an undersized resonator loaded by a lumped element shunt capacitor at each port. It may be matched by either an alternate line distributed transformer or again by a lowpass lumped element circuit. Still another means of miniaturizing the resonator is of course the introduction of radial magnetic walls. This chapter includes the development of the equivalence between a lowpass lumped element

The Stripline Circulator: Theory and Practice. By J. Helszajn

matching network and an alternate line transformer. It also brings together design equations and experimental results on a number of UHF circulators.

12.2 LUMPED ELEMENT CIRCULATOR

One circulator geometry that can readily be analyzed using the eigenvalue method is the lumped element configuration; Fig. 12.1 illustrates series and shunt possibilities. The symmetrical but nonreciprocal element can be wires or strips. The usual arrangement consists of a ferrite disk with three strips wound on it so that the radiofrequency (RF) magnetic fields of the strips are oriented at 120° with respect to each other. A direct magnetic field is applied normal to the plane of the disk. If the short-circuited striplines are electrically short, the energy within the disk geometry is essentially magnetic. The configuration dealt with here is the shunt structure. The degenerate resonances of the junction are established by connecting shunt capacitances at its terminals, while the impedance level of the device is fixed by suitably adjusting the gyrotropy. The voltage–current relationships at the terminals of the mesh structure are described by

$$\begin{bmatrix} V_1 \\ V_2 \\ V_3 \end{bmatrix} = j\omega L_0 \begin{bmatrix} \mu_i & -\dfrac{\mu_i}{2} & -\dfrac{\mu_i}{2} \\ -\dfrac{\mu_i}{2} & \mu_i & -\dfrac{\mu_i}{2} \\ -\dfrac{\mu_i}{2} & -\dfrac{\mu_i}{2} & \mu_i \end{bmatrix} \begin{bmatrix} I_1 \\ I_2 \\ I_3 \end{bmatrix} \tag{12.1}$$

Here L_0 is the inductance of each constituent mesh and μ_i is the initial permeability of the ferrite region.

The impedance eigenvalues of the network are

$$z_0 = 0 \tag{12.2a}$$

$$z_+ = j3\omega L_0 \mu_i/2 \tag{12.2b}$$

$$z_- = j3\omega L_0 \mu_i/2 \tag{12.2c}$$

The solution consists of one nondegenerate eigenvalue and a pair of degenerate ones. The corresponding admittance eigenvalues of this network are the reciprocal of the impedance ones:

$$y_0 = \infty \tag{12.3a}$$

$$y_+ = 2/j3\omega L_0 \mu_i \tag{12.3b}$$

$$y_- = 2/j3\omega L_0 \mu_i \tag{12.3c}$$

These eigenvalues can be adjusted to coincide with those of the first circulation condition by adding shunt capacitances at each port.

$$y_0 = \infty \tag{12.4a}$$

$$y_+ = j\omega C + 2/j3\omega L_0 \mu_i = 0 \tag{12.4b}$$

$$y_- = j\omega C + 2/j3\omega L_0 \mu_i = 0 \tag{12.4c}$$

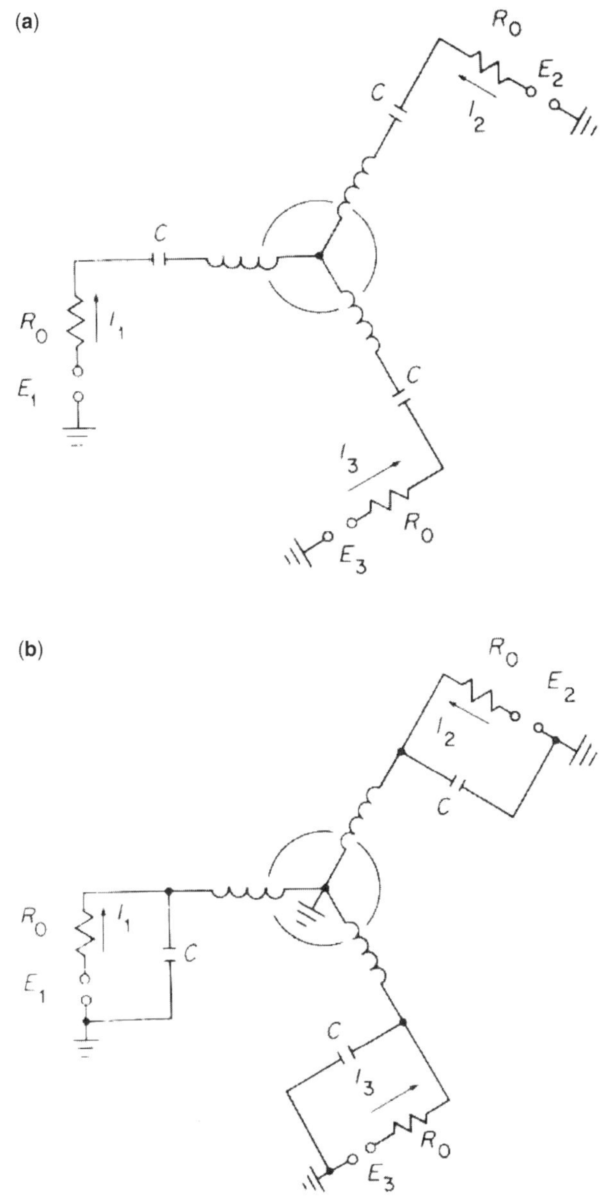

FIGURE 12.1 Schematic of lumped element circulator: (a) shunt network and (b) series network.

This adjustment establishes the isotropic resonances of the junction. The second circulation condition is now met by splitting the degenerate admittance eigenvalues with a suitable gyromagnetic material until these coincide with those of an ideal circulator. Making use of the scalar permeabilities defined in Chapter 2 gives

$$y_0 = \infty \tag{12.5a}$$
$$y_+ = j\omega C + 2/j3\omega L_0(\mu - \kappa) \tag{12.5b}$$
$$y_- = j\omega C + 2/j3\omega L_0(\mu + \kappa) \tag{12.5c}$$

The first and second circulation conditions are now established by recalling the boundary condition in the eigenvalue plane of the ideal circulator:

$$y_0 = \infty \tag{12.6a}$$
$$y_+ = -jY_0/\sqrt{3} \tag{12.6b}$$
$$y_- = jY_0/\sqrt{3} \tag{12.6c}$$

The real and imaginary parts of the complex gyrator admittance at midband are

$$Y_0 = \frac{\sqrt{3}}{\omega_0 L}\left(\frac{\kappa}{\mu}\right) \tag{12.7}$$

and

$$\omega_0^2 LC = 1 \tag{12.8}$$

where

$$L = \tfrac{3}{2}\mu_{\text{eff}} L_0 \tag{12.9}$$
$$\mu_{\text{eff}} = \frac{\mu^2 - \kappa^2}{\mu} \tag{12.10}$$

These two relationships fix the circulation conditions for this type of circulator.

12.3 COMPLEX GYRATOR CIRCUIT OF LUMPED ELEMENT CIRCULATOR

The equivalent circuit of a lumped element circulator near its circulation condition may be obtained by forming the input admittance of the circuit in terms of its eigenvalues,

$$Y_{\text{in}} = \left(\frac{y_{+1} + y_{-1}}{2}\right) + j\sqrt{3}\left(\frac{y_{+1} - y_{-1}}{2}\right) \tag{12.11}$$

which applies with $s_0 = -1$.

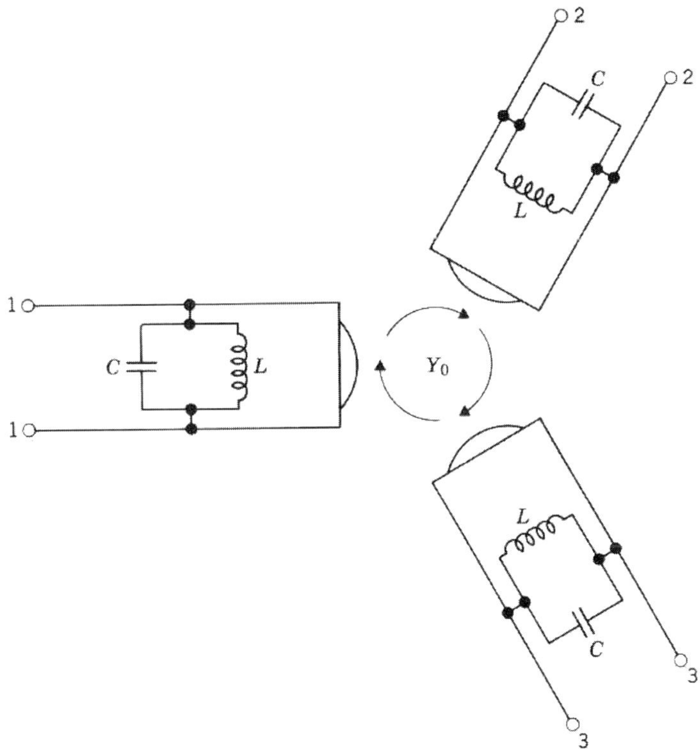

FIGURE 12.2 Equivalent circuit of lumped element circulator using frequency independent two-port gyrators.

The split eigenadmittances for the lumped element circulator are given from the previous section by

$$y_{+1} = j\left(\omega C - \frac{1}{\omega L}\right) - \frac{j}{\omega L} \cdot \frac{\kappa}{\mu} \tag{12.12a}$$

$$y_{-1} = j\left(\omega C - \frac{1}{\omega L}\right) + \frac{j}{\omega L} \cdot \frac{\kappa}{\mu} \tag{12.12b}$$

The result is

$$Y_{\text{in}} = \frac{\sqrt{3}}{\omega L} \cdot \frac{\kappa}{\mu} + j\left(\omega C - \frac{1}{\omega L}\right) \tag{12.13}$$

The imaginary and real parts of this equation are compatible with the first and second circulation conditions deduced in the previous section. The general form of the admittance of the network is therefore

$$Y_{\text{in}} = Y_0 + j\omega_0 C\left(\frac{\omega}{\omega_0} - \frac{\omega_0}{\omega}\right) \tag{12.14}$$

An approximate equivalent network for this equation is an ideal circulator available at any frequency with an admittance y_1 connected at each port. This is shown in Fig. 12.2.

12.4 GAIN-BANDWIDTH PRODUCT OF LUMPED ELEMENT CIRCULATOR

An important quantity of any circulator is its gain-bandwidth product. This quantity is fixed by the loaded quality factor of the complex gyrator admittance of the circulator. The task of this section is to determine this quantity. It is defined by

$$Q_L = B'/G \tag{12.15}$$

The susceptance slope parameter (B') is defined in terms of the imaginary part (B) of the complex gyrator admittance by

$$B' = \frac{\omega_0}{2} \frac{\partial B}{\partial \omega}\bigg|_{\omega=\omega_0} = \frac{1}{\omega_0 L} = \omega_0 C \tag{12.16}$$

Combining this equation with the real part condition of the complex gyrator admittance gives Q_L in terms of the magnetic variable:

$$Q_L = \frac{B'}{Y_0} = \frac{1}{\sqrt{3}}\left(\frac{\mu}{\kappa}\right) \tag{12.17}$$

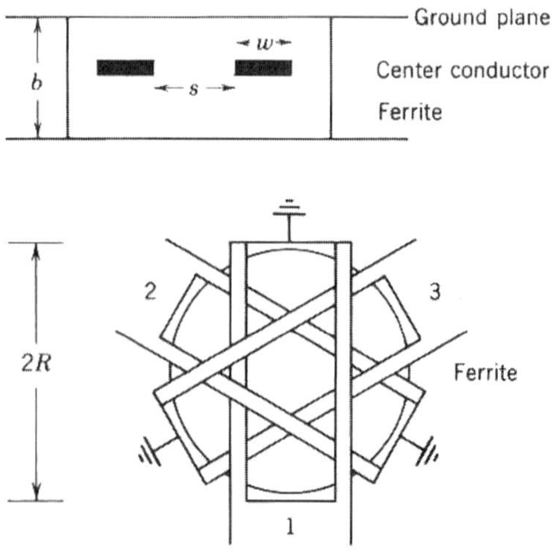

FIGURE 12.3 Geometry of lumped constant circulator using short-circuited striplines. (Reproduced with permission from Y. Konishi, Lumped element Y circulator, *IEEE Trans. Microwave Theory Tech.*, Vol. MTT-13, No. 6, pp. 852–864, 1965.)

The gyrotropy is related separately to the difference between the two split resonant frequencies of the junction by

$$\left(\frac{\kappa}{\mu}\right) = \frac{\omega_{+1} - \omega_{-1}}{\omega_0} \tag{12.18}$$

The loaded Q-factor of the junction is therefore also given in terms of the split frequencies ω_{+1} and ω_{-1} by

$$\frac{1}{\sqrt{3}Q_L} = \frac{\omega_{+1} - \omega_{-1}}{\omega_0} \tag{12.19}$$

which is a general result.

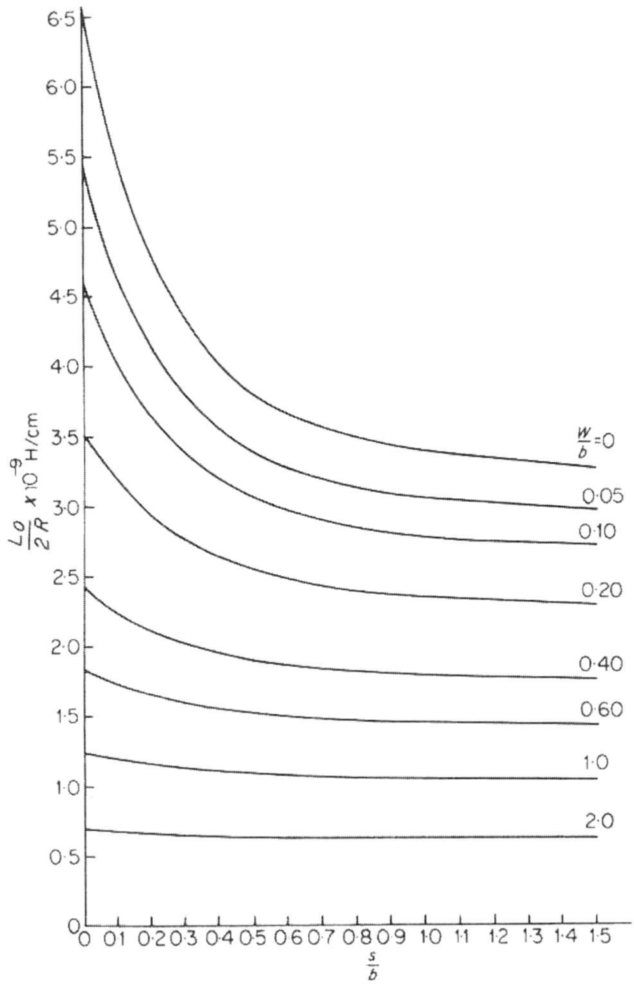

FIGURE 12.4 Junction inductance of stripline lumped element circulator for $t/b = 0.025$. (Reproduced with permission from J. Helszajn and M. McDermott, The inductance of a lumped constant circulator, *IEEE Trans. Microwave Theory Tech.*, Vol. MTT-18, No. 1, pp. 50–52, 1970.)

An expression for the Q-factor in terms of the junction specification may also be derived. It is given by

$$Q_{\mathrm{L}} = \frac{B'}{Y_0} = \frac{r-1}{2\delta_{\max}\sqrt{r}} \tag{12.20}$$

where r is the VSWR and $2\delta_{\max}$ is the full bandwidth $(\omega_2 - \omega_1)/\omega_0$.

The former quantity is deduced by making use of the connection between the reflection coefficient and the VSWR:

$$\mathrm{VSWR} = \frac{1+|\rho|}{1-|\rho|} \tag{12.21}$$

12.5 INDUCTANCE OF CONSTITUENT LUMPED ELEMENT RESONATOR

One possible planar configuration of the coil arrangement of the shunt arrangement is a mesh with elements insulated from each other. This geometry is shown in Fig. 12.3.

The inductance of a typical mesh may readily be calculated in terms of its geometry once its characteristic impedance is available. This impedance corresponds to half that of the even mode of two coupled striplines. It may be deduced in terms of the static capacitance per unit length of the geometry, provided the fields on the line may be assumed to be pure transverse ones.

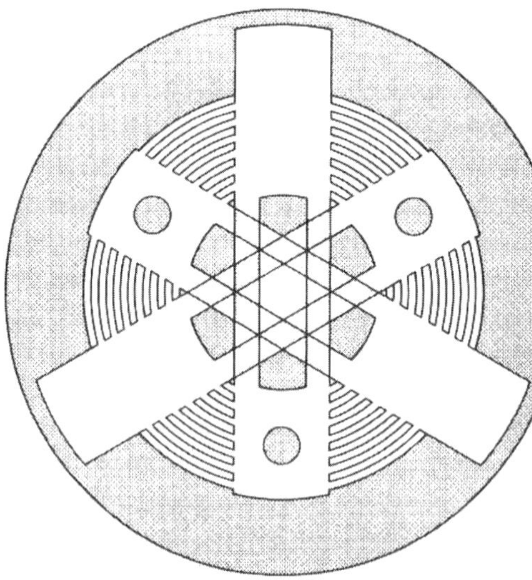

FIGURE 12.5 Lumped element circulator using interdigital capacitors. (Reproduced with permission from R. H. Knerr, A thin film lumped element circulator, *IEEE Trans. Microwave Theory Tech.*, Vol. MTT-17, No. 12, pp. 1152–1154, 1969.)

Figure 12.4 shows s/b versus $L_0/2R$ for parametric values of W/b, for the case where $t/b = 0.025$. For t/b small, the inductance is not strongly dependent on it. The mesh arrangement requires that t/b should be kept small; otherwise, it is not possible to interweave it. In the illustration R is the ferrite radius and L_0 is the inductance of the constituent resonator. Figure 12.5 illustrates a lumped element circulator using planar interdigital capacitors.

12.6 MAGNETIC VARIABLES OF LUMPED ELEMENT CIRCULATOR

At UHF frequencies the lumped element circulator is usually biased above the main resonance to prevent the onset of so-called low field loss, discussed in Chapter 2.

The gyrotropy and the effective permeability are given by

$$\frac{\kappa}{\mu} \approx \left(\frac{1}{\sigma}\right) \frac{p/\sigma}{1 + p/\sigma} \tag{12.22a}$$

and

$$\mu_e \approx 1 + p/\sigma \tag{12.22b}$$

provided $\sigma > 1$ and $p > 1$.

κ/μ is proportional to the center frequency, and μ_e is independent of it, under the same approximations. This means that the gyrator impedance is independent of

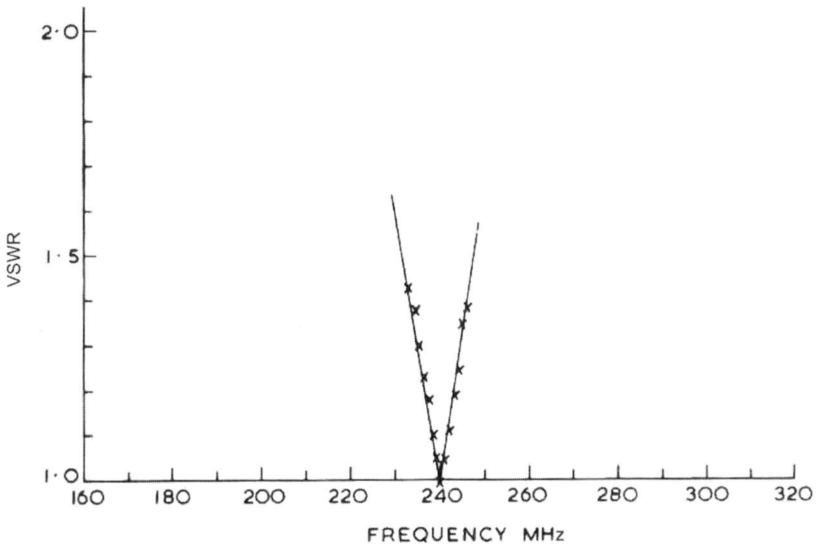

FIGURE 12.6 Performance of 240 MHz, degree-1 circulator. (Reproduced with permission from J. Helszajn and F. M. Aitken, U.H.F. techniques for lumped constant circulators, *Electron. Eng.*, pp. 53–59, Nov. 1973.)

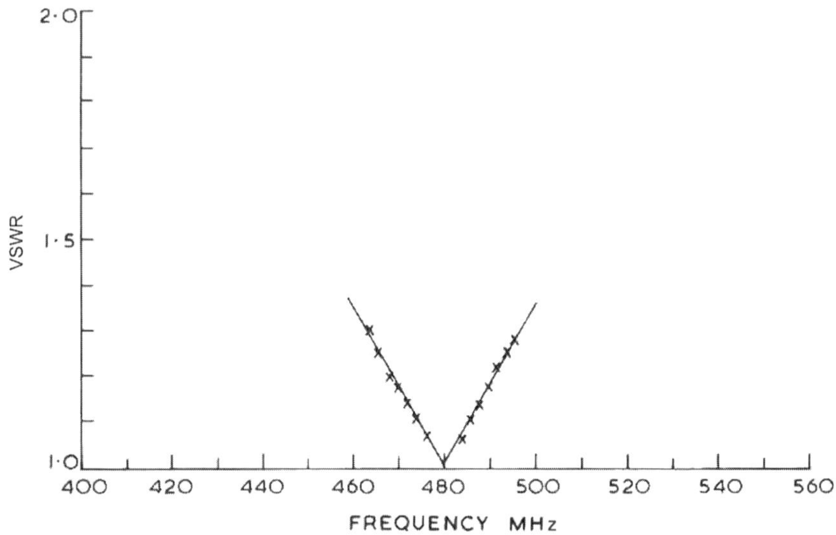

FIGURE 12.7 Performance of 480 MHz, degree-1 circulator. (Reproduced with permission from J. Helszajn and F. M. Aitken, U.H.F. techniques for lumped constant circulators, *Electron. Eng.*, pp. 53–59, Nov. 1973.)

frequency and that the center frequency of the device can readily be tuned by altering the lumped capacitance at the terminals of the junction. Another property of the device is that since the bandwidth is proportional to κ/μ the former is proportional to the center frequency.

Figure 12.6 depicts the experimental response of one circulator at 240 MHz. Figure 12.7 shows the same geometry retuned to 480 MHz by altering the terminal capacitances only. The experimental bandwidth at this frequency is approximately twice that obtained at 240 MHz (in keeping with the fact that the bandwidth is proportional to the center frequency of the device).

12.7 DEGREE-2 LUMPED ELEMENT CIRCULATOR

The bandwidth of the junction can be improved by connecting matching circuits at each port. One topology is obtained by embodying the gyrator network into a degree-2 bandpass filter in the manner indicated in Fig. 12.8. The details of this solution will now be deduced.

The unknowns in the degree-2 problem are the gyrator conductance and the quality factors of the series matching network and the complex gyrator circuit. The given quantities are the maximum VSWR and the character of the frequency response. The amplitude squared of the reflection coefficient is finite at the midband and bandedge frequencies and zero at finite frequencies. While the solution of this type of problem is a straightforward synthesis problem, the method employed here is the coefficient comparison method.

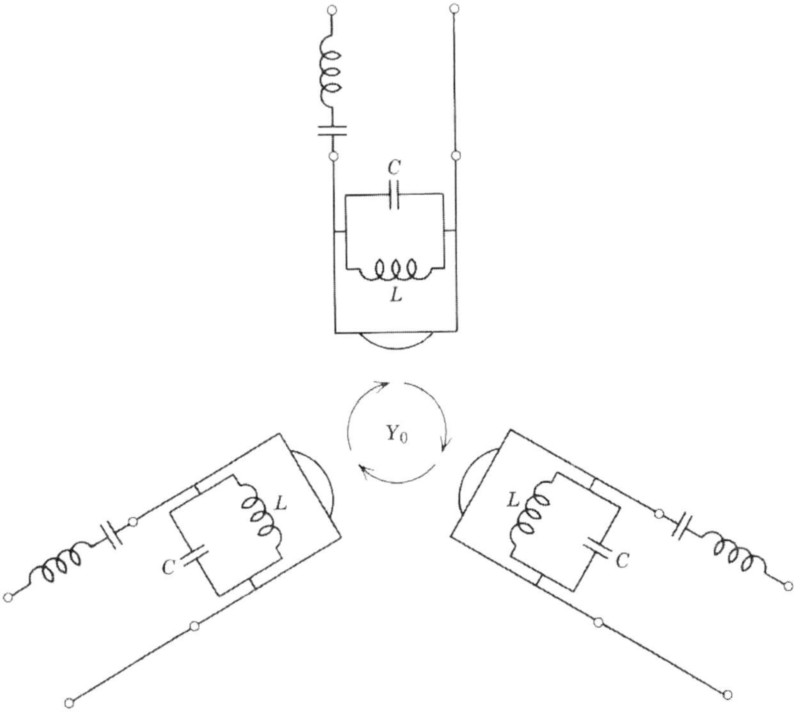

FIGURE 12.8 Schematic of degree-2 circulator.

The derivation of the required result starts with the description of the amplitude squared of the reflection coefficient:

$$|\rho_{\text{in}}|^2 = \frac{\varepsilon^2\, T_n^2(x)}{1 + \varepsilon^2 T_n^2(x)} \tag{12.23}$$

The ripple level ε is related to the VSWR by

$$\varepsilon = \frac{(\text{VSWR}) - 1}{2\sqrt{\text{VSWR}}} \tag{12.24}$$

The Chebyshev polynomial of degree-2 is

$$T_2(x) = 2x^2 - 1 \tag{12.25}$$

$|\rho_{\text{in}}|$ is a maximum at $x = 0$ and 1; it is zero at $x = 0.707$. This gives three conditions from which the unknowns of the problem may be deduced:

$$|\rho_{\text{in}}| = \frac{\text{VSWR} - 1}{\text{VSWR} + 1}, \quad x = 0 \tag{12.26a}$$

$$|\rho_{\text{in}}| = 0, \quad x = 0.707 \tag{12.26b}$$

$$|\rho_{in}| = \frac{\text{VSWR} - 1}{\text{VSWR} + 1}, \quad x = 1.0 \tag{12.26c}$$

The condition $|\rho_{in}| = 0$ may also be replaced by

$$\text{I.P.}\,(Z_{in}) = 0 \tag{12.27a}$$

$$\text{R.P.}\,(Z_{in}) = Z_0 \tag{12.27b}$$

The derivation now proceeds by forming the input impedance of the circuit,

$$Z_{in} = Z_1 + 1/Y_1 \tag{12.28}$$

where

$$Z_1 = j2\delta X_s' \tag{12.29}$$

$$Y_1 = G + j2\delta B_p' \tag{12.30}$$

and

$$\delta = \frac{\omega - \omega_0}{\omega_0} \tag{12.31}$$

X' and B' are the reactance and susceptance slope parameters of the series and parallel resonators of the filter circuit. The condition at $x = 0$ gives

$$Y_0/G = \text{VSWR} \tag{12.32}$$

The first impedance condition gives

$$Q_s = (\text{VSWR})Q_p \tag{12.33}$$

where

$$Q_s = \omega_0 L_s / Z_0 \tag{12.34}$$

$$Q_p = \omega_0 C_p / G \tag{12.35}$$

and

$$\omega_0^2 L_s C_s = 1 \tag{12.36}$$

$$\omega_0^2 L_p C_p = 1 \tag{12.37}$$

The second impedance condition readily gives

$$2\delta_{min} Q_p = (\text{VSWR} - 1)^{1/2} \tag{12.38}$$

The first minimum in the reflection coefficient occurs at

$$(2\delta_{\min}) = (2\delta_{\max})/\sqrt{2} \tag{12.39}$$

The gain-bandwidth product of the circuit is therefore specified by

$$2\delta_{\max} Q_p = \sqrt{2}\,(\mathrm{VSWR} - 1)^{1/2} \tag{12.40}$$

The amplitude squared of the reflection coefficient is now given by

$$|\rho_{in}|^2 = \frac{[(\mathrm{VSWR} - 1) - (\mathrm{VSWR})(2\delta Q_p)^2]^2 + [(\mathrm{VSWR} - 1)(2\delta Q_p)^2]^2}{[(\mathrm{VSWR} + 1) - (\mathrm{VSWR})(2\delta Q_p)^2]^2 + [(\mathrm{VSWR} + 1)(2\delta Q_p)^2]^2} \tag{12.41}$$

This quantity displays the appropriate frequency response of the filter specification.

Figure 12.9 indicates the response obtained with such a network at 480 MHz for the basic response in Fig 12.7. The bandwidth for VSWR = 1.22 is improved from about 5% to about 18%. Figure 12.10 illustrates the quality of one experimental complex gyrator circuit.

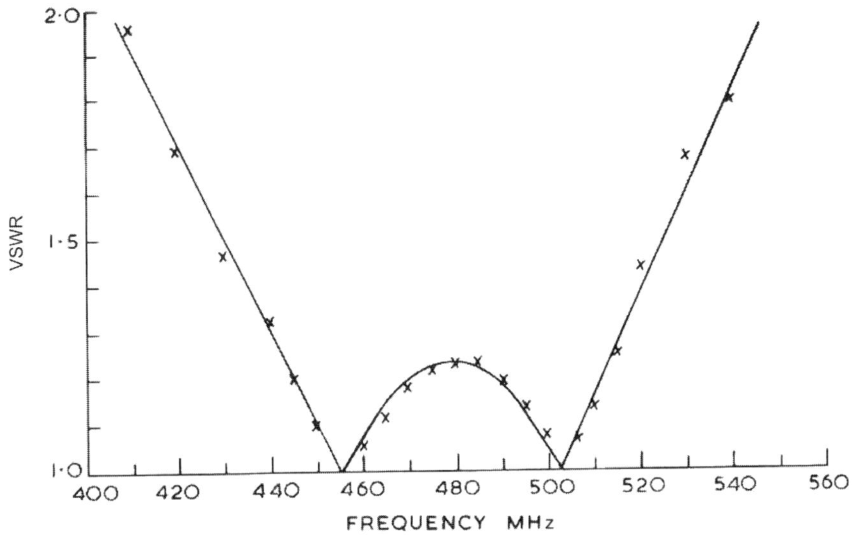

FIGURE 12.9 Performance of 480 MHz, degree-2 circulator. (Reproduced with permission from J. Helszajn and F. M. Aitken, U.H.F. techniques for lumped constant circulators, *Electron. Eng.*, pp. 53–59, Nov. 1973.)

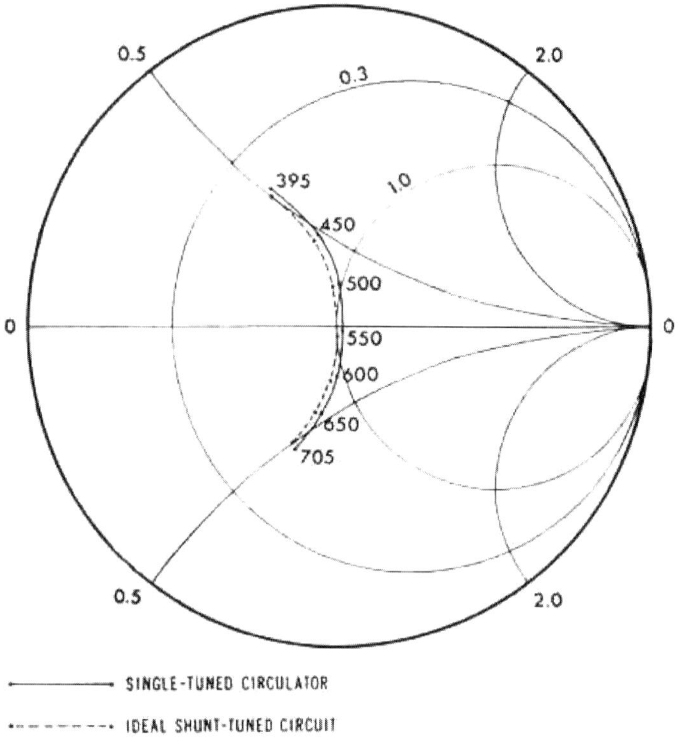

FIGURE 12.10 Schematic of degree-3 circulator.

12.8 DEGREE-3 LUMPED ELEMENT CIRCULATOR

A lumped element circulator with a degree-3 Chebyshev characteristic has also been realized. It consists of one series resonant circuit and one parallel resonant circuit connected in a first Cauer form in the manner indicated in Fig.12.11. The element values of this topology may be obtained from first principles or may be deduced by having recourse to a suitable lowpass to bandpass transformation. The results, based on the latter procedure, are

$$L_s = \left(\frac{2v^2}{v^2 + \frac{3}{4}}\right) R_0^2 C \tag{12.42}$$

$$C_s = 1/\omega^2 L_s \tag{12.43}$$

$$L_p = 1/\omega^2 C \tag{12.44}$$

$$C_p = C \tag{12.45}$$

$$v = (4/\varepsilon^2)^{1/4} - (4/\varepsilon^2)^{-1/4} \tag{12.46}$$

$$\varepsilon^2 = (r-1)^2/4r \tag{12.47}$$

r is the VSWR and ε is the ripple level.

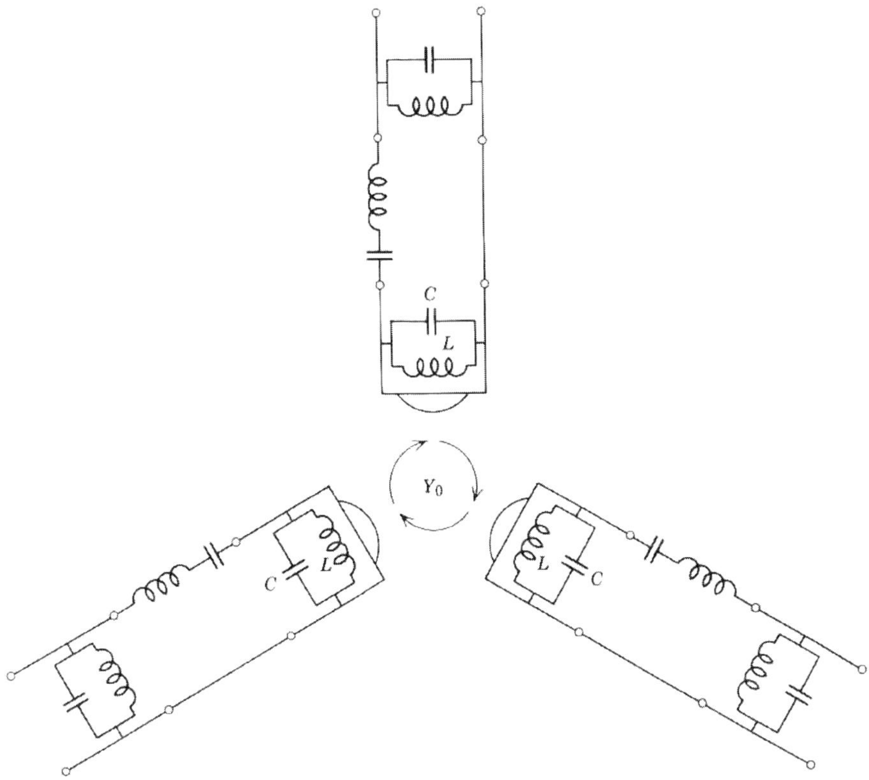

FIGURE 12.11 Complex gyrator of lumped element circulator. (Reproduced with permission from V. E. Dunn and R. W. Roberts, Miniature VHF and UHF circulators use lumped element design, *Microwaves*, pp. 46–47, Dec. 1963.)

In this situation the terminal impedance of the circulator is equal to that of the generator.

The bandwidth is increased by a factor of 4.25 for $r = 1.22$ and by a factor of 8.42 for $r = 1.065$.

12.9 QUASI LUMPED ELEMENT CIRCULATORS

A number of quasi lumped element circulators have also evolved over time in order either to overcome the restricted power rating of the lumped element circulator or to facilitate fabrication. One such geometry is a suitably magnetized undersized gyromagnetic region symmetrically loaded with quasi lumped element capacitors matched by an alternating impedance transformer. A similar or conventional distributed resonator matched by a lumped element circuit is another possibility. The general arrangement consists of a semidistributed resonator and a similar matching circuit. The equivalence between a lowpass filter circuit and the alternate line transformer is demonstrated separately.

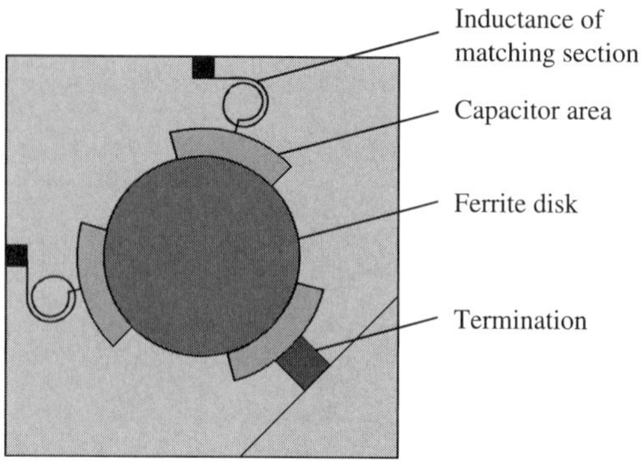

FIGURE 12.12 Miniature circulator using lumped element lowpass filter circuit. (Reproduced with permission from P. Barsony, Miniature strip-line circulators and isolators, International Conference On Microwave Ferrites, SMOLENICE, 1972.)

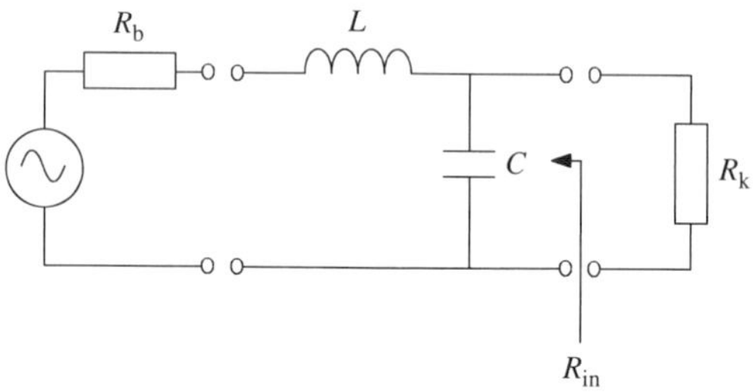

FIGURE 12.13 Equivalent circuit of miniature circulator using a lumped element lowpass filter circuit.

One example of a 1.0 GHz circulator built on a 1.0 mm by 25 × 25 mm alumina substrate is illustrated in Fig. 12.12. A ferrite disk with a magnetization $\mu_0 M_0$ equal to 0.1100 T is inserted into the substrate. The direct magnetic field intensity is not stipulated but a gyrator impedance of 180 Ω is given for the purpose of matching to the 50 Ω ports of the junction. The topology of the matching network is a lowpass lumped element impedance transformer. Its circuit is indicated in Fig. 12.13. The nature of this matching network may be understood by taking the reference terminals at the gyrator circuit instead of the more usual generator one and by separately making an equivalence between the lumped element circuit and the alternate line transformer.

12.10 LOWPASS MATCHING CIRCUIT

A lumped element matching circuit akin to the alternate line transformer is a lowpass filter. The required equivalence between the two is deduced by making a connection between an electrical short high impedance line and a series inductance and a low impedance line and a shunt capacitance.

The inductance of the high impedance line is determined in terms of its impedance by

$$\omega\mu_r L_0 = \beta(\mu_r \varepsilon_r) Z_{high} \tag{12.48}$$

The capacitance of the low impedance line is fixed by its impedance by

$$\omega\varepsilon_r C_0 = \beta(\mu_r \varepsilon_r)/Z_{low} \tag{12.49}$$

The phase constant is defined by

$$\beta(\mu_r \varepsilon_r) = k_0 \sqrt{\mu_r \varepsilon_r} \tag{12.50}$$

Z_{high} and Z_{low} may be obtained in terms of the overall specification of the device and the details of the gyrator circuit.

CHAPTER THIRTEEN

The Stripline Circulator Using a Gyromagnetic Planar Disk Resonator

13.1 INTRODUCTION

The classic stripline circulator consists of two ferrite planar disk resonators separated by a disk center conductor symmetrically coupled by three transmission lines. The gyromagnetic material is magnetized perpendicularly to the plane of the device by a static magnetic field. Its geometry is depicted in Fig. 13.1. An important property of this device is that this condition is met whenever all three ports are matched. For a three-port junction this requires two independent variables. Under certain simplifying conditions its adjustment can be described in terms of a standing wave pattern within the disk due to the interference of a pair of degenerate field patterns rotating in opposite directions. When the gyromagnetic junction is unmagnetized, the resonant frequencies of the two field patterns are identical. When it is magnetized, the degeneracy is removed, and the standing wave pattern is rotated. One circulation condition is established by operating between the two split frequencies. This requirement essentially fixes the radius of the gyromagnetic resonator. The second circulation condition is met by adjusting the splitting between the degenerate modes, until the standing wave pattern is rotated through 30°. From symmetry, port 3 is then situated at a null of the standing wave pattern and is therefore isolated. The junction then displays properties akin to that of a two-port transmission line resonator between the other two ports. This condition fixes the gyrotropy of the resonator. Figure 1.2a in Chapter 1 illustrates the field pattern in a demagnetized three-port junction while Fig. 1.2b depicts the same field pattern rotated through 30°.

The Stripline Circulator: Theory and Practice. By J. Helszajn

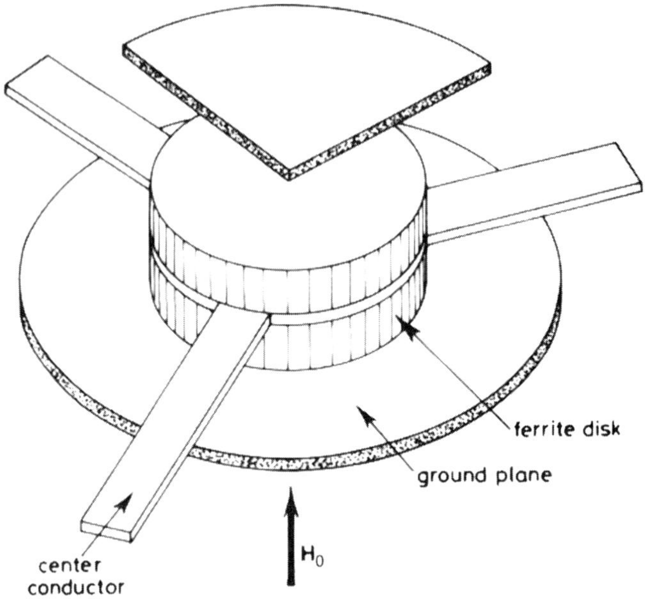

FIGURE 13.1 Schematic diagram of three-port stripline circulator.

There are in general six different ways in which circulator boundary conditions can be applied. This may be done in terms of the scattering, impedance, and admittance matrices and also in terms of the corresponding triplets of eigenvalues. In practice the choice is determined by the physical problem. Although it is always possible to construct the scattering matrix of any junction, it is not always possible to construct a corresponding impedance or admittance matrix. The method adopted in this chapter is based on the development of the impedance matrix.

13.2 MODE CHART OF GYROMAGNETIC DISK RESONATOR

A knowledge of the modes in a gyromagnetic resonator provides some valuable insight into the operation of junction circulators. If a magnetic field is applied along the axis of a planar disk, its two dominant counterrotating modes are no longer resonant at the same frequency. In addition, the standing pattern formed by the rotating field patterns is rotated within the disk. The derivation of the cutoff space and field patterns in a magnetized ferrite substrate proceeds in essentially the same way as for the isotropic problem, except that the permeability takes on a tensor form.

$$[\mu_r] = \begin{bmatrix} \mu & -j\kappa & 0 \\ j\kappa & \mu & 0 \\ 0 & 0 & 1 \end{bmatrix} \tag{13.1}$$

The magnetic field components H_r and H_ϕ in the plane of such a resonator are readily expressed in terms of the electric field E_z by having recourse to Maxwell's first curl equation, provided it is assumed that the alternating fields do not vary in the z-direction:

$$H_r = \omega\mu_0\mu_e j\left[\left(\frac{1}{r}\frac{\partial E_z}{\partial\phi}\right) - j\frac{\kappa}{\mu}\left(\frac{\partial E_z}{\partial \mathrm{r}}\right)\right] \tag{13.2}$$

$$H_\phi = \omega\mu_0\mu_e j\left[\left(\frac{\partial E_z}{\partial r}\right) + j\frac{\kappa}{\mu}\left(\frac{1}{r}\frac{\partial E_z}{\partial\phi}\right)\right] \tag{13.3}$$

where μ_e is the effective permeability of the gyromagnetic substrate and the other quantities have the usual meanings:

$$\mu_e = (\mu^2 - \kappa^2)/\mu \tag{13.4}$$

One solution for E_z that satisfies the wave equation is

$$E_z = A_n J_n(k_e r)\exp(jn\phi) \tag{13.5}$$

A_n is now different for n positive and negative. Combining the preceding relationships gives

$$H_r = -A_n\zeta_e\zeta_0\left[\frac{nJ_n(k_e r)}{k_e r} - \left(\frac{\kappa}{\mu}\right)J_n'(k_e r)\right]\exp(jn\phi) \tag{13.6}$$

$$H_\phi = -jA_n\zeta_e\zeta_0\left[J_n'(k_e r) - \left(\frac{\kappa}{\mu}\right)\frac{nJ_n(k_e r)}{k_e r}\right]\exp(jn\phi) \tag{13.7}$$

The wavenumber is defined by

$$k_e = \omega\sqrt{\varepsilon_0\varepsilon_r\mu_0\mu_e} \tag{13.8}$$

The relative wave admittance is given separately by

$$\zeta_e = 1/\eta_e = \sqrt{\varepsilon_r/\mu_e} \tag{13.9a}$$

The absolute admittance of free space is

$$\zeta_0 = 1/\eta_0 = \sqrt{\varepsilon_0/\mu_0} \tag{13.9b}$$

The cutoff space under consideration is now established by imposing a magnetic wall boundary condition at $r = R$:

$$H_\phi(R) = 0 \tag{13.10}$$

The ensuing characteristic equation fixes both the cutoff space and poles of the problem region:

$$J_n'(k_e R) - \left(\frac{\kappa}{\mu}\right)\frac{nJ_n(k_e R)}{k_e R} = 0 \tag{13.11}$$

Thus there are two roots for the planar magnetized disk resonator:

$$(k_e R)_{-nj} \tag{13.12a}$$

$$(k_e R)_{+nj} \tag{13.12b}$$

These define two resonant frequencies:

$$\omega_{+nj}\sqrt{\varepsilon_0 \varepsilon_r \mu_0 \mu_e}R = (k_e R)_{+nj} \tag{13.13}$$

$$\omega_{-nj}\sqrt{\varepsilon_0 \varepsilon_r \mu_0 \mu_e}R = (k_e R)_{-nj} \tag{13.14}$$

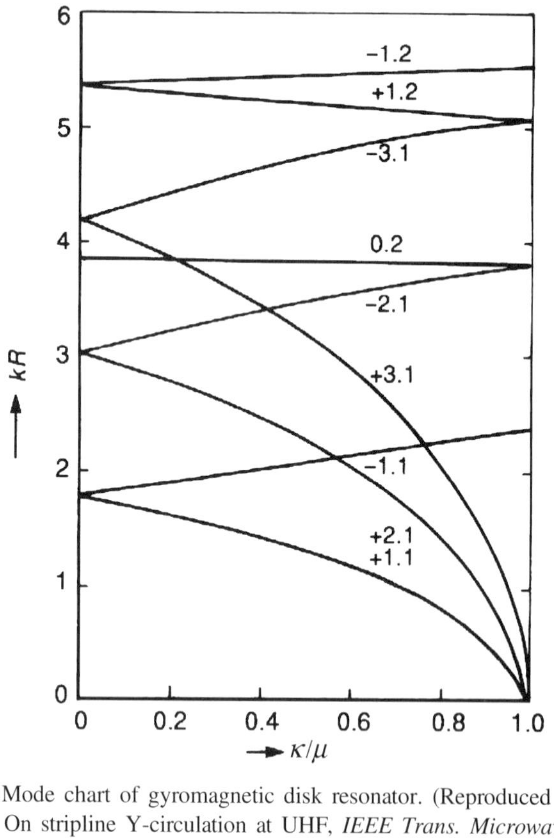

FIGURE 13.2 Mode chart of gyromagnetic disk resonator. (Reproduced with permission from H. Bosma, On stripline Y-circulation at UHF, *IEEE Trans. Microwave Theory Tech.*, Vol. MTT-12, pp. 61–72, 1964.)

The resonant frequency of the isotropic disk, for which κ/μ is zero, lies between the two split frequencies:

$$\omega_{nj}\sqrt{\varepsilon_0\varepsilon_r\mu_0\mu_e}R = (kR)_{nj} \tag{13.15}$$

Figure 13.2 indicates the mode chart of such a gyromagnetic disk resonator. One experimental chart is indicated in Fig. 13.3. The standing wave patterns of the first three modes in an isotropic disk resonator are illustrated in Fig. 13.4.

If the gyrotropy is small the difference between a pair of resonances is determined by

$$(\Delta k_e R)_{nj} \approx \frac{2n(k_e R)_{nj}}{(k_e R)_{nj}^2 - n^2} \cdot \left(\frac{\kappa}{\mu}\right) \tag{13.16}$$

This equation states that the splitting of a pair of resonant frequencies is proportional to the gyrotropy for κ/μ small.

Another important result of the arrangement under consideration is that the field patterns associated with counterrotating field patterns in a ferrite disk are rotated

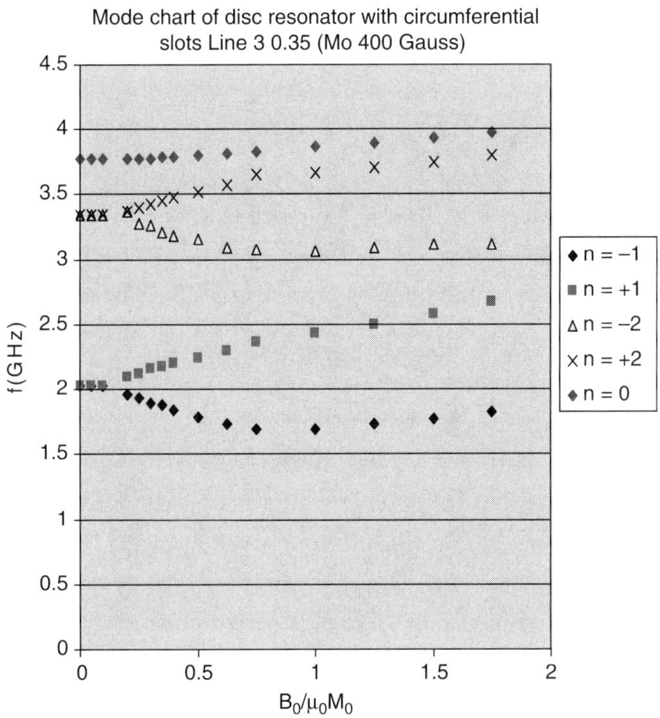

FIGURE 13.3 Experimental mode chart of below-resonance stripline circulator.

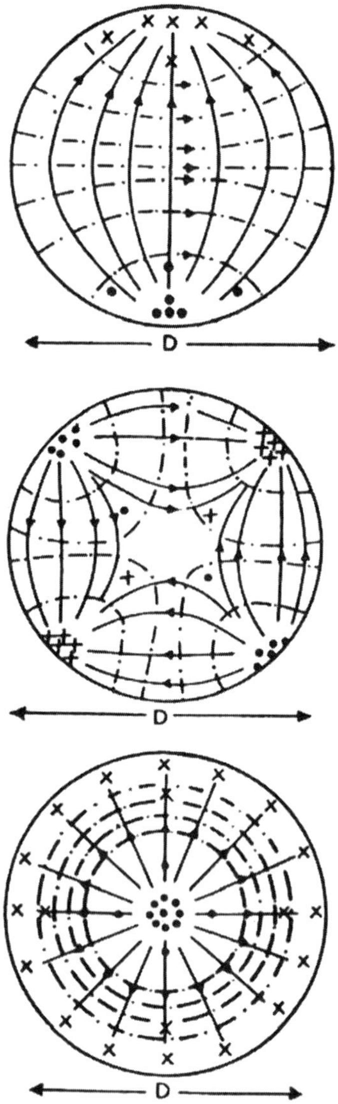

FIGURE 13.4 Standing wave patterns in planar disk resonator.

as the junction is magnetized. The direction in which this takes place is determined by that of the direct magnetic field intensity. If the dominant mode pattern is rotated by 30°, one of the ports is completely decoupled, and transmission occurs between the other two ports. In this configuration the three-port junction behaves as a circulator. This situation has already been noted in Chapter 1. The Bessel functions met in this work are defined for calculation purposes in Chapter 9.

13.3 IMPEDANCE MATRIX OF THREE-PORT JUNCTION CIRCULATOR

The derivation of the impedance matrix of the junction circulator requires a knowledge of the electromagnetic problem. The original boundary conditions adopted in the description of a planar circuit are that the magnetic field is a constant over the width of each microstrip or stripline and zero elsewhere. The electric fields are taken as the average values of E_z over the ports at $r = R$ and are assumed arbitrary elsewhere. The derivation of the required result starts with a statement of the magnetic fields of the problem region.

$$\begin{aligned} -\psi < \phi < \psi, &\quad H_\phi = H_1 \\ -120^\circ - \psi < \phi < \psi - 120^\circ, &\quad H_\phi = H_2 \\ 120^\circ - \psi < \phi < \psi + 120^\circ, &\quad H_\phi = H_3 \end{aligned}$$

elsewhere, $H_\phi = 0$.

ψ is known as the coupling angle of the circulator:

$$\sin\psi = W/2R \tag{13.17}$$

W is the width of the stripline at a typical port of the junction. It is assumed that the center conductor thickness t is zero. The schematic of the configuration studied is shown in Fig. 13.5.

The amplitude constant A_n in the description E_z is determined by expanding the magnetic field on the boundary into a Fourier series with respect to ϕ:

$$H_\phi = \sum_{n=-\infty}^{\infty} b_n \exp(jn\phi) \tag{13.18}$$

where

$$b_n = \frac{\sin n\psi}{\pi n}\left[H_1 + H_2 \exp\left(\frac{j2\pi n}{3}\right) + H_3 \exp\left(\frac{-j2\pi n}{3}\right)\right] \tag{13.19}$$

In obtaining this result the problem region in the construction of b_n has been broken up to correspond to the various disjointed sections of the boundary under consideration. The constant A_n can now be evaluated by comparing the two descriptions for H_ϕ at $r = R$.

The result is

$$A_n = \left(\frac{j\eta_e\eta_0 \sin\psi}{\pi n}\right) \frac{\left[H_1 + H_2 \exp\left(\frac{j2\pi n}{3}\right) + H_3 \exp\left(\frac{-j2\pi n}{3}\right)\right]}{\left[J_n'(k_e R) - \left(\frac{\kappa}{\mu}\right)\frac{n J_n(k_e R)}{k_e R}\right]} \tag{13.20}$$

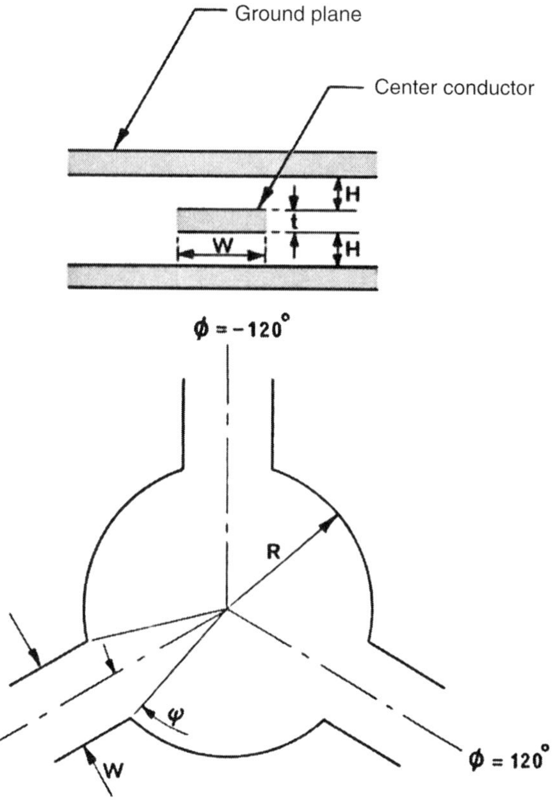

FIGURE 13.5 Schematic diagram of junction circulator using disk resonator.

The derivation of the required impedance matrix of the junction continues by imposing the boundary condition of the electric fields at the ports.

$$E_1 = \left(\frac{1}{2\Psi}\right) \int_{-\Psi}^{\Psi} E_z d\phi = \eta_{11} H_1 + \eta_{12} H_2 + \eta_{13} H_3 \tag{13.21a}$$

$$E_2 = \left(\frac{1}{2\Psi}\right) \int_{-2\pi/3-\Psi}^{-2\pi/3+\Psi} E_z d\phi = \eta_{13} H_1 + \eta_{11} H_2 + \eta_{12} H_3 \tag{13.21b}$$

$$E_3 = \left(\frac{1}{2\Psi}\right) \int_{2\pi/3-\Psi}^{2\pi/3+\Psi} E_z d\phi = \eta_{12} H_1 + \eta_{13} H_2 + \eta_{11} H_3 \tag{13.21c}$$

The wave impedance matrix of the circulator is therefore defined by

$$\overline{\eta} = \begin{bmatrix} \eta_{11} & \eta_{12} & \eta_{13} \\ \eta_{13} & \eta_{11} & \eta_{12} \\ \eta_{12} & \eta_{13} & \eta_{11} \end{bmatrix} \tag{13.22}$$

where

$$\eta_{11} = \sum_{n=-\infty}^{\infty} \frac{\eta_n}{3} \tag{13.23a}$$

$$\eta_{12} = \sum_{n=-\infty}^{\infty} \frac{\eta_n \exp(j2\pi n/3)}{3} \tag{13.23b}$$

$$\eta_{13} = \sum_{n=-\infty}^{\infty} \frac{\eta_n \exp(-j2\pi n/3)}{3} \tag{13.23c}$$

A typical pole is described by

$$\eta_n = \frac{j3\eta_e\eta_0\psi}{\pi}\left(\frac{\sin(n\psi)}{n\psi}\right)^2 \left[\frac{J_n'(k_eR)}{J_n(k_eR)} - \left(\frac{\kappa}{\mu}\right)\frac{n}{k_eR}\right]^{-1} \tag{13.24}$$

The impedance entering into the solution so far is that of the wave impedance of the junction. Its characteristic impedance is obtained by replacing the free space impedance η_0 by the characteristic impedance Z_r of the striplines.

$$Z_r = 30\pi \ln\left(\frac{W + t + 2H}{W + t}\right) \tag{13.25}$$

W is the width of a typical coupling line, H is the resonator thickness of each half-space, and t is the thickness of the center conductor. W, H, and t are in meters.

The preceding relationships are both necessary and sufficient for the description of the impedance matrix of the junction.

13.4 EIGENVALUE SOLUTION

A detailed scrutiny of the open-circuited parameters indicates that each may be broken up into linear combinations of in-phase and counterrotating eigenvalues or poles in keeping with Chapter 6. The relationships between the former two quantities are

$$\eta_{11} = \frac{\eta^0 + \eta^+ + \eta^-}{3} \tag{13.26a}$$

$$\eta_{12} = \frac{\eta^0 + \alpha\eta^+ + \alpha^2\eta^-}{3} \tag{13.26b}$$

$$\eta_{13} = \frac{\eta^0 + \alpha^2\eta^+ + \alpha\eta^-}{3} \tag{13.26c}$$

where $\alpha = \exp(-j2\pi/3)$.

The triplet of eigenvalues of the problem region may be revealed by constructing the classic linear combinations of the open-circuit parameters. This gives

$$\eta^0 = \sum_{n=-\infty}^{\infty} \eta_n \left[1 + \exp\left(-j\frac{2\pi n}{3}\right) + \exp\left(j\frac{2\pi n}{3}\right)\right] \tag{13.27a}$$

$$\eta^+ = \sum_{n=-\infty}^{\infty} \eta_n \left[1 + \alpha\exp\left(-j\frac{2\pi n}{3}\right) + \alpha^2\exp\left(j\frac{2\pi n}{3}\right)\right] \tag{13.27b}$$

$$\eta^0 = \sum_{n=-\infty}^{\infty} \eta_n \left[1 + \alpha^2\exp\left(-j\frac{2\pi n}{3}\right) + \alpha\exp\left(j\frac{2\pi n}{3}\right)\right] \tag{13.27c}$$

The in-phase impedance eigenvalue η^0 is zero for all n except $n = 0, \pm 3, \pm 6, \ldots$ Likewise the rotating eigenvalue η^+ is zero for all n except $n = 1, -2, 4, \ldots$ and the other rotating eigenvalue η^- is zero except for $n = -1, 2, -4, \ldots$. Thus

$$\eta^0 = \sum \eta_n, \quad n = 0, \pm 3, \pm 6, \ldots \tag{13.28a}$$

$$\eta^+ = \sum \eta_n, \quad n = 1, -2, 4, \ldots \tag{13.28b}$$

$$\eta^- = \sum \eta_n, \quad n = -1, 2, -4, \ldots \tag{13.28c}$$

Each of the 3 one-port reactances or eigenvalues η^0, η^+, and η^- appearing in the descriptions of the open-circuit parameters of a junction circulator may be expanded in terms of its poles in a first Foster or partial fractions form in the manner indicated in Fig. 13.6. This means that a knowledge of a typical pole is sufficient for the construction of any of the triplet eigenvalues.

A typical eigenvalue may also be directly deduced by replacing the arbitrary boundary conditions at the ports used so far by ones that coincide with each

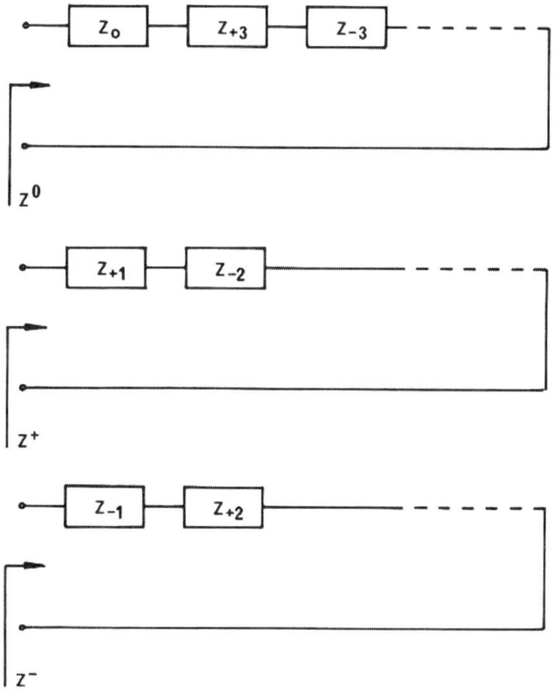

FIGURE 13.6 First Foster form realization of in-phase and counterrotating eigen-networks.

eigenvector one at a time. A typical pole may be obtained by constructing the self-impedance of the junction at port 1 with ports 2 and 3 open-circuited.

The scattering matrix of the circulator may be constructed once the corresponding eigenvalues are at hand. These are related to the impedance eigenvalues by the classic bilinear relationship between the two.

$$s_0 = \frac{\sum \eta_n - \eta_0}{\sum \eta_n + \eta_0}, \quad n = 0, \pm 3, \pm 6, \ldots \tag{13.29a}$$

$$s_+ = \frac{\sum \eta_n - \eta_0}{\sum \eta_n + \eta_0}, \quad n = 1, -2, 4, \ldots \tag{13.29b}$$

$$s_- = \frac{\sum \eta_n - \eta_0}{\sum \eta_n + \eta_0}, \quad n = -1, 2, -4, \ldots \tag{13.29c}$$

The use of these identities avoids the need to invert the impedance matrix in order to construct the scattering matrix.

13.5 COMPLEX GYRATOR CIRCUIT

The adjustment of the three-port junction circulator is often posed in terms of its midband cutoff number and its gyrator resistance or conductance. These two conditions coincide with the real and imaginary parts of the complex gyrator admittance of the device. This quantity is defined in terms of its open-circuit parameters by setting $V_3 = I_3 = 0$. The result is

$$Z_{in} = Z_{11} - Z_{12}^2/Z_{13} \tag{13.30}$$

The one-port equivalent circuit of a junction circulator using a resonator with a magnetic wall boundary is normally a shunt resonator in parallel with the gyrator conductance of the device. It is therefore usual to apply the boundary conditions in terms of its admittance.

$$Y_{in} = 1/Z_{in} \tag{13.31}$$

The two classic circulator boundary conditions are therefore given by

$$\mathrm{Im}(Y_{in}) = 0 \tag{13.32a}$$

$$\mathrm{Re}(Y_{in}) = Y_0 \tag{13.32b}$$

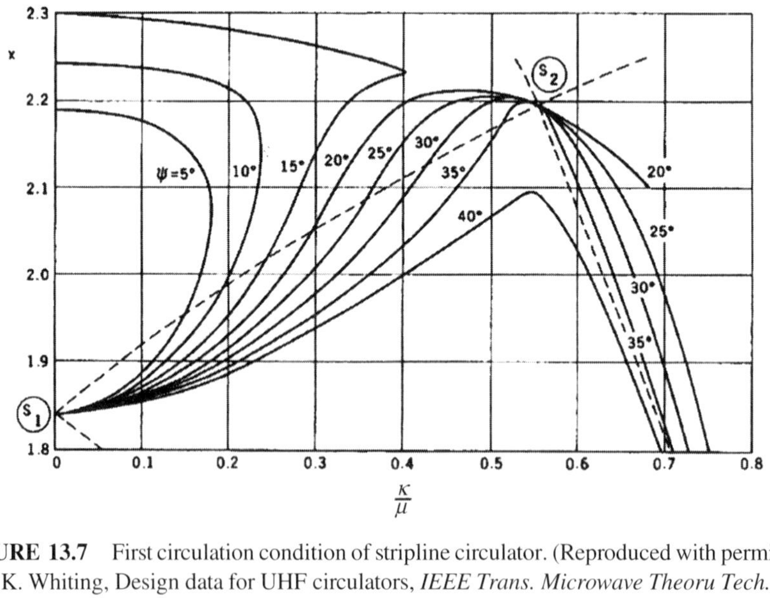

FIGURE 13.7 First circulation condition of stripline circulator. (Reproduced with permission from K. Whiting, Design data for UHF circulators, *IEEE Trans. Microwave Theoru Tech.*, Vol. MTT-15, pp. 195–198, Mar. 1967.)

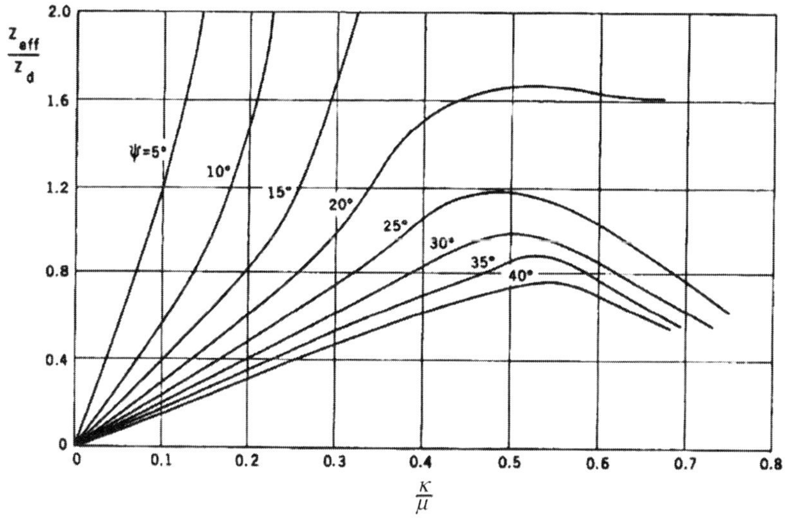

FIGURE 13.8 Second circulation condition of stripline circulator. (Reproduced with permission from K. Whiting, Design data for UHF circulators, *IEEE Trans. Microwave Theory Tech.*, Vol. MTT-15, pp. 195–198, Mar. 1967.)

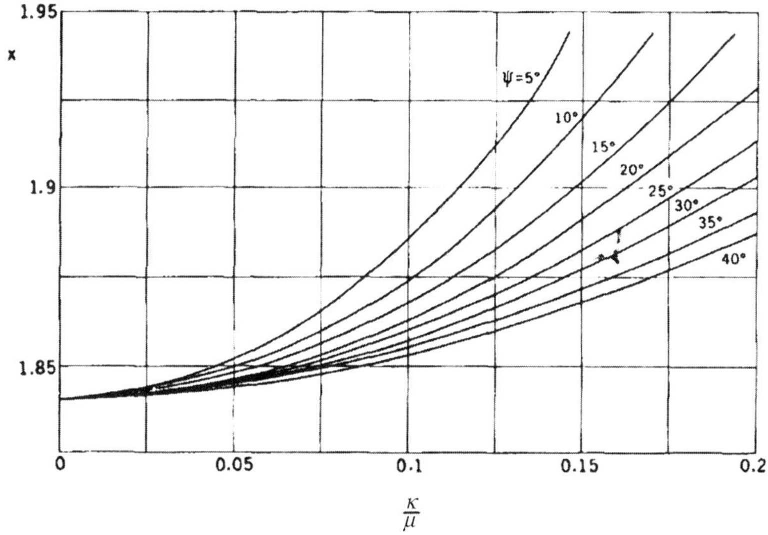

FIGURE 13.9 First circulation condition of weakly magnetized stripline circulator. (Reproduced with permission from K. Whiting, Design data for UHF circulators, *IEEE Trans. Microwave Theroy Tech.*, Vol. MTT-15, pp. 195–198, Mar. 1967.)

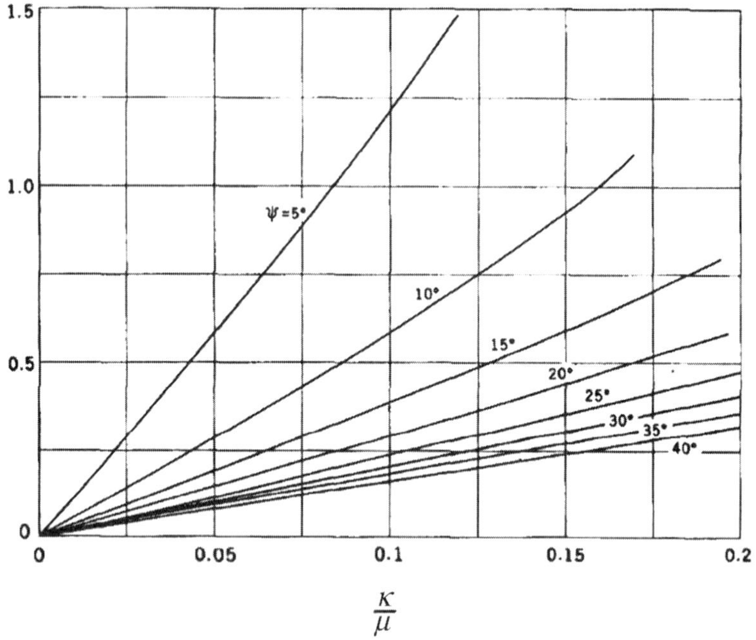

FIGURE 13.10 Second circulation condition of weakly magnetized stripline circulator. (Reproduced with permission from K. Whiting, Design data for UHF circulators, *IEEE Trans. Microwave Theory Tech.*, Vol. MTT-15, pp. 195–198, Mar. 1967.)

The first of these equations determines the center frequency of the junction whereas the second gives the gyrator conductance.

The field of solutions displayed by the circulator is at first sight quite complicated. In general it may be divided into a very weakly magnetized space, a weakly magnetized space, a moderately magnetized space, and a strongly magnetized space. It also contains a very strongly or edge mode solution in the case where the gyrotropy exceeds unity. Graphical solutions of these two conditions are depicted in Figs. 13.7 and 13.8. Expanded versions of the same solutions in the vicinity of the origin are indicated in Figs. 13.9 and 13.10.

13.6 SINGLE POLE CIRCULATION SOLUTION

The frequency responses of the first seven poles of a weakly magnetized junction are illustrated in Fig. 13.11. One feature of this illustration is that all degenerate poles except the symmetric ones are split by the gyromagnetic effect. Also, one equivalent circuit for the dominant in-phase pole is an open-circuited stub and the equivalent circuits for the dominant pair of split counterrotating ones are short-circuited ones as asserted.

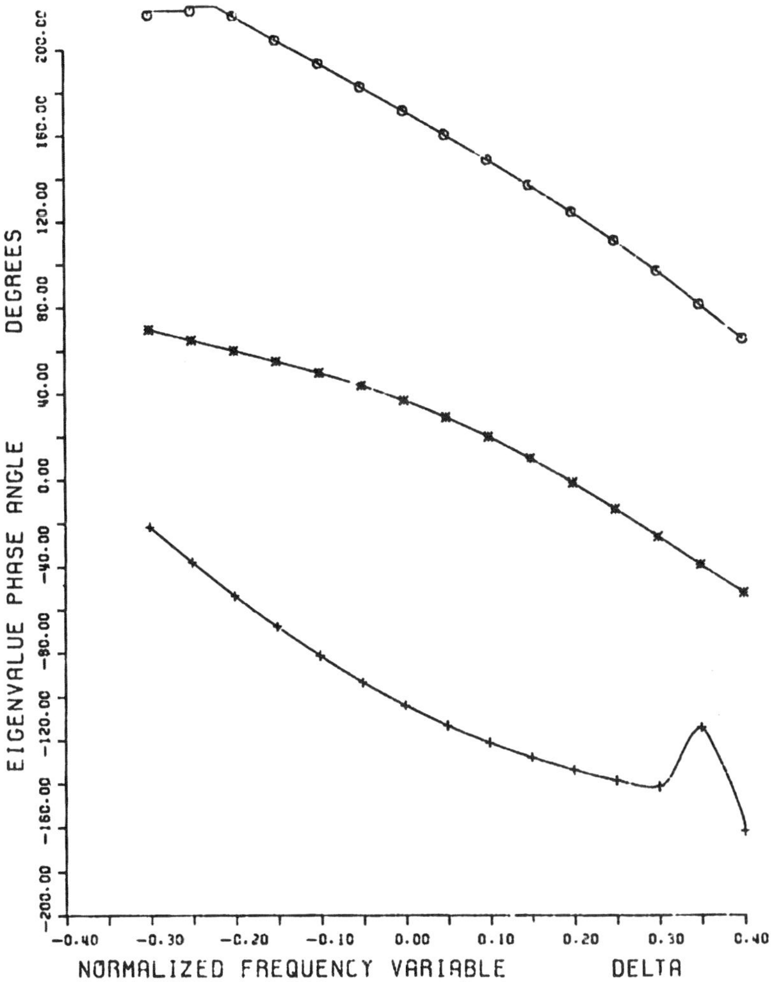

FIGURE 13.11 Frequency response of eigenvalues of three-port circulator ($\kappa = 0.30$).

Another classic feature of this result is that the in-phase eigen-network may sometimes be neglected compared to those of the split counterrotating ones.

One plausible solution in the vicinity of the first dominant pair of split modes is obtained by retaining single poles for the counterrotating modes and idealizing the in-phase mode by an impedance zero in the description of the complex gyrator circuit.

$$Z^0 \approx 0 \tag{13.33a}$$

$$Z^+ = Z_{+1} \tag{13.33b}$$

$$Z^- = Z_{-1} \tag{13.33c}$$

The first of these three conditions is consistent with the assumption employed in deriving the approximate nature of the complex gyrator admittance of the circulator.

$$Y_{\text{in}} = \left(\frac{Y_+ + Y_-}{2}\right) - j\sqrt{3}\left(\frac{Y_+ - Y_-}{2}\right) \tag{13.34}$$

The two circulation conditions of the device are now deduced by evaluating this quantity. Making use of the relationship between impedance and admittance on a one-port readily gives the required result.

$$Y_{\text{in}} = \frac{\pi\zeta_{\text{e}}Y_{\text{r}}}{\sqrt{3}(k_{\text{e}}R)\psi}\left(\frac{\psi}{\sin\psi}\right)^2\left(\frac{\kappa}{\mu}\right) - \frac{j\pi\zeta_{\text{e}}Y_{\text{r}}}{3\psi}\left(\frac{\psi}{\sin\psi}\right)^2\left[\frac{J_1'(k_{\text{e}}R)}{J_1(k_{\text{e}}R)}\right] \tag{13.35}$$

The first circulation condition is now obtained by setting the imaginary part of Y_{in} to zero:

$$J_1'(k_{\text{e}}R) = 0 \tag{13.36}$$

The first root is given by

$$(k_{\text{e}}R)_{1,1} = 1.84 \tag{13.37}$$

The second circulation condition is satisfied by setting the real part of Y_{in} to Y_0. The result is

$$Y_0 = \frac{\pi\zeta_{\text{e}}Y_{\text{r}}}{\sqrt{3}(1.84)\psi}\left(\frac{\psi}{\sin\psi}\right)^2\left(\frac{\kappa}{\mu}\right) \tag{13.38}$$

13.7 FREQUENCY RESPONSE OF WEAKLY MAGNETIZED CIRCULATOR

A complete characterization of a network also requires a knowledge of its frequency response. One possibility is to plot the real and imaginary parts of the complex gyrator circuit. Another is to form its susceptance slope parameter (B') and its loaded Q-factor (Q_{L}). The former quantity is defined by

$$B' = \frac{\omega_0}{2}\frac{dB}{d\omega}\bigg|_{\omega=\omega_0}. \tag{13.39}$$

The latter quantity is specified by

$$Q_{\text{L}} = B'/G \tag{13.40}$$

G is the real part of the complex gyrator admittance.

The evaluation of B' is facilitated by noting the following identity, which is in the vicinity of a typical root.

$$\frac{J_n'(x)}{J_n(x)} = \frac{1}{x}\left(\frac{x - x_{mn}}{x_{mn}}\right)\left(x_{mn}^2 - n^2\right)$$

n is the order of the Bessel function, m is its root.

The required results are given in terms of the original variables by having recourse to Eq. (13.35):

$$B' = \frac{\zeta_e Y_r}{\psi}\left(\frac{\psi}{\sin\psi}\right)^2\left[\frac{(k_e R)_{1,1}^2 - 1}{2(k_e R)_{1,1}}\right] \tag{13.41}$$

and

$$G = \frac{\pi\zeta_e Y_r}{\sqrt{3}(k_e R)\,\psi}\left(\frac{\psi}{\sin\psi}\right)^2\left(\frac{\kappa}{\mu}\right) \tag{13.42}$$

The loaded Q-factor is then given by

$$Q_L = \frac{1}{\sqrt{3}}\left[\frac{(k_e R)_{1,1}^2 - 1}{2}\right]\cdot\left(\frac{\mu}{\kappa}\right) \tag{13.43}$$

Two classic conditions met in the description of this class of device are now obtained by writing the preceding equation in terms of the two split frequencies of the gyromagnetic resonator. The required relationships are readily obtained by having recourse to Eq. (13.16):

$$\frac{1}{\sqrt{3}Q_L} = \frac{\omega_{+1,1} - \omega_{-1,1}}{\omega_{1,1}} \tag{13.44}$$

and

$$G = \sqrt{3}B'\left(\frac{\omega_{+1,1} - \omega_{-1,1}}{\omega_{1,1}}\right) \tag{13.45}$$

The normalized bandwidth $(2\delta_0)$ of this sort of circuit is related to its loaded Q-factor (Q_L) and the voltage standing wave ratio (VSWR) at its bandedges by

$$2\delta_0 = \frac{(\mathrm{VSWR} - 1)}{Q_L\sqrt{\mathrm{VSWR}}} \tag{13.46}$$

where

$$2\delta_0 = \frac{\omega_2 - \omega_1}{\omega_0} \tag{13.47}$$

$\omega_{1,2}$ are the bandedge frequencies, ω_0 is the midband frequency.

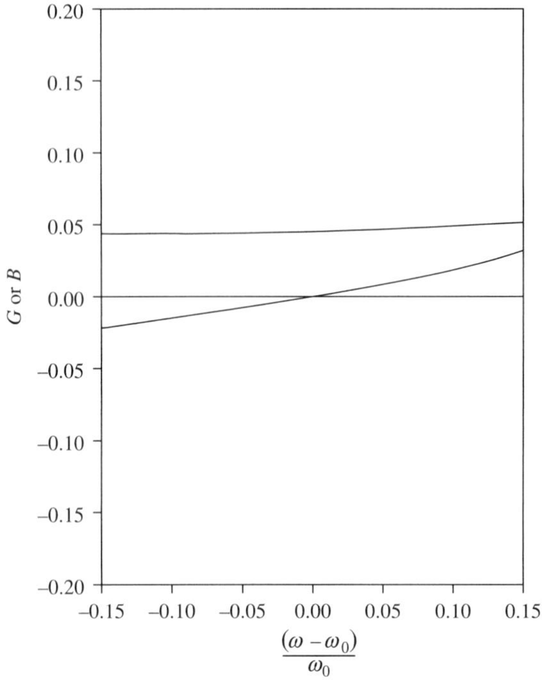

FIGURE 13.12 Frequency response of complex gyrator admittance ($\kappa = 0.30$).

13.8 VERY WEAKLY MAGNETIZED GYROMAGNETIC RESONATOR

The validity of the closed form description of a junction circulator employing a very weakly magnetized disk resonator is the topic of this section. It may be verified by comparing it with some numerical calculations using the first seven poles of the problem region. Such a calculation is summarized for parametric values of the coupling angle in Chapter 28. It indicates that the quality factor of the junction is independent of the coupling angle as long as the gyrotropy is bracketed between 0 and 0.25. It also suggests that the closed form description shows significant deterioration when the gyrotropy equals 0.35. The quality factor displayed by such a junction with a gyrotropy equal to 0.25 is about 2.30. This value is compatible with the synthesis of quarter-wave coupled junction circulators with 20 dB ripple levels over some 25% or with 26 dB ripple levels over 18%. Figure 13.12 illustrates one typical frequency response.

CHAPTER FOURTEEN

Green's Function Description of Junction Circulator

14.1 INTRODUCTION

The open-circuit parameters of m-port gyromagnetic circuits may be formed in one of a number of ways as is now understood. One classic means of doing so is to employ the Green's function of the problem region. This quantity may be constructed without difficulty once the eigenvectors and eigenvalues of the decoupled circuit are established. The latter quantities can always be deduced by having recourse to the finite element method. One gyromagnetic geometry for which a Green's function exists in closed form is that of a disk planar resonator. It has been employed by Bosma in his development of the theory of the classic three-port circulator. The procedure starts by defining the Green's matrix of the circuit in terms of the eigenfunctions of the homogeneous gyromagnetic circuit. It continues by calculating the electric fields at the ports in terms of this quantity. The construction of a typical open-circuit parameter is then obtained by forming the product of the line integrals of the fields over the ports under consideration. The calculation of the eigensolutions of an irregular gyromagnetic circuit using the finite element method is dealt with in the next chapter.

14.2 GREEN'S FUNCTION MATRIX OF JUNCTION CIRCULATOR

Before proceeding with the calculation of the open-circuit parameters of a gyromagnetic circuit using the Green's method, it is worthwhile to define a Green's matrix. It is also desirable to seek some possible relationships between its entries. For any

The Stripline Circulator: Theory and Practice. By J. Helszajn

three-port circuit this type of matrix may be written

$$\overline{\mathbf{G}}(\phi \mid \phi_0) = \begin{bmatrix} G_{11} & G_{12} & G_{13} \\ G_{21} & G_{22} & G_{23} \\ G_{31} & G_{32} & G_{33} \end{bmatrix} \tag{14.1}$$

The quantity ϕ refers to the observation point and ϕ_0 to the source point, both of which are assumed in this work to reside on the periphery of the circuit. Figure 14.1 indicates the situation.

A classic device based on a gyromagnetic resonator symmetrically coupled by three transmission lines is the junction circulator. If the ports are located at $-\pi/3$, $\pi/3$, and π in the manner indicated in Fig. 14.2, then the entries of the Green's

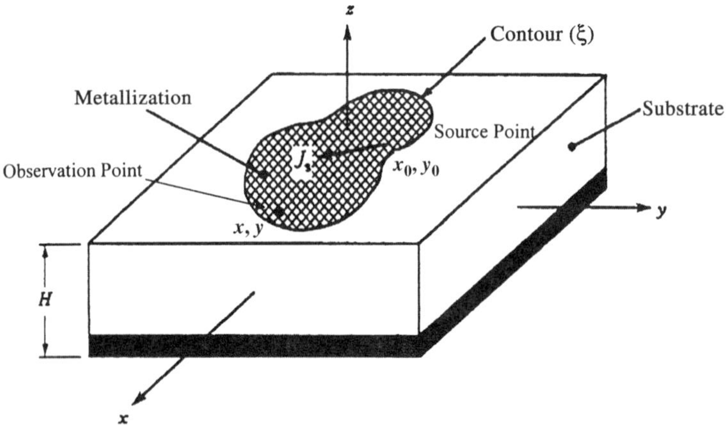

FIGURE 14.1 Definition of source and observation points in the definition of the Green's function.

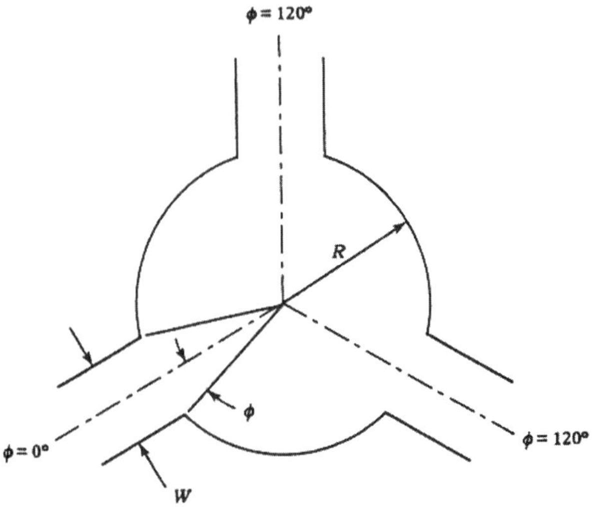

FIGURE 14.2 Schematic diagram of three-port junction circulator.

matrix are defined by

$$G_{11} = G\left(-\frac{\pi}{3}\middle| -\frac{\pi}{3}\right) \tag{14.2a}$$

$$G_{12} = G\left(-\frac{\pi}{3}\middle| \frac{\pi}{3}\right) \tag{14.2b}$$

$$G_{13} = G\left(-\frac{\pi}{3}\middle| \pi\right) \tag{14.2c}$$

$$G_{21} = G\left(\frac{\pi}{3}\middle| -\frac{\pi}{3}\right) \tag{14.2d}$$

$$G_{22} = G\left(\frac{\pi}{3}\middle| \frac{\pi}{3}\right) \tag{14.2e}$$

$$G_{23} = G\left(\frac{\pi}{3}\middle| \pi\right) \tag{14.2f}$$

$$G_{31} = G\left(\pi\middle| -\frac{\pi}{3}\right) \tag{14.2g}$$

$$G_{32} = G\left(\pi\middle| \frac{\pi}{3}\right) \tag{14.2h}$$

$$G_{33} = G\left(\pi\middle| \pi\right) \tag{14.2i}$$

In the absence of dissipation

$$G(\phi_0|\phi) = -G^*(\phi_0|\phi) \tag{14.3}$$

The cyclic symmetry of a three-port circulator requires the following additional condition to be fulfilled:

$$G\left(\phi + \frac{2\pi}{3}\middle| \phi_0 + \frac{2\pi}{3}\right) = G(\phi|\phi_0) \tag{14.4}$$

Making use of the preceding identities permits the Green's matrix of a symmetric three-port circulator to be described using only two entries,

$$\overline{\mathbf{G}}(\phi\,|\,\phi_0) = \begin{bmatrix} G_{11} & G_{12} & -G_{12}^* \\ -G_{12}^* & G_{11} & G_{12} \\ G_{12} & -G_{12}^* & G_{11} \end{bmatrix} \tag{14.5}$$

It is also useful in this type of problem to identify the eigenvalues of the Green's matrix. These are related to the entries of the matrix by

$$G^0(\phi|\phi_0) = G_{11}(\phi|\phi_0) + G_{12}(\phi|\phi_0) + G_{13}(\phi|\phi_0) \tag{14.6a}$$

$$G^+(\phi|\phi_0) = G_{11}(\phi|\phi_0) + \alpha G_{12}(\phi|\phi_0) + \alpha^2 G_{13}(\phi|\phi_0) \tag{14.6b}$$

$$G^-(\phi|\phi_0) = G_{11}(\phi|\phi_0) + \alpha^2 G_{12}(\phi|\phi_0) + \alpha G_{13}(\phi|\phi_0) \tag{14.6c}$$

where

$$\alpha = \exp(j(2\pi/3)) \tag{14.7a}$$

$$\alpha^2 = \exp(j(4\pi/3)) \tag{14.7b}$$

The entries of the Green's matrix are therefore linear combinations of its eigenvalues.

$$G_{11}(\phi\,|\,\phi_0) = \frac{G^0(\phi\,|\,\phi_0) + G^+(\phi\,|\,\phi_0) + G^-(\phi\,|\,\phi_0)}{3} \tag{14.8a}$$

$$G_{12}(\phi\,|\,\phi_0) = \frac{G^0(\phi\,|\,\phi_0) + \alpha G^+(\phi\,|\,\phi_0) + \alpha^2 G^-(\phi\,|\,\phi_0)}{3} \tag{14.8b}$$

$$G_{13}(\phi\,|\,\phi_0) = \frac{G^0(\phi\,|\,\phi_0) + \alpha^2 G^+(\phi\,|\,\phi_0) + \alpha G^-(\phi\,|\,\phi_0)}{3} \tag{14.8c}$$

$G^0(\phi|\phi_0)$ and $G^{\pm}(\phi|\phi_0)$ are related to the so-called in-phase and counterrotating eigen-networks of the circuit. A typical pole of the Green's function is then

$$G_n(\phi\,|\,\phi_0) = \tfrac{1}{3}[G^0(\phi\,|\,\phi_0) + G^+(\phi\,|\,\phi_0) + G^-(\phi\,|\,\phi_0)] \tag{14.9}$$

The boundary condition in a gyromagnetic circuit with a magnetic wall is given by

$$\frac{\partial G(\phi\,|\,\phi_0)}{\partial n} - j\frac{\kappa}{\mu}\frac{\partial G(\phi\,|\,\phi_0)}{\partial t} = 0 \tag{14.10}$$

For a reciprocal and symmetrical network

$$G_{13}(\phi\,|\,\phi_0) = G_{12}(\phi\,|\,\phi_0) \tag{14.11}$$

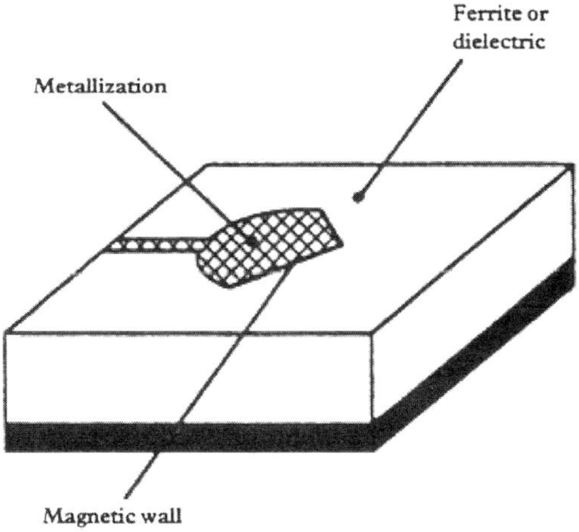

FIGURE 14.3 In-phase eigen-network.

and

$$G^{+}(\phi\,|\,\phi_0) = G^{-}(\phi\,|\,\phi_0) \tag{14.12}$$

The eigen-networks of the reciprocal circuit are shown in Figs. 14.3 and 14.4. The direct evaluations of these geometries provide an independent check on any calculations.

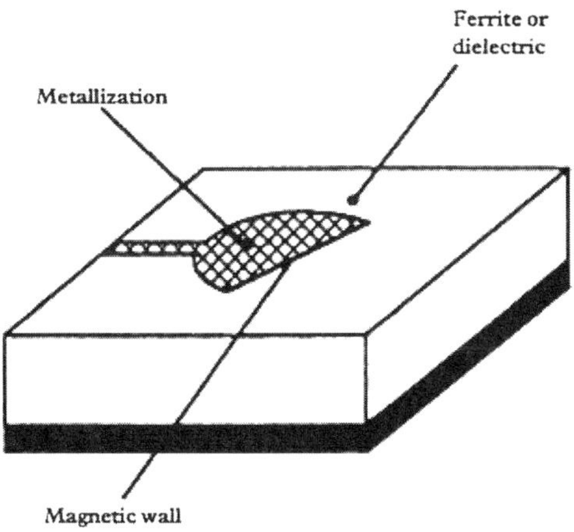

FIGURE 14.4 Degenerate counterrotating eigen-network.

14.3 WAVE IMPEDANCE MATRIX OF THREE-PORT CIRCULATOR

The open-circuit parameters of any gyromagnetic circuit may be calculated once its Green's function is available. Such a function exists for a gyromagnetic disk in closed form. The intention of this section is to deduce the open-circuit parameters of a symmetrical three-port junction circulator using such a resonator. This is done as a preamble to dealing with the general problem in Chapter 15. The electric field midway on any port at the observation point due to a distribution of unit point source is given by the superposition integral

$$E_z(R, \phi) = \int_C G(\phi \,|\, \phi_0) H_\phi(R, \phi_0)\, d\phi_0 \tag{14.13}$$

The required Green's function on the periphery of a gyromagnetic disk will not be derived here but may be shown to be given in terms of Bessel functions of the first kind of order n by

$$G(\phi \,|\, \phi_0) = \frac{-j\eta_e \eta_0}{2\pi} \sum_{-\infty}^{\infty} \left[\frac{J_n'(kR)}{J_n(kR)} - \left(\frac{\kappa}{\mu}\right) \frac{n}{kR} \right]^{-1} \exp[-jn(\phi - \phi_0)] \tag{14.14}$$

μ and κ are the usual diagonal and off-diagonal elements of the permeability tensor, η_e is the relative wave impedance in the magnetic insulator, and k_e is the wavenumber in the same medium.

The Green's function defined by this relationship assumes that the electric fields at the center of the conductors are constants there and that the magnetic fields over the center conductors are fixed by

$$H_\phi = H_1, \qquad -\pi/3 - \varphi < \phi < -\pi/3 + \varphi \tag{14.15a}$$

$$H_\phi = H_2, \qquad \pi/3 - \varphi < \phi < \pi/3 + \varphi \tag{14.15b}$$

$$H_\phi = H_3, \qquad \pi - \varphi < \phi < \pi + \varphi \tag{14.15c}$$

The coupling angle (φ) is determined by the width (W) of the center conductors and the radius (R) of the gyromagnetic disk by

$$\sin \varphi = W/2R \tag{14.16}$$

The nomenclature employed here is illustrated in Fig. 14.2.

The wave impedance matrix of the junction then has the symmetry of the Green's matrix

$$\overline{\boldsymbol{\eta}} = \begin{bmatrix} \eta_{11} & \eta_{12} & \eta_{13} \\ \eta_{13} & \eta_{11} & \eta_{12} \\ \eta_{12} & \eta_{13} & \eta_{11} \end{bmatrix} \tag{14.17}$$

A typical entry in this matrix may be defined by taking the constant magnetic term in Eq. (14.13) over the strips outside the integral sign:

$$\eta(\phi|\phi_0) = \int_C G(\phi|\phi_0)\, d\phi_0 \tag{14.18}$$

Evaluating this quantity at $\phi_0 - \pi/3$ between $\phi = \pi/3 \pm \phi$ by having recourse to Eq. (14.14) readily gives

$$\eta_{11} = \eta\left(\frac{\pi}{3}\middle|\frac{\pi}{3}\right) = \frac{j\eta_e\eta_0\psi}{\pi}\sum_{n=-\infty}^{n=\infty}\left(\frac{\sin n\psi}{n\psi}\right)\left[\frac{J_n'(k_eR)}{J_n(k_eR)} - \left(\frac{\kappa}{\mu}\right)\left(\frac{n}{k_eR}\right)\right]^{-1} \tag{14.19a}$$

Taking $\phi = \pi/3$, $\phi_0 = \pi/3$, and $\phi = \pi/3$, $\phi_0 = \pi$ gives the other two parameters.

$$\eta_{12} = \eta\left(\frac{\pi}{3}\middle|\frac{\pi}{3}\right) = \frac{j\eta_e\eta_0\psi}{\pi}\sum_{n=-\infty}^{n=\infty}\left(\frac{\sin n\psi}{n\psi}\right)\left[\frac{J_n'(k_eR)}{J_n(k_eR)} - \left(\frac{\kappa}{\mu}\right)\left(\frac{n}{k_eR}\right)\right]^{-1} \times \exp\left(\frac{j2n\pi}{3}\right) \tag{14.19b}$$

$$\eta_{13} = \eta\left(\frac{\pi}{3}\middle|\pi\right) = \frac{j\eta_e\eta_0\psi}{\pi}\sum_{n=-\infty}^{n=\infty}\left(\frac{\sin n\psi}{n\psi}\right)\left[\frac{J_n'(k_eR)}{J_n(k_eR)} - \left(\frac{\kappa}{\mu}\right)\left(\frac{n}{k_eR}\right)\right]^{-1} \times \exp\left(\frac{-j2\pi n}{3}\right) \tag{14.19c}$$

The two preceding equations are obtained by noting that, with the field source at either port 2 or 3, the integration around the periphery of the circuit is restricted to the port interval at port 1. The impedance poles appearing here differ from those met in connection with the mode matching approach by a factor of $\sin(n\psi)/n\psi$. This is due to the fact that the electric field is here taken as that midway between the ports, whereas it is taken as the average over the ports elsewhere.

A scrutiny of these relationships indicates that a typical pole in the description of the wave impedance eigenvalue is defined by

$$\eta_n = \frac{j3\eta_e\eta_0\psi}{\pi}\left(\frac{\sin n\psi}{n\psi}\right)\left[\frac{J_n'(k_eR)}{J_n(k_eR)} - \left(\frac{\kappa}{\mu}\right)\left(\frac{n}{k_eR}\right)\right]^{-1} \tag{14.20}$$

The mode chart of this function has been evaluated in Chapter 13 and will not be repeated here.

If η_{12} or η_{13} is expanded to reveal common factors 1, α, and α_2, then the open-circuit parameters may be expressed in terms of the eigenvalues of the matrix by

$$\eta_{11} = \frac{\eta^0 + \eta^+ + \eta^-}{3} \tag{14.21a}$$

$$\eta_{12} = \frac{\eta^0 + \alpha\eta^+ + \alpha^2\eta^-}{3} \tag{14.21b}$$

$$\eta_{13} = \frac{\eta^0 + \alpha^2\eta^+ + \alpha\eta^-}{3} \tag{14.21c}$$

The eigenvalues of the problem region are expanded in terms of its poles by separately collating the in-phase and counterrotating poles.

$$\eta^0 = \eta_0 + \eta_3 + \eta_{-3} + \eta_6 + \eta_{-6} + \cdots \tag{14.22a}$$

$$\eta^+ = \eta_1 + \eta_{-2} + \eta_4 + \eta_{-5} + \cdots \tag{14.22b}$$

$$\eta^- = \eta_{-1} + \eta_2 + \eta_{-4} + \eta_5 + \cdots \tag{14.22c}$$

The retention of the first six poles is usually assumed sufficient for the description of this class of problem.

It is preferable, for engineering purposes, to be able to describe the open-circuit parameters of the junction in terms of the characteristic impedance at the ports rather than by the wave impedance as already observed elsewhere in the text.

The relationship between the two matrix descriptions is again accomplished by replacing η_0, in any wave impedance function, by Z_r.

$$Z_e = \eta_e Z_r \tag{14.23}$$

The characteristic impedance of the junction is now obtained by introducing this substitution into the wave impedance description of the junction.

$$\overline{\mathbf{Z}} = \begin{bmatrix} Z_{11} & Z_{12} & Z_{13} \\ Z_{13} & Z_{11} & Z_{12} \\ Z_{12} & Z_{13} & Z_{11} \end{bmatrix} \tag{14.24}$$

Once the open-circuit parameters of any circuit are known, the other circuit quantities are given in the usual way. In particular, scattering parameters are given in terms of the respective eigenvalues by

$$\bar{\mathbf{S}} = \frac{\bar{\mathbf{I}} - \bar{\mathbf{Z}}}{\bar{\mathbf{I}} + \bar{\mathbf{Z}}} \tag{14.25}$$

and

$$S_{11} = \frac{s_0 + s_+ + s_-}{3} \tag{14.26a}$$

$$S_{12} = \frac{s_0 + \alpha s_+ + \alpha^2 s_-}{3} \tag{14.26b}$$

$$S_{13} = \frac{s_0 + \alpha^2 s_+ + \alpha s_-}{3} \tag{14.26c}$$

The reflection eigenvalues are related to the impedance ones by relationships similar to that in Eq. (14.25).

CHAPTER FIFTEEN

Finite Element Formulation of Junction Circulator

15.1 INTRODUCTION

The construction of the open-circuit parameters of regular isotropic and gyromagnetic m-port circuits based on a Green's function formulation has been dealt with in Chapter 14. It requires statements of the eigenvalues and eigenvectors of the problem region. The direct evaluation of these quantities using the finite element method is the topic of this chapter. It involves constructing an energy functional that, when discretized and extremized using the Rayleigh–Ritz procedure, produces the required eigensolutions of the geometry. One classic discretization involves partitioning the structure into triangular elements. The extremized fields at the nodes of the triangle associated with each eigenvalue represent the corresponding eigenvector. The classic three-port junction circulator is again taken by way of an example. The FE procedure outlined here is verified by reproducing some calculations on the first and second circulation conditions of a junction using a simple disk gyromagnetic resonator. This is done as a preamble to investigating the complex gyrator circuits of circulators using regular hexagonal and triangular resonators for which solutions based on separation of variables techniques are not possible. The task is completed once the equivalence between the closed form eigenfunctions dealt with so far and those based on the finite element method is established. A typical computer program consists of one suite that calculates the eigenvalues and eigenvectors of the homogeneous problem using the finite element method and a second one that calculates the open-circuit parameters by forming the line integrals of the Green's function over the ports of the network. Another numerical procedure that has been used to characterize the classic planar junction circulator is the contour integral method. It is appropriate provided the

The Stripline Circulator: Theory and Practice. By J. Helszajn

task is restricted to the description of the fields on the contour of the structure. This chapter also includes some calculations on the complex gyrator circuit of a junction using a triangular resonator based on this approach.

15.2 GREEN'S FUNCTION ANALYSIS USING FINITE ELEMENT METHOD

If the Green's function of an isotropic or gyromagnetic planar problem cannot be expanded in terms of closed form eigensolutions, it can always be constructed by having recourse to a variational method. The Green's function on the periphery of the circuit is related to the source coordinate ξ_0, which is assumed to reside on the periphery of the circuit, and that of the observer ξ point, which is also assumed to reside there, by

$$G(\xi|\xi_0) = j\omega\mu_0\mu_{\text{eff}}H\sum_{a=1}^{n}\frac{\phi_a^*(\xi)\phi_a(\xi_0)}{k_a^2 - k^2} \tag{15.1}$$

The factor in the denominator polynomial appearing in the definition of the Green's function may readily be revealed in the closed form descriptions of the open-circuit parameters of the circulator developed in Chapters 13 and 14. This may be done by expanding the ratio $xJ_n'(x)/J_n(x)$ in the vicinity of a typical pole:

$$\frac{xJ_n'(x)}{J_n(x)} = \left(\frac{x - x_{mn}}{x_{mn}}\right)\left(x_{mn}^2 - n^2\right)$$

The desired result is obtained by recognizing that the factor $J_n(x)/J_n'(x)$ rather than $J_n'(x)/J_n(x)$ appears in a typical pole in the first Foster expansion of Davis and Cohen or Bosma solutions,

$$\left[\frac{J_n'(x)}{J_n(x)}\right]^{-1} = \frac{1}{x_{mn}^2 - n^2}\left(\frac{2x_{mn}}{x^2 - x_{mn}^2}\right)$$

provided

$$x + x_{mn} \approx 2x_{mn}$$

In the variational method the eigensolutions coincide with the stationary values of the energy functional met in connection with the finite element method. The solution to this problem poses no particular difficulty. Provided the magnetic field is a constant, a typical open-circuit parameter between any two ports on the contour of the circuit is given by

$$Z_{ij} = \frac{1}{W_i W_j} \int_{W_j} \int_{W_i} G(\xi|\xi_0)\, dW_i\, dW_j \tag{15.2}$$

The nature of the impedance variable defined here differs from that employed in connection with the separation of variable solution in that the latter definition produced the wave impedance of the circuit whereas the variational approach employed here is in terms of the characteristic impedance at the ports of the problem region.

In order to proceed with the evaluation of the open-circuit parameters of the problem, it is useful to recognize the following equality:

$$\int_{W_j} \int_{W_i} \phi_a^*(\xi) \phi_a(\xi_0)\, dW_i\, dW_j = \int_{W_i} \phi_a^*(\xi)\, dW_i \int_{W_j} \phi_a(\xi_0)\, dW_j \tag{15.3}$$

where

$$\int_{W_i} \phi_a^*(\xi)\, dW_i \quad \text{and} \quad \int_{W_j} \phi_a(\xi_0)\, dW_j$$

are line integrals of the field variables over the ports i and j, respectively.

The derivation of the required matrix notation begins by expressing the field variables $\phi_a^*(\xi)$ and $\phi_a(\xi_0)$ over the ports in polynomial form:

$$\phi_a(\xi) = \sum_{k=1}^{n} \alpha_k u_k, \quad k = 1, 2, 3, \ldots, n \tag{15.4a}$$

$$\phi_a^*(\xi) = \sum_{k=1}^{n} \alpha_k u_k^*, \quad k = 1, 2, 3, \ldots, n \tag{15.4b}$$

In the finite element problem α_k are the basis functions that contain the spatial variation within each element of the region in question, while u_k are the field variables at the nodes.

Making use of this observation gives

$$\int_{W_i} \phi_a^*(\xi)\, dW_i = \sum_{k=1}^{n} u_k^* \int_{W_i} \alpha_k\, dW_i \tag{15.5a}$$

$$\int_{W_j} \phi_a(\xi_0)\, dW_j = \sum_{k=1}^{n} u_k \int_{W_j} \alpha_k\, dW_j \tag{15.5b}$$

In forming these quantities all nodes other than those on the ports in question are disregarded and the spatial variations of the fields over the strips are embodied in the shape functions only.

The development continues by noting the following additional equalities:

$$\int_{W_i} \phi_a^*(\xi)\, dW_i = \sum_{k=1}^{n} u_k^* v_k^i = (\overline{U}_a^*)^T \overline{V}^i \tag{15.6a}$$

$$\int_{W_j} \phi_a(\xi_0)\, dW_j = \sum_{k=1}^{n} u_k v_k^j = (\overline{U}_a)^T \overline{V}^j \tag{15.6b}$$

where

$$v_k^i = \int_{W_i} \alpha_k \, dW_i \tag{15.7a}$$

$$v_k^j = \int_{W_j} \alpha_k \, dW_j \tag{15.7b}$$

The desired relationship for a typical open-circuit parameter is now obtained by replacing the line integrals in the definition of this quantity by the preceding matrix operations. This gives

$$Z_{ij} = \frac{j\eta_e Z_r k_e}{W} \sum_{a=1}^{\infty} \frac{(\overline{\mathbf{U}}_a^*)^T \overline{\mathbf{V}}^i (\overline{\mathbf{U}}_a)^T \overline{\mathbf{V}}^j}{k_a^2 - k_e^2} \tag{15.8}$$

provided

$$W_i = W_j = W \tag{15.9}$$

Z_r is the free space characteristic impedance of the line, and η_e and k_e are defined in the usual way by

$$\eta_e = \sqrt{\mu_e / \varepsilon_f} \tag{15.10}$$

$$k_e = \omega \sqrt{\mu_0 \mu_e \varepsilon_0 \varepsilon_f} \tag{15.11}$$

The desired form for a typical open-circuit parameter is now obtained by having recourse to the following identity:

$$(\overline{\mathbf{U}}_a)^T \overline{\mathbf{V}}^j = (\overline{\mathbf{V}}^j)^T \overline{\mathbf{U}}_a \tag{15.12}$$

The required result is

$$Z_{ij} = \frac{j\eta_e Z_r k_e}{W} \sum_{a=1}^{n} \frac{(\overline{\mathbf{U}}_a^*)^T \overline{\mathbf{V}}^i (\overline{\mathbf{V}}^j)^T (\overline{\mathbf{U}}_a)}{k_a^2 - k_e^2} \tag{15.13}$$

For a stripline transmission line H in Eq. (15.1) is replaced by $2H$.

15.3 NORMALIZED EIGENFUNCTION

In order to proceed with a calculation of the open-circuit parameters of a planar circuit, it is also necessary to associate the u field variables at the nodes over the ports with a normalized eigenfunction,

$$\iint_S \phi_a(\overline{\mathbf{r}})\phi_a^*(\overline{\mathbf{r}})\, ds = 1 \tag{15.14}$$

Writing the eigenfunctions in polynomial form and noting that the spatial variation of the problem is represented by the basis functions gives

$$\sum_{i=1}^{n}\sum_{j=1}^{n} u_i u_j^* \iint_S (\alpha_i \alpha_j^*)\, ds = 1 \tag{15.15}$$

A typical surface integral appearing in the preceding equation is readily recognized as the B_{ij} element met in connection with the construction of the finite element method:

$$B_{ij} = \iint_S (\alpha_i \alpha_j^*)\, ds \tag{15.16}$$

Adopting this notation then gives

$$\sum_{i=1}^{n}\sum_{j=1}^{n} u_i B_{ij} u_j^* = 1 \tag{15.17}$$

If $\overline{\mathbf{U}} = \mathrm{diag}[u_i]$ then the required matrix statement is

$$\overline{\mathbf{U}}[B]\overline{\mathbf{U}}^* = \overline{\mathbf{I}} \tag{15.18}$$

The nature of this operation may readily be appreciated by way of an example by expanding a typical eigenfunction in polynomial form. If a degree-2 approximation problem is adopted to represent $\phi(x, y)$, then

$$\phi(x, y) = \alpha_1(x, y)u_1 + \alpha_2(x, y)u_2 \tag{15.19}$$

Introducing this approximation into the normalization condition ensures that the eigenfunctions are both orthogonal and suitably normalized.

$$\iint_S \{[\alpha_1(x, y)u_1 + \alpha_2(x, y)u_2][\alpha_1(x, y)u_1^* + \alpha_2(x, y)u_2^*]\}\, dx\, dy = 1 \tag{15.20}$$

The development of the desired insight continues by expanding the preceding equation:

$$\iint_S [u_1u_1^*\alpha_1(x, y)\alpha_1(x, y) + u_1u_2^*\alpha_1(x, y)\alpha_2(x, y) + u_2u_1^*\alpha_2(x, y)\alpha_1(x, y) + u_2u_2^*\alpha_2(x, y)\alpha_2(x, y)]\, dx\, dy = 1 \tag{15.21}$$

Since the spatial variation of $\phi(x, y)$ in each finite element is only contained in the definition of the basis functions, its coefficients or the field variables may be taken outside the integral sign. Making use of this consideration then gives

$$u_1\left[\iint_S [\alpha_1(x, y)\alpha_1(x, y)\, dx\, dy\right]u_1^* + u_1\left[\iint_S [\alpha_1(x, y)\alpha_2(x, y)\, dx\, dy\right]u_2^* + u_2\left[\iint_S [\alpha_1(x, y)\alpha_2(x, y)\, dx\, dy\right]u_1^* + u_2\left[\iint_S [\alpha_2(x, y)\alpha_2(x, y)\, dx\, dy\right]u_2^* = 1 \tag{15.22}$$

or

$$\sum_{i=1}^{2}\sum_{j=1}^{2} u_iB_{ij}u_j = 1 \tag{15.23}$$

as asserted.

A similar development indicates that the coefficients of the basis functions may be taken outside the line integrals appearing in the definition of the open-circuit parameters of the network.

15.4 FINITE ELEMENT PROCEDURE

One means of evaluating the cutoff space of a gyromagnetic resonator is to have recourse to a variational method. The finite element method is one of many attractive

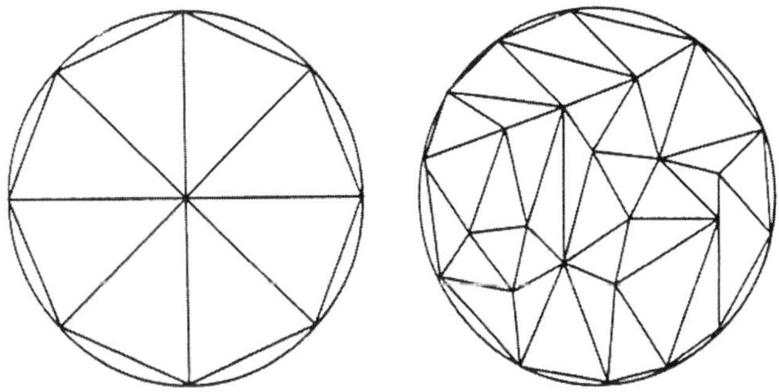

FIGURE 15.1 Discretization of hexagonal resonator.

procedures. Figure 15.1 illustrates a typical finite element segmentation in the case of a regular hexagonal resonator. The number of nodes within each finite element triangle is defined by the degree of the approximation problem. For degree-1 and degree-2, all such nodes lie on the boundary of the triangle, whereas for degrees greater than 2 some nodes lie within the triangle. For this sort of problem the solution to Maxwell's equations is based on a variational method, which consists of extremizing the energy functional of the problem region.

$$F(E_z) = \iint_S \left[|\nabla_t E_z|^2 - k^2 |E_z|^2 \right] ds - j\frac{\kappa}{\mu} \int_\xi E_z \frac{\partial E_z}{\partial t} dt \qquad (15.24)$$

This functional satisfies both the wave equation and a magnetic wall boundary condition on the geometry.

$$\frac{\partial E_z}{\partial n} - j\left(\frac{\kappa}{\mu}\right) \frac{\partial E_z}{\partial t} = 0 \qquad (15.25)$$

When the gyrotropy κ/μ is set equal to zero, the functional reduces to that of an isotropic resonator with an ideal magnetic side wall. The solution continues by approximating the true field E_z by the trial function expansion,

$$E_z' = \sum_{i=1}^{q} u_i \alpha_i(x, y) \qquad (15.26)$$

$\alpha_i(x, y)$ is a suitable set of real basis functions that contain the spatial variables of the problem and u_i are the complex coefficients that represent the electric fields at the nodes of the finite element mesh and are the unknowns of the problem. The functional of the problem region is now extremized by substitution of the trial function into the energy functional and applying the Rayleigh–Ritz condition:

$$\frac{\partial F(E_z')}{\partial u_i} = 0, \quad i = 1,\ 2, \ldots, q \tag{15.27}$$

The functional of the problem region has the nature of a quadratic form so that this amounts to separately minimizing each contribution. This step reduces the problem to a set of simultaneous equations, which may be expressed in a general matrix eigenvalue problem as

$$\left\{[D] + j\frac{\kappa}{\mu}[C]\right\}\overline{\mathbf{U}} = k^2[B]\overline{\mathbf{U}} \tag{15.28}$$

The matrices B_{mn}, C_{mn}, and D_{mn} are standard quantites in the literature:

$$D_{mn} = \iint_S (\nabla_t \alpha_m)(\nabla_t \alpha_n)\, ds \tag{15.29}$$

$$B_{mn} = \iint_S (\alpha_m \alpha_n)\, ds \tag{15.30}$$

$$C_{mn} = \int_\xi \left(\alpha_m \frac{\partial \alpha_n}{\partial t}\right) dt \tag{15.31}$$

The vector $\overline{\mathbf{U}}$ is a column matrix containing the trial fields at the nodes of the finite element mesh.

Once the basis or interpolation functions have been selected the general matrix eigenvalue problem may be solved for the q eigenvalues k and the q eigenvectors $\overline{\mathbf{U}}$ of the problem region. The eigenvalues are its normalized cutoff frequencies and the entries of the eigenvectors are the discrete values of the approximated field at its finite element nodes.

15.5 COMPLEX GYRATOR CIRCUIT OF JUNCTION CIRCULATOR

While the discussion so far is quite general, it will now be specialized to the calculation of the open-circuited parameters of a junction circulator. The geometry under consideration is a planar irregular gyromagnetic resonator with top and bottom electric walls and a magnetic side wall at the junction of three striplines.

Two circulation conditions are employed to describe a junction circulator. The classic solution is usually posed in terms of the complex gyrator impedance or admittance of the device introduced in Chapter 6 and this is the approach adopted here. One condition determines kR from the imaginary parts of either immittances. The other gives R or G from the real parts of either immittances. The frequency variation of the complex gyrator circuit is described in the vicinity of its midband frequency by its susceptance slope parameter (B') and loaded Q-factor (Q_L). The susceptance slope parameter is evaluated graphically in terms of its definition:

$$B' = \frac{\omega_0}{2} \frac{\partial B_{in}}{\partial \omega}\bigg|_{\omega_0} \tag{15.32}$$

The loaded Q-factor is defined separately in the usual way by

$$Q_L = B'/G \tag{15.33}$$

Figures 15.2 and 15.3 compare some results deduced using the finite element method with those based on the analytical solution in Chapter 13 in the case of a

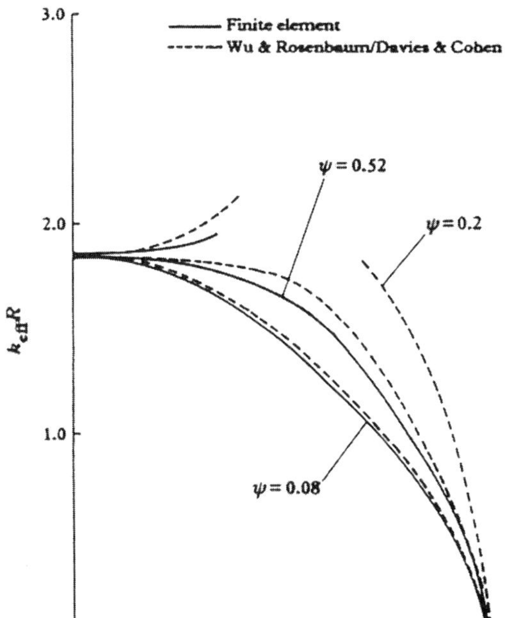

FIGURE 15.2 Comparison between finite element and closed form solutions for circulator using disk resonator (first circulation solution). (Reproduced with permission from R. Lyons and J. Helszajn, A finite element analysis of planar circulators using arbitrarily shaped resonators, *IEEE Trans. Microwave Theory Tech.*, Vol. MTT-30, pp. 1964–1974, 1982.)

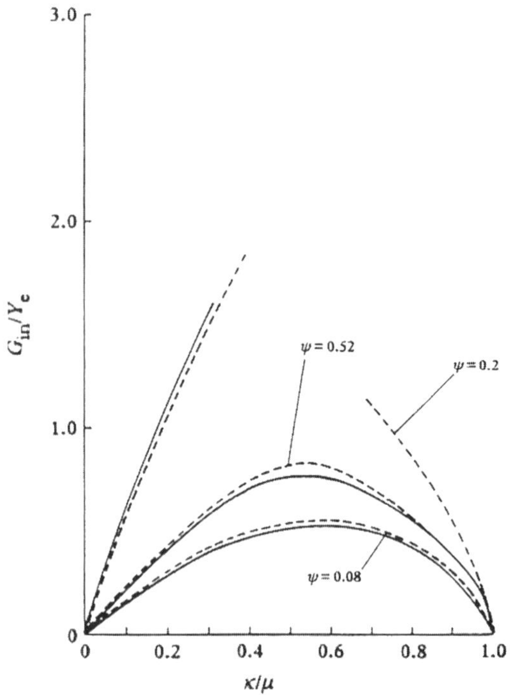

FIGURE 15.3 Comparison between finite element and closed form solutions for circulator using disk resonator (second circulation solution). (Reproduced with permission from R. Lyons and J. Helszajn, A finite element analysis of planar circulators using arbitrarily shaped resonators, *IEEE Trans. Microwave Theory Tech.*, Vol. MTT-30, pp. 1964–1974, 1982.)

junction using a classic disk resonator. The finite element result has been calculated by retaining the first 10 eigenfunctions of the problem. The agreement is best for large coupling angles and less good for small values. One explanation for this discrepancy is that smaller coupling angles excite higher order modes more strongly, and these higher order modes tend to be computed less accurately by the finite element program. The discretization utilized here is indicated in Fig. 15.4.

The first and second circulation solutions for circulators using disk, triangular, and regular hexagonal coupled resonators are tabulated in Tables 15.1–15.3. These tables are again computed by retaining the first 10 eigenfunctions of the geometry. The physical details entering into the description of these various shapes are summarized in Chapter 16 with A replaced by B and B replaced by A. It is observed that at certain points in the tables (e.g., a triangle $0.70 < \kappa/\mu < 0.85$) the values $k_{\text{eff}}R$ and $G_{\text{in}}/Y_{\text{f}}$ take the value 0. This indicates that no circulation condition was identified with $k_{\text{eff}}R < 3.0$. In certain cases, $G_{\text{in}}/Y_{\text{f}}$ takes a negative value. At these points, the lowest circulation condition represents rotation in the opposite direction.

$$q = 1$$
$$m = 48$$
$$n = 3$$
$$m \times n = 144$$
$$p = 37$$

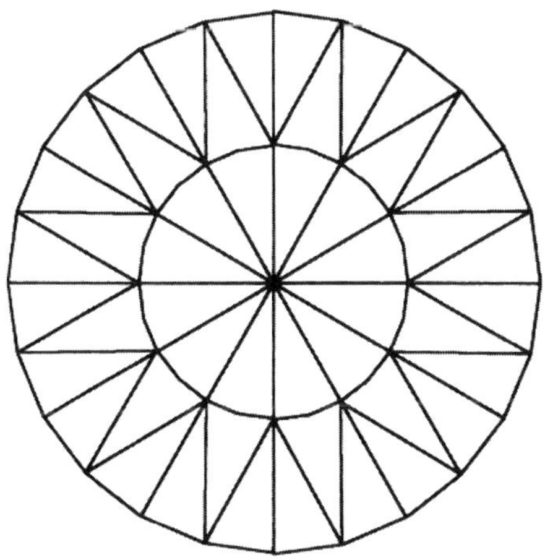

FIGURE 15.4 Discretization of disk resonators.

The values of G_{in} and B_{in} are normalized to Y_r. This quantity is defined by

$$Y_f = \sqrt{\varepsilon_f} Y_r \quad (15.34)$$

Y_r is the admittance of an air spaced planar transmission line whose width is defined by the coupling angle at the terminals of the resonator; ε_f is the relative dielectric constant of the gyromagnetic resonator.

15.6 CONTOUR INTEGRAL METHOD

An important means of establishing the open-circuit parameters of a planar circuit, besides the finite element method, is the contour integral method. It differs in that it involves segmenting the contour of the circuit instead of discretizing its surface. A shortcoming of the contour integral method is that it does not allow the fields inside the resonator to be calculated. The numerical approximation of the contour integral is obtained by dividing the periphery of the circuit into N segments, each of width W, and forming the field (electric) in each element one at a time by locating sources (magnetic) at every point along the contour.

TABLE 15.1 A First and Second Circulation Conditions for Circulator Using a Disk Resonator

First Circulation Solution for Disk Circulator							
ψ	0.200	0.300	0.400	0.500	0.600	0.700	0.800
κ/μ	$k_{eff} R$	$k_{eff} R$	$k_{eff} R$	$k_{eff} R$	$k_{eff} R$	$k_{eff} R$	$k_{eff} R$
0.10	1.862	1.859	1.857	1.854	1.851	1.848	1.846
0.15	1.864	1.859	1.852	1.845	1.839	1.833	1.828
0.20	1.869	1.859	1.847	1.834	1.821	1.810	1.801
0.25	1.880	1.861	1.840	1.818	1.798	1.780	1.766
0.30	1.903	1.869	1.833	1.798	1.767	1.741	1.720
0.35	1.968	1.889	1.826	1.772	1.729	1.693	1.665
0.40	2.206	1.968	1.823	1.741	1.681	1.636	1.600
0.45	2.216	2.180	1.842	1.711	1.630	1.572	1.529
0.50	2.228	2.224	2.192	1.664	1.564	1.497	1.448
0.55	0.000	0.000	0.000	1.605	1.487	1.414	1.361
0.60	2.193	2.187	2.171	1.532	1.402	1.325	1.270
0.65	2.162	2.136	2.071	1.443	1.308	1.230	1.175
0.70	2.518	2.064	1.884	1.305	1.204	1.133	1.078
0.75	2.342	1.941	1.692	1.190	1.088	1.023	0.972
0.80	2.328	1.762	1.469	1.068	0.969	0.907	0.859
0.85	2.375	1.519	1.235	0.927	0.836	0.779	0.736
0.90	1.973	1.214	0.976	0.756	0.679	0.631	0.594
0.95	0.904	0.812	0.654	0.530	0.477	0.442	0.415

Second Circulation Solution for Disk Circulator							
ψ	0.200	0.300	0.400	0.500	0.600	0.700	0.800
κ/μ	G_{in}/Y_f	G_{in}/Y_f	G_{in}/Y_f	G_{in}/Y_f	G_{in}/Y_f	G_{in}/Y_f	G_{in}/Y_f
0.10	0.495	0.335	0.255	0.210	0.181	0.162	0.149
0.15	0.750	0.505	0.383	0.314	0.271	0.242	0.223
0.20	1.010	0.676	0.511	0.417	0.359	0.320	0.295
0.25	1.282	0.850	0.637	0.517	0.443	0.394	0.363
0.30	1.580	1.029	0.760	0.612	0.522	0.463	0.427
0.35	1.980	1.223	0.882	0.701	0.593	0.526	0.484
0.40	3.011	1.500	1.001	0.780	0.656	0.580	0.534
0.45	2.685	1.728	1.079	0.837	0.706	0.623	0.572
0.50	2.828	1.803	1.350	0.893	0.749	0.660	0.606
0.55	−0.000	−0.000	−0.000	0.935	0.782	0.689	0.633
0.60	3.073	1.775	1.282	0.963	0.806	0.712	0.654
0.65	3.511	1.721	1.225	0.979	0.823	0.728	0.670
0.70	0.830	1.688	1.234	0.969	0.833	0.745	0.687
0.75	0.234	1.474	1.177	0.971	0.834	0.749	0.694
0.80	−0.429	1.356	1.154	0.972	0.839	0.755	0.700
0.85	−0.278	1.279	1.135	0.968	0.840	0.758	0.704
0.90	1.320	1.240	1.119	0.961	0.838	0.758	0.706
0.95	1.651	1.244	1.109	0.950	0.833	0.755	0.707

Source: R. Lyons and J. Helszajn, A finite element analysis of planar circulators using arbitrarily shaped resonators, *IEEE Trans. Microwave Theory Tech.*, Vol. MTT-30, pp. 1964–1974, 1982. Reproduced with permission.

TABLE 15.2 First and Second Circulation Conditions for Circulator Using a Triangular Resonator

First Circulation Solution for Triangular Circulator							
ψ	0.200	0.300	0.400	0.500	0.600	0.700	0.800
κ/μ	$k_{eff}R$	$k_{eff}R$	$k_{eff}R$	$k_{eff}R$	$k_{eff}R$	$k_{eff}R$	$k_{eff}R$
0.10	2.455	2.450	2.444	2.438	2.432	2.428	2.426
0.15	2.451	2.438	2.423	2.409	2.396	2.597	2.382
0.20	2.448	2.421	2.392	2.366	2.346	2.331	2.323
0.25	2.458	2.402	2.351	2.312	2.283	2.262	2.250
0.30	2.504	2.386	2.301	2.247	2.210	2.183	2.166
0.35	2.536	2.383	2.241	2.174	2.127	2.089	2.061
0.40	2.524	2.518	2.175	2.095	2.040	1.996	1.961
0.45	2.493	2.492	2.101	2.007	1.946	1.897	1.857
0.50	2.448	2.445	2.021	1.916	1.849	1.795	1.751
0.55	1.993	1.910	1.845	1.790	1.743	1.703	1.671
0.60	1.990	1.837	1.756	1.691	1.638	1.593	1.557
0.65	2.263	1.767	1.662	1.586	1.525	1.475	1.431
0.70	0.000	1.703	1.559	1.473	1.407	1.353	1.305
0.75	0.000	2.042	1.444	1.352	1.284	1.230	1.182
0.80	0.000	0.000	1.309	1.209	1.142	1.090	1.046
0.85	0.000	0.000	1.154	1.054	0.990	0.941	0.900
0.90	2.690	2.508	0.948	0.861	0.808	0.766	0.731
0.95	0.631	0.634	0.865	0.931	0.583	0.547	0.518
Second Circulation Solution for Triangular Circulator							
ψ	0.200	0.300	0.400	0.500	0.600	0.700	0.800
κ/μ	G_{in}/Y_f	G_{in}/Y_f	G_{in}/Y_f	G_{in}/Y_f	G_{in}/Y_f	G_{in}/Y_f	G_{in}/Y_f
0.10	1.706	1.095	0.780	0.583	0.448	0.347	0.268
0.15	2.562	1.629	1.146	0.849	0.649	0.503	0.391
0.20	3.444	2.144	1.476	1.079	0.820	0.637	0.500
0.25	4.431	2.640	1.755	1.262	0.956	0.745	0.590
0.30	5.703	3.142	1.974	1.398	1.057	0.829	0.663
0.35	7.066	3.797	2.134	1.497	1.143	0.916	0.754
0.40	7.323	5.322	2.243	1.566	1.200	0.969	0.804
0.45	−21.798	5.174	2.310	1.604	1.236	1.004	0.841
0.50	7.199	4.965	2.348	1.629	1.262	1.033	0.870
0.55	4.847	2.835	1.960	1.489	1.197	0.993	0.836
0.60	5.556	2.915	1.998	1.518	1.222	1.016	0.861
0.65	10.533	2.979	2.014	1.535	1.242	1.042	0.893
0.70	−0.000	3.130	2.025	1.542	1.253	1.057	0.913
0.75	−0.000	5.718	2.029	1.537	1.251	1.059	0.917
0.80	−0.000	−0.000	2.067	1.556	1.265	1.069	0.925
0.85	−0.000	−0.000	2.052	1.540	1.255	1.064	0.923
0.90	1.454	1.031	1.996	1.501	1.230	1.048	0.913
0.95	3.585	2.438	2.765	2.541	1.249	1.054	0.915

Source: R. Lyons and J. Helszajn, A finite element analysis of planar circulators using arbitrarily shaped resonators, *IEEE Trans. Microwave Theory Tech.*, Vol. MTT-30, pp. 1964–1974, 1982. Reproduced with permission.

TABLE 15.3 First and Second Circulation Conditions for Circulator Using a Regular Hexagonal Resonator

First Circulation Solution Using Regular Hexagon					
ψ	0.200	0.300	0.400	0.500	0.524
κ/μ	$k_{eff}R$	$k_{eff}R$	$k_{eff}R$	$k_{eff}R$	$k_{eff}R$
0.10	2.004	2.003	2.002	2.000	1.999
0.15	1.997	1.994	1.990	1.985	1.984
0.20	1.985	1.981	1.974	1.965	1.963
0.25	1.973	1.965	1.953	1.938	1.934
0.30	1.960	1.946	1.928	1.905	1.899
0.35	1.945	1.924	1.896	1.863	1.856
0.40	1.930	1.897	1.857	1.813	1.803
0.45	1.917	1.866	1.811	1.754	1.742
0.50	1.913	1.829	1.755	1.686	1.671
0.55	0.000	1.783	1.688	1.608	1.592
0.60	2.348	1.726	1.610	1.522	1.504
0.65	2.267	1.653	1.519	1.426	1.407
0.70	2.151	1.551	1.413	1.321	1.302
0.75	2.012	1.854	1.276	1.184	1.174
0.80	1.849	1.691	1.166	1.064	1.051
0.85	1.660	1.484	1.028	0.928	0.914
0.90	1.464	1.218	0.838	0.764	0.753
0.95	1.093	0.880	0.602	0.539	0.530
Second Circulation Solution Using Regular Hexagon					
ψ	0.200	0.300	0.400	0.500	0.524
κ/μ	G_{in}/Y_f	G_{in}/Y_f	G_{in}/Y_f	G_{in}/Y_f	G_{in}/Y_f
0.10	0.569	0.375	0.276	0.216	0.205
0.15	0.853	0.361	0.413	0.323	0.306
0.20	1.133	0.744	0.546	0.426	0.404
0.25	1.407	0.921	0.674	0.524	0.496
0.30	1.661	1.082	0.786	0.604	0.571
0.35	1.915	1.241	0.897	0.685	0.646
0.40	2.155	1.388	0.997	0.758	0.713
0.45	2.375	1.516	1.083	0.820	0.771
0.50	2.583	1.619	1.151	0.871	0.818
0.55	−0.000	1.693	1.199	0.910	0.856
0.60	3.079	1.739	1.229	0.938	0.884
0.65	2.881	1.759	1.245	0.957	0.903
0.70	2.739	1.753	1.248	0.968	0.915
0.75	2.462	1.720	1.231	0.955	0.910
0.80	2.374	1.648	1.239	0.961	0.915
0.85	2.350	1.617	1.237	0.964	0.918
0.90	2.314	1.623	1.223	0.960	0.916
0.95	2.018	1.577	1.237	0.962	0.914

Source: R. Lyons and J. Helszajn, A finite element analysis of planar circulators using arbitrarily shaped resonators, *IEEE Trans. Microwave Theory Tech.*, Vol. MTT-30, pp. 1964–1974, 1982. Reproduced with permission.

Subdividing the contour into n equal segments and partitioning the ports into p segments gives

$$\tan \phi = p \frac{3\sqrt{3}}{n}, \quad p \text{ and } n \text{ odd or } p \text{ and } n \text{ even}$$

In both cases W_n may be expressed in terms of the side A of the triangle by

$$A = nW_n/3$$

Some lower ordered segmentations are depicted in Fig. 15.5. Also tabulated on these illustrations are the values of coupling angle corresponding to some typical combinations of p and n. In selecting a lower bound for n, care must be taken to ensure not only that it is sufficient to represent the first pair of split modes in the magnetized resonator, but also that it is adequate to reproduce the second (and indeed third) pair of split modes of the resonator. Adopting a lower bound for n of 36 is equivalent to sampling a sine wave every 10° in the case of the dominant pair of degenerate modes and 20° and 30°, respectively, for the next two pairs of degenerate

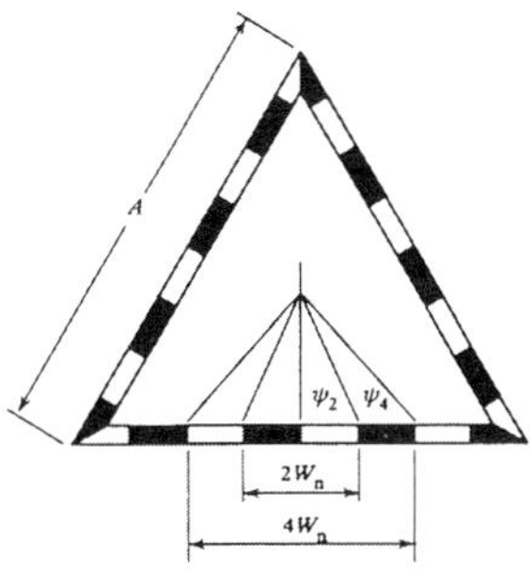

Definition of coupling angles for n odd and p=1 and 3

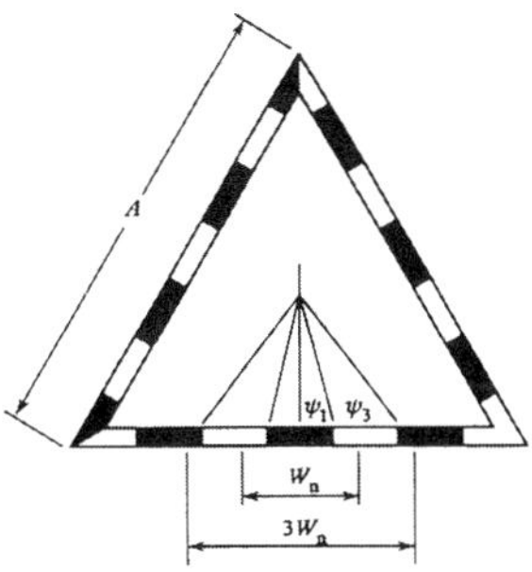

Definition of coupling angles for n even and p = 2 and 4

FIGURE 15.5 Definition of coupling angles for n even and $p = 2$ and 4. (Reproduced with permission from J. Helzsajn, Contour-integral equation formulation of complex gyrator admittance of junction circulators using triangular resonators, *IEE Proc.*, Vol. 32, Pt. H, No. 4, 1985.)

TABLE 15.4 Complex Gyrator Circuits of Circulators Using Triangular Resonators ($\psi = 0.52$, $\phi = 0.71$, $n = 18$)

κ	kA	G	B'	Q_L
0.05	4.073	0.227 Y_f	5.651 Y_f	24.89
0.10	4.057	0.450 Y_f	5.274 Y_f	11.72
0.20	3.989	0.857 Y_f	4.301 Y_f	5.02
0.30	3.856	1.169 Y_f	3.270 Y_f	2.80
0.40	3.654	1.345 Y_f	2.575 Y_f	1.91
0.50	3.378	1.403 Y_f	2.086 Y_f	1.47
0.60	3.037	1.413 Y_f	1.761 Y_f	1.25
0.67	2.760	1.442 Y_f	1.596 Y_f	1.11
0.80				
0.90				
1.00				

Source: J. Helszajn, Contour-integral equation formulation of complex gyrator admittance of junction circulators using triangular resonators, *IEE Proc.*, Vol. 32, Pt. H, No. 4, 1985.

modes. Inspection of Fig. 15.5 indicates that a representative range of coupling angles is obtained by taking n and p even, although n odd and p equal to 1 is appropriate for the realization of small values of ϕ.

The two circulation solutions and the complex gyrator parameters of a circulator employing a side-coupled triangular resonator, obtained using the contour integral method with $n = 18$ and 36, are tabulated in Tables 15.4 and 15.5. This is done for different values of the magnetic variable κ between zero and unity with a coupling angle equal to 0.52 rad.

The entries in these tables that have been left blank are those where either no solutions are located with $kA \leq 4.2$ (narrow coupling angles) or where the results

TABLE 15.5 Complex Gyrator Circuits of Circulators Using Triangular Resonators ($\psi = 0.52$, $\phi = 0.71$, $n = 36$)

κ	kA	G	B'	Q_L
0.05	4.067	0.138 Y_f	3.389 Y_f	24.56
0.10	4.045	0.275 Y_f	3.196 Y_f	11.61
0.20	3.938	0.529 Y_f	2.599 Y_f	4.91
0.30	3.748	0.720 Y_f	2.080 Y_f	2.89
0.40	3.490	0.836 Y_f	1.719 Y_f	2.06
0.50	3.184	0.899 Y_f	1.457 Y_f	1.62
0.60	2.839	0.934 Y_f	1.249 Y_f	1.34
0.67	2.598	0.943 Y_f	1.198 Y_f	1.27
0.80				
0.90				
1.00				

Source: J. Helszajn, Contour-integral equation formulation of complex gyrator admittance of junction circulators using triangular resonators, *IEE Proc.*, Vol. 32, Pt. H, No. 4, 1985.

were deemed untrustworthy. Detailed data on the susceptanceslope parameter and loaded Q-factor are not available in the literature for this device. Y_f in these tables has the meaning in Eq. (15.38).

The circulation data in these tables displays the classic relationship between the magnetic variables and the complex gyrator circuit for small κ. A similar correlation between the apex-coupled triangular resonator closed form and the contour integral method has also been mentioned. The weakly magnetized model of the device already exhibits significant deterioration at $\kappa = 0.30$, so that it is not adequate for everyday engineering. The full theory here is essential.

If all the independent variables may be freely chosen, then the loaded Q-factor may be fixed by the magnetic variables, the quality of the complex gyrator circuit may be set by the coupling angle, and the absolute level of the circuit can be adjusted by varying the ground-plane spacing.

CHAPTER SIXTEEN

Circulators Using Triangular and Irregular Hexagonal Planar Resonators

16.1 INTRODUCTION

An important gyromagnetic resonator met in the design of symmetrical three-port junction circulators is the triangular geometry. This sort of resonator may be coupled symmetrically by one of two possible triplets of ports. One purpose of this chapter is to provide a detailed description of its properties. Of import are its cutoff space, its mode nomenclature, and its dominant and higher order standing wave patterns. The stored energy of a typical solution is also of interest. Its circulation solution is developed separately in terms of this quantity and the power dissipated in port 2 under the assumptions that the conditions at the other two ports correspond to those of an ideal circulator. This is done by assuming single poles in the descriptions of the counterrotating eigen-networks and by separately neglecting the stored energy associated with the in-phase eigen-network. The standing wave solution of the ideal circulator is constructed by taking a suitable linear combination of those of the isotropic problem region under the assumption that the in-phase eigen-network may be idealized by a frequency independent short-circuit at the terminals of the junction. A perturbation description of the split frequencies of the resonator under consideration is also reproduced. A statement of this quantity and one of its quality factor is sufficient for the description of the complex gyrator circuit of the circulator. Calculations on the complex gyrator circuit of this class of circulator based on the contour integral method is available in Chapter 15. Another planar resonator, which has two triplets of ports with threefold symmetry, that may be employed in

The Stripline Circulator: Theory and Practice. By J. Helszajn

the construction of the three-port junction circulator is an irregular hexagonal geometry. It is also dealt with in detail. An important property of both the regular hexagonal and triangular resonators is that each of their possible triplet of ports displays different values of susceptance slope parameter. Thus there are altogether four discrete values available for design. Furthermore, the shape angle of the irregular hexagonal resonator may be used separately to interpolate between the two possible values of triangular geometry.

16.2 EIGENFUNCTIONS OF EQUILATERAL TRIANGLE

A planar circuit for which a circulation solution may be established in closed form is that of the equilateral triangle illustrated in Fig. 16.1. One solution for the electric field that satisfies the homogeneous Helmholtz differential equation inside the resonator is

$$E_z(u, v, w) = A_{m,n,\ell} T(u, v, w) \tag{16.1}$$

The function $T(u, v, w)$ is defined as

$$\begin{aligned} T(u, v, w) = {} & \cos \frac{4\pi\ell}{\sqrt{3}A}\left(\frac{u}{2} + \frac{A}{2\sqrt{3}}\right) \cos\left(\frac{2\pi(m-n)(v-w)}{3\sqrt{3}A}\right) \\ & + \cos \frac{4\pi m}{\sqrt{3}A}\left(\frac{u}{2} + \frac{A}{2\sqrt{3}}\right) \cos\left(\frac{2\pi(n-\ell)(v-w)}{3\sqrt{3}A}\right) \\ & + \cos \frac{4\pi n}{\sqrt{3}A}\left(\frac{u}{2} + \frac{A}{2\sqrt{3}}\right) \cos\left(\frac{2\pi(\ell-m)(v-w)}{3\sqrt{3}A}\right) \end{aligned} \tag{16.2}$$

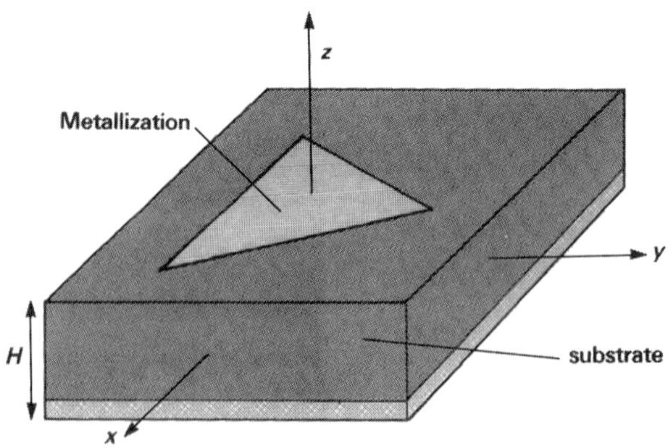

FIGURE 16.1 Schematic diagram of equilateral triangle.

$T(u, v, w)$ satisfies the wave equation, as is readily verified:

$$\left(\frac{\partial^2}{\partial x^2} + \frac{\partial^2}{\partial y^2} + k_{m,n,\ell}^2\right) T(u, v, w) = 0 \tag{16.3}$$

where

$$k_{m,n,\ell} = (4\pi/3A)\sqrt{m^2 + mn + n^2} \tag{16.4}$$

A is the side of the triangle, and

$$m + n + \ell = 0 \tag{16.5}$$

It is observed that the interchange of the three digits m, n, ℓ leaves the cutoff number $k_{m,n,\ell}$ and field patterns unchanged.

The coordinate system employed here is indicated in Fig. 16.2. One possible solution is described with $\alpha = 0$, $\beta = 120^\circ$, and $\gamma = 240^\circ$ by

$$u = x \tag{16.6a}$$

$$v = -\frac{x}{2} + \frac{\sqrt{3}}{2}y \tag{16.6b}$$

$$w = -\frac{x}{2} - \frac{\sqrt{3}}{2}y \tag{16.6c}$$

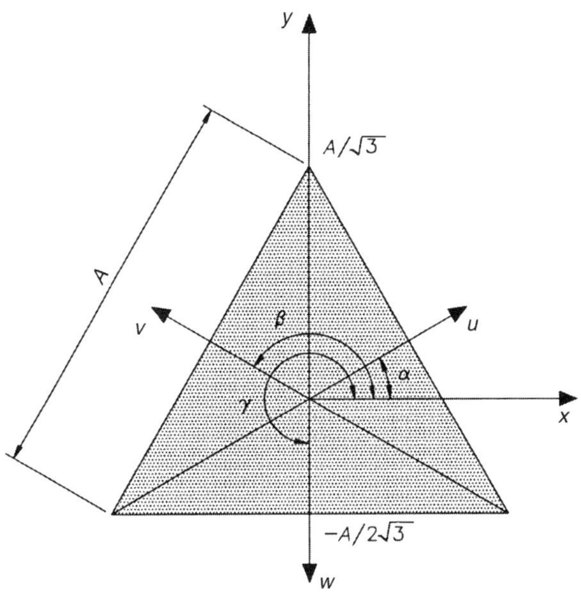

FIGURE 16.2 Coordinate system of planar equilateral triangle. (Reproduced with permission from X. Schelkunoff, *Electromagnetic Waves*. New York: Van Nostrand, 1943, p. 393.)

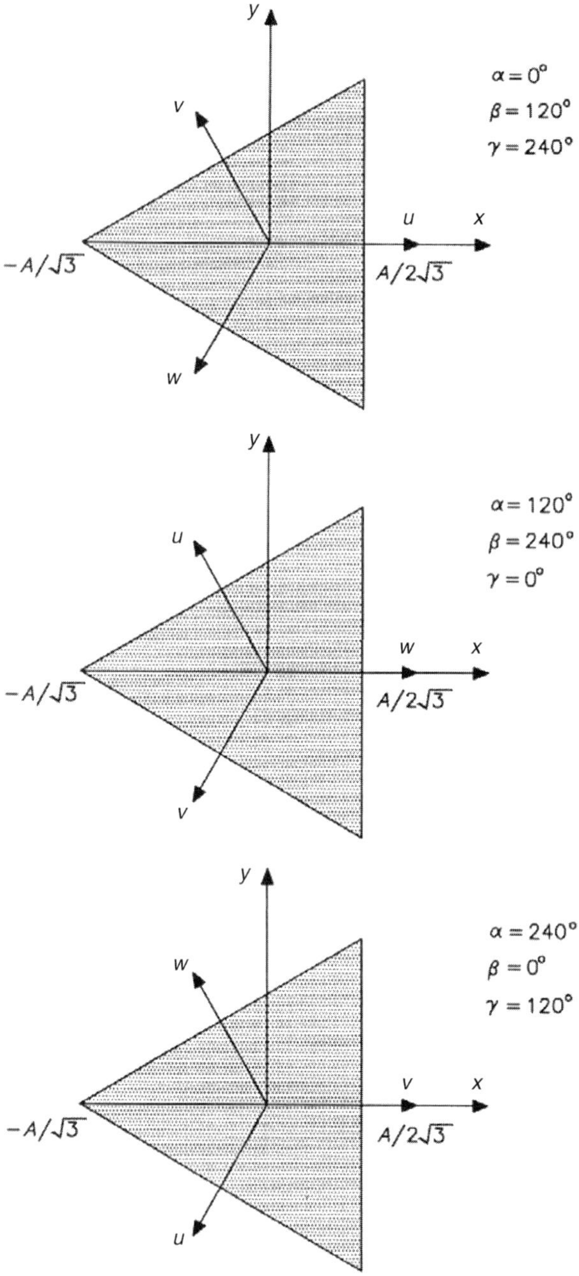

FIGURE 16.3 Coordinate transformations of equilateral triangle. (Reproduced with permission from X. Schelkunoff, *Electromagnetic Waves*. New York: Van Nostrand, 1943, p. 393.)

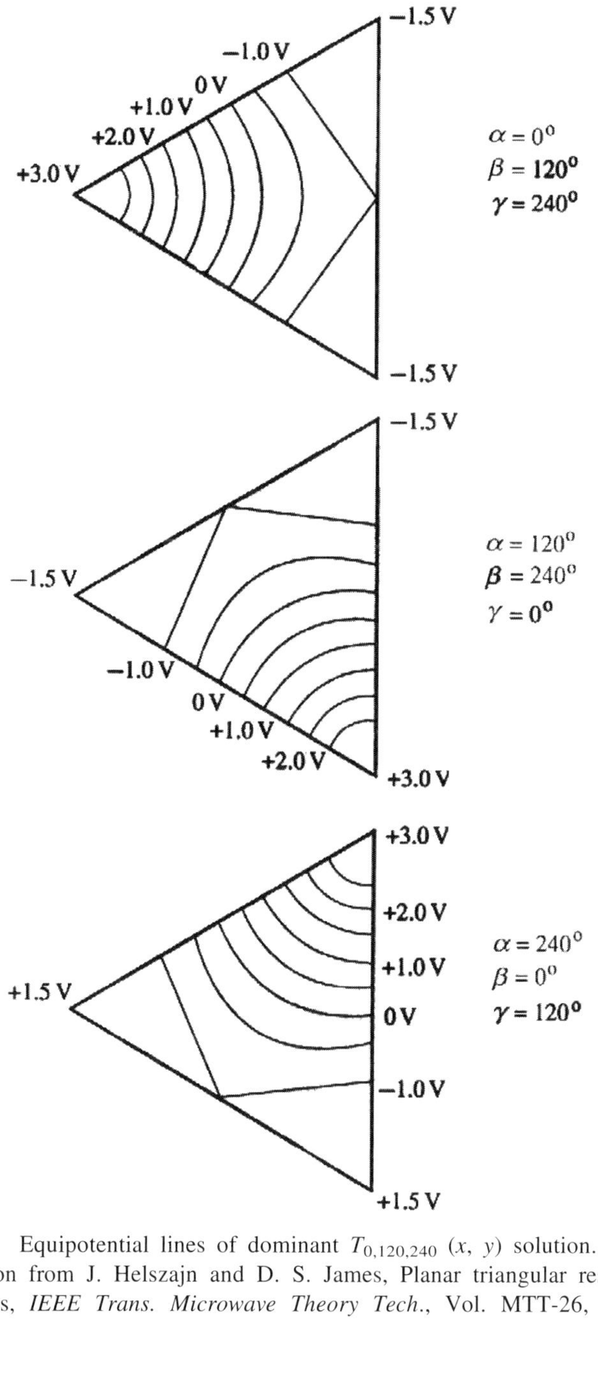

FIGURE 16.4 Equipotential lines of dominant $T_{0,120,240}$ (x, y) solution. (Reproduced with permission from J. Helszajn and D. S. James, Planar triangular resonators with magnetic walls, *IEEE Trans. Microwave Theory Tech.*, Vol. MTT-26, pp. 95–100, Feb. 1978.)

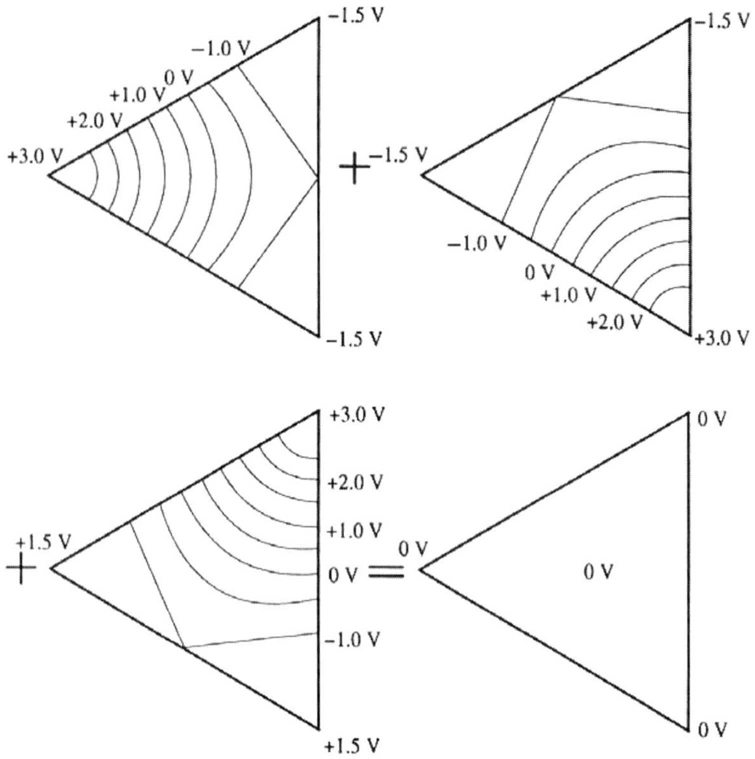

FIGURE 16.5 Linear combination of $T_{1,-1,0}$ (x, y) dominant solutions.

A second solution is obtained with $\alpha = 120°$, $\beta = 240°$, and $\gamma = 360°$, giving

$$u = -\frac{x}{2} + \frac{\sqrt{3}}{2}y \tag{16.7a}$$

$$v = -\frac{x}{2} - \frac{\sqrt{3}}{2}y \tag{16.7b}$$

$$w = x \tag{16.7c}$$

The third solution coincides with $\alpha = 240°$, $\beta = 360°$, and $\gamma = 120°$ but is not necessary for the present discussion.

$$u = -\frac{x}{2} - \frac{\sqrt{3}\,y}{2} \tag{16.8a}$$

$$v = x \tag{16.8b}$$

$$w = -\frac{x}{2} + \frac{\sqrt{3}\,y}{2} \tag{16.8c}$$

The three possible orientations of the equilateral triangle are shown in Fig. 16.3. The dominant solution coincides with $m = 1$, $n = 1$, $\ell = 0$. It is illustrated in Fig. 16.4.

It may be demonstrated separately that the solution $m = 1$, $n = -1$, $\ell = 0$ satisfies

$$T_{0,120,240}(x, y) + T_{120,240,360}(x, y) + T_{240,360,120}(x, y) = 0 \tag{16.9}$$

This relationship is summarized graphically in Fig. 16.5.

16.3 TM FIELD PATTERNS OF TRIANGULAR PLANAR RESONATOR

The TM-mode field components in a triangular dielectric resonator having no variation of the field patterns along the thickness of the resonator are given by

$$E_z = A_{m,n,\ell} T(x, y) \tag{16.10a}$$

$$H_x = \frac{j}{\omega\mu_0\mu_e}\frac{\partial E_z}{\partial y} \tag{16.10b}$$

$$H_y = \frac{-j}{\omega\mu_0\mu_e}\frac{\partial E_z}{\partial x} \tag{16.10c}$$

$$H_z = E_x = E_y = 0 \tag{16.10d}$$

where $A_{m,n,\ell}$ is a constant. Figure 16.1 shows the geometry of the planar resonator under discussion.

$T(x, y)$, in the case of a planar resonator with top and bottom electric walls and magnetic side walls, may be obtained by duality from its existing description met in connection with a TE mode with electric side walls. Aligning the orientation of the triangle with $\alpha = 0$, $\beta = 120^\circ$, $\gamma = 360^\circ$ gives

$$\begin{aligned} T(x, y) = {} & \cos\left[\left(\frac{2\pi x}{\sqrt{3}A} + \frac{2\pi}{3}\right)\ell\right] \cos\left(\frac{2\pi(m-n)y}{3A}\right) \\ & + \cos\left[\left(\frac{2\pi x}{\sqrt{3}A} + \frac{2\pi}{3}\right)m\right] \cos\left(\frac{2\pi(n-\ell)y}{3A}\right) \\ & + \cos\left[\left(\frac{2\pi x}{\sqrt{3}A} + \frac{2\pi}{3}\right)n\right] \cos\left(\frac{2\pi(\ell-m)y}{3A}\right) \end{aligned} \tag{16.11}$$

The complete standing wave solution is

$$E_z = A_{m,n,\ell} T(x, y) \tag{16.12a}$$

$$\begin{aligned} H_x = \frac{-jA_{m,n,\ell}}{\omega\mu_0\mu_e} \Bigg\{ & \frac{2\pi(m-n)}{3A} \cos\left[\left(\frac{2\pi x}{\sqrt{3}A} + \frac{2\pi}{3}\right)\ell\right] \sin\left(\frac{2\pi(m-n)y}{3A}\right) \\ & + \frac{2\pi(n-\ell)}{3A} \cos\left[\left(\frac{2\pi x}{\sqrt{3}A} + \frac{2\pi}{3}\right)m\right] \sin\left(\frac{2\pi(n-\ell)y}{3A}\right) \\ & + \frac{2\pi(\ell-m)}{3A} \cos\left[\left(\frac{2\pi x}{\sqrt{3}A} + \frac{2\pi}{3}\right)n\right] \sin\left(\frac{2\pi(\ell-m)y}{3A}\right) \Bigg\} \end{aligned} \tag{16.12b}$$

$$\begin{aligned} H_y = \frac{-jA_{m,n,\ell}}{\omega\mu_0\mu_e} \Bigg\{ & \frac{2\pi\ell}{\sqrt{3}A} \sin\left[\left(\frac{2\pi x}{\sqrt{3}A} + \frac{2\pi}{3}\right)\ell\right] \cos\left(\frac{2\pi(m-n)y}{3A}\right) \\ & + \frac{2\pi m}{\sqrt{3}A} \sin\left[\left(\frac{2\pi x}{\sqrt{3}A} + \frac{2\pi}{3}\right)m\right] \cos\left(\frac{2\pi(n-\ell)y}{3A}\right) \\ & + \frac{2\pi n}{\sqrt{3}A} \sin\left[\left(\frac{2\pi x}{\sqrt{3}A} + \frac{2\pi}{3}\right)n\right] \cos\left(\frac{2\pi(\ell-m)y}{3A}\right) \Bigg\} \end{aligned} \tag{16.12c}$$

The interchange of the three digits m, n, ℓ leaves the cutoff number $k_{m,n,\ell}$ unchanged; similarly, the field patterns are retained, without rotation.

16.4 $TM_{1,0,-1}$ FIELD COMPONENTS OF TRIANGULAR PLANAR RESONATOR

The field patterns of the dominant mode in a planar triangular resonator are given by Eq. (16.12) with $m = 1$, $n = 0$, $\ell = -1$. The result is

$$E_z = A_{1,0,-1}\left[2\cos\left(\frac{2\pi x}{\sqrt{3}A} + \frac{2\pi}{3}\right)\cos\left(\frac{2\pi y}{3A}\right) + \cos\left(\frac{4\pi y}{3A}\right)\right] \tag{16.13a}$$

$$H_x = -jA_{1,0,-1}\zeta_e\left[\cos\left(\frac{2\pi x}{\sqrt{3}A} + \frac{2\pi}{3}\right)\sin\left(\frac{2\pi y}{3A}\right) + \sin\left(\frac{4\pi y}{3A}\right)\right] \tag{16.13b}$$

$$H_y = j\sqrt{3}A_{1,0,-1}\zeta_e\left[\sin\left(\frac{2\pi x}{\sqrt{3}A} + \frac{2\pi}{3}\right)\cos\left(\frac{2\pi y}{3A}\right)\right] \tag{16.13c}$$

where

$$k_{1,0,-1} = 4\pi/3A \tag{16.14}$$

$$\zeta_e = \sqrt{\varepsilon_0\varepsilon_r/\mu_0\mu_e} \tag{16.15}$$

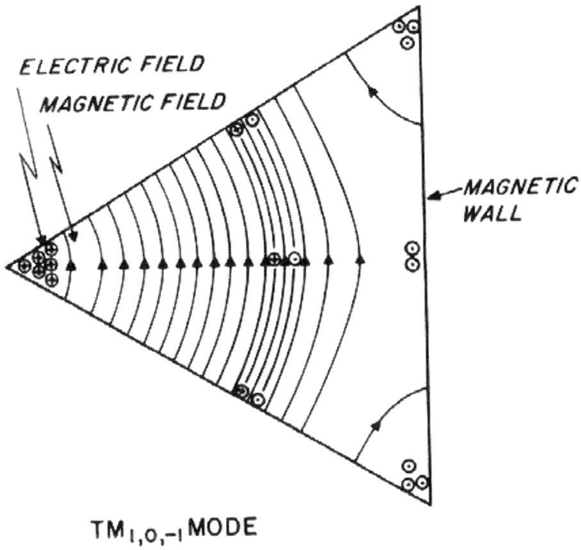

FIGURE 16.6 $TM_{1,0,-1}$ dominant mode field pattern in triangular resonator with magnetic walls. (Reproduced with permission from J. Helszajn and D. S. James, Planar triangular resonators with magnetic walls, *IEEE Trans. Microwave Theory Tech.*, Vol. MTT-26, pp. 95–100, Feb. 1978.)

Figure 16.6 is a sketch of the magnetic and electric fields for the dominant $TM_{1,0,-1}$ mode in a triangular resonator.

16.5 $TM_{1,1,-2}$ FIELD COMPONENTS OF TRIANGULAR PLANAR RESONATOR

In addition to the dominant mode the next higher order mode in a planar triangular resonator has also been investigated. This mode is a symmetrical or in-phase one for which the field pattern is given in Fig. 16.7. It is obtained with $m = 1$, $n = 1$, $\ell = -2$; $T(x, y)_{m,n,\ell}$, and $k_{m,n,\ell}$ are in this instance described by

$$T(x, y)_{1,1,-2} = \cos 2\left(\frac{2\pi x}{\sqrt{3}A} + \frac{2\pi}{3}\right) + 2\cos\left(\frac{2\pi x}{\sqrt{3}A} + \frac{2\pi}{3}\right)\cos\left(\frac{2\pi y}{A}\right) \qquad (16.16)$$

$$k_{1,1,-2} = 4\pi/\sqrt{3}A \qquad (16.17)$$

A property of this mode is that it has a maximum at the origin, unlike the $T(x, y)_{1,0,-1}$ mode, which has a zero there. This maximum is given by

$$T(0, 0)_{1,1,-2} = -3/2 \qquad (16.18)$$

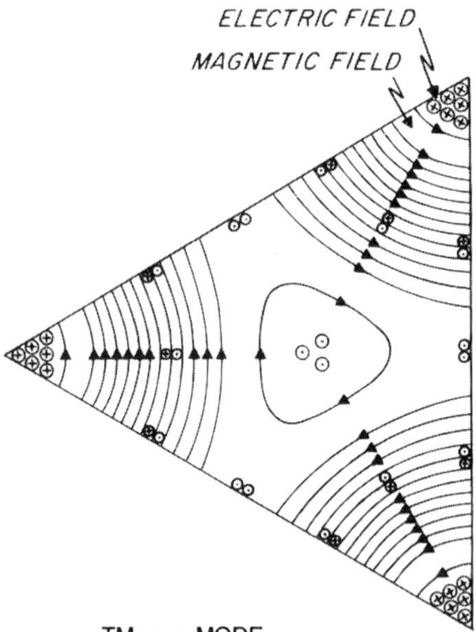

FIGURE 16.7 $TM_{1,1,-2}$ field pattern in triangular resonator with magnetic walls. (Reproduced with permission from J. Helszajn and D. S. James, Planar triangular resonators with magnetic walls, *IEEE Trans. Microwave Theory Tech.*, Vol. MTT-26, pp. 95–100, Feb. 1978.)

It is easily verified that this mode is symmetrical:

$$\begin{aligned} T(0, A/2\sqrt{3})_{1,1,-2} &= T(-A/4\sqrt{3}, A/4)_{1,1,-2} \\ &= T(-A/4\sqrt{3}, -A/4)_{1,1,-2} \end{aligned} \tag{16.19}$$

16.6 ELECTRIC FIELD PATTERN IN CIRCULATOR USING TRIANGULAR RESONATOR

The geometry of a triangular resonator makes it difficult to visualize rotation of the $TM_{1,0,-1}$ mode in order to satisfy the boundary conditions of an ideal circulator. One possible way to construct the electric field distribution is by taking a linear combination of two $TM_{1,0,-1}$ standing wave solutions with one of them rotated through 120° with respect to the other. The demonstration of this construction starts with a statement of that for the dominant mode in the u, v, w plane:

$$T(u, v, w) = 2\cos\left(\frac{2\pi u}{\sqrt{3}A} + \frac{2\pi}{3}\right)\cos\left(\frac{2\pi(v-w)}{3\sqrt{3}A}\right) + \cos\left(\frac{4\pi(v-w)}{3\sqrt{3}A}\right) \tag{16.20}$$

One field pattern in the x, y plane is given with $\alpha = 0$, $\beta = 120°$, and $\gamma = 240°$:

$$T(x, y)^{0,120,240} = 2\cos\left(\frac{2\pi x}{\sqrt{3}A} + \frac{2\pi}{3}\right)\cos\left(\frac{2\pi y}{3A}\right) + \cos\left(\frac{4\pi y}{3A}\right) \tag{16.21}$$

A second solution is obtained with $\alpha = 120°$, $\beta = 240°$, and $\gamma = 360°$:

$$\begin{aligned} T(x, y)^{120,240,360} = {} & 2\cos\left[\frac{2\pi}{\sqrt{3}A}\left(-\frac{x}{2} + \frac{\sqrt{3}y}{2}\right) + \frac{2\pi}{3}\right] \\ & \times \cos\left[\frac{2\pi}{3\sqrt{3}A}\left(-\frac{3x}{2} + \frac{\sqrt{3}y}{2}\right)\right] \\ & + \cos\left[\frac{4\pi}{3\sqrt{3}A}\left(-\frac{3x}{2} + \frac{\sqrt{3}y}{2}\right)\right] \end{aligned} \tag{16.22}$$

The third field pattern is obtained with $\alpha = 240°$, $\beta = 360°$, and $\gamma = 120°$ but is not necessary for the present discussion.

One standing wave solution that is in keeping with a phenomenological description of a circulator using a triangular resonator is given by

$$T'(x, y) = T(x, y)^{0,120,240} - T(x, y)^{120,240,360} \tag{16.23}$$

Figure 16.8 illustrates the construction of the standing wave solution for the circulator using the equipotential diagram of the isotropic resonator. The standing wave solution defined by Eq. (16.23) can also be written in terms of counterrotating variables as

$$T'(x, y) = T^{+}(x, y)\exp(j\pi/6) + T^{-}(x, y)\exp(-j\pi/6) \tag{16.24}$$

where

$$\begin{aligned} T'(x, y)^{\pm} = {} & \tfrac{1}{3}[T(x, y)^{0,120,240} + T(x, y)^{120,240,360}\exp(\pm j120) \\ & + T(x, y)^{240,360,120}\exp(\pm j240)] \end{aligned} \tag{16.25}$$

The result for the $TM^{\pm}_{1,0,-1}$ mode is

$$T^{\pm}(x, y)_{1,0,-1} = \frac{1}{2}\exp\left(\mp j\frac{4\pi y}{3A}\right) + \cos\left(\frac{2\pi x}{\sqrt{3}A} + \frac{2\pi}{3}\right)\exp\left(\mp\frac{j2\pi y}{3A}\right) \tag{16.26}$$

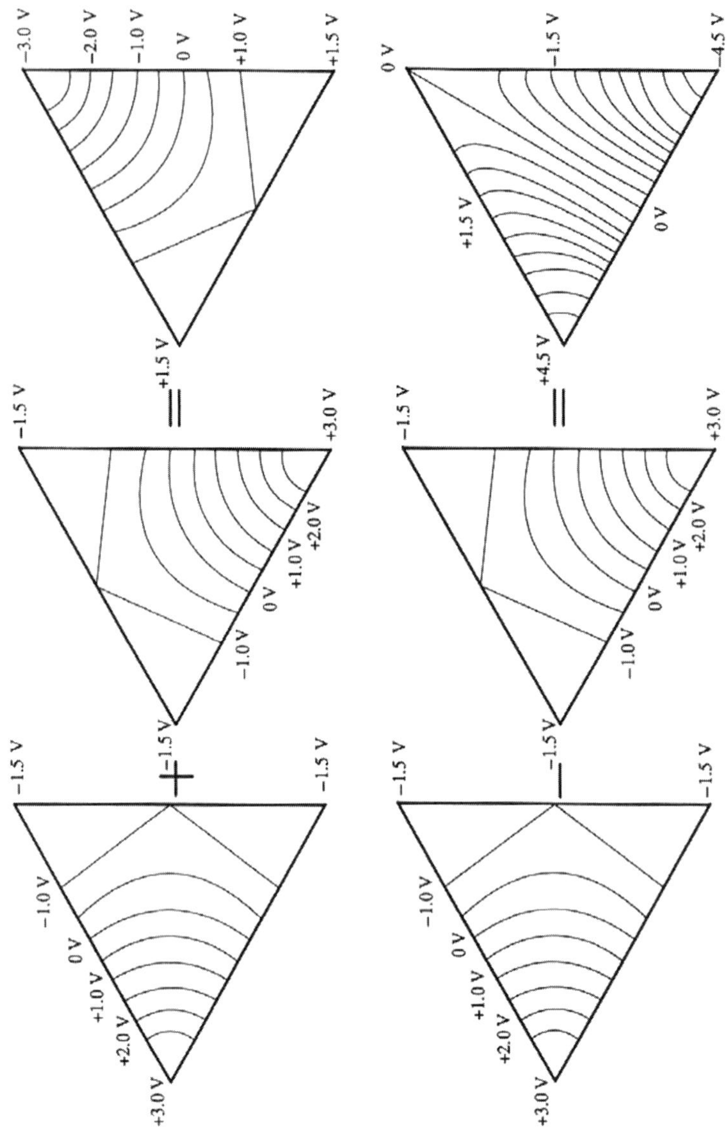

FIGURE 16.8 Sum and difference of dominant $T_{0,120,240}(x, y)$ and $T_{240,0,120}(x, y)$ solutions.

It is easily verified that

$$T^{\pm}\left(-A/\sqrt{3}, 0\right)_{1,0,-1} = \frac{3}{2} \tag{16.27}$$

$$T^{\pm}\left(A/2\sqrt{3}, A/2\right)_{1,0,-1} = \frac{3}{2}\exp\left(\mp j120\right) \tag{16.28}$$

$$T^{\pm}\left(A/2\sqrt{3}, -A/2\right)_{1,0,-1} = \frac{3}{2}\exp\left(\mp j240\right) \tag{16.29}$$

The in-phase mode is constructed in a similar way in terms of the $TM_{1,1,-2}$ nomenclature as

$$T(x, y)_{1,1,-2} = \frac{1}{3}\left[T(x, y)^{0,120,240} + T(x, y)^{120,240,360} + T(x, y)^{240,360,120}\right] \qquad (16.30)$$

The meanings of these field settings are illustrated in Fig. 16.9.

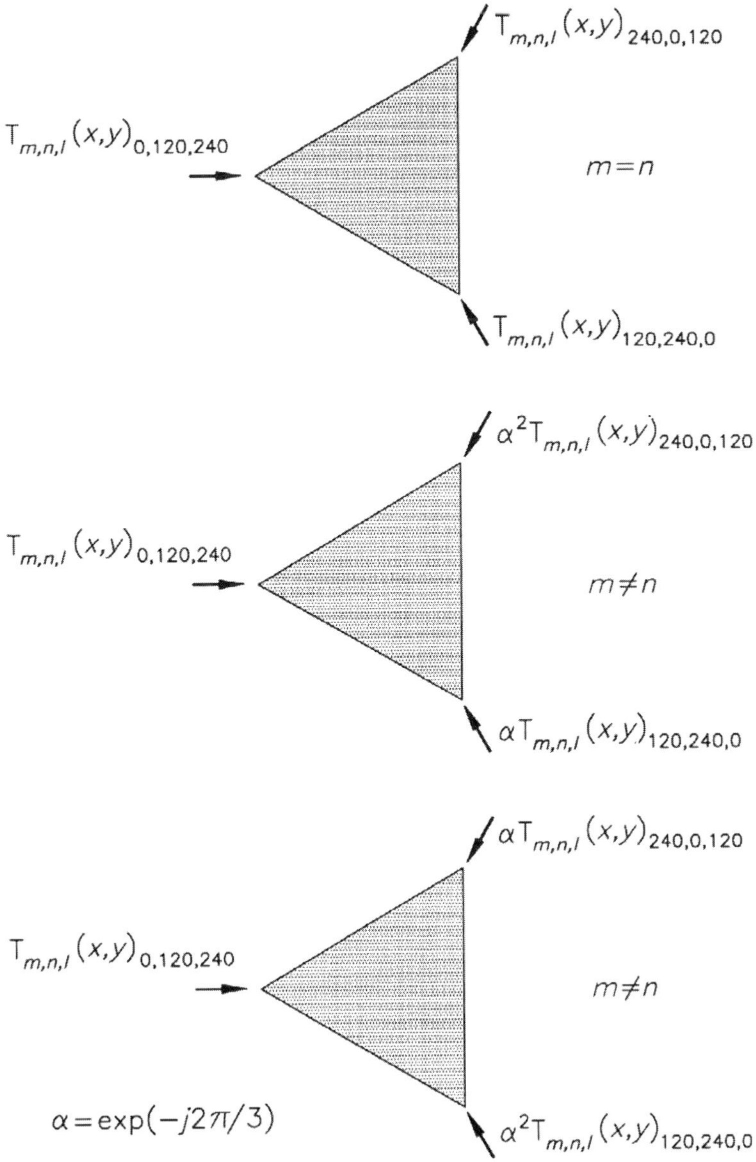

FIGURE 16.9 Definitions of $T^0(x, y)$, $T^+_{m,n,l}(x, y)$, and $T^-_{m,n,l}(x, y)$ spatial functions.

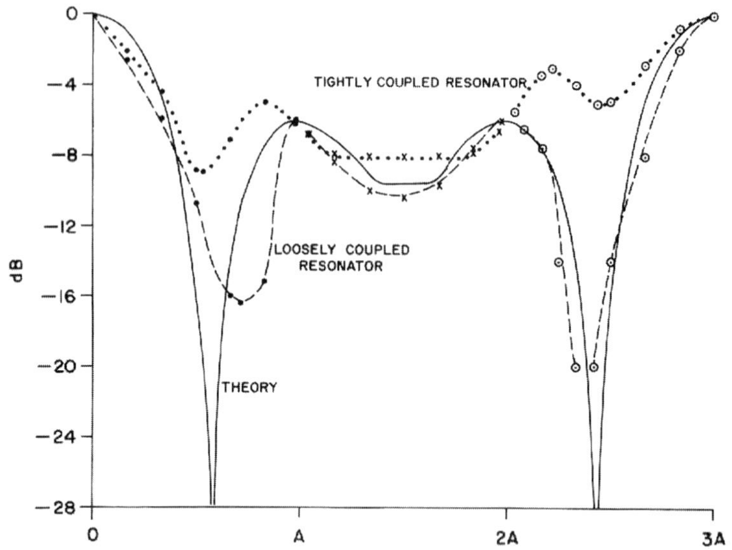

FIGURE 16.10 Electric field on the periphery of circulator using triangular resonator. (Reproduced with permission from J. Helszajn and D. S. James, Planar triangular resonators with magnetic walls, *IEEE Trans. Microwave Theory Tech.*, Vol. MTT-26, pp. 95–100, Feb. 1978.)

Figure 16.10 indicates the theoretical and experimental electric field distribution at the edge of a loosely coupled circulator.

The constructions of the standing wave solutions in an irregular hexagonal resonator are akin to those of the related triangular geometries.

16.7 STORED ENERGY

An important quantity in the description of any resonator is its stored energy. It is defined by

$$U_0 = 2H \int\int_s \frac{1}{2} \varepsilon_0 \varepsilon_f E^2 \, ds \tag{16.31}$$

E may be written in terms of the linear combinations of the T functions in Eq. (16.23) or (16.24). Adopting the latter possibility gives

$$E = A_{1,0,-1}[T^+(x, y)\exp(j\pi/6) + T^-(x, y)\exp(-j\pi/6)] \tag{16.32}$$

where $T^\pm(x)$ is stipulated in Eq. (16.26).

TABLE 16.1 $TM_{m,n,l}$ Modes and Stored Energies for Triangular Resonators

Sequence Number	m, n, ℓ	$\int_s T(x, y)^2_{m,n,\ell}\, ds$
1 (dominanat)	1, 0, −1	$(9\sqrt{3})/2$
2 (symmetric)	1, 1, −2	$(9\sqrt{3})/2$
3	2, −2, 0	$(9\sqrt{3})/2$
4	1, 2, −3	$(9\sqrt{3})/4$
5	3, −3, 0	$(9\sqrt{3})/2$
6	2, 2, −4	$(9\sqrt{3})/2$
7	1, 3, −4	$(9\sqrt{3})/4$
8	2, 3, −5	$(9\sqrt{3})/4$
9	1, 4, −5	$(9\sqrt{3})/4$

$$W_{m,n,\ell} = \frac{A^2_{m,n,\ell}\varepsilon_0\varepsilon_r A^2 H}{24}\int_S [T(x, y)_{m,n,\ell}]^2\, ds$$

The stored energy, after some algebra, is given by

$$U_0 = \varepsilon_0\varepsilon_f A_{1,0,-1} H \int_{-A/2\sqrt{3}}^{A/2\sqrt{3}} \int_{(-x/\sqrt{3})-(A/3)}^{(x/\sqrt{3})+(A/3)} \left[\cos\left(\frac{4\pi y}{3A} - \frac{\pi}{6}\right) + 2\cos\left(\frac{2\pi x}{\sqrt{3}A} + \frac{2\pi}{3}\right) \times \cos\left(\frac{2\pi y}{3A} + \frac{\pi}{6}\right)\right]^2 dy\, dx \tag{16.33}$$

The required result is now obtained by evaluating the right-hand side of Eq. (16.33):

$$U_0 = 0.64\varepsilon_0\varepsilon_f A_{1,0,-1} A^2 H$$

where H is the thickness of each half-space of the triangular region. This solution has been verified using a conversational numerals multiple-integration package available as a standard time-sharing utility on a Burrough's B5700 digital computer. The computed answer is

$$U_0 = 0.649519 \tag{16.34}$$

Table 16.1 summarizes the energies for various higher order modes of the resonator.

16.8 THE IRREGULAR HEXAGONAL RESONATOR

The topology of the irregular hexagonal resonator is illustrated in Fig. 16.11. One triplet of ports reduces the problem region to a side-coupled triangular resonator and the other maps it into an apex-coupled triangular one. The existing literature

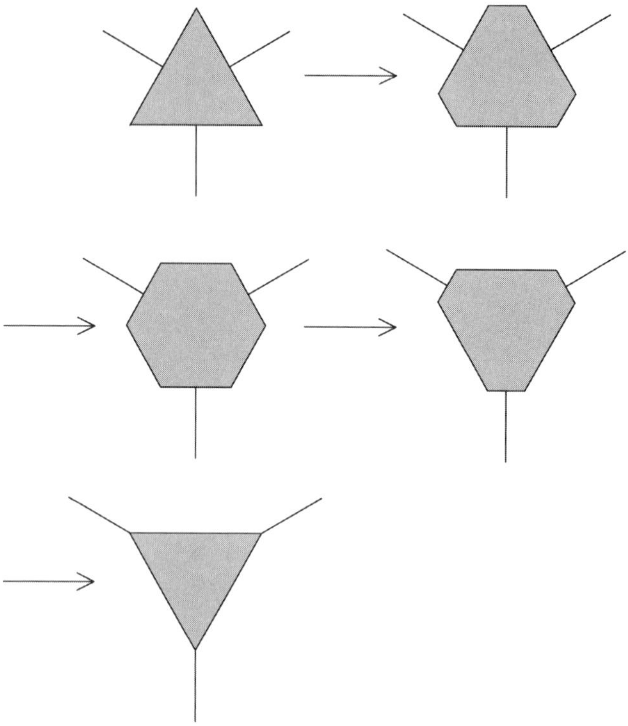

FIGURE 16.11 Mapping between regular and irregular hexagonal resonators.

contains papers that give the frequency response of some specific geometries based on the contour integral method or the point matching one, papers that give statements on the two circulation conditions by having recourse to the finite element method.

The physical variables entering into the description of this sort of geometry are the shape angle of the resonator, ϕ; the side dimensions of the resonator, A, B; and its circumscribed radius, r, respectively. The shape angle takes the value 0° for a triangle and the value 60° for a regular hexagonal one.

The details of a typical coupling port are described by the width (W) of the coupling strip, by the coupling angle (ψ), and by the radius (R) subtended by the coupling angle on the periphery of the resonator. The required relationships between these physical variables for the two possible triplets of ports of the problem region are

$$W = 2r\cos(\phi/2)\tan\psi, \text{ apex-coupled resonator} \tag{16.35}$$

$$W = 2r\cos(60 - \phi/2)\tan\psi, \text{ side-coupled resonator} \tag{16.36}$$

respectively.

The physical variables entering into the description of the two possible triplets of ports are indicated in Fig. 16.12.

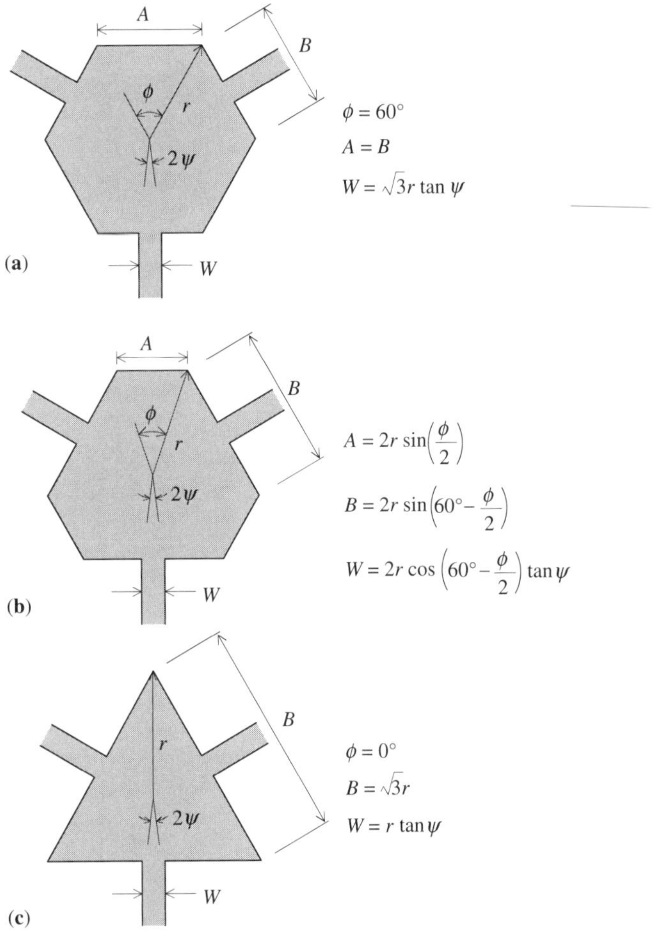

FIGURE 16.12 Physical variables of irregular hexagonal resonator.

16.9 CUTOFF SPACE OF PLANAR IRREGULAR HEXAGONAL RESONATOR

One means of deducing the cutoff space of a planar irregular hexagonal resonator is to have recourse to the finite element method. The case where $\phi = 0^\circ$ is a triangle and that for which $\phi = 60^\circ$ is a regular hexagon. The modes of the irregular hexagon resonator are designated $TM_{m,n,\ell}$ limit modes. This nomenclature is consistent with that used to describe the modes in a triangular resonator. The cutoff numbers for the first three regular hexagon modes are

$$(kr)_{1,0,-1} = 2.00$$
$$(kr)_{2,-2,0} = 3.35$$
$$(kr)_{1,1,-2} = 4.20$$

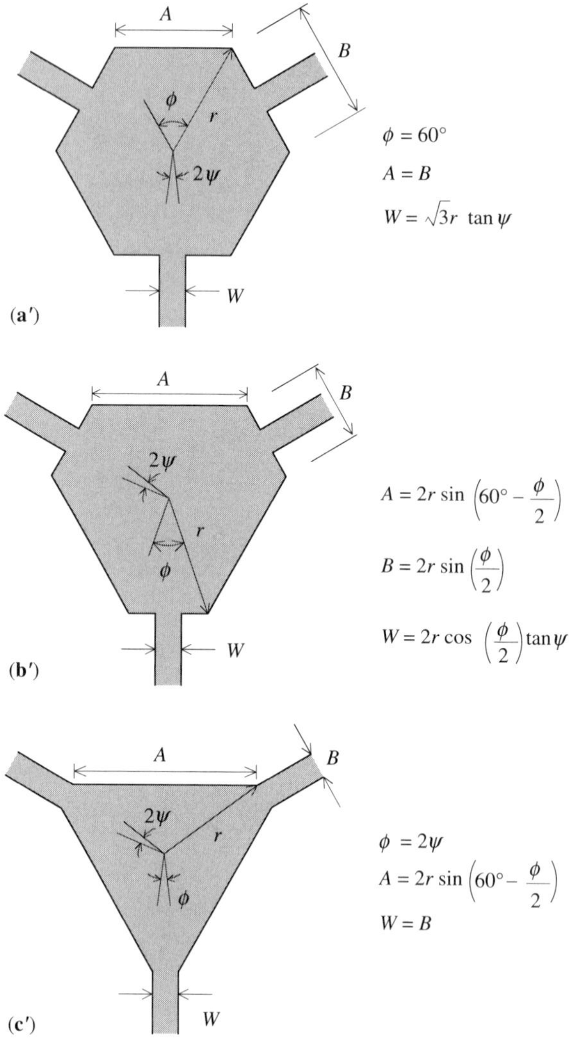

FIGURE 16.12 *Continued.*

and for the triangle resonator;

$$(kr)_{1,0,-1} = 2.45$$
$$(kr)_{1,1,-2} = 4.15$$
$$(kr)_{2,-2,0} = 4.85$$

m, n, ℓ satisfy

$$m + n + \ell = 0$$

A graph of the cutoff numbers of the first three modes versus the apex angle of the smaller triangle contained within the hexagon is illustrated in Fig. 16.13. It indicates that the symmetric $TM_{1,1,-2}$ limit mode is degenerate with the $TM_{2,-2,0}$ limit mode. One polynomial representation of the cutoff space of the irregular resonator is

$$kr = a_0 + a_1\phi + a_2\phi^2 + a_3\phi^3 \tag{16.37}$$

where

$$a_0 = 2.45$$
$$a_1 = -2.85 \times 10^{-2}$$
$$a_2 = 6.5 \times 10^{-4}$$
$$a_3 = -5.0 \times 10^{-6}$$

and k has the usual meaning.

The details of the discretization employed in the case of the regular isotropic hexagonal resonator illustrated in Fig. 16.14 are

$$p = 2$$
$$n = 6$$
$$m = 90$$
$$n \times m = 540$$
$$q = 211$$

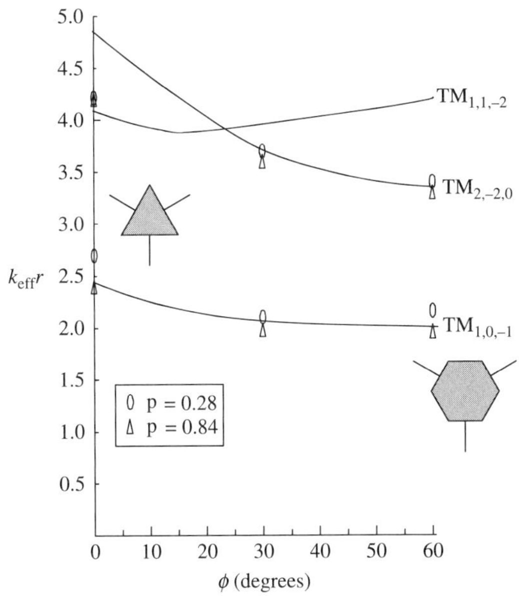

FIGURE 16.13 Mode chart of regular hexagonal resonator.

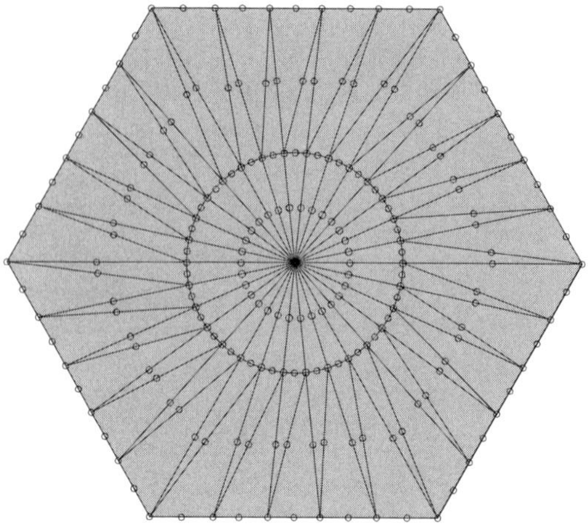

FIGURE 16.14 Discretization of regular hexagonal resonator.

p is the degree of the polynomial within each finite element triangle, n is the number of nodes inside each finite element triangle, m is the number of triangles, $n \times m$ is the total number of nodes before assembly of the finite element mesh, and q is the number of nodes after assembly of the mesh.

The equipotential lines of the TM modes of the regular hexagonal resonator with a perfect magnetic wall boundary condition at its periphery reduce to those of the triangular geometry and are not reproduced here.

16.10 SPLIT FREQUENCIES OF IRREGULAR HEXAGONAL RESONATOR

The split frequencies of the regular gyromagnetic hexagonal resonator are depicted in Fig. 16.15. The details of the finite element mesh utilized in this calculation are specified by

$$\begin{aligned} p &= 2 \\ n &= 6 \\ m &= 24 \\ n \times m &= 148 \\ q &= 73 \end{aligned}$$

One polynomial approximation for the split frequencies of this sort of weakly magnetized resonator based on the finite element solution of six different shape angles (ϕ) is given by

$$\left(\frac{\omega_+ - \omega_-}{\omega_0}\right)\left(\frac{\mu}{\kappa}\right) = a_0 + a_1\phi + a_2\phi^2 + a_3\phi^3 + a_4\phi^4 + a_5\phi^5, \; 0 \leq \frac{\kappa}{\mu} \leq 0.30 \tag{16.38}$$

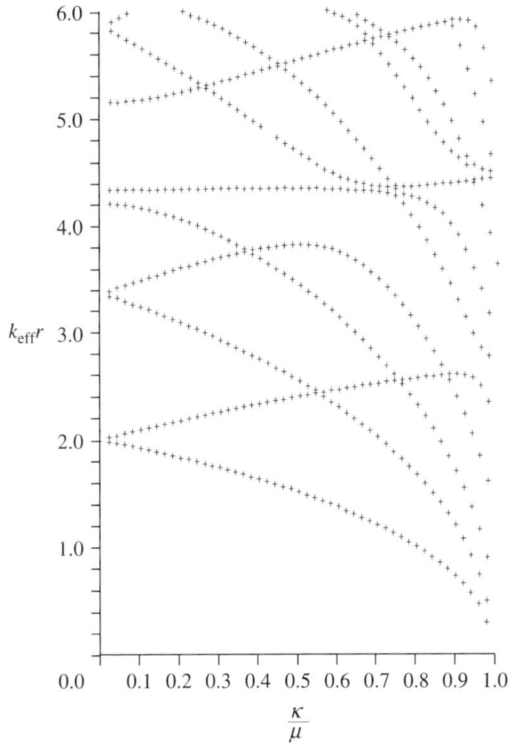

FIGURE 16.15 Split frequencies of regular hexagonal gyromagnetic planar resonators using FE process. (Reproduced with permission from J. Helszajn and R. W. Lyon, Mode charts of magnetised and demagnetised planar hexagonal resonators on ferrite substrates using finite elements, *IEE Proc.*, Pt. H, pp. 420–422, Dec. 1984.)

where

$$a_0 = 0.55$$

$$a_1 = 12.64 \times 10^{-3}$$

$$a_2 = -12.98 \times 10^{-4}$$

$$a_3 = 65.15 \times 10^{-6}$$

$$a_4 = -12.59 \times 10^{-7}$$

$$a_5 = 8.29 \times 10^{-9}$$

and ϕ is in degrees.

The right-hand side of this polynomial approximation reduces to 0.55 in the case of the triangular resonator and to 0.838 in the case of the regular hexagonal geometry.

The result for a regular hexagonal resonator is given by

$$\left(\frac{\omega_+ - \omega_-}{\omega_0}\right)\left(\frac{\mu}{\kappa}\right) = 0.84, \ 0 \le \frac{\kappa}{\mu} \le 0.30 \tag{16.39}$$

The relationship between the split frequencies of the triangular resonator and its gyrotropy has previously been evaluated using perturbation theory. The result is

$$\left(\frac{\omega_+ - \omega_-}{\omega_0}\right)\left(\frac{\mu}{\kappa}\right) = \frac{\sqrt{3}}{\pi}, \ 0 \le \frac{\kappa}{\mu} \le 0.30 \tag{16.40}$$

A comparison between these two relationships indicates that the split frequencies of the hexagonal resonator are somewhat larger than those of the equilateral one. This observation is in keeping with measurements.

The tie between the split frequencies and the coupling angle of a regular hexagonal resonator is indicated in Fig. 16.16. This relationship is plotted here in terms of the

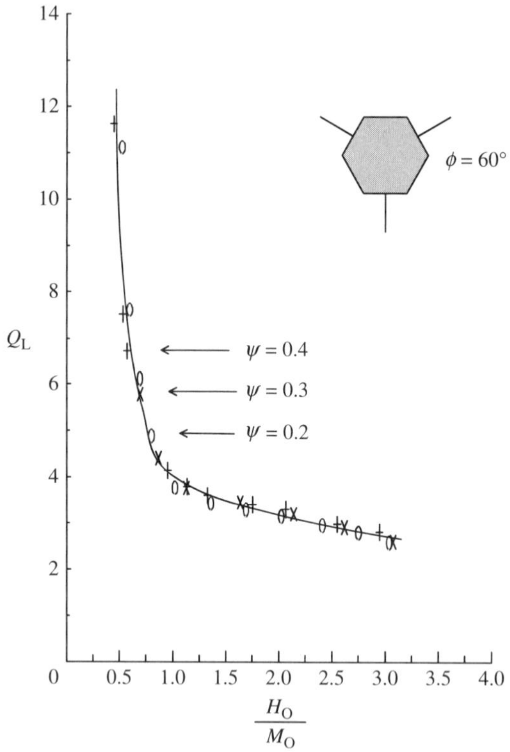

FIGURE 16.16 Quality factor of coupling angle irregular hexagonal resonator for parametric values of $\psi(\phi = 60°)$. (Reproduced with permission from J. Helszajn, M. Mckay, and D. J. Lynch, Complex gyrator circuit of a junction circulator using weakly magnetised planar irregular hexagonal resonator, *IEE Proc. Microwave Antennas Propag.*, Vol. 143, pp. 532–538, Dec. 1996.)

quality factor of the complex gyrator circuit instead of its split frequencies by making use of the connection between the two. This condition is given in Eq. (16.51) and elsewhere in the text. It suggests that the coupling angle at the terminals of the resonator has only a second order effect on the opening between the split frequencies and the gyrotropy of the magnetic insulator.

16.11 QUALITY FACTOR OF CIRCULATOR USING APEX-COUPLED TRIANGULAR RESONATOR

The approach utilized here to obtain the loaded Q-factor of a weakly magnetized junction follows that employed by Fay and Comstock. This quantity is defined by

$$Q_{\mathrm{L}} = \frac{\omega\,(\text{stored energy})}{\text{power dissipated} + \text{power radiated}} \tag{16.41}$$

The quality factor deduced in this way in terms of the physical geometry of the junction must be compatible with that obtained in terms of the split frequencies of the gyromagnetic resonator but must not be confused with it.

In the case of a three-port circulator,

$$Q_{\mathrm{L}} = \frac{\omega U_0}{P_{\mathrm{r1}} + P_{\mathrm{r2}} + P_{\mathrm{r3}}} \tag{16.42}$$

U_0 is the stored energy in the two disks and $P_{\mathrm{r}i}$ is the power radiated out into a typical stripline. For an ideal circulator

$$P_{\mathrm{r1}} = P_{\mathrm{r3}} = 0$$

The total radiated power at port 2 is defined by

$$P_{\mathrm{r2}} = \frac{V_{\mathrm{r2}}^2}{R_0} \tag{16.43}$$

or

$$P_{\mathrm{r2}} = E_{\mathrm{r2}}^2 H^2 Y_0 \tag{16.44}$$

Y_0 is the conductance at the output stripline and also that looking into the resonator circuit, and E_{r2} is the electric field amplitude at the output port. The electric field for an apex-coupled equilateral gyromagnetic triangle is given at $x = A/2\sqrt{3}$, $y = -A/2$ by

$$E_{\mathrm{r2}} = (3\sqrt{3}/2)A_{1,0,-1} \tag{16.45}$$

The power at the same port is then given by

$$P_{\mathrm{r2}} = (27/4)A_{1,0,-1}^2 \cdot H^2 Y_0 \tag{16.46}$$

The quality factor is evaluated in this instance as

$$Q_{\rm L} = 0.095\varepsilon_0\varepsilon_{\rm f}\omega A^2/Y_0 H \tag{16.47}$$

This may be compared with the value for a simple disk, which is described by

$$Q_{\rm L} = 1.48\varepsilon_0\varepsilon_{\rm f}\omega R^2/Y_0 H \tag{16.48}$$

16.12 CIRCULATOR USING SIDE-WALL-COUPLED TRIANGULAR RESONATOR

The symmetry of the triangular resonator indicates that it also supports a triplet of ports along the side walls of the triangle. Inspection of Fig. 16.9 suggests that a combination of $TM_{1,0,-1}$ modes also satisfies this coupling arrangement. Its solution relies on the same field patterns as the apex-coupled one and has the same operating frequency.

The evaluation of the Q-factor for this geometry proceeds as in the previous section. The field pattern and the stored energy are identical to those of the apex-coupled arrangements, but the electric field at port 2 is different. Its value at $(A/2\sqrt{3}, 0)$ is

$$E_{\rm r2} = (\sqrt{3}/2)A_{1,0,-1} \tag{16.49}$$

The corresponding quality factor is

$$Q_{\rm L} = 0.855\varepsilon_0\varepsilon_{\rm f}\omega A^2/Y_0 H \tag{16.50}$$

The Q-factor of this geometry is 9 times that of the apex-coupled circulator and 3 times that of the conventional disk arrangement.

16.13 GYRATOR CONDUCTANCE

A complete description of a junction circulator also requires one of the gyrator conductance. Its derivation proceeds by developing one more description of the quality factor of the junction. One such formulation in terms of the split frequencies of the gyromagnetic resonator is

$$\frac{1}{Q_{\rm L}} = \sqrt{3}\left(\frac{\omega_+ - \omega_-}{\omega_0}\right) \tag{16.51}$$

Combining the two formulations in the case of the apex-coupled geometry produces the required circulation condition and fixes the gyrotropy of the junction.

$$\sqrt{3}\left(\frac{\omega_+ - \omega_-}{\omega_0}\right) = \frac{Y_0 H}{0.855\varepsilon_0\varepsilon_f \omega A^2} \tag{16.52}$$

The gyrator conductance is now obtained by solving the preceding relationship for Y_0. The result is

$$Y_0 = \frac{\sqrt{3}\,0.855(k_e A)^2 \zeta_0 \zeta_{eff}}{k_e H}\left(\frac{\omega_+ - \omega_-}{\omega_0}\right) \tag{16.53}$$

where

$$\zeta_0\zeta_{eff} = \sqrt{\varepsilon_0\varepsilon_f/\mu_0\mu_{eff}} \tag{16.54}$$

The solution for the side-coupled geometry is

$$Y_0 = \frac{\sqrt{3}\;0.095(k_e A)^2 \zeta_0 \zeta_{eff}}{k_e H}\left(\frac{\omega_+ - \omega_-}{\omega_0}\right) \tag{16.55}$$

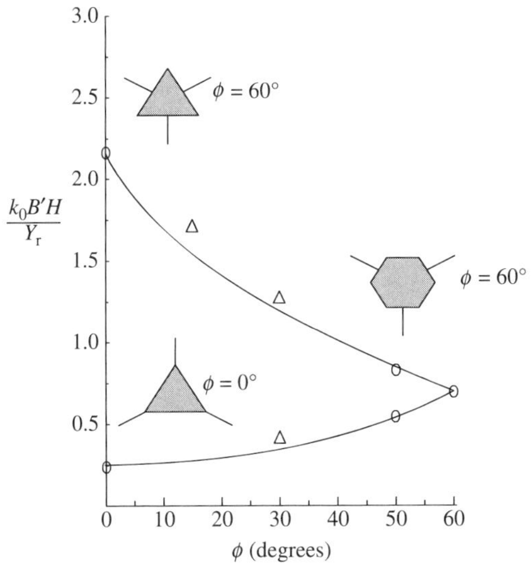

FIGURE 16.17 Relationships between normalized susceptance slope parameter B'/Y_{eff} product and aspect ratio for irregular resonator, $p = 0.28$. (Reproduced with permission from J. Helszajn, M. Mckay, and D. J. Lynch, Complex gyrator circuit of a junction circulator using weakly magnetised planar irregular hexagonal resonator, *IEE Proc. Microwave Antennas Propag.*, Vol. 143, pp. 532–538, Dec. 1996.)

The difference between the split frequencies of a triangular resonator based on a perturbation formulation is given by Eq. (16.40).

The corresponding result in the case of a junction using a disk gyromagnetic resonator is

$$Y_0 = \frac{\sqrt{3}\, 0.740 (k_e R)^2 \zeta_0 \zeta_{eff}}{k_e H} \left(\frac{\omega_+ - \omega_-}{\omega_0} \right) \tag{16.56}$$

16.14 SUSCEPTANCE SLOPE PARAMETERS OF DISK AND TRIANGULAR RESONATORS

Equations (16.47) and (16.48) may be used to extract the absolute susceptance slope parameter of the two geometries:

$$B' = Q_L Y_0 = \frac{0.855 (k_e A)^2 \zeta_0 \zeta_{eff}}{k_e H} \tag{16.57}$$

$$B' = Q_L Y_0 = \frac{0.095 (k_e A)^2 \zeta_0 \zeta_{eff}}{k_e H} \tag{16.58}$$

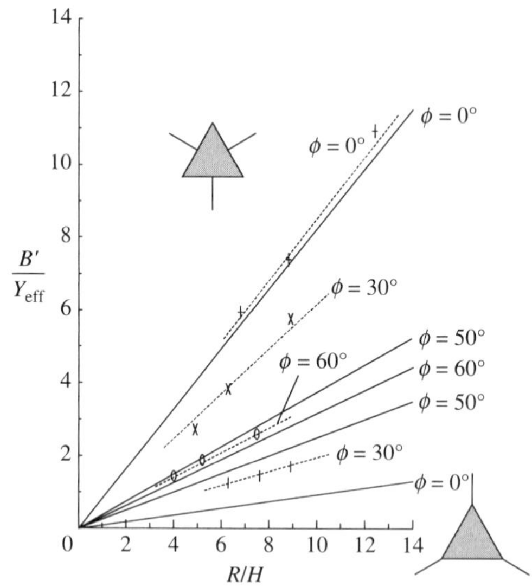

FIGURE 16.18 Relationships between normalized susceptance slope parameter B'/Y_{eff} product and aspect ratio for irregular resonator, $p = 0.28$. (Reproduced with permission from J. Helszajn, M. Mckay, and D. J. Lynch, Complex gyrator circuit of a junction circulator using weakly magnetised planar irregular hexagonal resonator, *IEE Proc. Microwave Antennas Propag.*, Vol. 143, pp. 532–538, Dec. 1996.)

The corresponding result for a disk is

$$B' = Q_L Y_0 = \frac{0.740(k_e R)^2 \zeta_0 \zeta_{\text{eff}}}{k_e H} \tag{16.59}$$

The latter result due to Fay and Comstock reduces to that due to Bosma in Chapter 11 in the limit of wide strips. This may be done by recalling that the ground plane spacing of the junction must be consistent with the definition of the characteristic impedance Z_r of the strips at the terminals of the junction.

A scrutiny of the expressions in the preceding equations indicates that the susceptance slope parameters of irregular planar resonators vary with both the ground plane spacing and the shape angle. The connection between the susceptance slope parameter and the shape factor of an irregular hexagonal resonator is indicated in Fig. 16.17.

Some data on the susceptance slope parameter of a number of circulators using disk-, side-, and apex-coupled triangular resonators with different ground plane spacings are summarized in Fig. 16.18. It compares the calculations deduced here with some experimental ones obtained by measuring the 20 dB bandwidth of different circulators. The case of the disk resonator, due to Bosma, is also plotted. The agreement between theory and experiment suggests the susceptance slope parameters of disk and triangular resonators may be calculated with reasonable accuracy.

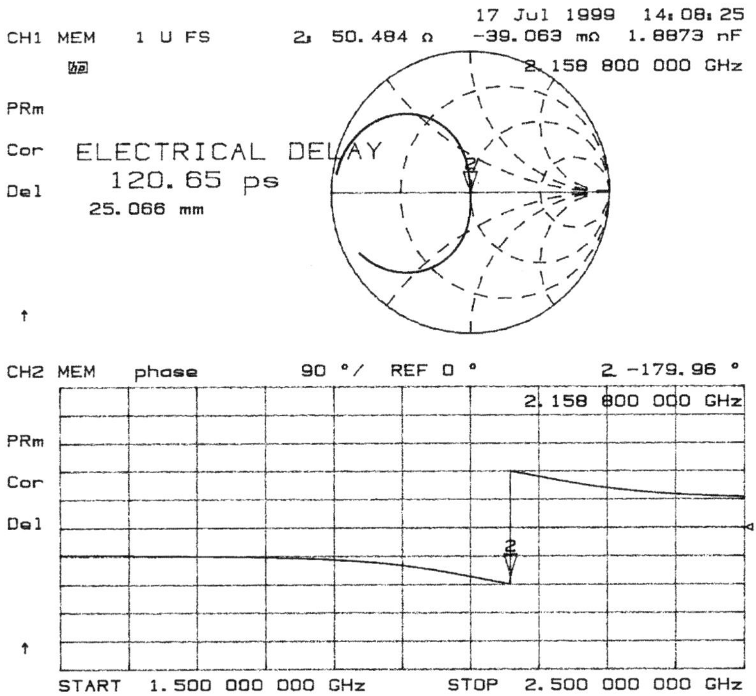

FIGURE 16.19 Definition of reference terminals of junction circulator.

The A dimension, in the case of the triangular resonator, was held constant at 28 mm and its thickness was varied between 1.27 and 3.175 mm. Its center frequency was approximately 2.25 GHz. The radius, in the case of the disk resonator, was held constant at 9.9 mm and its thickness was again varied between 1.27 and 3.175 mm. The center frequency of this arrangement was 2.8 GHz. The data on the side-coupled resonators is restricted to the 3.175 mm substrate as it was not possible to obtain a reliable measurement on the other substrates because of the relatively large values of susceptance slope parameter displayed by these geometries. The material used in these circulators had a magnetization $\mu_0 M_0$ equal to 0.0400 T and a dielectric constant of ε_f equal to15.3.

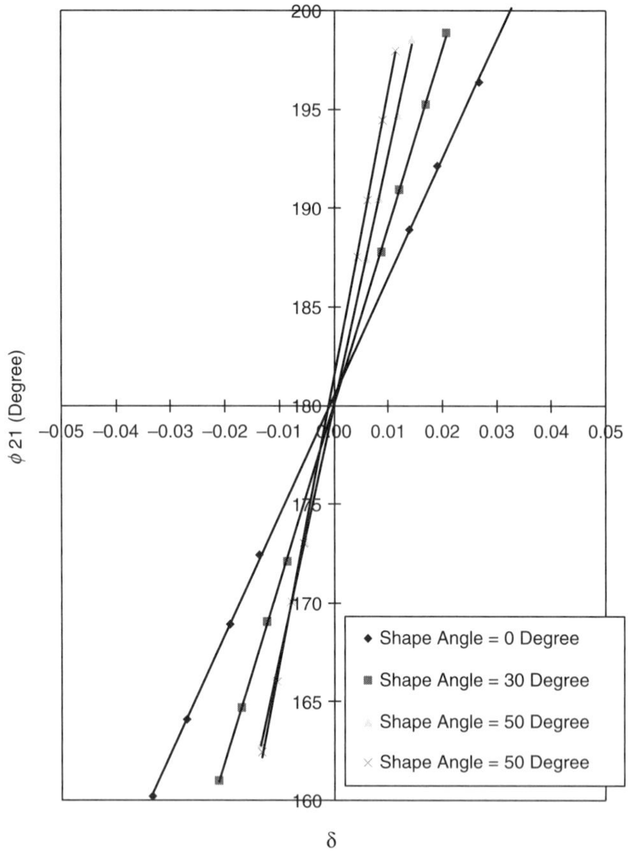

FIGURE 16.20 Experimental relationships between transmission phase and frequency for parametric values of shape angle of irregular hexagonal resonator. (Reproduced with permission from J. Helszajn, W. D'Orazio, and M. Caplin, Insertion phase and phase slope parameter of microwave junction circulators, *IEE Proc. Microwaves, Antennas and Propagation*, Vol. 151, pp. 54–60, Feb. 2004.)

16.15 TRANSMISSION PHASE ANGLE OF DEGREE-1 CIRCULATOR

The insertion phase of a junction circulation using an irregular hexagonal gyromagnetic resonator is the topic of this paper.

The geometry employed here is a degree-1 stripline circulator using an irregular hexagonal gyromagnetic resonator operating at about 2.0 GHz. This sort of topology may be described by one radius (r) and one shape angle (ϕ) in the manner indicated in Fig. 16.12. The shape angles employed in this work are 0°, 30°, 50°, and 60°, respectively. The circumscribed radii are 16.54, 13.84, 13.72, and 13.54 mm and the thickness is 1.65 mm. The saturation magnetization of the material $\mu_0 M_0$ is equal to 0.0475 T.

One important aspect in the calibration of the insertion phase of a two-port network is the determination of its reference terminals. In practice, the electrical reference terminals do not coincide with the physical planes. The calibration procedure

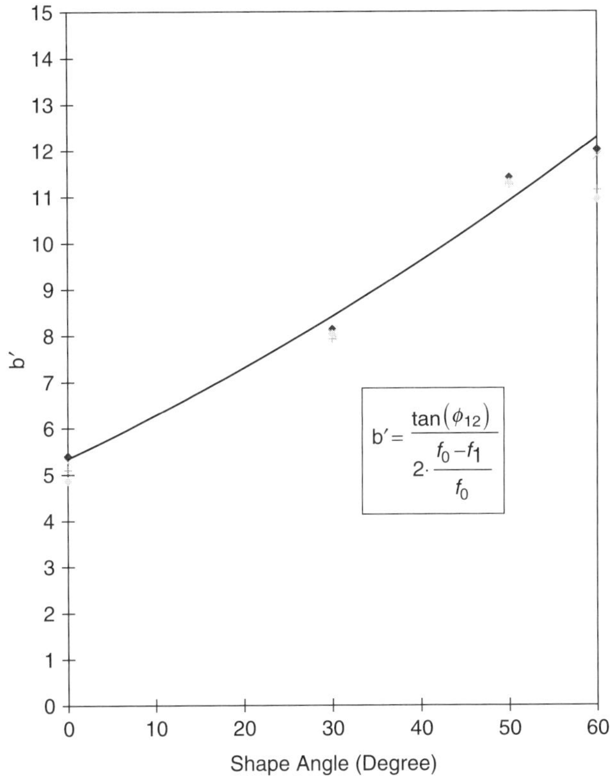

FIGURE 16.21 Experimental relationship between susceptance slope parameter and shape angle of irregular hexagonal resonator. (Reproduced with permission from J. Helszajn, W. D'Orazio, and M. Caplin, Insertion phase and phase slope parameter of microwave junction circulators, *IEE Proc. Microwaves, Antennas and Propagation*, Vol. 151, pp. 54–60, Feb. 2004.)

employed here, in order to obtain the insertion phase between ports 1 and 2 of the junction, commences by assuming that this is the case. The required condition is then established by aligning the reflection coefficient at port 1 with a constant conductance circle. Figure 16.19 depicts one typical situation.

Figure 16.20 indicates the relationship between the insertion phase and the frequency for parametric values of the shape angle of the irregular hexagonal resonator. The one between the susceptance slope parameter and the shape angle of the resonator is illustrated separately in Fig. 16.21. Separate measurements were undertaken at the 17 dB, 20 dB, and 23 dB points of the return loss.

CHAPTER SEVENTEEN

Operation of the Tracking and Semitracking Stripline Circulators

17.1 INTRODUCTION

An important stripline circulator solution, which has a nearly frequency independent resistance over an octave band, is the tracking one. Its first circulation condition coincides with the frequency at which the two counterrotating eigen-networks exhibit complex conjugate immittances and the in-phase eigen-network may be idealized by a nearly frequency independent short-circuit. This boundary condition produces a unique coupling angle and is satisfied by describing the eigenvalue of the in-phase eigen-network in terms of the $n = 0, \pm 3$ modes or poles and those of the counterrotating ones in terms of the $n = +1, -2$, and $n = -1, +2$ modes or poles. The second circulation condition merely involves the calculation of the gyrator resistance using the counterrotating eigenvalues. Two other essential requirements for the operation of this sort of circulator are that the ferrite material is saturated and that its gyrotropy is bracketed by $0.50 \leq \kappa \leq 1.0$ over the frequency interval in question. In a circulator, using a weakly magnetized resonator, the in-phase eigen-network is adequately described by the $n = 0$ mode or pole (although it is often idealized by a frequency independent short-circuit) and the counterrotating ones by the $n = +1$ and -1 modes or poles. The eigenvalue problem for which the eigen-networks are restricted to single resonator modes or poles is readily extended to the situation for which higher order modes or poles are necessary by recognizing that these may be realized in a first Foster form expansion of the resonator poles. Using the rotational properties of the eigenvectors it is possible, by inspection, to

The Stripline Circulator: Theory and Practice. By J. Helszajn

distribute the poles of the problem region among the counterrotating and in-phase eigen-networks. This notation has the advantage that existing literature, which has been formulated in terms of eigen-networks, supporting single resonator modes, is readily extended in the case where higher order modes of the junction are required to satisfy the boundary conditions at terminals. One important field of semitracking solutions in the vicinity of the tracking has also been formulated. It is given special attention.

17.2 EIGENVALUES OF TRACKING CIRCULATORS

The boundary conditions of a symmetrical three-port junction circulator may be described either in terms of its scattering, impedance, or admittance matrices, or in terms of the corresponding reflection, impedance, or admittance eigenvalues. The three sets of eigenvalues are the scalar variables of three one-port networks known as the eigen-networks of the junction. A physical understanding of the junction requires the identification of the eigenvalue problem.

The derivation of the tracking solution of the three-port junction circulator is facilitated by segregating its in-phase and counterrotating poles between the impedance eigenvalues of the junction. Inspection of the electromagnetic problem indicates that this may be done by realizing the junction eigen-networks in a first Foster

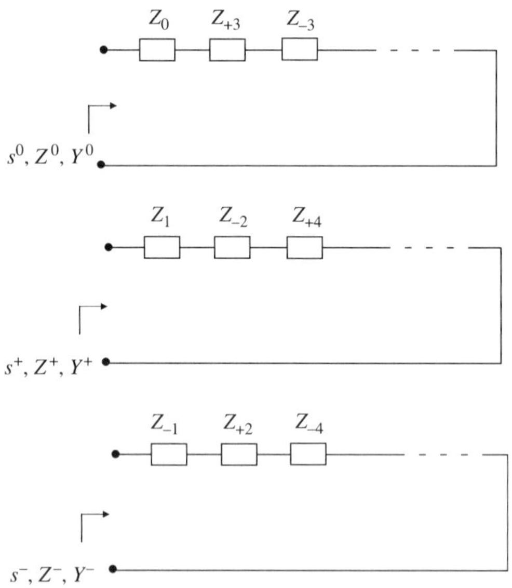

FIGURE 17.1 First Foster form realization of eigen-networks of three-port junction circulator.

form (rather than in a second Foster form) one-port reactance network in the manner illustrated in Fig. 17.1.

$$Z^0 = \sum Z_n, \quad n = 0,\ \pm 3,\ \pm 6,\ \pm 9, \ldots \tag{17.1}$$

$$Z^+ = \sum Z_n, \quad n = +1,\ -2,\ +4,\ -5,\ +7, \ldots \tag{17.2}$$

$$Z^- = \sum Z_n, \quad n = -1,\ +2,\ -4,\ +5,\ -7, \ldots \tag{17.3}$$

This formulation has the advantage that once the required terms in each summation have been determined, the other eigenvalues and matrix relations are readily evaluated in the normal way.

$$Y^q = 1/Z^q \tag{17.4}$$

$$S^q = (Y_0 - Y^q)/(Y_0 + Y^q) \tag{17.5}$$

with $q = 0, +, -$.

The open-circuit impedance parameters of the three-port junction are expanded in terms of the eigenvalues in the usual way.

$$Z_{11} = \frac{Z^0 + Z^+ + Z^-}{3} \tag{17.6}$$

$$Z_{12} = \frac{Z^0 + \alpha Z^+ + \alpha^2 Z^-}{3} \tag{17.7}$$

$$Z_{13} = \frac{Z^0 + \alpha^2 Z^+ + \alpha Z^-}{3} \tag{17.8}$$

The admittance matrix does not normally exist at the terminals of the junction in the case of a conventional stripline circulator.

In the case of a disk resonator the poles of Z^+, Z^-, and Z^0 are defined by the coupling angle ψ and the thickness H of each half-resonator in keeping with the development in Chapter 13:

$$Z_n = \frac{j3\sqrt{\mu_{\mathrm{eff}}} R_{\mathrm{f}} \psi}{\pi} \left(\frac{\sin n\psi}{n\psi}\right)^2 \left[\frac{J_{n-1}(kR)}{J_n(kR)} - n\left(\frac{1 + \kappa/\mu}{kR}\right)\right]^{-1} \tag{17.9}$$

where

$$R_{\mathrm{f}} = R_{\mathrm{r}}/\sqrt{\varepsilon_{\mathrm{f}}} \tag{17.10}$$

$$R_{\mathrm{r}} = 30\pi \ln\left(\frac{W + t + 2H}{W + t}\right) \tag{17.11}$$

$$k = 2\pi\sqrt{\varepsilon_{\mathrm{f}}\mu_{\mathrm{eff}}}/\lambda_0 \tag{17.12}$$

$$\sin\psi = W/2R \tag{17.13}$$

and the other variables have the usual meaning. The schematic diagram of the classic planar circulator using a simple disk resonator discussed here is depicted in Chapter 13.

A scrutiny of the poles entering into the description of the eigenvalues of the problem region indicates that all the poles have a zero at the origin except for the $n = 0$, which has a pole there. This may readily be appreciated by noting the asymptotic form of $J_n(x)$ as x goes to zero.

$$J_n(x) \approx \frac{1}{n!}\left(\frac{x}{2}\right)^n \tag{17.14}$$

An essential prerequisite for the operation of the tracking circulator is that the ferrite material is saturated. This implies that

$$\kappa = \omega_{\mathrm{m}}/\omega \tag{17.15a}$$

$$\mu = 1 \tag{17.15b}$$

and

$$\mu_{\mathrm{eff}} = 1 - (\omega_{\mathrm{m}}/\omega)^2 \tag{17.15c}$$

where

$$\omega_{\mathrm{m}} = \gamma M_0$$

$\gamma = 2.21 \times 10^5$ (rad/s per A/m), M_0 is in A/m, ω is in rad/s, $\mu_0 = 4\pi \times 10^{-7}$ H/m, and $\mu = 1$ in a saturated material.

Making use of the preceding relationships allows Eq. (17.12) to be expressed as

$$k = \left(\frac{\omega_m \sqrt{\varepsilon_f \mu_{eff}}}{c\kappa} \right) = k_m \frac{\sqrt{\varepsilon_f \mu_{eff}}}{\kappa} \tag{17.16}$$

where $k_m = \omega_m/c$, and c is the free space velocity.

17.3 COMPLEX GYRATOR CIRCUIT

The nature of the gyrator impedance of the circulator is the topic of this section. It is defined in Chapter 6 by

$$Z_{in} = Z_{11} - Z_{12}^2/Z_{13} \tag{17.17}$$

This equation is deduced by writing $V_3 = I_3 = 0$ in obtaining Z_{in} in terms of the open-circuit parameters.

In what follows, it is useful to express the real and imaginary parts of the preceding condition in terms of the eigenvalues of the problem region. The results are

$$R_{in} = \frac{jB(B^2 - 3A^2)}{3(A^2 + B^2)} \tag{17.18a}$$

$$X_{in} = \left(Z^0 + \frac{A(B^2 - 3A^2)}{3(A^2 + B^2)} \right) \tag{17.18b}$$

where

$$A = (-1/2)(Z^+ + Z^-) + Z^0 \tag{17.19}$$

$$B = (\sqrt{3}/2)(Z^+ - Z^-) \tag{17.20}$$

The eigenvalues Z^0, Z^-, and Z^+ are pure imaginary numbers, so that the imaginary part in the expression of the complex gyrator impedance is the gyrator resistance, and the real part is the reactance part of Z_{in}.

Figures 17.2 and 17.3 illustrate the two standard circulation conditions obtained by retaining the first seven poles of the eigenvalues ($n = 0, \pm 1, \pm 2, \pm 3$) in the description of the gyrator impedance according to the scheme in Eqs. (17.1)–(17.3). The illustrations are obtained by setting the imaginary part of the standard complex gyrator impedance (or admittance) equal to zero and computing the corresponding real part.

The classic circulation solution is approximately defined by $|\kappa/\mu|$ in the interval 0 to 0.50 on these illustrations (with ψ variable) and its three eigen-networks shown in

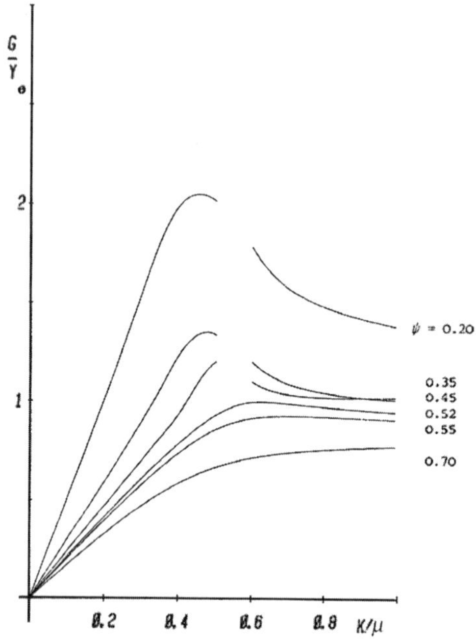

FIGURE 17.2 First circulation solution of junction circulator using disk resonator. (Reproduced with permission from J. Helszajn, Theory of tracking circulators, *IEEE Trans. Microwave Theory Tech.*, Vol. MTT-29, pp. 700–707, July 1981.)

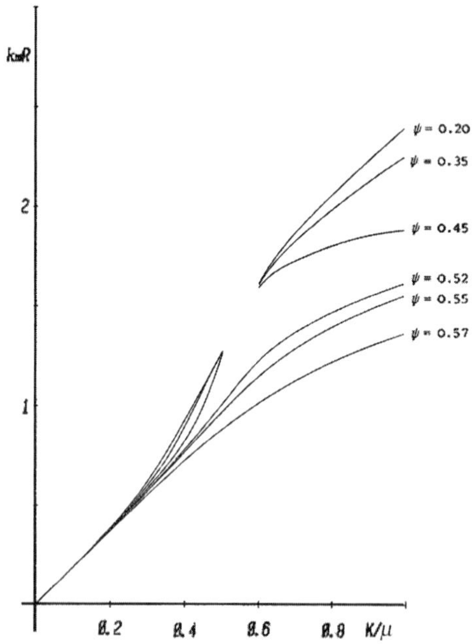

FIGURE 17.3 Second circulation solution of junction circulator using disk resonator. (Reproduced with permission from J. Helszajn, Theory of tracking circulators, *IEEE Trans. Microwave Theory Tech.*, Vol. MTT-29, pp. 700–707, July 1981.)

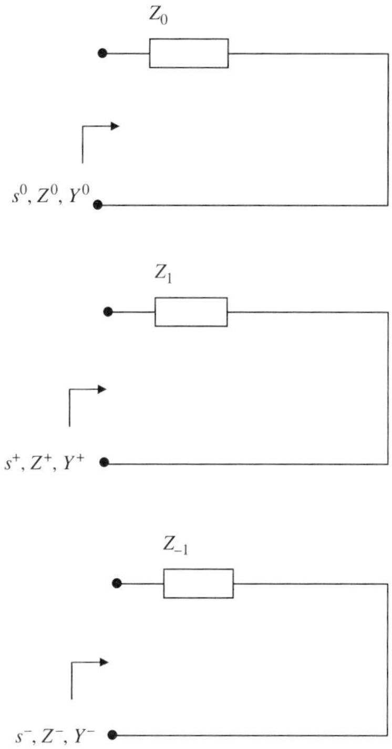

FIGURE 17.4 Eigen-networks of stripline circulator with off-diagonal entry κ of permeability tensor in vicinity of zero.

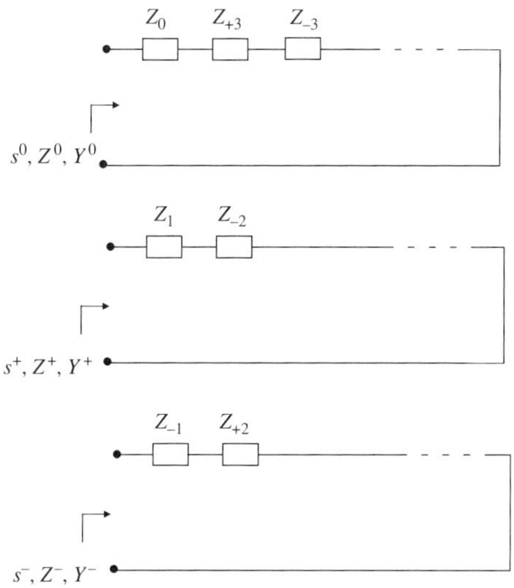

FIGURE 17.5 Eigen-networks of stripline circulator with off-diagonal entry κ of permeability tensor in tracking circulator.

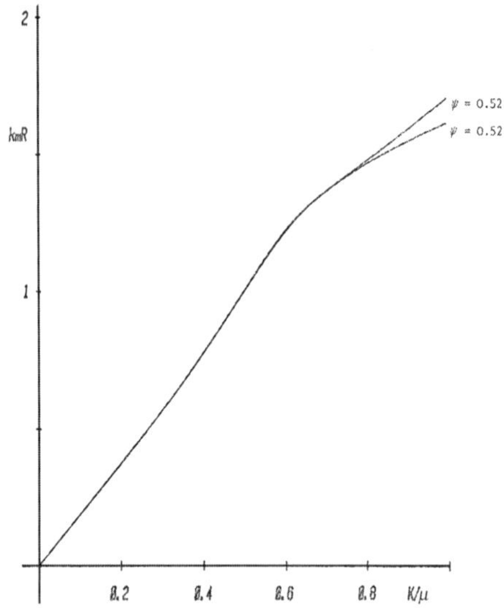

FIGURE 17.6 Comparison between exact and approximate first circulation solutions of junction circulator using disk resonator. (Reproduced with permission from J. Helszajn, Theory of tracking circulators, *IEEE Trans. Microwave Theory Tech.*, Vol. MTT-29, pp. 700–707, July 1981.)

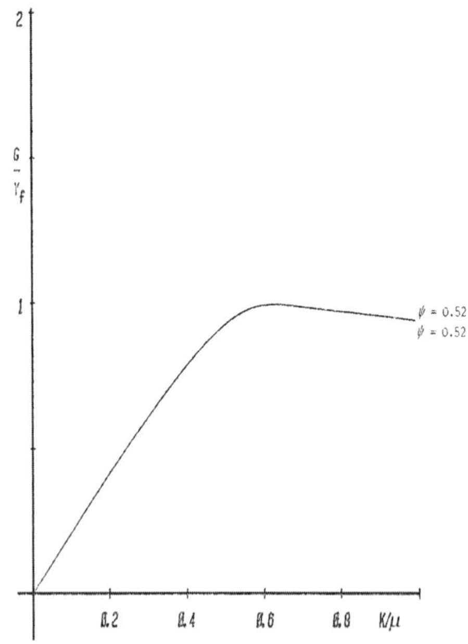

FIGURE 17.7 Comparison between exact and approximate second circulation solutions of junction circulator using disk resonator. (Reproduced with permission from J. Helszajn, Theory of tracking circulators, *IEEE Trans. Microwave Theory Tech.*, Vol. MTT-29, pp. 700–707, July 1981.)

Fig. 17.4 support the $n = 0, +1, -1$ resonances of the disk resonator. These are the Bosma and Fay and Comstock solutions. The Wu and Rosenbaum tracking solution requires the $n = 0, \pm 1, \pm 2$, and ± 3 modes for its description and operates with the magnetic variable $|\kappa|$ between 0.5 and 1 with the coupling angle $\psi \approx 0.55$, but no closed form representation is available for it. For $0 \leq \psi \leq 0.50$, the solution of $k_m R$ is ill defined for $\kappa/\mu \approx 0.50$, and this region is left blank in Figs. 17.2 and 17.3. Figure 17.5 depicts the corresponding eigen-networks.

An approximate equivalent circuit of the complex gyrator has been derived in Chapter 11. It is verified here. This has been done by expressing the imaginary part of the input impedance in terms of its real part as a preamble to neglecting the in-phase term in the description of the real part.

The agreements between the two circulation conditions, for $\psi = 0.52$, based on the approximate real and imaginary parts of the complex gyrator circuit and the exact quantities are depicted in Figs. 17.6 and 17.7.

17.4 THREE EIGEN-NETWORK THEORY OF TRACKING CIRCULATOR

The purpose of this section is to show that the first Wu and Rosenbaum boundary condition coincides with the frequency at which the two counterrotating eigen-networks have complex conjugate immittances, and the in-phase eigen-network has a short-circuit boundary condition. The second boundary condition is approximately satisfied by assuming that the counterrotating eigen-networks support the $n = +1, -2$ and $n = -1, +2$ resonator modes.

The first condition may be demonstrated by simultaneously satisfying

$$Z^0 = Z_0 + Z_{+3} + Z_{-3} = 0 \tag{17.21a}$$

$$Z^+ + Z^- = (Z_{+1} + Z_{-2}) + (Z_{-1} + Z_{+2}) = 0 \tag{17.21b}$$

The first equation ensures that Z^0 displays a short-circuit boundary condition at the input terminals of the device, and the second one ensures that the operating frequency coincides with that at which the two counterrotating eigen-networks exhibit complex conjugate immittances. The simultaneous solution of these two conditions is a unique solution to Eq. (17.18). The first seven entries in Table 17.1 give the required result over the whole field of variables. It is apparent from this data that the Wu and Rosenbaum solution given by $\psi \approx 0.522$ rad, $\kappa/\mu \approx 0.67$, and $kR \approx 1.465$ is a compatible with the three eigen-network theory articulated here.

Once the first circulation (frequency) is satisfied, the second one (gyrator level) may be calculated by having recourse to the real part of the complex gyrator impedance,

$$R_{\text{in}} = j\left(\frac{Z^+ - Z^-}{2\sqrt{3}}\right) \tag{17.22}$$

TABLE 17.1 Circulation Conditions for Tracking Circulators

κ/μ	ψ	kR	R	Z	G	B'	Q
0.10	0.5428	1.8371	$5.0056\,R_f$	$-33.6200\,R_f$	$0.1998\,Y_f$	$1.341S\,Y_f$	6.7165
0.20	0.5449	1.8238	$2.5381\,R_f$	$-7.8322\,R_f$	$0.3940\,Y_f$	$1.2155\,Y_f$	3.0852
0.30	0.5473	1.7984	$1.7566\,R_f$	$-3.0590\,R_f$	$0.5758\,Y_f$	$1.0144\,Y_f$	1.7615
0.40	0.5483	1.7550	$1.3586\,R_f$	$-1.4011\,R_f$	$0.7360\,Y_f$	$0.7590\,Y_f$	1.0313
0.50	0.5449	1.6338	$1.1590\,R_f$	$-0.6638\,R_f$	$0.8628\,Y_f$	$0.4942\,Y_f$	0.5728
0.60	0.5344	1.5727	$1.0556\,R_f$	$-0.3095\,R_f$	$0.9474\,Y_f$	$0.2777\,Y_f$	0.2932
0.67	0.5224	1.4650	$1.0181\,R_f$	$-0.1803\,R_f$	$0.9823\,Y_f$	$0.1740\,Y_f$	0.1771
0.70	0.5163	1.4103	$1.0079\,R_f$	$-0.1436\,R_f$	$0.9922\,Y_f$	$0.1413\,Y_f$	0.1424
0.80	0.4928	1.1833	$0.9898\,R_f$	$-0.0706\,R_f$	$1.0103\,Y_f$	$0.0721\,Y_f$	0.0714
0.90	0.4663	0.8552	$0.9851\,R_f$	$-0.0406\,R_f$	$1.0152\,Y_f$	$0.0418\,Y_f$	0.0412
0.95	0.4526	0.6105	$0.9847\,R_f$	$-0.0336\,R_f$	$1.0156\,Y_f$	$0.0346\,Y_f$	0.0341

Source: J. Helszajn, Operation of tracking circulators, *IEEE Trans. Microwave Theory Tech.*, Vol. MTT-29, pp. 700–707, July 1981. Reproduced with permission.

where

$$Z^{+} \approx Z_{+1} + Z_{-2} \tag{17.23a}$$

$$Z^{-} \approx Z_{-1} + Z_{+2} \tag{17.23b}$$

Evaluating Z_n, the counterrotating poles of the problem region, with $\psi = 0.52244$, $kR = 1.46503$, and $\kappa/\mu = 0.67$, yields

$$\begin{aligned} Z_{+1} &= -j1.88869R_{\mathrm{f}} \\ Z_{-1} &= +j0.45920R_{\mathrm{f}} \\ Z_{+2} &= +j1.30413R_{\mathrm{f}} \\ Z_{-2} &= +j0.12537R_{\mathrm{f}} \end{aligned}$$

The corresponding eigenvalues are

$$\begin{aligned} Z^{+} &= -j1.76332R_{\mathrm{f}} \\ Z^{-} &= +j1.76333R_{\mathrm{f}} \end{aligned}$$

and the gyrator conductance is

$$R_{\mathrm{in}} = 1.01805R_{\mathrm{f}}$$

in agreement with the appropriate entry in column 7 of Table 17.1. Evaluating R_{f} in Eq. (17.10) with $\varepsilon_{\mathrm{f}} = 15.3$ and $R_{\mathrm{r}} = 50\,\Omega$ gives the gyrator resistance of the circulator as

$$R_{\mathrm{in}} = 13.00\,\Omega$$

For completeness, the magnitudes of the in-phase poles are

$$\begin{aligned} Z_{0} &= -j0.35593R_{\mathrm{f}} \\ Z_{+3} &= +j0.30928R_{\mathrm{f}} \\ Z_{-3} &= +j0.04666R_{\mathrm{f}} \end{aligned}$$

The in-phase eigenvalue is therefore equal to zero as asserted:

$$Z^{0} = 0$$

The eigenvalues and poles at the circulation conditions are tabulated in Table 17.2.

TABLE 17.2 Circulation Poles of Tracking Circulators

κ/μ	Z_0	Z_3	Z_{-3}	Z_1	Z_{-1}	Z_2	Z_{-2}	Z^0	Z^+	Z^-
0.10	$-0.2822\ R_f$	$0.1577\ R_f$	$0.1246\ R_f$	$-9.0644\ R_f$	$8.1430\ R_f$	$0.5273\ R_f$	$0.3949\ R_f$	$0.00001\ R_f$	$-8.6695\ R_f$	$8.6703\ R_f$
0.20	$-0.2858\ R_f$	$0.1763\ R_f$	$0.1095\ R_f$	$-4.7372\ R_f$	$3.7826\ R_f$	$0.6138\ R_f$	$0.3414\ R_f$	$0.00001\ R_f$	$-4.3959\ R_f$	$4.3965\ R_f$
0.30	$-0.2923\ R_f$	$0.1972\ R_f$	$0.0951\ R_f$	$-3.3006\ R_f$	$2.2870\ R_f$	$0.7206\ R_f$	$0.2926\ R_f$	$0.00000\ R_f$	$-3.0081\ R_f$	$3.0077\ R_f$
0.40	$-0.3027\ R_f$	$0.2213\ R_f$	$0.0815\ R_f$	$-2.5995\ R_f$	$1.5004\ R_f$	$0.8529\ R_f$	$0.2464\ R_f$	$0.00000\ R_f$	$-2.3531\ R_f$	$2.3533\ R_f$
0.50	$-0.3182\ R_f$	$0.2498\ R_f$	$0.0684\ R_f$	$-2.2085\ R_f$	$0.9963\ R_f$	$1.0111\ R_f$	$0.2101\ R_f$	$0.00000\ R_f$	$-2.0074\ R_f$	$2.0074\ R_f$
0.60	$-0.3390\ R_f$	$0.2834\ R_f$	$0.0557\ R_f$	$1.9845\ R_f$	$0.6436\ R_f$	$1.1847\ R_f$	$0.1562\ R_f$	$-0.00000\ R_f$	$-1.8283\ R_f$	$1.8283\ R_f$
0.67	$\sim 0.3559\ R_f$	$0.3093\ R_f$	$0.0467\ R_f$	$-1.8887\ R_f$	$0.4592\ R_f$	$1.3041\ R_f$	$0.1254\ R_f$	$0.00001\ R_f$	$-1.7633\ R_f$	$1.7633\ R_f$
0.70	$-0.3635\ R_f$	$0.3208\ R_f$	$0.0427\ R_f$	$-1.8581\ R_f$	$0.3929\ R_f$	$1.3528\ R_f$	$0.1124\ R_f$	$0.00001\ R_f$	$-1.7456\ R_f$	$1.7457\ R_f$
0.80	$-0.3884\ R_f$	$0.3593\ R_f$	$0.0290\ R_f$	$-1.7855\ R_f$	$0.2157\ R_f$	$1.4936\ R_f$	$0.0711\ R_f$	$-0.00001\ R_f$	$-1.7144\ R_f$	$1.7144\ R_f$
0.90	$-0.4111\ R_f$	$0.3964\ R_f$	$0.0147\ R_f$	$-1.7378\ R_f$	$0.0902\ R_f$	$1.6160\ R_f$	$0.0335\ R_f$	$0.00001\ R_f$	$-1.7063\ R_f$	$1.7062\ R_f$
0.95	$-0.4211\ R_f$	$0.4138\ R_f$	$0.0076\ R_f$	$-1.7218\ R_f$	$0.0415\ R_f$	$1.6640\ R_f$	$0.0162\ R_f$	$0.00001\ R_f$	$-1.7056\ R_f$	$1.7055\ R_f$

Source: J. Helszajn, Operation of tracking circulators, *IEEE Trans. Microwave Theory Tech.*, Vol. MTT-29, pp. 700–707, July 1981. Reproduced with permission.

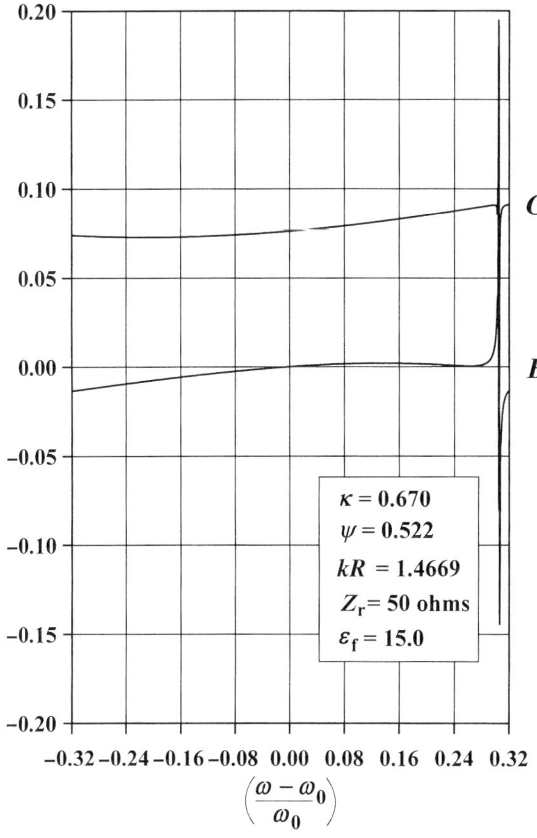

$G = 0.0762, B' = 0.0133, Q_L = 0.1746$

FIGURE 17.8 Variation of real and imaginary parts of tracking circulator with $\kappa/\mu = 0.67$, $kR = 1.46$, $\psi = 0.52$. (Reproduced with permission from J. Helszajn, Theory of tracking circulators, *IEEE Trans. Microwave Theory Tech.*, Vol. MTT-29, pp. 700–707, July 1981.)

Figure 17.8 indicates the variation of the real and imaginary parts of Z_{in} with frequency for this boundary condition. It shows that its equivalent circuit in this case is well behaved, as asserted. It is of note that the loaded Q-factor obtained graphically from this illustration is in excellent agreement with the value tabulated in Table 17.1. The gyromagnetic space of the tracking solution is indicated by the shaded area in Fig. 17.9.

17.5 SYNTHESIS OF SEMITRACKING STRIPLINE JUNCTION CIRCULATORS

The tracking gyrator circuit is characterized by a nearly frequency independent conductance over approximately an octave frequency band. However, it is not an

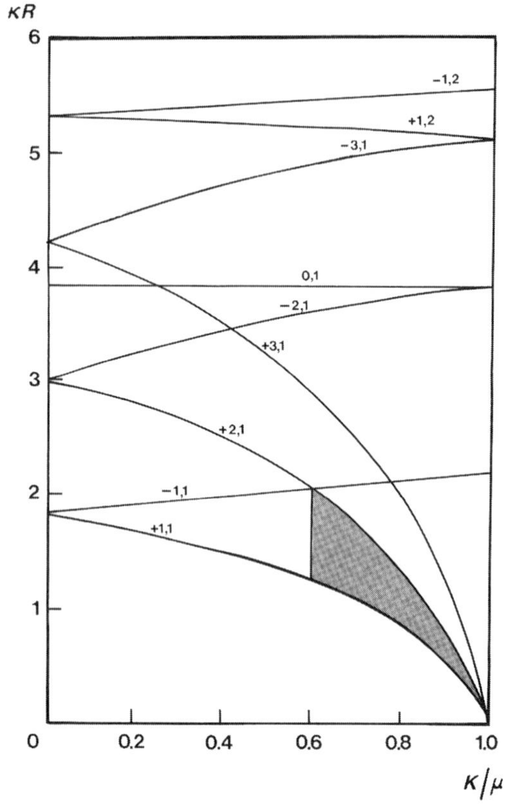

FIGURE 17.9 Gyromagnetic mode chart of the tracking circulator.

optimum solution in so far as the network problem is concerned in that a complex rather than a real gyrator circuit is to be preferred. A field of semitracking solutions can in fact be realized in the vicinity of the tracking one by perturbing either the coupling angle of the junction or its magnetization. The two most important parameters in the synthesis of a semitracking circulator are again the quality factor of the junction and the frequency interval over which its complex gyrator circuit has a nearly frequency independent conductance and a nearly constant susceptance slope. Once these quantities are fixed (by the coupling angle of the resonator and its gyrotropy), it is merely necessary to adjust the absolute levels of the real and imaginary parts of the junction (with the aid of the ground plane spacing) to meet the required network specification. A description of the junction in terms of its complex gyrator admittance and loaded Q-factor is therefore again necessary and sufficient. Although no closed form result is in general available, it may be derived graphically from a knowledge of the frequency response of the device. A detailed investigation of this problem indicates that a host of semitracking solutions may be realized in the vicinity of the tracking one by properly adjusting the details of the junction. One possibility is the design of a degree-3 equal ripple frequency response over one octave. The bandwidth of semitracking circulators is restricted to about 75% unless some precautions are taken to avoid so-called low-field loss. One way to avoid this type of loss is to ensure that every region within the ferrite resonator is equally magnetized. A feature of quarter-wave coupled circulators using semitracking junctions is that the relative dielectric constant of the transformer regions is small compared to that of the ferrite material. This is a somewhat fortunate situation in that it ensures that the usual assumed magnetic wall boundary condition between the circulator ports is consistent with practice and ensures fair agreement between theory and experiment.

17.6 SEMITRACKING CIRCULATION SOLUTIONS

Circulator solutions using semitracking gyrator circuits may be constructed by varying the coupling angle (ψ) and/or the gyrotropy (κ/μ) of the resonator about the values given by the Wu and Rosenbaum tracking solution. A complete description of this class of device requires a knowledge of the complex gyrator circuit and loaded Q-factor of the junction. These quantities are evaluated in this section by having recourse to a calculation of the complex gyrator admittance of the circulator in the neighborhood of the first circulation condition of Davies and Cohen. It is also necessary to investigate to what extent these variables hold over the desired frequency interval of the device. This may be done numerically by expanding the two circulation conditions in the vicinity of the first one. This expansion is also necessary in order to analyze the frequency response of quarter-wave coupled devices.

The complex gyrator circuit in the admittance plane is described in terms of its real and imaginary parts in Eqs. (17.18a) and (17.18b) by

$$G_{\text{in}} = \frac{R_{\text{in}}}{R_{\text{in}}^2 + X_{\text{in}}^2} \tag{17.24a}$$

$$B_{\text{in}} = \frac{-jX_{\text{in}}}{R_{\text{in}}^2 + X_{\text{in}}^2} \tag{17.24b}$$

The second of the two conditions defines the planar circuit in terms of ψ, κ/μ, and kR, and the first fixes the absolute conductance level in terms of the free space admittance (Y_r) defined by the resonator terminals.

The susceptance slope parameter is determined graphically from a knowledge of B_{in} in the vicinity of the first circulation condition:

$$B' = \frac{k_0R}{2}\left[\frac{B_{\text{in}}(\delta) - B_{\text{in}}(-\delta)}{k_0R(1+\delta) - k_0R(1-\delta)}\right] = \frac{B_{\text{in}}(\delta) - B_{\text{in}}(-\delta)}{4\delta} \tag{17.25}$$

δ is the frequency deviation about ω_0:

$$\delta = (\omega - \omega_0)/\omega_0 \tag{17.26}$$

Finally, the quality factor (Q_L) of the circuit is calculated by forming

$$Q_L = B'/G_0 \tag{17.27}$$

Table 17.3 depicts semitracking solutions with κ/μ and ψ in the neighborhood of the tracking one. Such semitracking solutions are particularly attractive for the design of octave band devices.

Intermediate values of ψ and Q_L may be approximated in terms of the first two terms of a Taylor expansion or by some more elaborate interpolation procedure.

$$\psi_i = \psi_{i-1} + \frac{\psi_{i-1} - \psi_{i+1}}{Q_{i-1} - Q_{i+1}}(Q_i - Q_{i-1}) \tag{17.28}$$

i indicates the required quantities; $i-1$ and $i+1$ refer to the known quantities.

The physical variables in the tables may be understood by constructing an example. One possibility is defined by

$$\begin{aligned} \kappa &= \omega_m/\omega \approx 0.6 \\ \mu &= 1 \\ \psi &\approx 0.7\ \text{rad} \\ kR &= 1.35\ \text{rad} \end{aligned}$$

The absolute gyrator conductance and susceptance slope parameter displayed by the above variables are

$$\begin{aligned} G &= 0.7089Y_f \\ B' &= 0.4445Y_f \end{aligned}$$

where

$$Y_f = Y_r\sqrt{\varepsilon_f}$$

The corresponding quality factor is

$$Q_L = 0.6263$$

Y_r is the free space conductance of a typical port at the junction. In a directly coupled junction it is usually chosen as 0.02 mhos, but in a quarter-wave coupled one, with $(n-1)$ transformer sections, it is used to adjust the gyrator conductance by varying the ground plane spacing H in order to satisfy the network problem. The dielectric

TABLE 17.3 Circulation Conditions for Semitracking Circulators

ψ	kR	G	B'	Q_L
		$\kappa = 0.525$, $\mu = 1$		
0.543*	1.6610	0.8880 Y_f	0.4333 Y_f	0.4880
0.550*	1.6453	0.8727 Y_f	0.4480 Y_f	0.5138
0.575*	1.6018	0.8238 Y_f	0.4855 Y_f	0.5894
0.600*	1.5685	0.7841 Y_f	0.5074 Y_f	0.6472
0.625	1.5413	0.7506 Y_f	0.5197 Y_f	0.6924
0.650	1.5183	0.7219 Y_f	0.5253 Y_f	0.7275
0.675	1.4985	0.6971 Y_f	0.5267 Y_f	0.7543
0.700	1.4810	0.6763 Y_f	0.5233 Y_f	0.7739
		$\kappa = 0.575$, $\mu = 1$		
0.538*	1.6062	0.9302 Y_f	0.3486 Y_f	0.3486
0.550	1.5727	0.9011 Y_f	0.3628 Y_f	0.4021
0.575	1.5241	0.8507 Y_f	0.4118 Y_f	0.4842
0.600	1.4879	0.8098 Y_f	0.5446 Y_f	0.5446
0.625	1.4589	0.7757 Y_f	0.5905 Y_f	0.5905
0.650	1.4347	0.7466 Y_f	0.6254 Y_f	0.6254
0.675	1.4139	0.7216 Y_f	0.4670 Y_f	0.6513
0.700	1.3956	0.7002 Y_f	0.4689 Y_f	0.6697
		$\kappa = 0.600$, $\mu = 1$		
0.535*	1.5740	0.947 Y_f	0.2772 Y_f	0.2927
0.550*	1.5290	0.9093 Y_f	0.3305 Y_f	0.3634
0.575*	1.4794	0.8598 Y_f	0.3834 Y_f	0.4453
0.600	1.4428	0.8192 Y_f	0.4138 Y_f	0.5051
0.625	1.4136	0.7852 Y_f	0.4319 Y_f	0.5500
0.650	1.3892	0.7561 Y_f	0.4416 Y_f	0.5839
0.675	1.3682	0.7313 Y_f	0.4452 Y_f	0.6089
0.700	1.3498	0.7098 Y_f	0.4445 Y_f	0.6263
		$\kappa = 0.625$, $\mu = 1$		
0.531*	1.5384	0.9537 Y_f	0.2433 Y_f	0.2552
0.550	1.4806	0.9151 Y_f	0.3053 Y_f	0.3336
0.575	1.4313	0.8666 Y_f	0.3581 Y_f	0.4133
0.600	1.3949	0.8267 Y_f	0.3897 Y_f	0.4714
0.625	1.3658	0.7930 Y_f	0.4083 Y_f	0.5150
0.650	1.3414	0.7642 Y_f	0.4185 Y_f	0.5475
0.675	1.3205	0.7396 Y_f	0.4224 Y_f	0.5712
0.700	1.3021	0.7183 Y_f	0.4220 Y_f	0.5876
		$\kappa = 0.650$, $\mu = 1$		
0.550	1.4281	0.9818 Y_f	0.2853 Y_f	0.3107
0.575	1.3799	0.8715 Y_f	0.3372 Y_f	0.3870

(Continued)

TABLE 17.3 *Continued*

ψ	kR	G	B'	Q_L
0.600	1.3442	0.8324 Y_f	0.3685 Y_f	0.4427
0.625	1.3155	0.7993 Y_f	0.3872 Y_f	0.4844
0.650	1.2915	0.7708 Y_f	0.3972 Y_f	0.5153
0.675	1.2708	0.7465 Y_f	0.4013 Y_f	0.5376
0.700	1.2525	0.7255 Y_f	0.4009 Y_f	0.5527
		$\kappa = 0.670,\ \mu = 1$		
0.522	1.4650	0.9830 Y_f	0.1735 Y_f	0.1765
0.550	1.3834	0.9200 Y_f	0.2726 Y_f	0.2963
0.575	1.3367	0.8744 Y_f	0.3228 Y_f	0.3692
0.600	1.3017	0.8359 Y_f	0.3533 Y_f	0.4227
0.625	1.2737	0.8034 Y_f	0.3717 Y_f	0.4626
0.650	1.2501	0.7755 Y_f	0.3816 Y_f	0.4921
0.675	1.2296	0.7513 Y_f	0.3855 Y_f	0.5131
0.700	1.2115	0.7303 Y_f	0.3850 Y_f	0.5272

*The results in Table 17.3 marked by an asterisk are not suitable for the design of such devices in that the functions defined by these boundary conditions exhibit a reversal in the direction of circulating at the high frequency end of the band.

Source: J. Helszajn, Synthesis of octave-band quarterwave coupled semi-tracking junction circulators, *IEEE Trans. Microwave Theory Tech.*, Vol. MTT-43, pp. 573–578, Mar. 1995. Reproduced with permission.

constant ε_t of the dielectric region adjacent to the junction may be employed to set the admittance Y_{n-1} of the quarter-wave transformer adjacent to the junction.

$$Y_{n-1} = \sqrt{\varepsilon_t} Y_r \tag{17.29}$$

A knowledge of the quality factor although mandatory is not sufficient, however, in that it is also necessary to ensure that the complex gyrator circuit is well behaved over the frequency interval of interest. Figure 17.10 illustrates the frequency response of the semitracking solution under consideration. It indicates that the frequency characteristics of semitracking solutions using disk resonators are exceptionally well behaved and are indeed appropriate with the design of octave-band devices.

The results in Table 17.3 marked by an asterisk are not suitable for the design of such devices in that the junctions defined by these boundary conditions have a reversal in the direction of the circulation at the high frequency end of the band.

17.7 SYNTHESIS OF DEGREE-3 CIRCULATOR

The general matching theory in which VSWR(min) is allowed to deviate from unity suggests that a precise realization of the susceptance slope parameter may not be as critical as historically supposed. One engineering decision is to fix VSWR(min) in terms of VSWR(max) by

$$\text{VSWR(min)} = \sqrt{\text{VSWR(max)}} \tag{17.30}$$

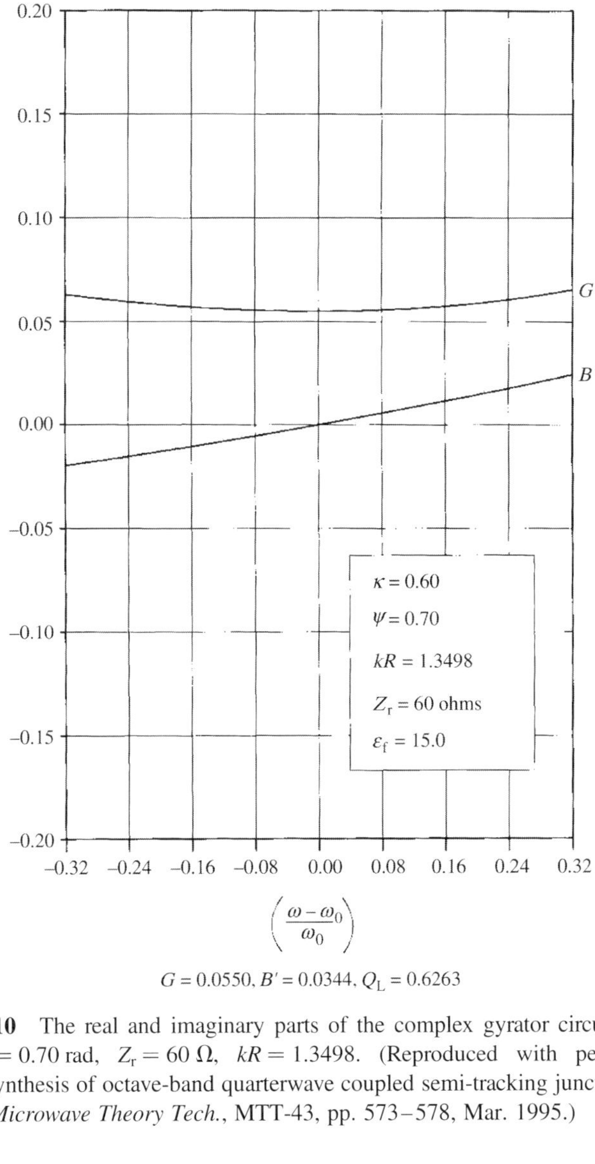

FIGURE 17.10 The real and imaginary parts of the complex gyrator circuit with $\mu = 1$, $\kappa = 0.60$, $\psi = 0.70$ rad, $Z_r = 60\ \Omega$, $kR = 1.3498$. (Reproduced with permission from J. Helszajn, Synthesis of octave-band quarterwave coupled semi-tracking junction circulators, *IEEE Trans. Microwave Theory Tech.*, MTT-43, pp. 573–578, Mar. 1995.)

A typical solution is described by VSWR(max) $= 1.04$ and $2\delta_0 = 0.66$. Its network solution is

$$G = 0.0661$$
$$B' = 0.0291$$
$$Q_L = 0.4407$$
$$Y_1 = 0.0249$$
$$Y_2 = 0.0453$$

Table 17.3 indicates that the boundary conditions required to satisfy this value of Q_L are not unique. One solution is

$$\begin{aligned} \kappa &= 0.650 \\ \psi &= 0.600 \\ kR &= 1.3442 \\ G &= 0.8324\, Y_f \\ B' &= 0.3685\, Y_f \end{aligned}$$

Y_f may be evaluated using either G or B'. In the procedure adopted here it is obtained by reconciling the two statements for G.

Taking ε_f as 14.7 and making use of the definition of Y_f leads to

$$G = 3.206\, Y_r$$

Equating the values of G determined by the network and EM problems gives the following relationship:

$$0.06612 = 3.406\, Y_r$$

Solving this equation for Y_r yields

$$Y_r = 0.0206$$

An important quantity in the design of a junction circulator is the value of the dielectric constant of the region adjacent to the resonator. Unfortunately, this quantity is not an independent variable. In practice it is dependent on both the overall specification of the device and the semitracking solution adopted for its realization. The evaluation of this quantity will now be undertaken. It is determined from a knowledge of Y_r and Y_{n-1}. The result is

$$\varepsilon_d = (Y_2/Y_r)^2 = 4.8$$

The nearest integer value for ε_d is 5. If ε_d does not correspond to an integer or commercial value the design must be repeated with a different ripple level or with a different choice of semitracking solution in Table 17.3.

The width of the striplines adjacent to the resonator is calculated next in terms of ψ once the radius of the resonator has been formed from a knowledge of the center frequency of the device and kR. Finally, the thickness H of the resonators is evaluated from a knowledge of W and Y_r.

One feature of the ideal semitracking solution of circulators using disk resonators is that the dielectric constant of the transformer region adjacent to the resonator lies approximately between 3 and 6. The use of such low values of dielectric constant adjacent to the resonator is helpful in reproducing the assumed magnetic wall boundary condition between the ports of the junction. Failure to accurately do so leads to some uncertainty in the definition of the effective coupling angle of the junction (defined by the transmission lines and the resonator circuit) and of the radius of the resonator. There is also some corresponding modification in the susceptance slope parameter and to a lesser extent in the conductance of the complex gyrator circuit. Fortunately, the field of solutions of the semitracking subspace outlined here permits some laxity in the definition of the former parameters and the network problem can accommodate some uncertainty in the latter quantities if the minima in the reflection coefficient are not forced to pass through zero.

17.8 FREQUENCY RESPONSE OF QUARTER-WAVE COUPLED CIRCULATORS

The frequency response of a quarter-wave coupled circulator may be traced by forming the reflection coefficient at the input terminals of its one-port complex gyrator network. The topology in question is illustrated in Fig. 17.11a. The frequency

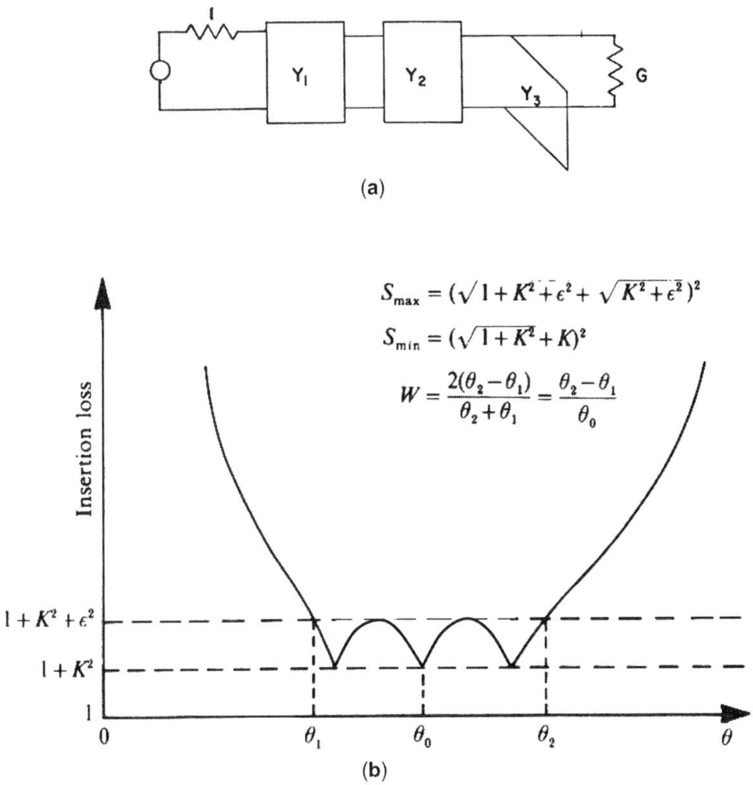

FIGURE 17.11 (a) Topology of quarter-wave coupled complex gyrator circuit. (b) Frequency response of $n = 3$ network with nonzero minima in the reflection coefficient.

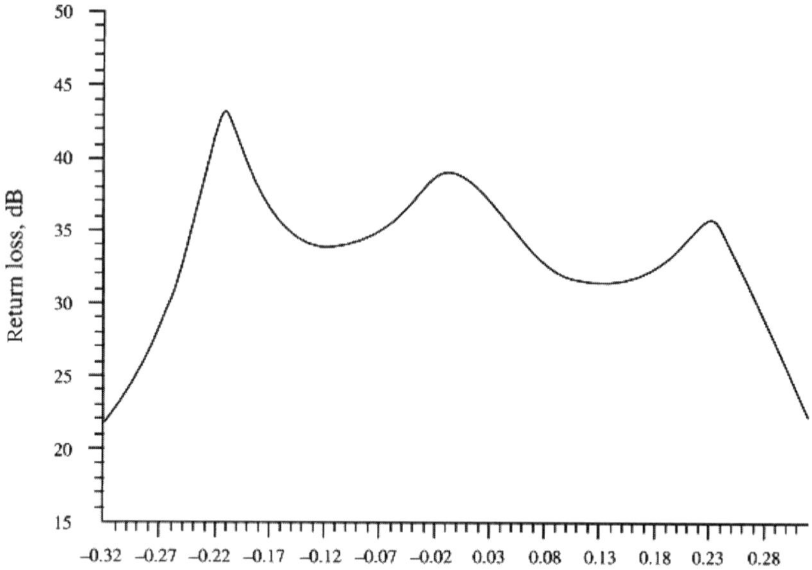

FIGURE 17.12 Theoretical frequency response of two section quarter-wave coupled semi-tracking junction with $\mu = 1$, $\kappa = 0.67$, $\psi = 0.70$, $G = 0.077$, $B' = 0.0404$, $Q_L = 0.527$, $Z_r = 36.50\ \Omega$, $\varepsilon_d = 3.50$. (Reproduced with permission from J. Helszajn, Synthesis of octave-band quarterwave coupled semi-tracking junction circulators, *IEEE Trans. Microwave Theory Tech.*, MTT-43, pp. 573–578, Mar. 1995.)

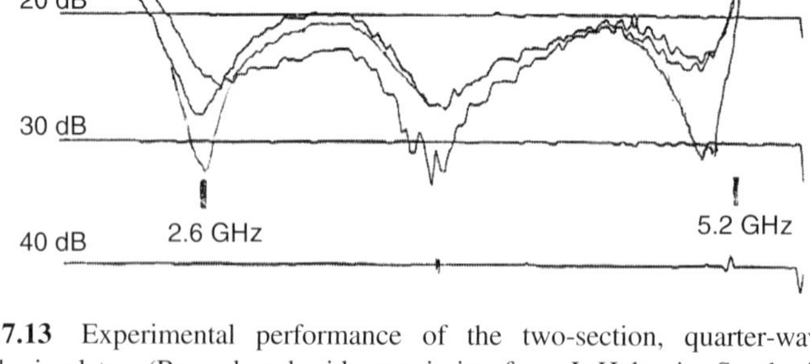

FIGURE 17.13 Experimental performance of the two-section, quarter-wave coupled, octave-band, circulator. (Reproduced with permission from J. Helszajn, Synthesis of octave-band quarterwave coupled semi-tracking junction circulators, *IEEE Trans. Microwave Theory Tech.*, MTT-43, pp. 573–578, Mar. 1995.)

response of the degree-3 solution is shown in Fig. 17.11b. This allows the frequency response of the quarter-wave coupled junction to be displayed without difficulty:

$$|\Gamma_{\text{in}}|^2 = \frac{[AR_{\text{in}} - Z_0(D - CX_{\text{in}})]^2 + [B + AX_{\text{in}} - Z_0CR_{\text{in}}]^2}{[AR_{\text{in}} + Z_0(D - CX_{\text{in}})]^2 + [B + AX_{\text{in}} + Z_0CR_{\text{in}}]^2} \tag{17.31}$$

A, *B*, *C*, and *D* are the parameters of the overall region. The *ABCD* parameters of a typical transformer region are

$$A = \cos\theta \tag{17.32a}$$

$$B = Y_t \sin\theta \tag{17.32b}$$

$$C = (\sin\theta)/Y_t \tag{17.32c}$$

$$D = \cos\theta \tag{17.32d}$$

θ is the electrical length of the transformer:

$$\theta = (\pi/2)(1 + \delta) \tag{17.33}$$

δ is the normalized radian frequency variable in Eq. (17.26).

Y_0 is the characteristic impedance of the generator circuit and Y_t is that of the quarter-wave-long transformer:

$$Y_t = Y_r\sqrt{\varepsilon_t} \tag{17.34}$$

X_{in} and R_{in} are the imaginary and real parts of the complex gyrator impedance, and ε_t is the relative permittivity of the transformer impedance adjacent to the junction.

The frequency response of the $n = 3$ network is obtained with

$$A = \cos^2\theta - (y_{02}/y_{01})\sin^2\theta \tag{17.35a}$$

$$B = (1/y_{01} + 1/y_{02})\sin\theta\cos\theta \tag{17.35b}$$

$$C = (y_{01} + y_{02})\sin\theta\cos\theta \tag{17.35c}$$

$$D = (-y_{01}/y_{02})\sin^2\theta + \cos^2\theta \tag{17.35d}$$

Figure 17.12 illustrates a typical calculation.

The performance of a 2.6–5.2 GHz device developed prior to the development of the full theory is indicated in Fig. 17.13.

CHAPTER EIGHTEEN

Complex Gyrator Circuit of Negative Permeability Tracking and Semitracking Circulators

18.1 INTRODUCTION

The shortcoming of either the tracking or semitracking solution of the junction circulator is that the bandwidth of the former is restricted to about 66% and the latter to about 80%. These sorts of circulators are associated with a gyrotropy that varies between 0.50 and 1.0 and with an effective permeability that is positive over the full frequency range. In order to extend the bandwidth beyond these values, it is necessary to allow the gyrotropy to vary between 0.50 and 1.50. One difficulty at first sight with this approach is that the effective permeability of the gyromagnetic resonator becomes negative when the gyrotropy exceeds unity; another is that low-field loss exists unless special attention is given to the profile of the internal direct field. The first feature poses no particular mathematical difficulty and the second can be avoided in practice or at least minimized by ensuring that the ferrite or garnet material is saturated. One way to avoid the latter shortcoming is to embody semispherical pole pieces in the magnetic circuit with the same material as that of the ferrite resonator. In this way the bandwidth of the junction can be extended from about 80% to 100%. The purpose of this chapter is to develop the complex gyrator circuit of the device. It involves replacing the usual Bessel functions of a typical pole by ones with imaginary argument under the assumption that the resonator is uniformly magnetized. The substitution of the usual Bessel functions by modified ones permits the analysis of this type of circulator to proceed without difficulty. When the effective

The Stripline Circulator: Theory and Practice. By J. Helszajn

permeability μ_{eff} approaches zero and κ approaches μ, the pole descriptions in the two Bessel planes approach the same value for $\mu_{\text{eff}} < 0$ and $\mu_{\text{eff}} > 0$.

A common 100% device is a 6–18 GHz circulator. In the design of this sort of junction the negative permeability region extends from 6 to 9 GHz and the positive one covers the 9–18 GHz band. The center frequency is 12 GHz and coincides with $\kappa = 0.75$. This chapter reproduces the tracking complex gyrator half-space defined by the positive permeability region and separately investigates that defined by the negative one. The semitracking rather than the tracking solution is again the optimum one.

Another gyromagnetic resonator of some interest is made up of an inner disk with one value of magnetization and an outer sleeve with a lower value. The effective permeability of the inner region may be negative or positive; that of the outer region is restricted to a positive value.

18.2 NEGATIVE PERMEABILITY GYROMAGNETIC RESONATOR

The effective permeability in a saturated magnetic insulator is positive when the normalized magnetization (p) is less than unity, zero when it is equal to unity, and negative provided the normalized magnetization exceeds unity. It is again positive above the Kittel line. The previous situation implies that, for a given magnetization, the permeability is negative at low frequencies, zero at a specific frequency determined by the magnetization of the material, and positive above this frequency. A similar picture is obtained at a single frequency if the magnetization is kept constant and the normalized direct magnetic field intensity (σ) is varied.

To characterize the cutoff space of the negative permeability gyromagnetic resonator, it is necessary to replace the Bessel functions of the first kind appearing in its characteristic equation by modified ones. It is appropriate in the derivation of the required result to rewrite the existing characteristic equation in the following form:

$$\frac{J_{n-1}(\pm jk_e R)}{J_n(\pm jk_e R)} - \left(\frac{n}{\pm jk_e R}\right)\left(1 + \frac{\kappa}{\mu}\right) = 0, \quad \mu_{\text{eff}} \leq 0 \tag{18.1}$$

The wavenumber in the preceding equation is defined by

$$k_e = k_o\sqrt{\varepsilon_f(-\mu_{\text{eff}})}, \quad \mu_{\text{eff}} < 0 \tag{18.2}$$

The required characteristic equation is now deduced by making use of the connection between the Bessel functions of the first kind and the related modified ones. The result is

$$(\mp j)\left[\frac{I_{n-1}(k_e R)}{I_n(k_e R)} - \left(\frac{n}{k_e R}\right)\left(1 + \frac{\kappa}{\mu}\right)\right] = 0, \quad \mu_{\text{eff}} < 0 \tag{18.3}$$

The $(\mp j)$ factor is retained in order not to lose information in the development of the poles of the problem region, which are dealt with separately.

The modified Bessel function is related to the standard Bessel one by

$$I_n(x) = J_n(jx)/(j)^n$$
$$I_{-n}(x) = I_n(x)$$

Useful polynomial approximations for the first two modified Bessel functions are

$$I_0(x) = 1 + 2.249\,999\,7\left(\frac{x}{3}\right)^2 + 1.265\,620\,8\left(\frac{x}{3}\right)^4 + 0.316\,386\,6\left(\frac{x}{3}\right)^6 + 0.044\,447\,9\left(\frac{x}{3}\right)^8 + 0.003\,944\,4\left(\frac{x}{3}\right)^{10} + 0.000\,210\,0\left(\frac{x}{3}\right)^{12}$$

$$I_1(x) = x\left[0.50 + 0.562\,499\,85\left(\frac{x}{3}\right)^2 + 0.210\,935\,73\left(\frac{x}{3}\right)^4 + 0.039\,542\,89\left(\frac{x}{3}\right)^6 + 0.004\,433\,19\left(\frac{x}{3}\right)^8 + 0.000\,317\,61\left(\frac{x}{3}\right)^{10} + 0.000\,011\,09\left(\frac{x}{3}\right)^{12}\right]$$

The description of the modified higher order Bessel functions can be deduced by having recourse to the recurrence condition:

$$\left(\frac{2n}{x}\right)I_n(x) = I_{n-1}(x) - I_{n+1}(x)$$

It is also of note that

$$\frac{I(-x)_{\text{odd}}}{I(-x)_{\text{even}}} = (-1)\frac{I(x)_{\text{odd}}}{I(x)_{\text{even}}}$$

A similar identity applies to the ratio of even to odd polynomials.

The effective permeability (μ_{eff}) is specified in terms of the diagonal (μ) and off-diagonal (κ) elements of the permeability tensor in the usual way.

$$\mu_{\text{eff}} = \mu\left[1 - (\kappa/\mu)^2\right] \tag{18.4}$$

This quantity is positive provided the gyrotropy (κ/μ) is less than unity:

$$(\kappa/\mu)^2 \leq 1 \tag{18.5a}$$

It is zero provided it is equal to unity:

$$(\kappa/\mu)^2 = 1 \tag{18.5b}$$

It is negative when it is larger than unity:

$$(\kappa/\mu)^2 \geq 1 \tag{18.5c}$$

μ_{eff} and κ/μ are, in general, functions of both p and σ in keeping with the development in Chapter 2.

$$\mu_{\text{eff}} = \frac{(p+\sigma)^2 - 1}{\sigma(p+\sigma) - 1} \tag{18.6}$$

$$\left(\frac{\kappa}{\mu}\right) = \frac{p}{\sigma(p+\sigma) - 1} \tag{18.7}$$

where

$$p = \omega_m/\omega \tag{18.8a}$$

$$\sigma = \frac{\gamma(H_0 - N_z M_0)}{\omega} \tag{18.8b}$$

The only case considered here is that of a magnetic insulator at saturation below the Kittel line

$$\sigma = 0 \tag{18.9}$$

The corresponding effective permeability and gyrotropy are

$$\mu = 1 \tag{18.10}$$

$$\mu_{\text{eff}} = 1 - \kappa^2 \tag{18.11}$$

$$\kappa = p \tag{18.12}$$

γ is the gyromagnetic ratio (2.21×10^5 rad/s per A/m), M_0 is the magnetization (A/m), ω is the radian frequency (rad/s), and μ_0 is the free-space permeability ($4\pi \times 10^{-7}$ H/m).

Figure 18.1 indicates the mode chart obtained in this way with the polarity of the gyrotropy positive. It is identical for each of the two possible roots of -1 introduced

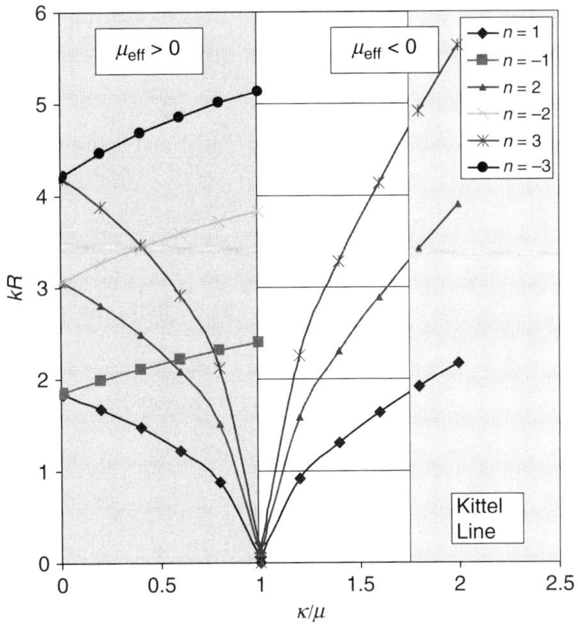

FIGURE 18.1 Cutoff space of gyromagnetic resonator in positive and negative permeability half-spaces (after A. P. Gibson).

in connection with the definition of the wavenumber. An interesting feature of the frequency of $TM_{1,1,0}$ and the $TM_{2,1,0}$ modes is that these decrease for κ between 0.50 and 1.0 but increase thereafter.

Recall that the mode chart in Fig. 18.1 applies to a decoupled gyromagnetic resonator. In the presence of ports, the actual situation is somewhat more complicated in that it also involves the coupling angle as discussed elsewhere.

18.3 IMPEDANCE POLES OF VERY STRONGLY MAGNETIZED JUNCTION

In order to proceed with a calculation of the complex gyrator circuit in the negative permeability region of a strongly magnetized gyromagnetic function, it is necessary to construct a typical pole of the problem region. The required impedance pole in the modified Bessel plane is

$$Z_n = \frac{-j3\eta_{\text{eff}} Z_r \psi}{\pi} \left(\frac{\sin n\psi}{n\psi}\right)^2 \left[\frac{I_{n-1}(k_e R)}{I_n(k_e R)} - \left(\frac{n}{k_e R}\right)\left(1 + \frac{\kappa}{\mu}\right)\right]^{-1} \tag{18.13}$$

The wavenumber has the meaning met in connection with the development of the cutoff space. The wave impedance is similarly defined by

$$\eta_{\mathrm{eff}} = \sqrt{\frac{(-\mu_{\mathrm{eff}})}{\varepsilon_{\mathrm{f}}}} \tag{18.14}$$

The change in sign of the pole in the modified Bessel plane is of note. The eigenvalues, open-circuit parameters, and the two circulation conditions of the junction may be constructed once the poles of the problem region have been evaluated.

For calculation purposes, it is usual to introduce the following substitution in the pole formulation of the problem region:

$$\eta_{\mathrm{eff}} Z_{\mathrm{r}} = \sqrt{(-\mu_{\mathrm{eff}})} Z_{\mathrm{f}} \tag{18.15}$$

where

$$Z_{\mathrm{f}} = Z_{\mathrm{r}} / \sqrt{\varepsilon_{\mathrm{f}}} \tag{18.16}$$

Z_{r} is the characteristic impedance of the strips at the termination of the resonator, ε_{f} is the relative dielectric constant of the ferrite material.

18.4 CIRCULATION SOLUTION USING NONDEGENERATE RESONATOR MODES

An interesting feature of the circulation solutions summarized in Table 17.1 in Chapter 17 is that the gyrator resistance, with the variable κ in the vicinity of unity, is essentially made up by a pair of counterrotating poles of the junction that are not degenerate in the demagnetized state. In this it differs from the usual result met with κ in the vicinity of the origin for which the gyrator resistance involves a pair of split degenerate poles. The state of affairs under consideration may be understood by perusing Table 17.2, which gives the amplitudes of the impedance poles and eigenvalues over the complete range of the gyrotropy. It is readily apparent from this data that the gyrator level in the tracking interval is mainly established by the $\mathrm{TM}_{1,1,0}$ and $\mathrm{TM}_{2,1,0}$ modes of the resonator. This is all the more so with the variable κ in the neighborhood of unity. The eigen-networks for the latter situation are depicted in Fig. 18.2. The intersection between the $n = -1$ and $n = +2$ branches of the magnetized cutoff space of the gyromagnetic resonator in Fig. 18.1 may be taken as the boundary between the two types of solutions. It is separately observed from Table 17.2 that the idealization of the in-phase eigen-network is satisfied for each value of the off-diagonal entry κ provided the angle ψ subtended by the stripline at a typical port of the resonator is equal to about 0.52 rad. The purpose of this work is to examine the solution in the case for which κ exceeds unity. In order for the transition between the two regions to be seamless, it is necessary that the

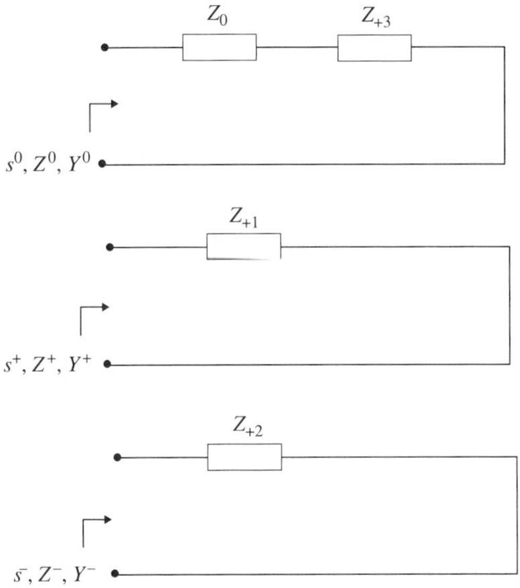

FIGURE 18.2 Eigen-networks of junction circulator using nondegenerate counterrotating poles (κ_0 in vicinity of unity).

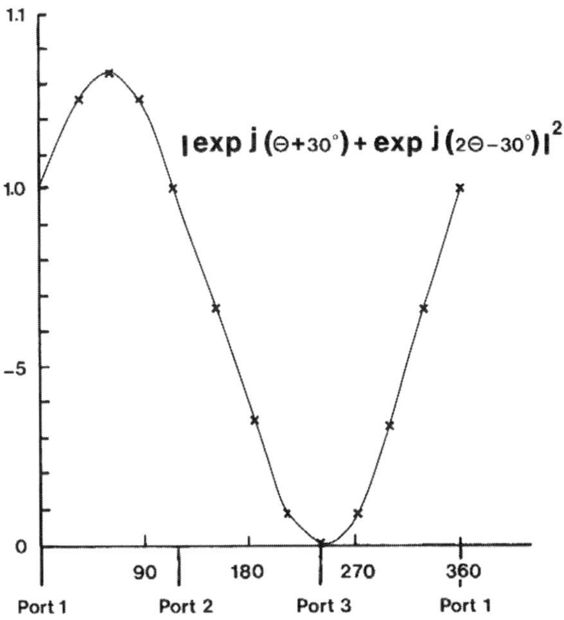

FIGURE 18.3 Standing wave solution of junction circulator using nondegenerate $TM_{1,1,0}$ and $TM_{2,1,0}$ counterrotating modes.

counterrotating Z_{+1} and Z_{+2} poles in the first Foster expansion of the counterrotating eigenvalues should again be the dominant poles. It will be seen that this is the case.

18.5 STANDING WAVE SOLUTION OF CIRCULATOR USING NONDEGENERATE RESONATOR MODES

The fact that the $TM_{n,1,0}$ modes with $n = +1$ and $+2$ may be used to construct a circulation solution may be appreciated by recognizing that the split $TM_{n,m,0}$ modes produce a circulator that circulates in the opposite direction to that using $n = \pm 2$ modes with $n = \pm 1$. This feature implies that the lower branch of the split ± 1 modes rotates in the opposite direction to that of the $n = \pm 2$ and may therefore be employed to construct an acceptable circulator solution. This possibility may be verified by checking that the corresponding eigenadmittances form a complex conjugate set, and that a standing wave solution can indeed be constructed using only the $TM_{1,1,0}$ and $TM_{2,1,0}$ modes. The construction of such a standing wave solution is indicated in Fig. 18.3.

18.6 NEGATIVE PERMEABILITY TRACKING SOLUTION

The tracking solution developed in the previous chapter ensures that the in-phase eigen-network displays an electric wall at the terminals of the resonator and that the counterrotating ones have complex conjugate immitances there.

The two tracking circulation conditions with the gyrotropy on either side of unity are determined in terms of the eigenvalues of the problem region by

$$Z^0 = 0 \tag{18.17a}$$

$$Z^+ + Z^- = 0 \tag{18.17b}$$

and

$$R = j(Z^+ - Z^-)/2\sqrt{3} \tag{18.18}$$

The number of poles necessary to reproduce the boundary conditions at the ports is an open question. The choice adopted here corresponds to that employed in the description of the positive permeability half-space, except that the $TM_{m,n,0}$ modes with negative values of the integer n have been omitted. While the latter modes do not at first sight reveal poles in the negative permeability cutoff space of interest, these still have finite reactances there that have to be considered. At first sight the omission of these terms does not produce any significant variation of the result.

The poles in the expansion of the eigenvalues in the present work are

$$Z^0 \approx Z_{+3} + Z_0 \quad (18.19a)$$

$$Z^+ \approx Z_{+1} \quad (18.19b)$$

$$Z^- \approx Z_{+2} \quad (18.19c)$$

The first tracking condition fixes the relationship between the coupling angle (ψ) and the radial wavenumber (kR) for a given value of the off-diagonal element κ. It produces a unique value of the coupling angle. The second sets the gyrator resistance.

The solutions of the first condition of a circulator using a negative permeability gyromagnetic resonator produce unique values of kR and ψ for each value of κ. Taking $\kappa = 1.10$ by way of example gives $kR = 0.879$ and $\psi = 0.416$. The corresponding poles are

$$\begin{aligned} Z_0 &= -j0.443\,R_f \\ Z_{+1} &= -j1.728\,R_f \\ Z_{+2} &= j1.728\,R_f \\ Z_{+3} &= j0.443\,R_f \end{aligned}$$

The in-phase eigenvalue is equal to zero as asserted.

The second circulation condition merely involves the calculation of the gyrator resistance. It is usual to represent the real part condition by the gyrator conductance instead of its resistance.

$$G = 0.957Y_f$$

The two circulation conditions of the tracking solution for kR in the negative permeability space bracketed by κ between 1.0 and 1.50 are tabulated in Table 18.1. The first solution lies approximately halfway between the $TM_{1,1,0}$ and the $TM_{2,1,0}$ poles of the problem region. The reactances of the poles are equal in amplitude with opposite sign under the same condition. These are indicated in Table 18.2. These two tables in conjunction with Tables 17.1 and 17.2 provide the complete

TABLE 18.1 Circulation Conditions of Negative Permeability Tracking Junction Circulator

κ	ψ	kR	R	G	B'	Q
1.1	0.416	0.879	0.9976 R_f	1.0024 Y_f	0.0097 Y_f	0.0097
1.2	0.395	1.256	1.0089 R_f	0.9911 Y_f	0.0231 Y_f	0.0233
1.3	0.376	1.553	1.0187 R_f	0.9816 Y_f	0.0384 Y_f	0.0391
1.4	0.358	1.811	1.0261 R_f	0.9746 Y_f	0.0558 Y_f	0.0573
1.5	0.342	2.046	1.0340 R_f	0.9671 Y_f	0.0752 Y_f	0.0777

TABLE 18.2 Impedance Poles of Negative Permeability Tracking Junction Circulator

κ/μ	Z_0	Z_1	Z_2	Z_3
1.05	$-j0.4427\ R_f$	$-j1.7195\ R_f$	$j1.7206\ R_f$	$j0.4416\ R_f$
1.10	$-j0.4529\ R_f$	$-j1.7311\ R_f$	$j1.7248\ R_f$	$j0.4522\ R_f$
1.15	$-j0.4627\ R_f$	$-j1.7386\ R_f$	$j1.7359\ R_f$	$j0.4626\ R_f$
1.20	$-j0.4723\ R_f$	$-j1.7465\ R_f$	$j1.7488\ R_f$	$j0.4725\ R_f$
1.25	$-j0.4809\ R_f$	$-j1.7548\ R_f$	$j1.7560\ R_f$	$j0.4813\ R_f$
1.30	$-j0.4898\ R_f$	$-j1.7654\ R_f$	$j1.7634\ R_f$	$j0.4895\ R_f$
1.35	$-j0.4975\ R_f$	$-j1.7721\ R_f$	$j1.7718\ R_f$	$j0.4974\ R_f$
1.40	$-j0.5043\ R_f$	$-j1.7786\ R_f$	$j1.7757\ R_f$	$j0.5044\ R_f$
1.45	$-j0.5115\ R_f$	$-j1.7849\ R_f$	$j1.7851\ R_f$	$j0.5115\ R_f$
1.50	$-j0.5177\ R_f$	$-j1.7899\ R_f$	$j1.7920\ R_f$	$j0.5181\ R_f$

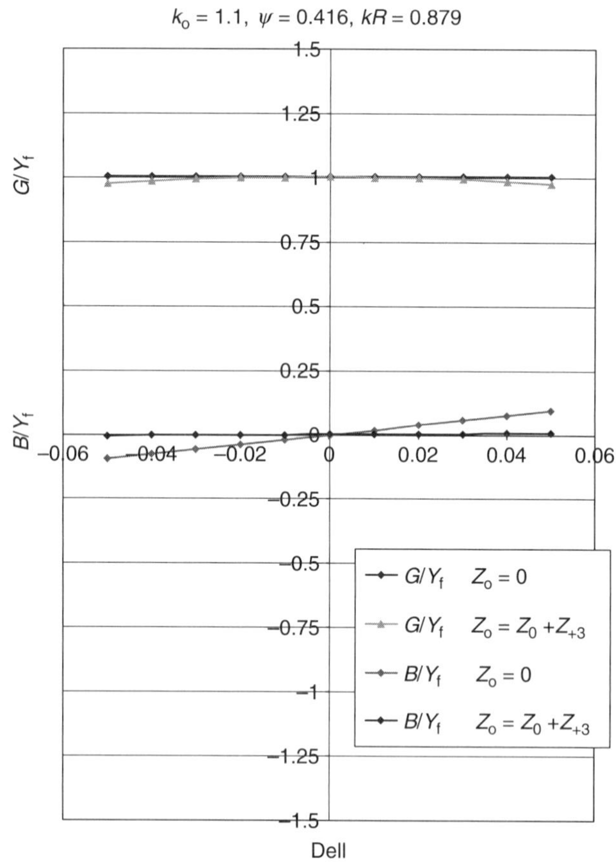

FIGURE 18.4 Effect of in-phase poles on complex gyrator circuit in negative permeability half-space.

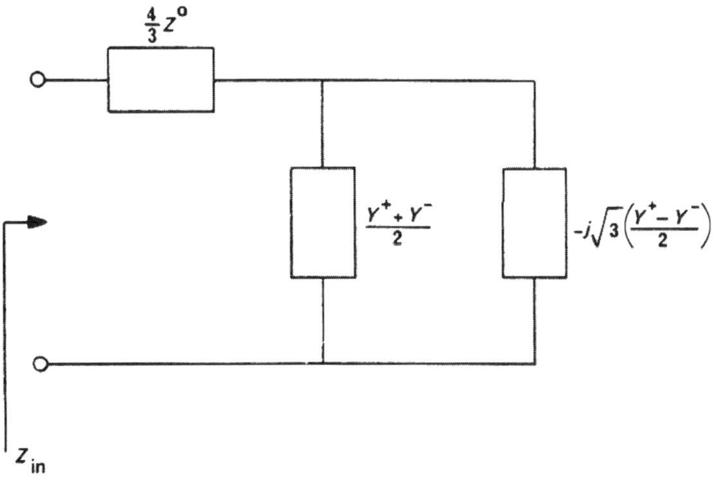

FIGURE 18.5 Effect of in-phase poles on complex gyrator circuit of junction circulator.

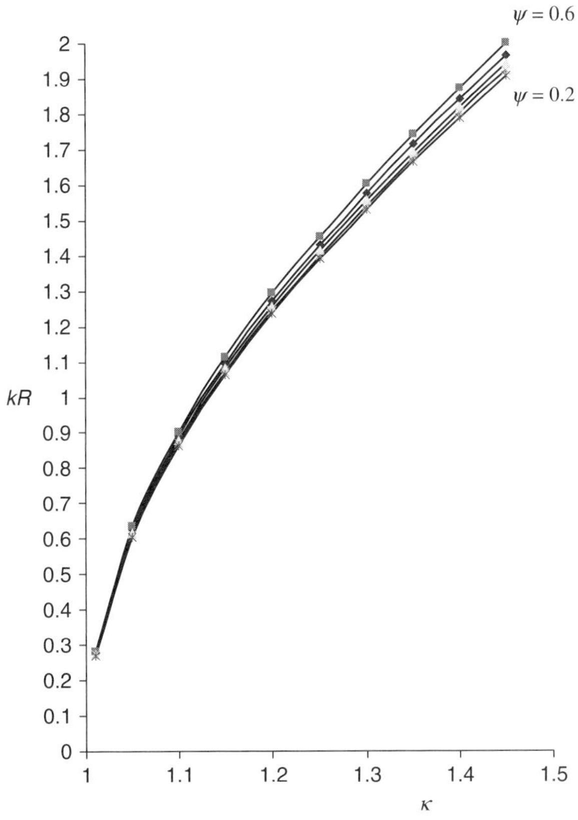

FIGURE 18.6 First circulation condition of nontracking negative permeability junction circulator.

solution of the poles, eigenvalues, and complex gyrator circuit of the 100% band tracking junction circulator.

One useful approximation met in the derivation of the frequency response of the complex gyrator circuit of the junction circulator is obtained by idealizing the in-phase eigenvalue in the descriptions of both its real and imaginary parts. While the former approximation is valid for the case under discussion, the latter one is not. Figure 18.4 indicates one solution with and without the in-phase term. The equivalent circuit, obtained by retaining the in-phase term in the description of the imaginary part of the gyrator circuit, has a series element of value $4Z^0/3$ in cascade with the usual complex gyrator circuit involving simple linear combinations of Y^+ and Y^-. It is illustrated in Fig. 18.5. Figure 18.6 indicates the connection between kR and κ for parametric values of ψ obtained by disregarding the tracking condition placed on Z^0. The corresponding relationship between the gyrator conductance and the gyrotropy is illustrated separately in Fig. 18.7.

It may occasionally be useful to construct tracking circulators with narrow coupling angles. In order to do so it is desirable to independently adjust the in-phase eigen-network. One independent variable that allows Z_0 to be tuned without perturbing the

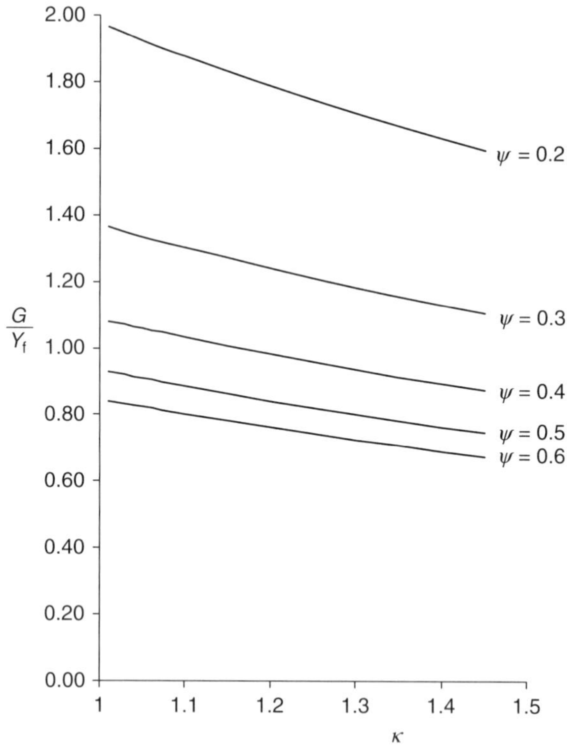

FIGURE 18.7 Second circulation condition of nontracking negative permeability junction circulator.

other resonator modes is a thin metal post through the center of the resonator or the introduction of circumferential slots in the disk metallization in the manner outlined in Chapter 10.

18.7 FREQUENCY RESPONSE

The exact circulation conditions, in the nontracking situation, are obtained by constructing G_{in} and B_{in} in terms of the real and imaginary parts of the complex gyrator impedance.

$$G_{in} = \frac{R_{in}}{R_{in}^2 + X_{in}^2} \tag{18.20}$$

and

$$B_{in} = \frac{-jX_{in}}{R_{in}^2 + X_{in}^2} \tag{18.21}$$

R_{in} and X_{in} are defined in terms of the impedance eigenvalues in the usual way in the previous chapter.

In order to display the frequency response of a typical pole or that of the complex gyrator circuit, it is necessary to write the frequency dependent magnetic variables in its description as

$$\kappa = \kappa_0(\omega_0/\omega) \tag{18.22}$$

$$|\mu_{eff}| = |1 - \kappa^2| \tag{18.23}$$

$$k_e R \rightarrow k_e R \sqrt{\frac{|1 - \kappa^2|}{|1 - \kappa_0^2|}\left(\frac{\omega}{\omega_0}\right)} \tag{18.24}$$

It is also useful to define a normalized frequency variable

$$(\omega/\omega_0) = 1 + \delta \tag{18.25}$$

where

$$\delta = (\omega - \omega_0)/\omega_0 \tag{18.26}$$

The midband variables are

$$\kappa_0 = (\omega_m/\omega_0) \tag{18.27}$$

$$|\mu_{eff}| = 1 - \kappa_0^2 \tag{18.28}$$

$$k_e R = k_0 \sqrt{|1 - \kappa_0^2| \varepsilon_f} R \tag{18.29}$$

The susceptance slope parameter and the quality factor of the complex gyrator circuit may readily be evaluated once the frequency response of the circuit is established. These quantities are included in Table 18.1. A typical frequency response is depicted in Fig. 18.8. The frequency variation of the poles is illustrated in Fig. 18.9. The midband gyrotropy is κ_0 equal to 1.25.

The calculation undertaken here on the frequency response of the complex gyrator circuit of the tracking solution in the negative permeability half-space indicates that

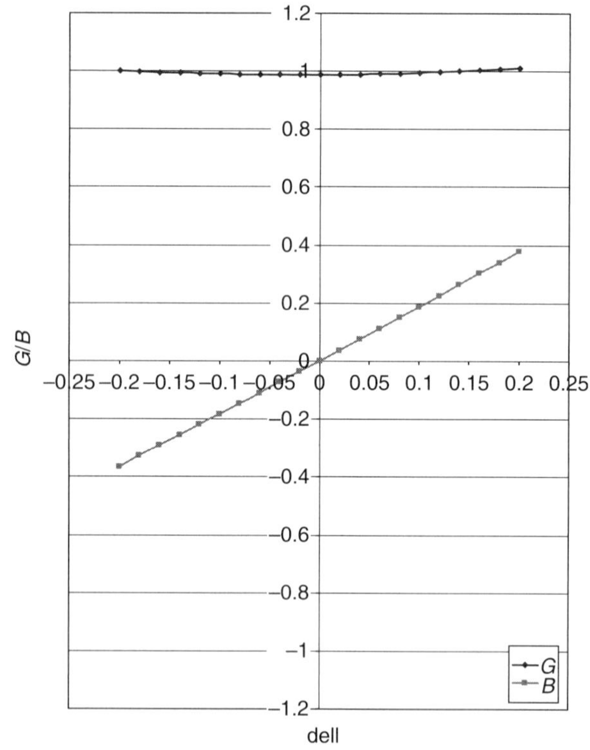

FIGURE 18.8 Frequency response of negative permeability tracking circulator ($\kappa_0 = 1.25$).

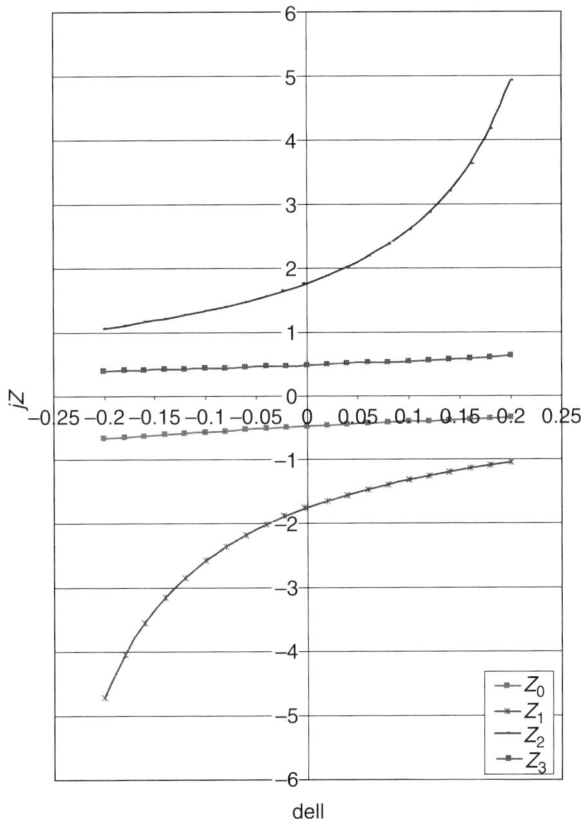

FIGURE 18.9 Frequency responses of impedance poles of negative permeability tracking circulator ($\kappa_0 = 1.25$).

the ideal in-phase condition is not only satisfied at the midband frequency but also elsewhere. Figure 18.9 indicates that this is surprisingly the case.

18.8 COMPLEX GYRATOR CIRCUIT OF 100% CIRCULATOR

The design elements of the complex gyrator circuit of the 100% junction circulator are summarized in this section. The midband frequency of the device under consideration is taken as 12 GHz and the lower and upper bandedge frequencies are at 6 and 18 GHz, respectively. The permeability of the gyromagnetic resonator is negative between 6 and 9 GHz and positive between 9 and 18 GHz. The corresponding gyrotropy varies between 1.50 and 0.50. The situation under consideration is depicted in

μ_{eff}		$\mu_{eff} < 0$		$\mu_{eff} > 0$	
Freq		6 GHz	9 GHz	12 GHz	18 GHz
κ/μ		1.5	1.0	0.75*	0.5

FIGURE 18.10 Frequencies and gyrotropies of 6–18 GHz circulator. (*Design frequency is 12 GHz.)

Fig. 18.10. The permeabilities at 6, 12, and 18 GHz of the 100% circulator are illustrated in Fig. 18.11. The gyrotropy at 12 GHz is

$$\kappa_0 = 0.75$$

The corresponding magnetization is given by

$$\frac{\gamma M_0}{\omega} = 0.75$$

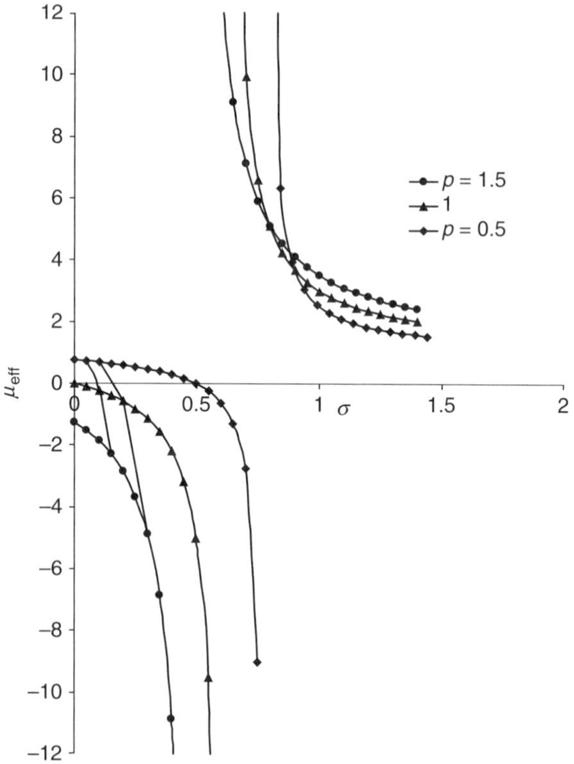

FIGURE 18.11 Effective permeabilities of 6–18 GHz circulator.

The tracking geometry of the microwave circuit is uniquely defined by

$$kR = 1.2968 \quad \text{and} \quad \psi = 0.5045$$

The complex gyrator circuit is specified separately by

$$\begin{aligned} B' &= 0.1067Y_f \\ G &= 1.0012Y_f \\ Q_L &= 0.1069 \end{aligned}$$

The absolute values of G and B' are fixed by Y_r once the topology of the matching network is selected and the gain-bandwidth product is specified by Q_L.

The frequency responses of the tracking solution and its poles are illustrated in Figs. 18.12 and 18.13. The glitch in this frequency response coincides with the

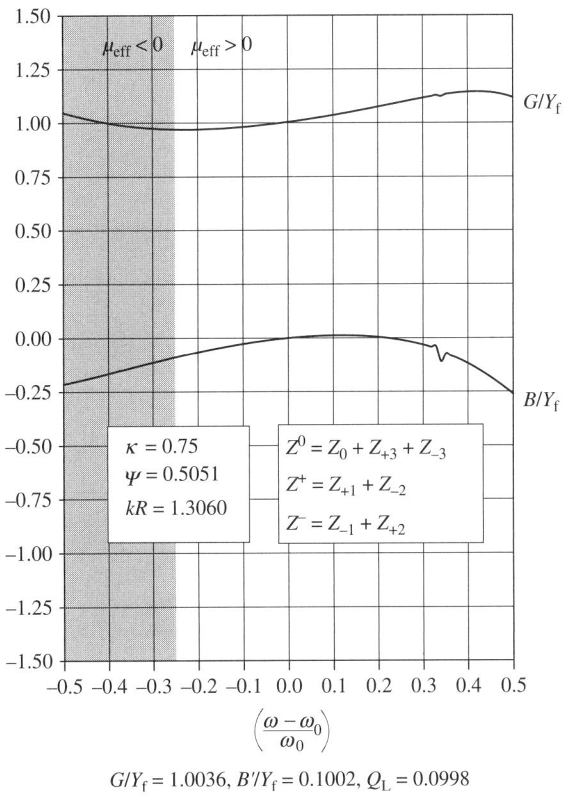

FIGURE 18.12 Frequency response of tracking circulator ($\psi = 0.5051$ rad, $kR = 1.3060$ rad, $\kappa_0 = 0.75$).

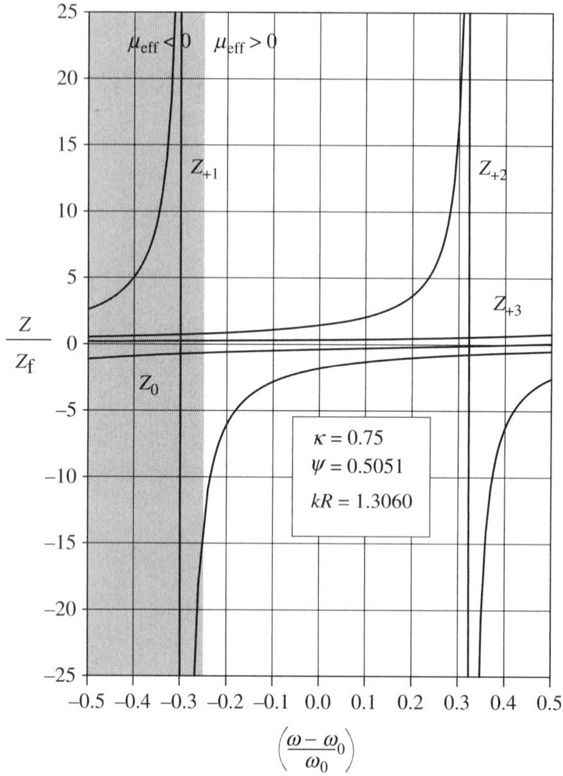

FIGURE 18.13 Frequency responses of impedance poles of tracking circulator ($\psi = 0.5051$ rad, $kR = 1.3060$ rad, $\kappa_0 = 0.75$).

degeneracy between the Z_{-1} and Z_{+2} in the positive permeability half-space. In practice the frequency responses of the semitracking solutions rather than the tracking one are preferable. Figures 18.14 and 18.15 indicate the corresponding solutions. The complex gyrator circuit of this solution is described by

$$B' = 0.3279Y_f$$
$$G = 0.7882Y_f$$
$$Q_L = 0.4160$$

One realizable ripple specification is

$$\text{VSWR(max)} \approx 1.15$$
$$\text{VSWR(min)} \approx 1.06$$

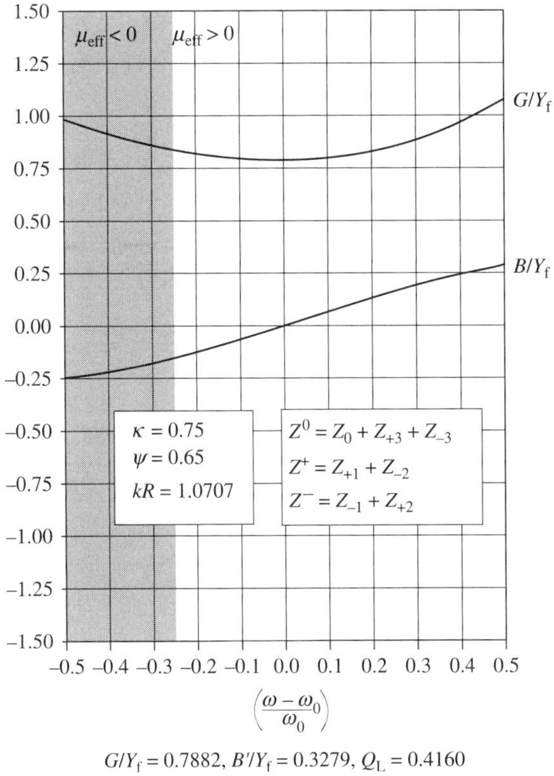

FIGURE 18.14 Frequency response of semitracking solution ($\psi = 0.65$ rad, $kR = 1.0707$ rad, $\kappa_0 = 0.75$).

In order to avoid low-field loss in the negative permeability region, it is necessary to embody semispherical pole pieces in the magnetic circuit with the same material as that of the ferrite resonator. Figure 18.16 indicates the arrangement. The possibility of using a ferrite ring with a lower magnetization around the ferrite disk has been mentioned in this context.

18.9 COMPOSITE GYROMAGNETIC RESONATOR

Another means of ensuring that every part of the gyromagnetic resonator is saturated is to use a composite resonator consisting of a ferrite disk with one value of magnetization, surrounded by a ferrite ring or sleeve with a lower value. The structure under consideration is depicted in Fig. 18.17. The derivation of its cutoff space is a straightforward process. The case where one or the other region displays a negative permeability must be treated separately from that for which both regions have positive

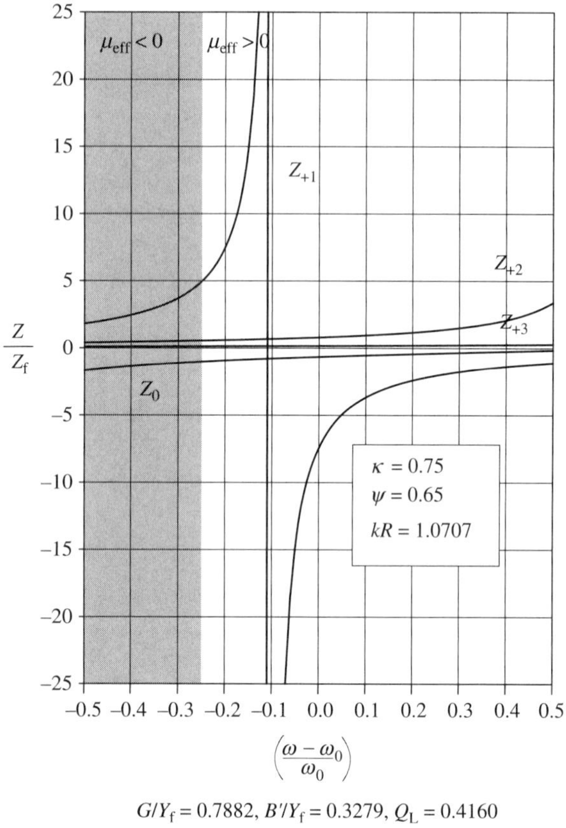

FIGURE 18.15 Frequency responses of impedance poles of semitracking solution ($\psi = 0.65$ rad, $kR = 1.0707$ rad, $\kappa_0 = 0.75$).

permeabilities. The most simple case is that for which the permeability is positive in each region. The development of its characteristic equation starts by writing down the electric field inside the ferrite disk:

$$E_z = \sum_{n=-\infty}^{\infty} A_n J_n(k_1 r) \exp(jn\phi) \tag{18.30}$$

The ϕ component of the magnetic field H_ϕ, is

$$H_\phi = \frac{-j}{\omega\mu_0\mu_{e1}}\left(\frac{\partial E_z}{\partial \mathrm{r}} + j\frac{\kappa_1}{\mu_1}\frac{1}{r}\frac{\partial E_z}{\partial \phi}\right) \tag{18.31}$$

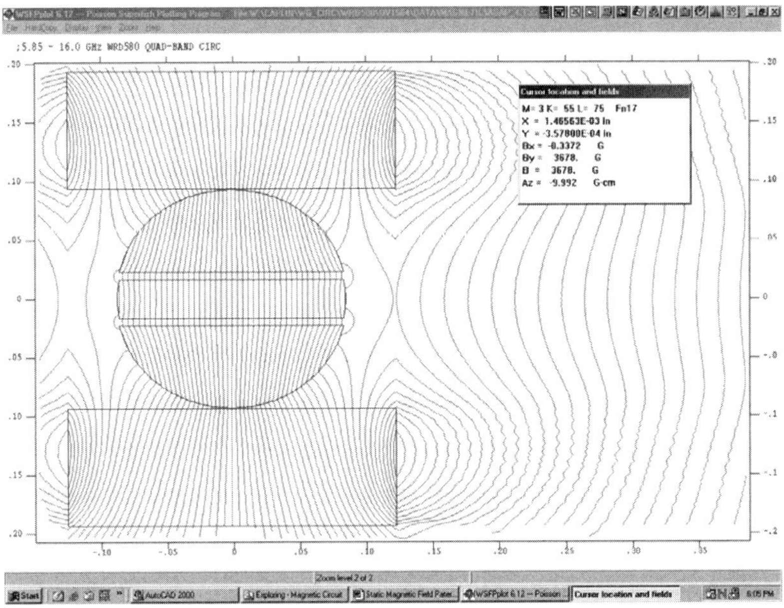

Ferrite Material: TT1-3000
Magnet: SmCo 18 – Dia. 0.250 in. × 0.100 in.
Air gap spacing between disk and dome: 0.006 in.
B_y at edge of ferrite disk (at the plane of symmetry): 4017 G

FIGURE 18.16 Gyromagnetic resonator using semispherical pole pieces. (Reproduced with permission from E. F. Schloemann, *Proc. IEEE*, Vol. 76, No. 2, Feb. 1988.)

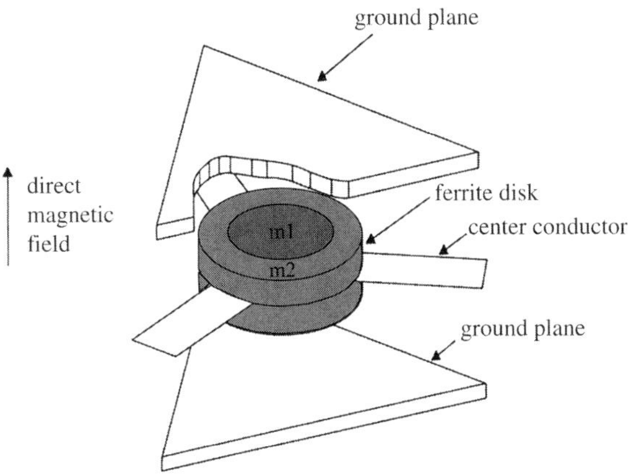

FIGURE 18.17 Schematic diagram of composite resonator.

The required result in terms of the original variables is

$$H_\phi = \sum_{n=-\infty}^{\infty} -jA_n\zeta_0\zeta_1\left(J_n'(k_1r) - \frac{\kappa_1}{\mu_1}\frac{nJ_n(k_1r)}{k_1r}\right)\exp(jn\phi) \tag{18.32}$$

where

$$k_1 = k_0\sqrt{\varepsilon_f\mu_{e1}} \tag{18.33a}$$

$$\zeta_1 = \sqrt{\varepsilon_f/\mu_{e1}} \tag{18.33b}$$

The solution for the electric field E_z in the ferrite ring is given by the following expression:

$$E_z = \sum_{n=-\infty}^{\infty} A_n[B_nJ_n(k_2r) + C_nY_n(k_2r)]\exp(jn\phi) \tag{18.34}$$

where

$$k_2 = k_0\sqrt{\varepsilon_r\mu_{e2}} \tag{18.35a}$$

$$\zeta_2 = \sqrt{\varepsilon_f/\mu_{e2}} \tag{18.35b}$$

where B_n and C_n are constants, and $Y_n(kr)$ and $J_n(kr)$ denote Bessel functions of order n of the first and second kinds, respectively.

In the most simple case the direct magnetic field in the outside ring is in the same sense as that of the inner disk but its magnetization is different. The ϕ component of the magnetic field, H_ϕ, in the ferrite ring is

$$H_\phi = \sum_{n=-\infty}^{\infty} -jA_n\xi_0\xi_2\left[\left(J_n'(k_2r) - \frac{\kappa_2}{\mu_2}\frac{nJ_n(k_2r)}{k_2r}\right)B_n + \left(Y_n'(k_2r) - \frac{\kappa_2}{\mu_2}\frac{nY_n(k_2r)}{k_2r}\right)C_n\right]\exp(jn\phi) \tag{18.36}$$

The constants B_n and C_n can be expressed in terms of the physical variables by using the conditions that E_z and H_ϕ are continuous at $r = r_i$:

$$B_n J_n(k_2 r_i) + C_n Y_n(k_2 r_i) = J_n(k_1 r_i) \tag{18.37}$$

$$B_n\left(J'_n(k_2 r_i) - \frac{\kappa_2}{\mu_2}\frac{nJ_n(k_2 r_i)}{k_2 r_i}\right) + C_n\left(Y'_n(k_1 r_i) - \frac{\kappa_2}{\mu_2}\frac{nY_n(k_1 r_i)}{k_2 r_i}\right)$$
$$= \left(J'_n(k_1 r_i) - \frac{\kappa_1}{\mu_1}\frac{nJ_n(k_1 r_i)}{k_1 r_i}\right) \tag{18.38}$$

The two families of roots for the uncoupled resonator are deduced in the usual way by imposing a magnetic wall at $r = r_0$:

$$H_\phi(r = r_0) = 0 \tag{18.39}$$

This gives

$$\left(J'_n(k_2 r_0) - \frac{\kappa_2}{\mu_2}\frac{nJ_n(k_2 r_0)}{k_2 r_0}\right)B_n + \left(Y'_n(k_2 r_0) - \frac{\kappa_2}{\mu_2}\frac{nY_n(k_2 r_0)}{k_2 r_0}\right)C_n \tag{18.40}$$

The description of the cutoff space in the case for which the permeability of the inner region is negative proceeds by replacing the Bessel function of the first kind in that of the magnetic field intensity by modified ones.

A typical pole is

$$Z_n = -ja_n^2\eta_e Z_r\left[\frac{[B_n J_n(k_2 r) + C_n Y_n(k_2 r)]}{\left[\left(J'_n(k_2 r) + \frac{\kappa_2}{\mu_2}\frac{nJ_n(k_2 r)}{k_2 r}\right)B_n + \left(Y'_n(k_2 r) + \frac{\kappa_2}{\mu_2}\frac{nY_n(k_2 r)}{k_2 r}\right)C_n\right]}\right]^{-1} \tag{18.41}$$

a_n is a term ratio connecting the terminal of the eigen-resonator to those of the eigen-network.

$$a_n^2 = \left(\frac{3\psi}{\pi}\right)\left(\frac{\sin n\psi}{n\psi}\right)^2 \tag{18.42}$$

CHAPTER NINETEEN

Synthesis of Wideband Planar Circulators Using Narrow Coupling Angles

19.1 INTRODUCTION

A drawback of the classic tracking solution of the junction circulator using a planar disk resonator is that it requires a wide coupling angle for its realization. While many practical devices appear to respect this solution, many more apparently do not. One way to cater for this discrepancy is to place open instead of magnetic walls at the boundaries between the ports of the circuit. One approximation that goes some way to reconcile theory and practice is obtained by loading the complex gyrator terminals with lumped element capacitances. The nature of the lumped element variable is not unique and one way to realize it is to make use of the fringing capacitance at the boundary of a dielectric resonator with one value of relative dielectric constant and a substrate with a higher value. Such a hybrid resonator displays a ψ, κ, kR, ω, Q_L, B', G (coupling angle, off-diagonal element of the permeability tensor, radial wavenumber, frequency, loaded Q-factor, susceptance slope parameter, gyrator conductance) space that contains circulation solutions that are outside those of the conventional field, are exceptionally well-behaved, have narrow coupling angles, and exhibit values of loaded Q-factors that are compatible with the synthesis of quarter-wave coupled devices. The magnitude of this fringing capacitance is dependent on both the thickness of the resonator and the relative dielectric constant of the substrate of the transformer region. In practice both these quantities can be varied by altering the topology of the matching network and the specification of the overall device. This means that a range of practical solutions is in fact available. The saving grace in many

The Stripline Circulator: Theory and Practice. By J. Helszajn

experimental efforts is of course the fact that the network problem accommodates some uncertainty in the absolute level of the gyrator immittance provided the minimum ripple level of the specification is not tied down.

19.2 COMPLEX GYRATOR CIRCUITS OF CIRCULATORS USING RADIAL/LUMPED ELEMENT RESONATORS

The space described by the classic solution of a junction circulator using a simple planar disk resonator is usually fixed by two independent variables and one dependent variable. The two independent ones are the coupling angle (ψ) and the value of the off-diagonal element (κ) of the permeability tensor, the dependent variable is normally the radial wavenumber (kR). This classic situation contains a multiplicity of solutions, some of which are well behaved over some frequency interval while others are not. One result that is rather ill behaved is obtained with $\psi = 0.25$ rad (say) and $\kappa = 0.67$ (say) for which the independent variable $kR = 1.99$. The frequency response of its complex gyrator circuit, illustrated in Fig. 19.1, has a characteristic resonance within its passband and therefore, at first sight, is of little value in the design of commercial devices. Although its character may be altered by either varying the coupling angle or the magnetic variables, the ensuing response

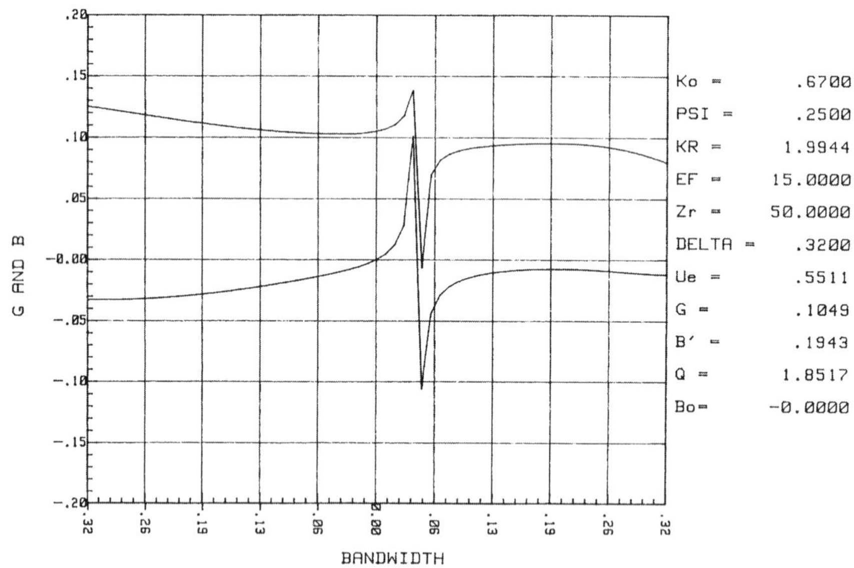

FIGURE 19.1 Frequency responses of complex gyrator circuit for $\psi = 0.25$, $\kappa = 0.67$, $kR = 1.99$, $Z_r = 50\ \Omega$, $\varepsilon_f = 15$, and $\omega_0 C = 0$. (Reproduced with permission from J. Helszajn and R. D. Baars, Synthesis of wide-band planar circulators using narrow coupling angles and undersized disk resonators, *IEEE Trans. Microwave Theory Tech.*, Vol. MTT-39, pp. 1681–1687, Oct. 1991.)

more often than not will still be inadequate. Scrutiny of this problem indicates that for $\psi < 0.26$ rad, kR only exists for $\kappa \leq 0.40$. For $\psi > 0.26$ rad, it exists for all values of κ between 0 and 1.

One way to alter this situation is to replace the distributed resonator by a mixed distributed radial/lumped element arrangement. This geometry displays a new useful field of solutions and permits the radial wavenumber (kR) to be employed as an independent instead of a dependent variable. It is satisfied by replacing the conventional first circulation condition, obtained by setting the imaginary part of the complex gyrator immittance to zero,

$$B_{\text{in}} = 0 \tag{19.1}$$

with

$$B_{\text{in}} + j\omega_0 C = 0 \tag{19.2}$$

B_{in} is the susceptance of the distributed disk circuit, and $j\omega_0 C$ is that of a lumped capacitance at each port. In this instance, the topology of the gyrator circuit is indicated in Fig. 19.2.

Figure 19.3 indicates the real and imaginary parts of the distributed complex gyrator admittance for the situation employed in obtaining the frequency response in Fig. 19.1 but with kR equal to 1.30. This equivalent circuit is well behaved over a considerable frequency interval but its resonant frequency is outside the range under consideration. However, it may now be readily centered by adding a shunt lumped element susceptance at each port of the device in keeping with the condition in Eq. (19.2).

Figure 19.4 displays the frequency response of one solution. It is symmetrical about its center frequency and both its gyrator conductance and susceptance slope parameter are exceptionally well behaved over the frequency response under consideration. It should therefore be of value in the synthesis of quarter-wave coupled devices. In obtaining the data in Figs. 19.1, 19.3, and 19.4, the first six poles of the circuit have been retained.

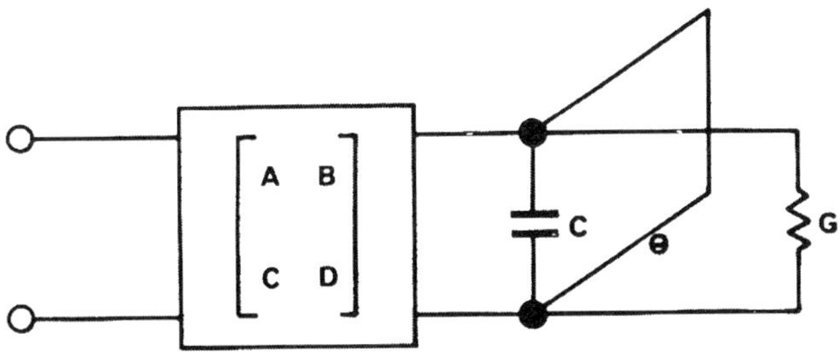

FIGURE 19.2 Schematic diagram complex gyrator circuit using distributed radial/lumped element resonator.

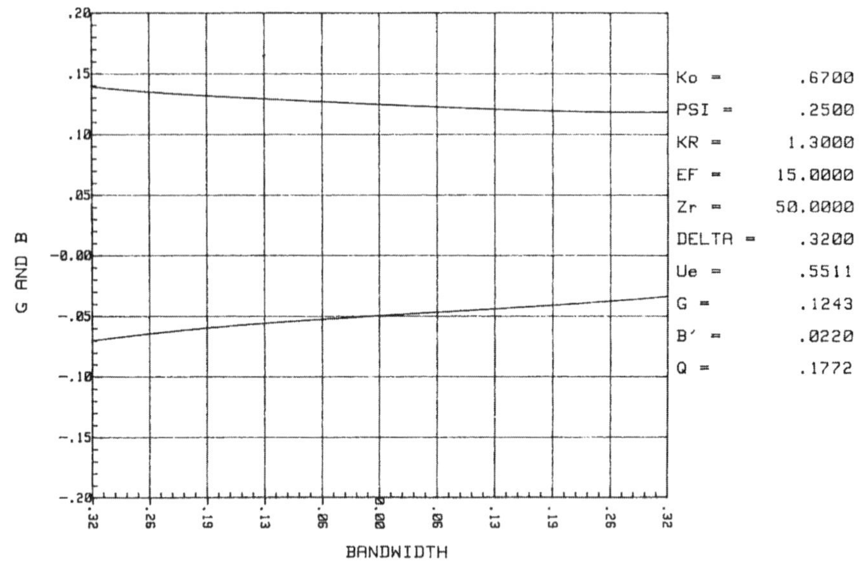

FIGURE 19.3 Frequency response of complex gyrator circuit for $\psi = 0.25$, $\kappa = 0.67$, $kR = 1.30$, $Z_r = 50\ \Omega$, $\varepsilon_f = 15$, and $\omega_0 C = 0$. (Reproduced with permission from J. Helszajn and R. D. Baars, Synthesis of wide-band planar circulators using narrow coupling angles and undersized disk resonators, *IEEE Trans. Microwave Theory Tech.*, Vol. MTT-39, pp. 1681–1687, Oct. 1991.)

In Eq. (19.2) $j\omega_0 C$ is written separately as

$$j\omega_0 C(1 + \delta) \tag{19.3}$$

where δ is the normalized frequency variable

$$\delta = (\omega - \omega_0)/\omega_0 \tag{19.4}$$

ω is the usual frequency variable (rad/s) and ω_0 is the center frequency (rad/s).

Table 19.1 gives some (but not necessarily the best) circulation solutions for one choice of radial wavenumber. The first four entries in this table determine the boundary conditions of the junction and fix the first circulation condition of the circuit; the next three entries define the complex gyrator circuit of the device for the situation for which $Z_r = 50\ \Omega$. The absolute value of the lumped element susceptance required to meet the first circulation condition is primarily dependent on the choice of the radial wavenumber. The most significant parameter in the description of this kind of circuit is of course the value of the loaded Q-factor displayed by the solution. This quantity fixes the gain bandwidth of the device, in keeping with the material in Chapter 25. The values displayed by the solutions tabulated here are in fact all suitable for the

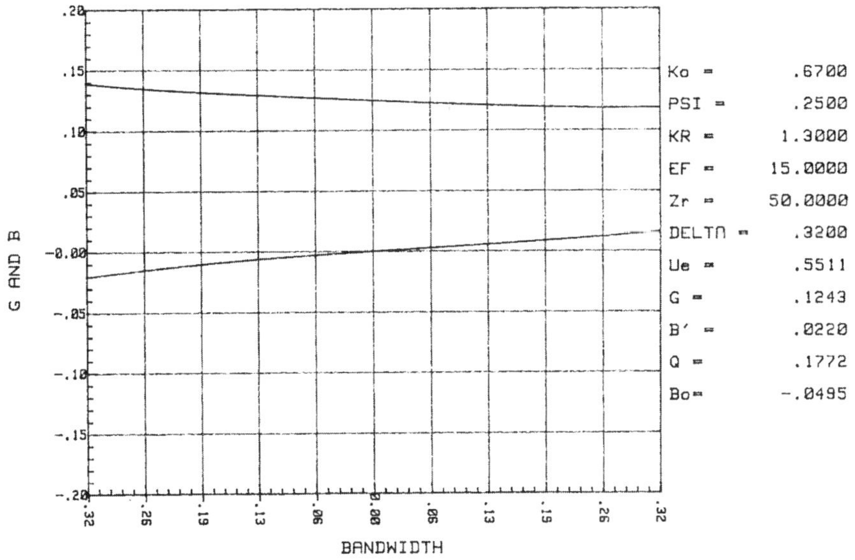

FIGURE 19.4 Frequency response of complex gyrator circuit for $\psi = 0.25$, $\kappa = 0.67$, $kR = 1.30$, $Z_r = 50\ \Omega$, $\varepsilon_f = 15$, and $\omega_0 C = 0.0495$. (Reproduced with permission from J. Helszajn and R. D. Baars, Synthesis of wide-band planar circulators using narrow coupling angles and undersized disk resonators, *IEEE Trans. Microwave Theory Tech.*, Vol. MTT-39, pp. 1681–1687, Oct. 1991.)

synthesis of devices with degree-3 equal ripple frequency responses over an octave band or more. Of some interest is the fact that the quality factor of the circuit increases as kR decreases. This would suggest that only a modest decrease in the radial wavenumber is sufficient to significantly alter the character of the complex gyrator circuit of the circulator. Such solutions are also associated with small values for the lumped

TABLE 19.1 Circulation Solutions for One Choice of Radial Wavenumber

κ_0	ψ	kR	B_0	G	B'	Q_L
0.500	0.200	1.3	0.086	0.123	0.089	0.724
0.500	0.250	1.3	0.067	0.100	0.074	0.733
0.500	0.300	1.3	0.055	0.089	0.058	0.752
0.525	0.200	1.3	0.082	0.128	0.080	0.628
0.525	0.250	1.3	0.064	0.105	0.067	0.638
0.525	0.300	1.3	0.052	0.090	0.058	0.652
0.550	0.200	1.3	0.079	0.133	0.071	0.536
0.550	0.250	1.3	0.061	0.109	0.060	0.547
0.550	0.300	1.3	0.049	0.093	0.052	0.562
0.575	0.200	1.3	0.076	0.138	0.062	0.449
0.575	0.250	1.3	0.058	0.113	0.052	0.460
0.575	0.300	1.3	0.046	0.097	0.046	0.476

element circuits. Note also that the values of the gyrator conductance displayed in this table are essentially independent of the radial wavenumber.

The complex gyrator admittance of one experimental junction using an undersized disk resonator at the junction of three 50 Ω air lines over the frequency interval 1.0–2.0 GHz and parametric values of the direct magnetic field is indicated in Fig. 19.5. The topology of this geometry has a coupling angle (ψ) of 0.30 rad, an off-diagonal element (κ) with a nominal value of 0.70, and a radial wavenumber (kR) of 1.21. The theoretical value of κ applies to a saturated material—a condition seldom established in practice. This result was obtained by decoupling port 3 from port 1 by placing a variable mismatch at port 2. The gyrator conductance and the susceptance slope parameter of this junction are nearly constants over the full experimental frequency interval. It is also apparent that the magnitude of the real part intersects the theoretical value at approximately the direct field required to saturate the garnet material. The agreement between the theoretical and experimental values of the

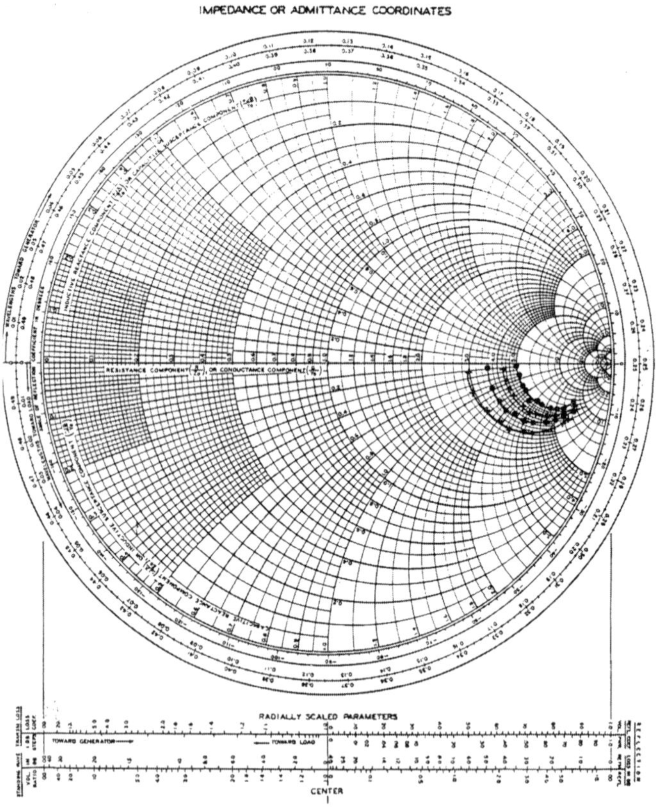

FIGURE 19.5 Experimental Smith chart of undersized resonator ($kR = 1.2$, $\psi = 0.30$, $\kappa = 0.67$, $Z_r = 50$ Ω). (Reproduced with permission from J. Helszajn and R. D. Baars, Synthesis of wide-band planar circulators using narrow coupling angles and undersized disk resonators, *IEEE Trans. Microwave Theory Tech.*, Vol. MTT-39, pp. 1681–1687, Oct. 1991.)

susceptance slope parameters (and therefore loaded Q-factors) while less good is still gratifying. The value of the lumped element susceptance required to center the experimental data in Fig. 19.5 is likewise in keeping with that calculated in Table 19.1. The slight deterioration in the complex gyrator circuit at the low frequency extremity of the band may in part be attributed to the onset of low-field loss.

19.3 SYNTHESIS OF QUARTER-WAVE COUPLED JUNCTION CIRCULATORS USING DISTRIBUTED/LUMPED ELEMENT RESONATORS

The physical design of quarter-wave coupled junction circulators with equal ripple frequency responses is not complete until both the network specification and the topology of the junction are stipulated. The network problem fixes the characteristic impedance (Z_r) at the terminals of the resonator and the relative dielectric constant (ε_t) of the transformer region in the normal way from a knowledge of the loaded Q-factor and the relationship between the ripple specification and the bandwidth. The aspect ratio of the resonator (R/H) is also fixed once the coupling angle (ψ) and the impedance (Z_r) at the resonator terminals are stipulated. If the thickness of the center conductor is neglected then the required relationship is

$$\frac{R}{H} = \frac{1}{\sin(\psi)\,[\text{anti}\ln(Z_r/30\pi) - 1]} \tag{19.5}$$

Table 19.2 summarizes some network solutions corresponding to the entries in Table 19.1 for the case of a topology employing two quarter-wave long impedance transformers. The first three entries are therefore common in the two tables; the next two entries specify the relative dielectric constant of the transformer region

TABLE 19.2 Network Solutions

κ_0	ψ	kR	B_0	ε_t	R/H	W
0.500	0.200	1.3	0.086	15.9	4.85	0.760
0.500	0.250	1.3	0.067	10.4	5.35	0.755
0.500	0.300	1.3	0.055	7.5	5.72	0.748
0.525	0.200	1.3	0.082	20.0	3.58	0.819
0.525	0.250	1.3	0.064	13.1	3.99	0.812
0.525	0.300	1.3	0.052	9.4	4.32	0.804
0.550	0.200	1.3	0.079	24.9	2.20	0.884
0.550	0.250	1.3	0.061	16.4	2.57	0.875
0.550	0.300	1.3	0.049	11.7	2.88	0.864
0.575	0.200	1.3	0.076	30.5	1.43	0.954
0.575	0.250	1.3	0.058	20.1	1.74	0.945
0.575	0.300	1.3	0.046	14.4	2.02	0.931

and the aspect ratio of the resonator. These quantities are not unique but are dependent on the network specification. The entries in this table satisfy the network problem with $S(\min) = 1.05$ and $S(\max) = 1.15$ over one octave band or more. Fairly large values for the relative dielectric constant of the transformer region, narrow coupling angles (in contrast with the conventional solution), and practical ground plane spacings are features of these solutions. Scrutiny of the network problem indicates that quite small changes in the ripple specification can produce fairly large variations in both these quantities. The last entry in Table 19.2 is the bandwidth for the network problem adopted. The problem of establishing the required lumped element susceptance is of separate issue. One possibility is to make use of the fringing capacitance at the terminals of the semi-ideal magnetic wall of the open gyromagnetic resonator. If this is the approach adopted, then the only entries that need to be considered in these tables are those for which R/H leads to practical ground plane spacings (say, $1.5 \leq R/H \leq 4$) and for which the lumped element susceptance can be realized.

Figures 19.6 and 19.7 depict the frequency responses of the gyrator circuit over an 80% band for one situation without and with lumped element susceptances at each port. Figure 19.8 indicates the frequency response of a degree-3 quarter-wave coupled device using this solution. The aspect ratio of this resonator is 2.75, the radial wavenumber is 1.40, and the value of the lumped susceptance is 0.0557 S. The relative dielectric constant of the transformer region is 16.7. The classic solution, in the absence of any fringing effects, is described by $\kappa = 0.67$, $\psi = 0.53$, and $kR = 1.57$.

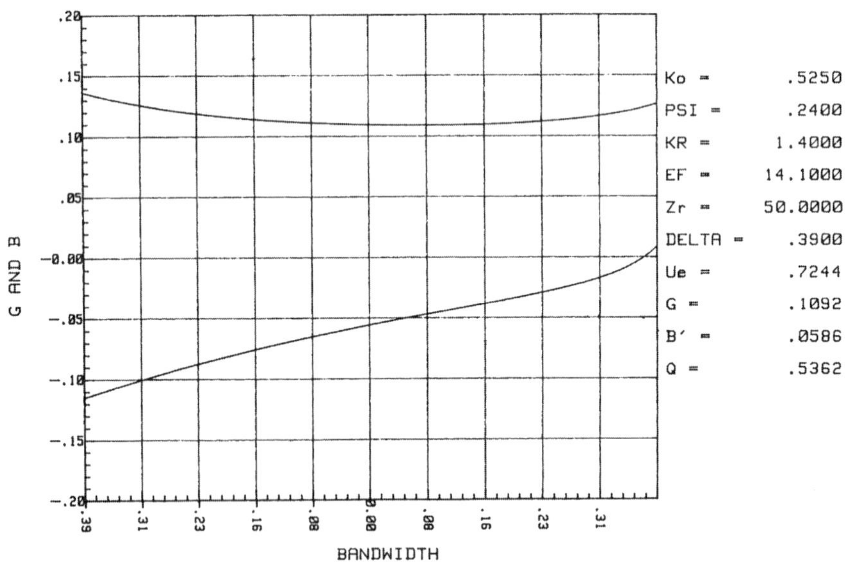

FIGURE 19.6 Frequency responses of complex gyrator circuit for $\psi = 0.24$, $\kappa = 0.525$, $kR = 1.40$, $Z_r = 50\ \Omega$, and $\omega_0 C = 0$. (Reproduced with permission from J. Helszajn and R. D. Baars, Synthesis of wide-band planar circulators using narrow coupling angles and undersized disk resonators, *IEEE Trans. Microwave Theory Tech.*, Vol. MTT-39, pp. 1681–1687, Oct. 1991.)

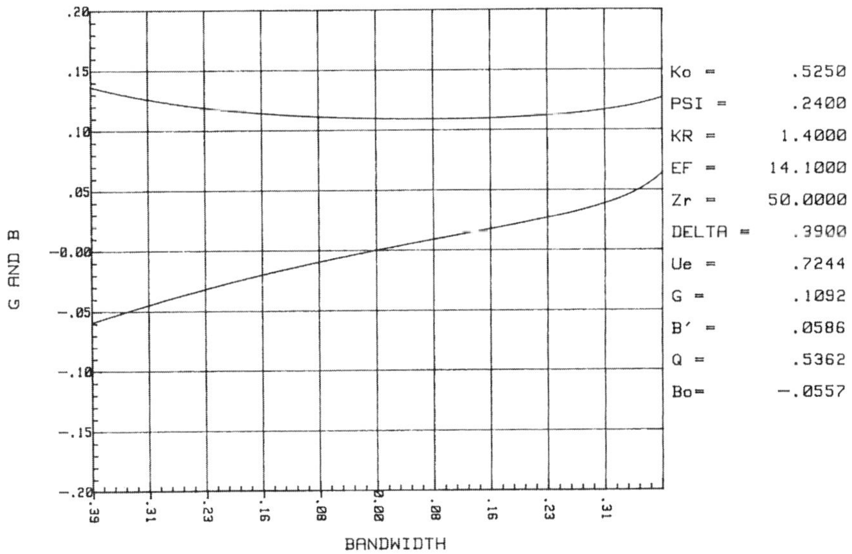

FIGURE 19.7 Frequency responses of complex gyrator circuit for $\psi = 0.24$, $\kappa = 0.525$, $kR = 1.40$, $Z_r = 50\,\Omega$, and $\omega_0 C = 0.056$. (Reproduced with permission from J. Helszajn and R. D. Baars, Synthesis of wide-band planar circulators using narrow coupling angles and undersized disk resonators, *IEEE Trans. Microwave Theory Tech.*, Vol. MTT-39, pp. 1681–1687, Oct. 1991.)

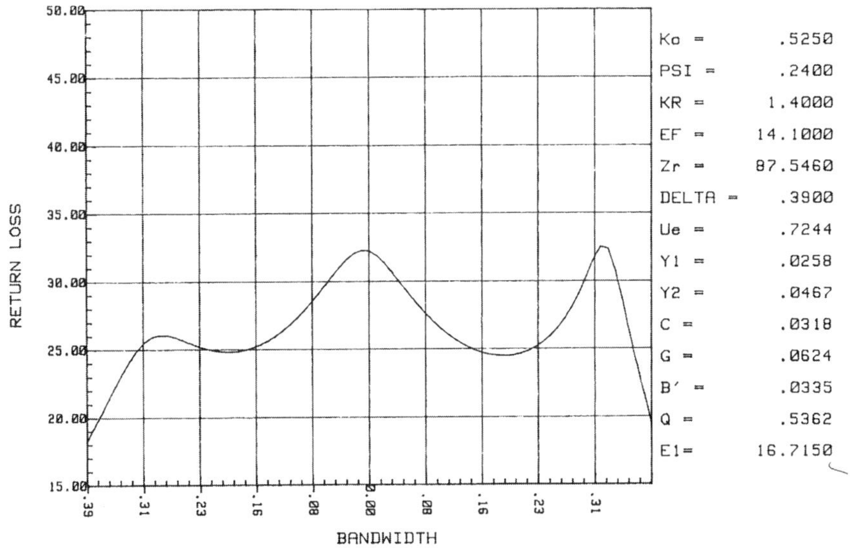

FIGURE 19.8 Frequency response of degree-3 junction for $\psi = 0.24$, $\kappa = 0.525$, $kR = 1.40$, and $\omega_0 C = 0.056$. (Reproduced with permission from J. Helszajn and R. D. Baars, Synthesis of wide-band planar circulators using narrow coupling angles and undersized disk resonators, *IEEE Trans. Microwave Theory Tech.*, Vol. MTT-39, pp. 1681–1687, Oct. 1991.)

19.4 MIXED DISTRIBUTED RADIAL/LUMPED ELEMENT RESONATOR

The required form of the lumped element discontinuity at the resonator terminals is not unique and may be obtained in a number of different ways. Here we make use of the naturally occurring fringing capacitance at the edge of a resonator embedded in a suitable dielectric substrate (Fig. 19.9). The value of this capacitance is a function of both the aspect ratio (R/H) of the resonator and the relative dielectric constant of the substrate at the boundary of the open resonator. The equivalent circuit of this arrangement consists of an undersized radial resonator loaded by a lumped element capacitance. This geometry in fact arises naturally in the design of quarter-wave coupled devices, since the impedance level of the transformer section adjacent to the junction is usually realized by loading the free space impedance defined by the resonator terminals with a material with an appropriate dielectric constant. In practice, it is a complicated function of all these variables and a solution to the characteristic equation, which is compatible with the network problem, is therefore not at first sight assured. It also differs for each of the three eigen-networks of the junction—a problem that awaits resolution. A preliminary scrutiny of this problem and the practical device described in the next section suggests that by itself it can essentially provide, as asserted, a means of realizing mixed distributed radial/lumped element resonators that are compatible with the construction of the class of octave-band circulators discussed here. This is especially so if the relative dielectric constant of the substrate material is comparable or larger than that of the garnet resonator and if the aspect ratio of the resonator is bracketed by $1.5 \leq R/H \leq 4.0$.

Figure 19.10 depicts the radial wavenumber $k_0 R\sqrt{\varepsilon_f \mu_d}$ of the dominant mode of a loosely coupled garnet resonator embedded in a medium with a relative dielectric constant ε_d for parametric values of R/H. The demagnetized permeability in this quantity has the meaning defined in Chapter 2. It is furthermore assumed that the ratio of the demagnetized and effective radial wavenumbers is on the order of unity.

Scrutiny of the experimental frequency response in Fig. 19.5 indicates that its radial wavenumber must be lowered by a factor of about 0.75 in order to center it. One way this may be achieved is by introducing a lumped susceptance equal to about 0.06 (S) at each of its terminals. Another possibility is to make use of the natural fringing capacitance at the boundary between a disk resonator and a suitable

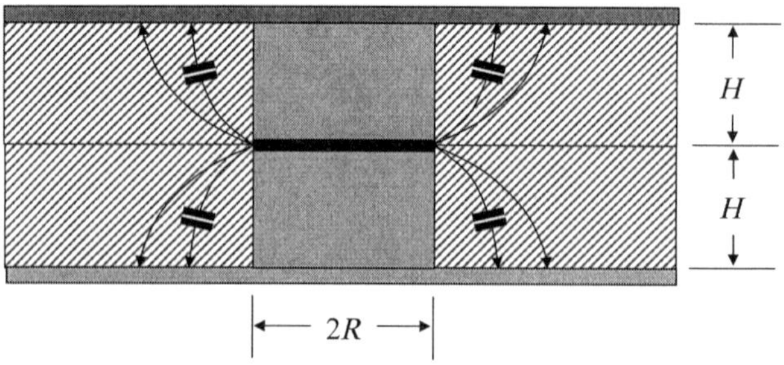

FIGURE 19.9 Equivalent circuit of disk resonator embedded in a dielectric region.

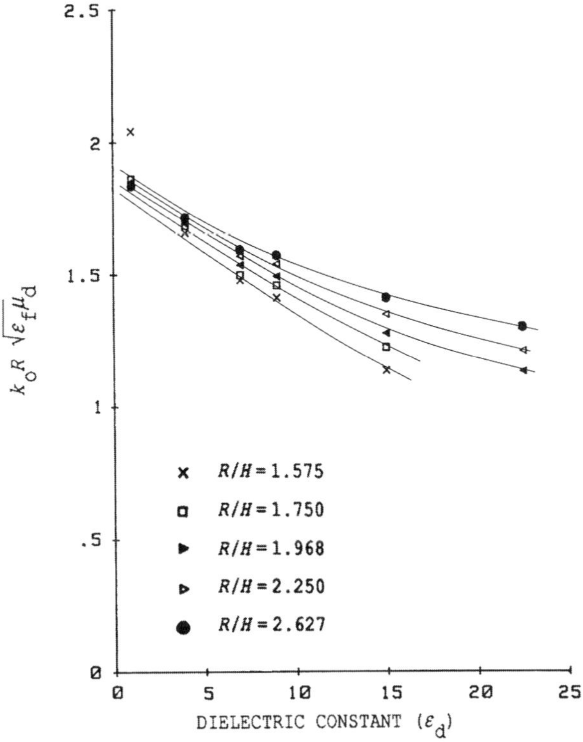

FIGURE 19.10 Resonant frequency of loosely coupled stripline resonator embedded in dielectric substrate. (Reproduced with permission from J. Helszajn and R. D. Baars, Synthesis of wide-band planar circulators using narrow coupling angles and undersized disk resonators, *IEEE Trans. Microwave Theory Tech.*, Vol. MTT-39, pp. 1681–1687, Oct. 1991.)

dielectric substrate. A solution to this problem is not assured but inspection of Fig. 19.10 indicates that constant R/H curves between 1.75 and 2.62 provide a number of such solutions for ε_d between 15 and 25. Any discrepancy between the required discontinuity and that displayed by the junction may in practice be accommodated by loading the junction with appropriate magnetic walls.

Scrutiny of this graph suggests that the required fringing susceptance specified in Eq. (19.2) is nearly naturally met by the solution summarized by Figs. 19.6–19.8. In practice, not only is the dominant mode modified by the presence of open walls but the other higher order modes of the problem region are also. The effect is to perturb the overall gyromagnetic space of the geometry.

19.5 1–2 GHz DEVICE

A 1–2 GHz device based on the topology outlined here, using the junction described in Fig. 19.5, has in fact been partly experimentally developed prior to the theory

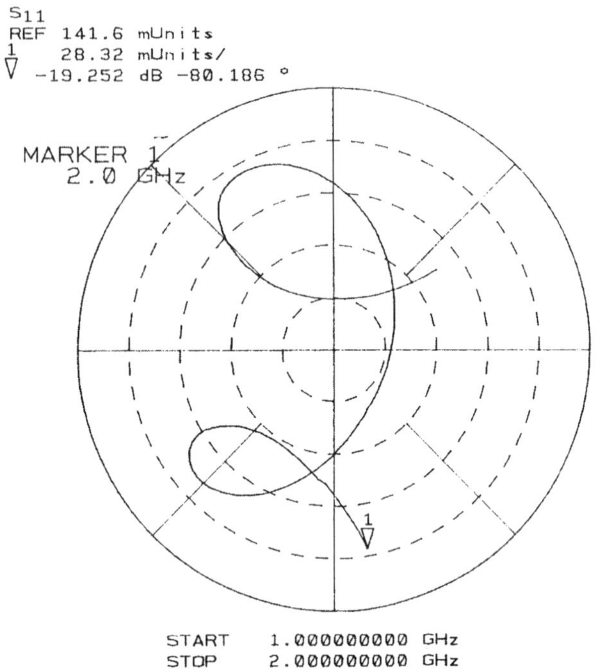

FIGURE 19.11 Frequency response of quarter-wave long transformer coupled junction using mixed distributed radial/lumped element resonators. (Reproduced with permission from J. Helszajn and R. D. Baars, Synthesis of wide-band planar circulators using narrow coupling angles and undersized disk resonators, *IEEE Trans. Microwave Theory Tech.*, Vol. MTT-39, pp. 1681–1687, Oct. 1991.)

outlined here and optimized thereafter. It is described by

$$\psi = 0.27$$
$$\kappa = 0.70$$
$$kR = 1.21$$
$$R/H = 2.77$$
$$\varepsilon_d = 25$$

Some possible theoretical entries that are suitable for the design of a wideband device, which straddle these physical parameters except for the value of κ, are given by the entries in row seven in Table 19.1. A perusal of Figs. 19.5 and 19.10 separately indicates that this solution approximately satisfies the first circulation solution of the circulator.

The discrepancy between the two values of the normalized magnetization (κ) of the garnet material may be understood by recognizing that its value at magnetic saturation (κ) represents in practice an upper bound on its actual value κ_p. The two

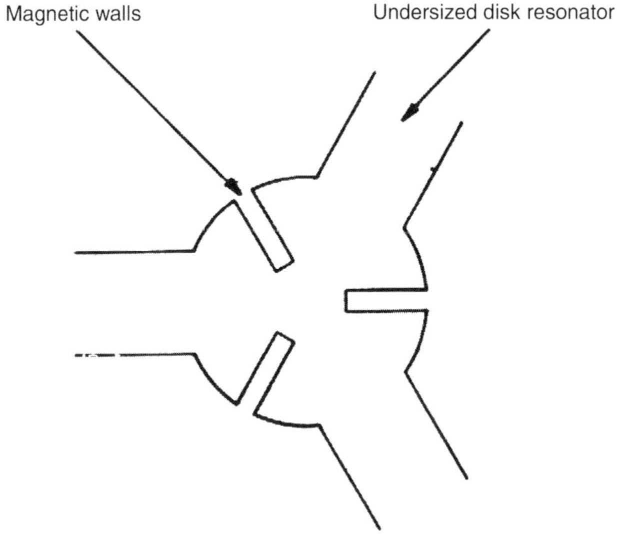

FIGURE 19.12 Schematic diagram of disk resonator loaded with magnetic walls.

values of the normalized magnetization are related by

$$\kappa_p = \left(\frac{M}{M_0}\right)\kappa \tag{19.6}$$

where M is the actual magnetization of the material, and M_0 is its value at magnetic saturation. If M is taken as the remanence value of the hysteresis of the material loop, then

$$M/M_0 = 0.70 \tag{19.7}$$

and the value of κ_p in this example is consistent with those appearing in Table 19.1.

Figure 19.11 illustrates a typical response at one port of the device. Some fine trimming of the junction and or the fringing capacitance with the aid of transverse magnetic walls in the manner indicated in Fig. 19.12 was employed in obtaining this result; otherwise all the variables are within 15% of the theoretical values. The insertion loss between any two ports is less than 0.35 dB over the full frequency interval.

19.6 SYNTHESIS OF WEAKLY MAGNETIZED UNDERSIZED JUNCTIONS

In a weakly magnetized junction for which the in-phase eigen-network can be idealized by a short-circuit boundary condition, the situation is fairly simple. The first circulation condition is determined by the sum of the counterrotating eigenvalues and the second one involves the difference between the two. In this simplified situation the characteristic equation takes the form in Eq. (19.2) with the phenomenological dynamical capacitance replaced by that associated with the dominant mode of the resonator; the gyrator conductance may be assumed to be unaffected by the fringing capacitance provided the dynamical relative dielectric constant of the resonator is not too different from that of the ferrite or garnet material. The complex gyrator circuit is invariably better behaved as a consequence of this type of offset. A simple correction to the radius of the resonator therefore goes some way to satisfy the first two circulation conditions of the device as is sometimes assumed. However, it does leave something to be desired insofar as its susceptance slope parameter is concerned. This is especially the case if the junction is not weakly magnetized, as is more often than not the case.

CHAPTER TWENTY

Complex Gyrator Circuit of Three-Port Circulators Using Gyromagnetic Resonators with Sixfold Symmetry

20.1 INTRODUCTION

A common resonator met in the design of three-port circulators is an undersized circular gyromagnetic disk symmetrically loaded by six stubs. This sort of junction may be reduced to a symmetrical three-port circulator by either open- or short-circuiting three of the stubs. The description of the port relationships is usually expressed in terms of either its open-circuit or scattering matrices or its impedance or scattering eigenvalues. A knowledge of any one of these formulations is sufficient for that of any of the other ones. The procedure adopted in this chapter is based on a formulation of the open-circuit parameters and the complex gyrator circuit of the open-circuit arrangement. It begins by constructing the open-circuit parameters of the six-port network in terms of its eigenvalues by having recourse to symmetry considerations only. The required parameters of the three-port circulator are then obtained from those of the six-port one by terminating a symmetrical triplet of ports by magnetic walls. The derivation is completed by recalling the description of a typical eigenvalue of a symmetrical m-port gyromagnetic plate. The important condition for which the in-phase eigen-network lies midway between the split counterrotating ones is given special attention. While a solution to the problem is not assured, some such solutions are indeed displayed by this geometry. If the resonator has sixfold symmetry then the maximum coupling

The Stripline Circulator: Theory and Practice. By J. Helszajn

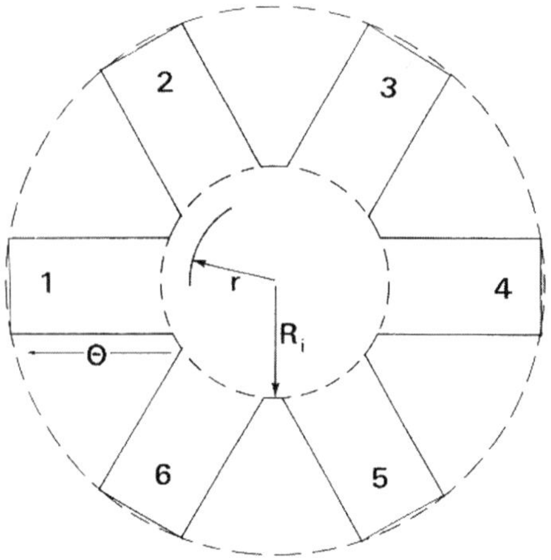

FIGURE 20.1 Schematic diagram of six-port planar gyromagnetic circuit with six-fold symmetry.

angle (ψ) at the circular region is restricted to $\pi/6$ rad. This value is half that met in the design of the classic planar resonator using a simple disk resonator. The geometry under consideration provides one means of adjusting the quality factor of the classic tracking solution using a simple gyromagnetic disk. The topology of the resonator under consideration is illustrated in Fig. 20.1.

20.2 EIGENVALUES AND EIGENVECTORS OF *m*-PORT SYMMETRIC PLANAR JUNCTION CIRCULATOR

If the eigenvectors and eigenvalues of an m-port junction are available, it is possible to recover the coefficients of the related matrix using a simple mathematical technique. This may be done by having recourse to the similarity transformation in Chapter 5.

The diagonalization procedure will now be developed in the case of an ideal m-port symmetric circuit by constructing the square matrices $\overline{\mathbf{U}}$ and $(U^*)^{\mathrm{T}}$ in terms of the eigenvectors of the problem region. The eigenvectors of an m-port symmetrical junction can be derived from symmetry conditions only. The required result is

$$(U_n)_q = \frac{\exp\left[-j2\pi n(q-1)\right]/m}{\sqrt{m}} \tag{20.1}$$

m is the number of ports, q is the port location, and n refers to the mode nomenclature of the resonator.

If $m = 6$ then n and q are given by

$$n = 0,\ 1,\ -1,\ 2,\ -2,\ 3 \tag{20.2}$$

$$q = 1,\ 2,\ 3,\ 4,\ 5,\ 6 \tag{20.3}$$

The eigenvectors of a six-port symmetric junction are therefore described by

$$\mathbf{U}_{+2} = \frac{1}{\sqrt{6}}\begin{bmatrix}1\\ \beta^5\\ \beta^4\\ \beta^3\\ \beta^2\\ \beta\end{bmatrix}, \quad \mathbf{U}_{+1} = \frac{1}{\sqrt{6}}\begin{bmatrix}1\\ \alpha^5\\ \alpha^4\\ \alpha^3\\ \alpha^2\\ \alpha\end{bmatrix}, \quad \mathbf{U}_{0} = \frac{1}{\sqrt{6}}\begin{bmatrix}1\\ 1\\ 1\\ 1\\ 1\\ 1\end{bmatrix} \tag{20.4a}$$

$$\mathbf{U}_{-1} = \frac{1}{\sqrt{6}}\begin{bmatrix}1\\ \alpha\\ \alpha^2\\ \alpha^3\\ \alpha^4\\ \alpha^5\end{bmatrix}, \quad \mathbf{U}_{-2} = \frac{1}{\sqrt{6}}\begin{bmatrix}1\\ \beta\\ \beta^2\\ \beta^3\\ \beta^4\\ \beta^5\end{bmatrix}, \quad \mathbf{U}_{3} = \frac{1}{\sqrt{6}}\begin{bmatrix}1\\ \gamma^5\\ \gamma^4\\ \gamma^3\\ \gamma^2\\ \gamma\end{bmatrix} \tag{20.4b}$$

where

$$\alpha = \exp(j60) \tag{20.5a}$$

$$\beta = \exp(j120) \tag{20.5b}$$

$$\gamma = \exp(j180) \tag{20.5c}$$

The square matrix $[U]$ with the eigenvectors of the problem region along its columns is given by

$$[U] = \frac{1}{\sqrt{6}}\begin{bmatrix}1 & 1 & 1 & 1 & 1 & 1\\ 1 & \alpha & \alpha^2 & -1 & (\alpha^*)^2 & \alpha^2\\ 1 & \alpha^2 & (\alpha^*)^2 & 1 & \alpha^2 & (\alpha^*)^2\\ 1 & -1 & 1 & 1 & 1 & -1\\ 1 & (\alpha^*)^2 & \alpha^2 & 1 & (\alpha^*)^2 & \alpha^2\\ 1 & \alpha^* & (\alpha^*)^2 & -1 & \alpha^2 & \alpha\end{bmatrix} \tag{20.6}$$

In forming the matrix $\overline{\mathbf{U}}$ in terms of the m eigenvectors of the problem region, care has been taken to ensure that the resultant matrix is symmetrical.

The diagonal matrix containing the impedance eigenvalues is

$$\overline{\mathbf{Z}} = \begin{bmatrix}Z_0 & 0 & 0 & 0 & 0 & 0\\ 0 & Z_{-1} & 0 & 0 & 0 & 0\\ 0 & 0 & Z_{-2} & 0 & 0 & 0\\ 0 & 0 & 0 & Z_3 & 0 & 0\\ 0 & 0 & 0 & 0 & Z_2 & 0\\ 0 & 0 & 0 & 0 & 0 & Z_1\end{bmatrix} \tag{20.7}$$

Since the matrix $[U]$ is symmetric it satisfies

$$[U]^{-1} = [U^*]^{\mathrm{T}} \tag{20.8}$$

The required open-circuit parameters are now obtained by having recourse to the similarity transformation in Chapter 5. The results are

$$Z_{11}(6) = \frac{(Z_0 + Z_3) + (Z_{+1} + Z_{-2}) + (Z_{-1} + Z_{+2})}{6} \tag{20.9a}$$

$$Z_{12}(6) = \frac{(Z_0 - Z_3) + (Z_2 - Z_{-1})\beta + (Z_{-2} - Z_1)\beta^2}{6} \tag{20.9b}$$

$$Z_{13}(6) = \frac{(Z_0 + Z_3) + (Z_1 + Z_{-2})\beta + (Z_{-1} + Z_2)\beta^2}{6} \tag{20.9c}$$

$$Z_{14}(6) = \frac{(Z_0 - Z_3) + (Z_2 - Z_{-1}) + (Z_{-2} - Z_1)}{6} \tag{20.9d}$$

$$Z_{15}(6) = \frac{(Z_0 + Z_3) + (Z_1 + Z_{-2})\beta^2 + (Z_{-1} + Z_2)\beta}{6} \tag{20.9e}$$

$$Z_{16}(6) = \frac{(Z_0 - Z_3) + (Z_2 - Z_{-1})\beta^2 + (Z_{-2} - Z_1)\beta}{6} \tag{20.9f}$$

20.3 OPEN-CIRCUIT PARAMETERS OF A THREE-PORT CIRCULATOR WITH THREE-FOLD SYMMETRY

The open-circuit parameters of the required three-port circulator formed by open-circuiting ports 2, 4, and 6 may be deduced by introducing the appropriate boundary conditions at the ports in question.

$$I_2 = I_4 = I_6 = 0 \tag{20.10}$$

The voltage–current relationships at the various ports are then specified by

$$\begin{bmatrix} V_1 \\ V_2 \\ V_3 \\ V_4 \\ V_5 \\ V_6 \end{bmatrix} = \begin{bmatrix} Z_{11} & Z_{12} & Z_{13} & Z_{14} & Z_{15} & Z_{16} \\ Z_{16} & Z_{11} & Z_{12} & Z_{13} & Z_{14} & Z_{15} \\ Z_{15} & Z_{16} & Z_{11} & Z_{12} & Z_{13} & Z_{14} \\ Z_{14} & Z_{15} & Z_{16} & Z_{11} & Z_{12} & Z_{13} \\ Z_{13} & Z_{14} & Z_{15} & Z_{16} & Z_{11} & Z_{12} \\ Z_{12} & Z_{13} & Z_{14} & Z_{15} & Z_{16} & Z_{11} \end{bmatrix} \begin{bmatrix} I_1 \\ 0 \\ I_3 \\ 0 \\ I_5 \\ 0 \end{bmatrix} \tag{20.11}$$

The port variables at ports 1, 3, and 5 are given without ado by

$$\begin{bmatrix} V_1 \\ V_3 \\ V_5 \end{bmatrix} = \begin{bmatrix} Z_{11} & Z_{13} & Z_{15} \\ Z_{15} & Z_{11} & Z_{13} \\ Z_{13} & Z_{15} & Z_{11} \end{bmatrix} \begin{bmatrix} I_1 \\ I_3 \\ I_5 \end{bmatrix} \tag{20.12}$$

It is convenient, in what follows, to relabel the ports to conform to those of a standard three-port circulator. If this is done, then the new open-circuited parameters are

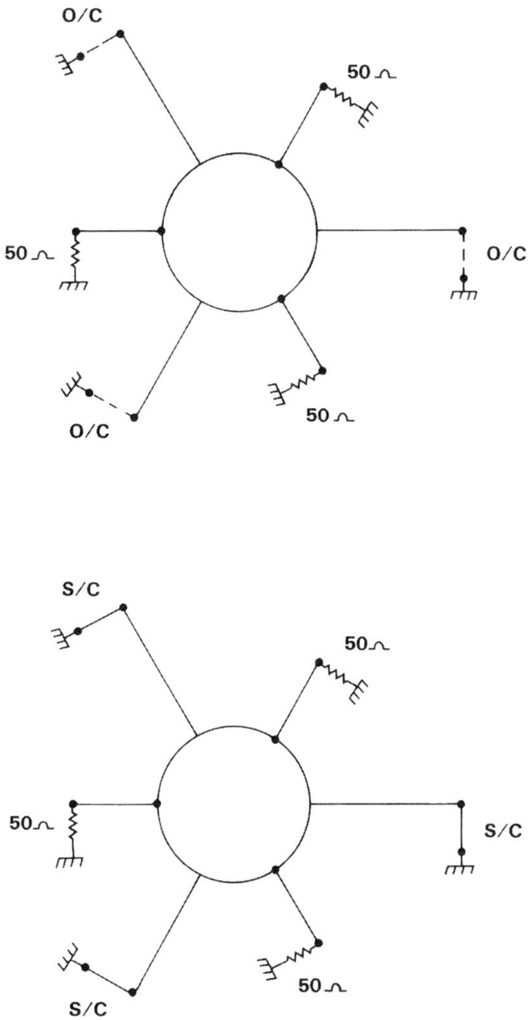

FIGURE 20.2 Schematic diagram of wye resonator using short- and open-circuit stubs.

described in terms of the old ones:

$$Z_{11}(3) = Z_{11}(6) \tag{20.13a}$$

$$Z_{12}(3) = Z_{13}(6) \tag{20.13b}$$

$$Z_{13}(3) = Z_{15}(6) \tag{20.13c}$$

Scrutiny of the open-circuit parameters of the three-port circuit indicates that if the poles of the six-port with the same coefficients are collated then

$$Z_{11}(3) = \frac{Z^0(6) + Z^+(6) + Z^-(6)}{6} \tag{20.14a}$$

$$Z_{12}(3) = \frac{Z^0(6) + \beta Z^+(6) + \beta^2 Z^-(6)}{6} \tag{20.14b}$$

$$Z_{13}(3) = \frac{Z^0(6) + \beta^2 Z^+(6) + \beta Z^-(6)}{6} \tag{20.14c}$$

where

$$Z^0(6) = Z_0(6) + Z_3(6) \tag{20.15a}$$

$$Z^+(6) = Z_{+1}(6) + Z_{-2}(6) \tag{20.15b}$$

$$Z^-(6) = Z_{-1}(6) + Z_{+2}(6) \tag{20.15c}$$

and β has the meaning previously noted.

The possibility of short-circuiting ports 2, 4, and 6 instead of open-circuiting these ports is also understood. Figure 20.2 indicates this situation.

20.4 EIGENVALUES OF SYMMETRICAL *m*-PORT GYROMAGNETIC RESONATOR

The poles in the description of the eigenvalues of the open-circuited parameters of a symmetrical m-port gyromagnetic disk are a standard result in the literature. A typical pole is described by

$$Z_n(m) = \frac{jm\eta_e Z_r \psi}{\pi} \left(\frac{\sin n\psi}{n\psi}\right)^2 \left[\frac{J_n'(kR)}{J_n(kR)} - \left(\frac{\kappa}{\mu}\right)\left(\frac{n}{kR}\right)\right]^{-1} \tag{20.16}$$

where $m = 6$ for a six-port circuit.

Some useful identities between the Bessel functions for the purposes of calculation are available in Chapter 9.

ψ is the coupling angle at a typical port of the terminals of the junction:

$$\psi = W/2R \tag{20.17}$$

where

$$\eta_e = \sqrt{\mu_e/\varepsilon_f} \tag{20.18}$$

and

$$Z_r = 30\pi \ln\left(\frac{W + t + 2H}{W + t}\right) \tag{20.19}$$

W is the width of the center conductor (in meters), H is the thickness of each garnet resonator (meters), R is the radius (meters), and t is the thickness of the center conductor (meters).

A feature of some import in the description of this class of circuit is the dependence of the gyrotropy of the gyromagnetic resonator on the frequency. For a saturated material

$$\mu = 1 \tag{20.20}$$

$$\mu_{eff} = 1 - \kappa^2 \tag{20.21}$$

$$\kappa = \left(\frac{\omega_m}{\omega_0}\right)\left(\frac{\omega_0}{\omega}\right) \tag{20.22}$$

where

$$\omega_m = \gamma M_0 \tag{20.23}$$

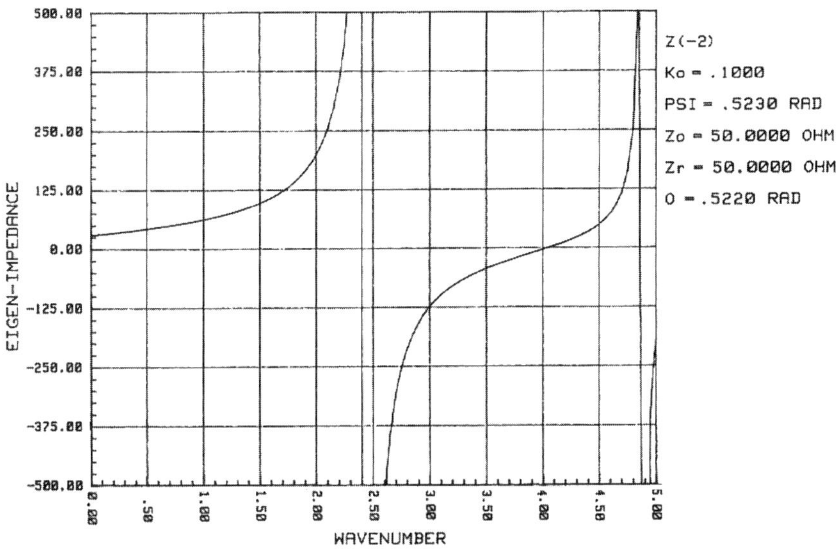

FIGURE 20.3 Typical impedance pole of gyromagnetic resonator with six-fold symmetry.

γ is the gyromagnetic ratio (2.21×10^5 rad/s per A/m), M_0 is the saturation magnetization (A/m), ω_0 is the midband frequency (rad/s), and ω is the frequency (rad/s).

Figure 20.3 indicates the frequency variation of a typical impedance pole met in connection with a symmetrical junction with sixfold symmetry.

20.5 AUGMENTED EIGENVALUES

The resonator under consideration consists of a radial gyromagnetic region symmetrically loaded by uniform gyromagnetic transmission lines. The open-circuit parameters at the terminals of the junction can be calculated once the eigenvalues of the problem region are established. In this instance, the eigenvalues at the outer set of terminals are related to those of the inner set in a straightforward way. This may be done by mapping each inner eigen-network through a section of transmission line of electrical length θ and characteristic impedance $\eta_e Z_r$. The required transformation is

$$Z_L' = \frac{AZ_L + B}{CZ_L + D} \tag{20.24}$$

where

$$A = \cosh\,\theta \tag{20.25a}$$

$$B = \eta_e Z_r \sinh\,\theta \tag{20.25b}$$

$$C = \frac{\sinh\theta}{\eta_e Z_r} \tag{20.25c}$$

$$D = \cosh\,\theta \tag{20.25d}$$

This transformation is illustrated in Fig. 20.4. Introducing this bilinear transformation between the eigenvalues at the two planes under consideration gives

$$Z_0'(6) = \eta_e Z_r \frac{Z_0(6) + \eta_e Z_r \tanh\,\theta}{\eta_e Z_r + Z_0(6)\tanh\,\theta} \tag{20.26a}$$

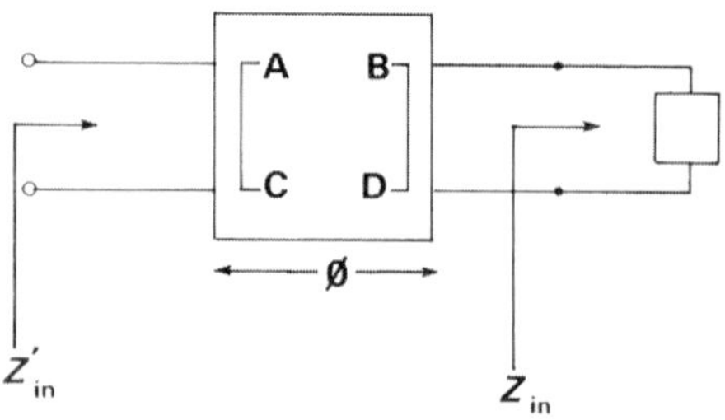

FIGURE 20.4 Mapping between impedances at input and output planes of uniform transmission line.

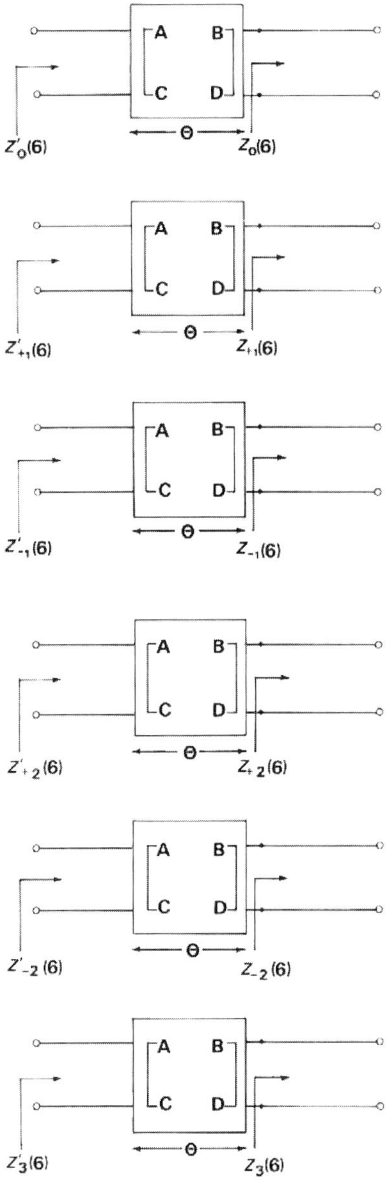

FIGURE 20.5 Eigen-networks of six-port planar gyromagnetic circuit with six-fold symmetry.

$$Z'_{-1}(6) = \eta_e Z_r \frac{Z_{-1}(6) + \eta_e Z_r \tanh \theta}{\eta_e Z_r + Z_{-1}(6) \tanh \theta} \tag{20.26b}$$

$$Z'_{+1}(6) = \eta_e Z_r \frac{Z_{+1}(6) + \eta_e Z_r \tanh \theta}{\eta_e Z_r + Z_{+1}(6) \tanh \theta} \tag{20.26c}$$

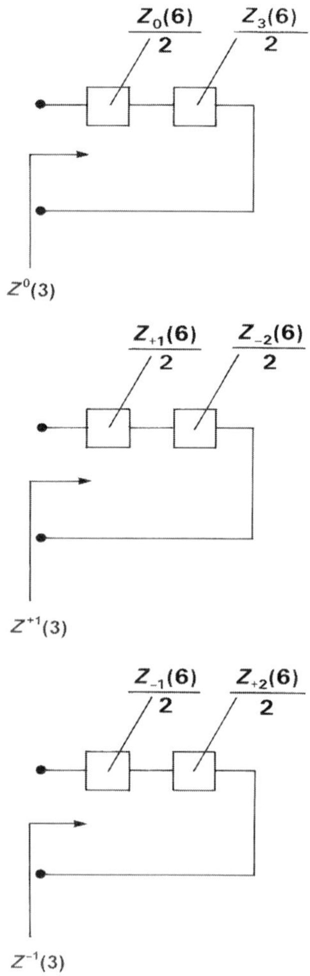

FIGURE 20.6 Eigen-networks of three-port planar gyromagnetic circuit with six-fold symmetry.

$$Z'_{-2}(6) = \eta_e Z_r \frac{Z_{-2}(6) + \eta_e Z_r \tanh\,\theta}{\eta_e Z_r + Z_{-2}(6)\tanh\,\theta} \tag{20.26d}$$

$$Z'_{+2}(6) = \eta_e Z_r \frac{Z_{+2}(6) + \eta_e Z_r \tanh\,\theta}{\eta_e Z_r + Z_{+2}(6)\tanh\,\theta} \tag{20.26e}$$

$$Z'_{3}(6) = \eta_e Z_r \frac{Z_{3}(6) + \eta_e Z_r \tanh\,\theta}{\eta_e Z_r + Z_{3}(6)\tanh\,\theta} \tag{20.26f}$$

θ is the electrical length of the unit element (UE):

$$\theta = jk_0\sqrt{\varepsilon_f \mu_{eff}}\,\ell \tag{20.27}$$

and Z_r is its characteristic impedance and η_e has the meaning met in connection with the definition of the eigenvalues. Figure 20.5 indicates the eigen-networks of the six-port network. Figure 20.6 gives those of the three-port circuit.

20.6 TRACKING SOLUTION

A circulation condition that is always of some interest in the design of wideband circulators is the tracking solution dealt with in Chapter 17. Its first circulation condition coincides with that for which the in-phase eigen-network lies between the two split counterrotating ones. It is satisfied provided

$$Z^0(3) = 0 \tag{20.28}$$

$$Z^+(3) + Z^-(3) = 0 \tag{20.29}$$

If ψ is fixed by the topology of the circuit as $\pi/6$, then the available physical variables for the adjustment of this boundary condition are κ, θ, or kR. While a solution to this problem is not assured, one possibility is described by

$$\kappa = 0.66, \quad \theta = 0.2074 \text{ rad}, \quad kR = 1.126 \text{ rad}$$

20.7 COMPLEX GYRATOR CIRCUIT

The complex gyrator immittance is again defined in terms of the open-circuit parameters of the three-port circuit in the usual way by

$$Y_{in} = 1/Z_{in} \tag{20.30}$$

$$Z_{in} = Z_{11} - Z_{12}^2/Z_{13} \tag{20.31}$$

The real and imaginary parts of Y_{in} are described in terms of R_{in} and X_{in} in Chapter 17.

The frequency response obtained in this way for $\delta_0 = \pm 0.33$ for the tracking solution stipulated in the previous section is displayed in Fig. 20.7.

In this instance, the complex gyrator circuit is described by

$$G = 0.0221\ \Omega^{-1}, \quad B' = 0.0003, \text{ and } Q_L = 0.0158$$

A feature of the resonator under consideration is that the loaded Q-factor of its semi-tracking solution may be adjusted by varying the length of the stubs. One possible solution that is compatible with the synthesis of an equal ripple degree-3 octave band circulator is

$$\kappa = 0.62, \quad \theta = 0.279 \text{ rad}, \text{ and } kR = 1.379 \text{ rad}$$

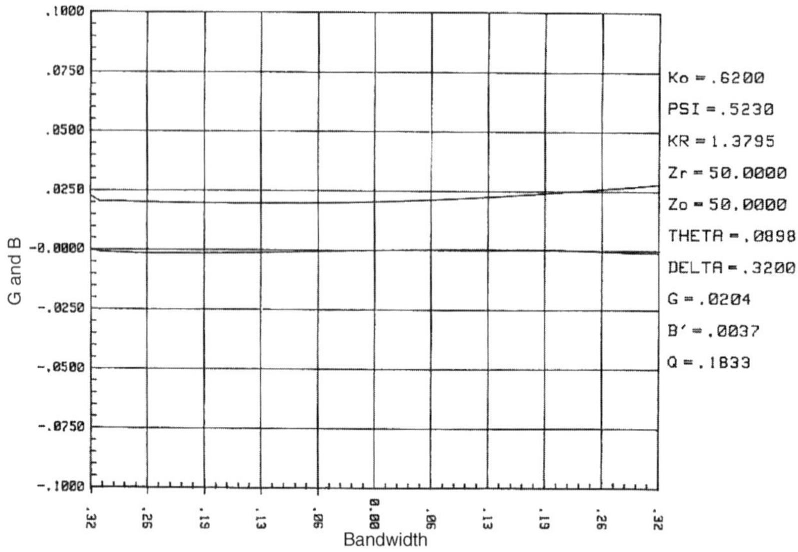

FIGURE 20.7 Complex gyrator circuit of planar junction using resonator with six-fold symmetry ($\kappa = 0.66$, $\theta = 0.2074$ rad, $\psi = \pi/6$ rad, $kR = 1.126$ rad). (Reproduced with permission from J. Helszajn and Y. Lapointe, Complex gyrator circuit of 3-port circulators using gyromagnetic resonators with six-fold symmetry, *IEE Proc. Microwaves Antennas Propag.*, Vol. 148, pp. 318–322, Oct. 2001.)

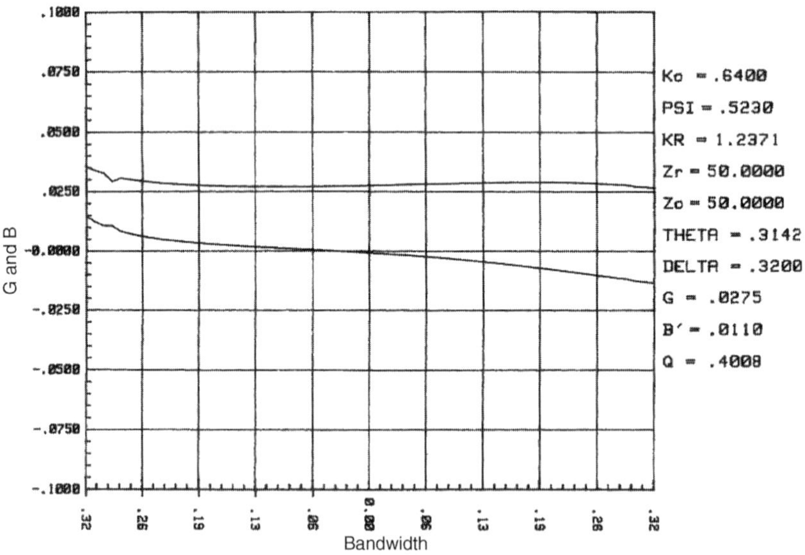

FIGURE 20.8 Complex gyrator circuit of planar junction using resonator with six-fold symmetry ($\kappa = 0.62$, $\theta = 0.2790$ rad, $\psi = \pi/6$ rad, $kR = 1.379$ rad). (Reproduced with permission from J. Helszajn and Y. Lapointe, Complex gyrator circuit of 3-port circulators using gyromagnetic resonators with six-fold symmetry, *IEE Proc. Microwaves Antennas Propag.*, Vol. 148, pp. 318–322, Oct. 2001.)

In this instance, the absolute elements of the complex gyrator circuit are specified by

$$G = 0.278\ \Omega^{-1},\ B' = 0.0139,\ \text{and}\ Q_{\mathrm{L}} = 0.5020$$

Figure 20.8 depicts the solution in this situation.

A property of the complex gyrator immittance of a junction circulator is that it is not a positive real function in that it contains gyrator circuits in addition to the usual positive elements. This means that the variation between its real and imaginary parts is not constrained to positive slopes.

CHAPTER TWENTY-ONE

Open-Circuit Parameters of Circulators Using Side-Coupled Wye Resonators: An Impedance Pole Approach

21.1 INTRODUCTION

The circulator described in this chapter consists of a 6-port nonresonant gyromagnetic disk with three open-circuited gyromagnetic stubs and three coupling ports with two degrees of threefold symmetry. Figure 21.1 shows two typical arrangements. A feature of either geometry is that it may readily be quarter-wave coupled to produce a compact commercial package. This type of function may be analyzed by forming a 6×6 impedance matrix for the central nonresonant disk region defined by the three open-circuited stubs of the resonator circuit and the three coupling intervals of the terminals. A careful scrutiny of its open-circuited parameters suggests that these quantities may be expressed in terms of linear combinations of the poles met in connection with each symmetrical sextet of ports. The boundary conditions at the terminals of the three stubs are subsequently used to reduce the 6×6 matrix to the required 3×3 impedance matrix at the ports of the circulator. The complex gyrator impedance of the circulator is then constructed with the help of this matrix and the standard circulator conditions thereafter satisfied. The impedance poles associated with each constituent junction are the classic result so that all calculations may proceed without difficulty.

The geometry under consideration reduces to the classic symmetrical three-port junction when the coupling angle of a typical stub is made equal to zero. It

The Stripline Circulator: Theory and Practice. By J. Helszajn

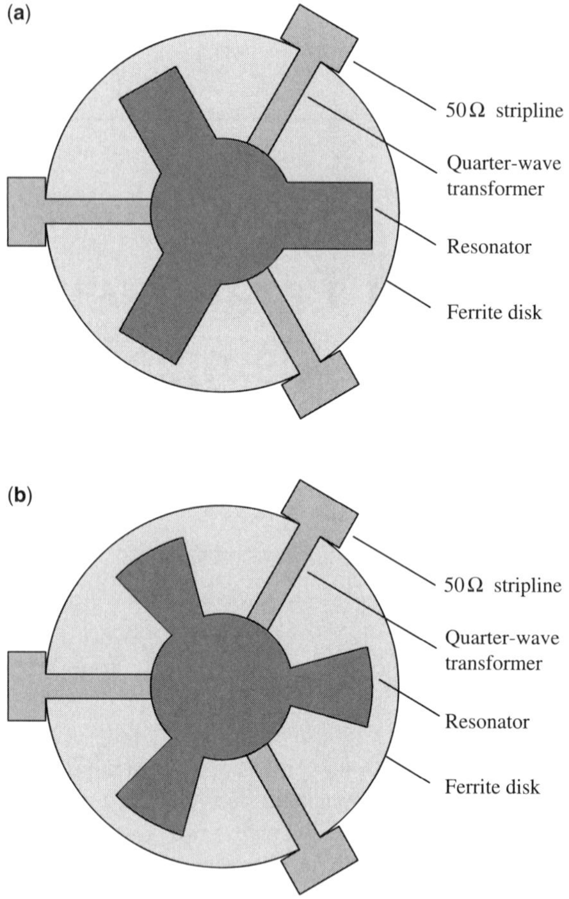

FIGURE 21.1 Quarter-wave coupled wye resonator using (a) stubs and (b) radial stubs.

coincides with the classic symmetrical six-port arrangement when the coupling angles of the circulator ports and stubs are made equal. The class of planar junction circulators discussed in this chapter may also be handled by having recourse to various numerical techniques. The finite element method, the contour integral method, the point matching technique, and the mode matching approach have been employed.

The method outlined in this chapter, in conjunction with an optimization subroutine, provides perhaps the most efficient way of constructing this type of circulator.

21.2 STANDING WAVE SOLUTION

The gyromagnetic wye resonator is but one of a number of geometries with the necessary symmetry for the realization of a three-port junction circulator. The

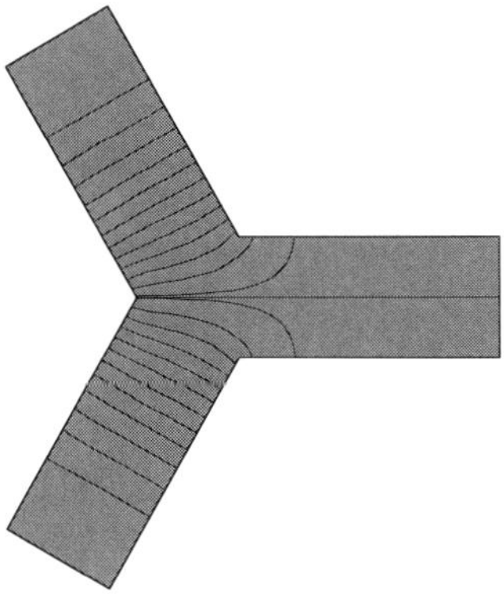

FIGURE 21.2 Construction of standing wave solution of circulator using wye resonator.

nature of this resonator structure may be appreciated by assuming that it consists of three symmetrically connected quarter-wave long open-circuited stubs. The equipotential standing wave solution of its dominant mode based on a finite element method calculation in such an isotropic resonator is given in Chapter 9. The difficulty of visualizing rotation of this standing wave pattern has also been dealt with by taking a linear combination of two isotropic field patterns. Figure 21.2 reproduces the standing wave pattern obtained in this way. A scrutiny of this illustration suggests that an ideal circulation condition can be realized by coupling the resonator at either of the two triplets of ports, as is now understood. Figure 21.3 depicts the two possibilities.

21.3 SYMMETRY PROPERTIES OF CIRCULATORS

A circulator using planar wye shaped resonators may be described approximately by subdividing it into a six-port, central, nonresonant circular region, defined by the triplet open-circuited stubs of the resonator circuit and the three coupling intervals of the circulator terminals. The intermediate intervals are assumed open-circuited. A six-port impedance matrix for the central disk region may be defined in terms of the voltages and currents at these terminals. A schematic diagram of the junction under consideration is illustrated in Fig. 21.4. The port nomenclature is chosen such that the numbering of the circulator corresponds to that of a conventional three-port device.

The total number of independent entries in the scattering or immittance matrices representing the six-port circuit under consideration may be deduced by having recourse to its symmetry. This may be done by satisfying the commutation relationship between the scattering or immittance matrices and the symmetry operators of the junction. The symmetry operators are matrices that indicate how the terminal fields transform under reflections and rotations, which map the junction on itself. The

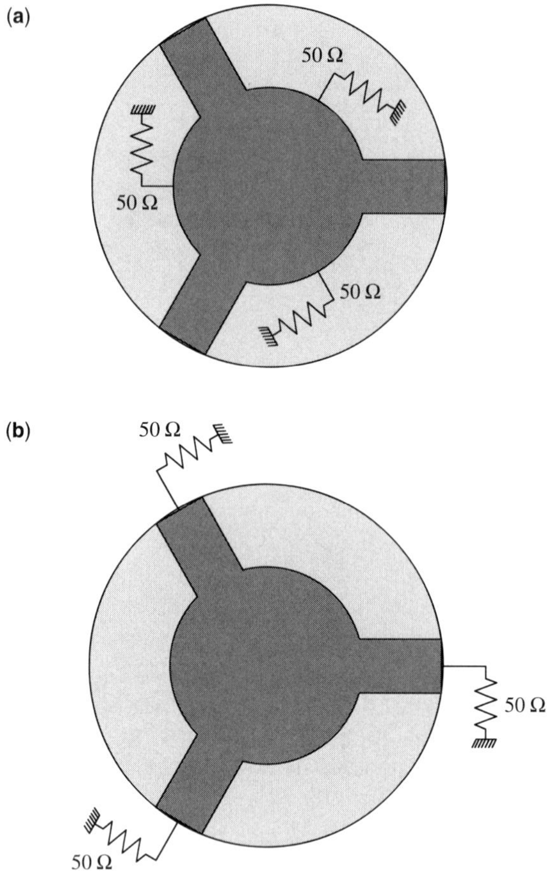

FIGURE 21.3 (a) High Q connection of junction resonator using a planar disk resonator. (b) Low Q connection of circulator using a planar wye resonator.

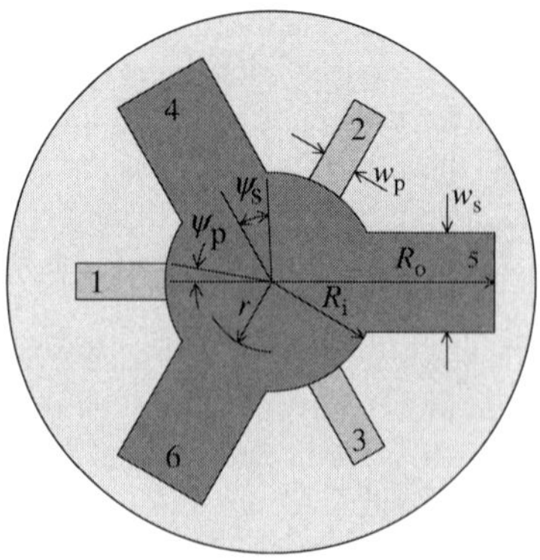

FIGURE 21.4 Coordinate system of circulator using a planar disk resonator.

symmetry operator for a junction rotation of 120° is

$$\mathbf{F}_1 = \begin{bmatrix} 0 & 1 & 0 & 0 & 0 & 0 \\ 0 & 0 & 1 & 0 & 0 & 0 \\ 1 & 0 & 0 & 0 & 0 & 0 \\ 0 & 0 & 0 & 0 & 1 & 0 \\ 0 & 0 & 0 & 0 & 0 & 1 \\ 0 & 0 & 0 & 1 & 0 & 0 \end{bmatrix} \tag{21.1}$$

The other two rotational operators are specified by

$$\mathbf{F}_2 = \mathbf{F}_1^2 \tag{21.2a}$$

$$\mathbf{F}_3 = \mathbf{F}_1^3 \tag{21.2b}$$

$\mathbf{F}_2$ and $\mathbf{F}_3$ represent junction rotations of 240° and 360°, respectively.

The general form of the impedance matrix of a six-port network is

$$\mathbf{Z} = \begin{bmatrix} Z_{11} & Z_{12} & Z_{13} & Z_{14} & Z_{15} & Z_{16} \\ Z_{21} & Z_{22} & Z_{23} & Z_{24} & Z_{25} & Z_{26} \\ Z_{31} & Z_{32} & Z_{33} & Z_{34} & Z_{35} & Z_{36} \\ Z_{41} & Z_{42} & Z_{43} & Z_{44} & Z_{45} & Z_{46} \\ Z_{51} & Z_{52} & Z_{53} & Z_{54} & Z_{55} & Z_{56} \\ Z_{61} & Z_{62} & Z_{63} & Z_{64} & Z_{65} & Z_{66} \end{bmatrix} \tag{21.3}$$

Making use of the fact that the symmetry and impedance matrices commute gives

$$\mathbf{F}_1\mathbf{Z} = \mathbf{Z}\mathbf{F}_1 \tag{21.4}$$

The general expression for the impedance matrix of a six-port junction with two degrees of threefold symmetry is therefore described by

$$\mathbf{Z} = \begin{bmatrix} Z_{11} & Z_{12} & Z_{13} & Z_{14} & Z_{15} & Z_{16} \\ Z_{13} & Z_{11} & Z_{12} & Z_{16} & Z_{14} & Z_{15} \\ Z_{12} & Z_{13} & Z_{11} & Z_{15} & Z_{16} & Z_{14} \\ Z_{41} & Z_{42} & Z_{43} & Z_{44} & Z_{45} & Z_{46} \\ Z_{43} & Z_{41} & Z_{42} & Z_{46} & Z_{44} & Z_{45} \\ Z_{42} & Z_{43} & Z_{41} & Z_{45} & Z_{46} & Z_{44} \end{bmatrix} \tag{21.5}$$

Meeting the commutation relations for the other symmetry operators produces a similar result.

21.4 OPEN-CIRCUIT PARAMETERS OF A JUNCTION WITH TWO DEGREES OF THREEFOLD SYMMETRY

The derivation of the 6×6 impedance matrix for the central circular region defined by the terminals of the triplet of open-circuited stubs and the circulator ports is the topic of this section. Perfect magnetic wall boundary conditions are assumed to exist at the periphery of the circular region other than at the two triplets of ports. All of the transmission lines are assumed to propagate a fundamental TEM mode.

Assuming no z dependence and $\exp(j\omega t)$ time variation, the electric field E_z and transverse magnetic field H_ϕ in the central circular region are

$$E_z = \sum_{-\infty}^{+\infty} A_n J_n(kr) \exp(jn\phi), \ \ 0 \le r \le R_i \tag{21.6}$$

$$H_\phi = \frac{-j}{\eta_e \eta_0} \sum_{-\infty}^{+\infty} A_n \left(J_n'(kr) - \frac{\kappa}{\mu} \frac{n J_n(kr)}{(kr)} \right) \exp(jn\phi), \ \ 0 \le r \le R_i \tag{21.7}$$

and

$$k = (2\pi/\lambda_0)\sqrt{\varepsilon_f \mu_e} \tag{21.8}$$

$$\eta_0 = \sqrt{\mu_0/\varepsilon_0} \tag{21.9a}$$

$$\eta_e = \sqrt{\mu_e/\varepsilon_f} \tag{21.9b}$$

λ_0 is the free space wavelength (meters), ω is the frequency (rad/s), ε_f is the relative dielectric constant, and μ_e is the effective permeability of the ferrite material.

$$\mu_e = \mu^2 - \kappa^2/\mu \tag{21.10}$$

μ and κ are the entries of the permeability tensor of the ferrite material.

In order to deduce the unknown coefficient A_n it is necessary to have one other description of either E_z or H_ϕ. Following Bosma, it is assumed that H_ϕ is constant over the width of the connecting striplines and stubs and zero elsewhere. The boundary condition at the inner radius of the circular gyromagnetic region is depicted in

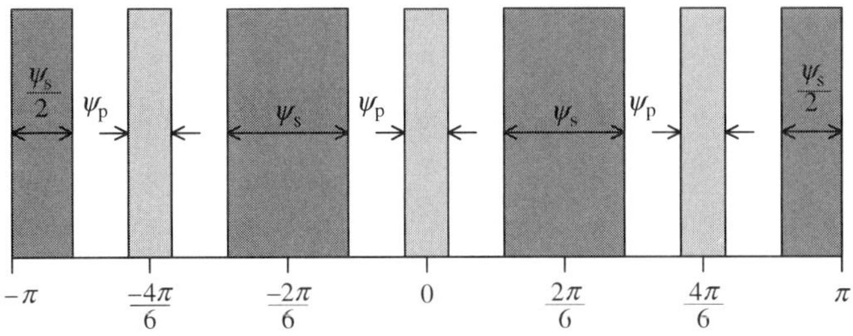

FIGURE 21.5 Boundary conditions of circulator.

Fig. 21.5 and specified by

$$(-\psi_p) < \phi < (\psi_p),\ H_\phi = H_1$$

$$(2\pi/3 - \psi_p) < \phi < (2\pi/3 + \psi_p),\ H_\phi = H_2$$

$$(4\pi/3 - \psi_p) < \phi < (4\pi/3 + \psi_p),\ H_\phi = H_3$$

$$(\pi/3 - \psi_s) < \phi < (\pi/3 + \psi_s),\ H_\phi = H_4$$

$$(\pi - \psi_s) < \phi < (\pi + \psi_s),\ H_\phi = H_5$$

$$(5\pi/3 - \psi_s) < \phi < (5\pi/3 + \psi_s),\ H_\phi = H_6$$

The coupling angles ψ_p and ψ_s are defined by

$$\psi_p = \sin^{-1}(W_p/2R_i) \tag{21.11}$$

$$\psi_s = \sin^{-1}(W_s/2R_i) \tag{21.12}$$

W_p is the width of the connecting stripline, W_s is the width of the open-circuited stub, and R_i is the radius of the central circular gyromagnetic region.

The unknown amplitude constant A_n may be found by expanding H_ϕ into an exponential Fourier series with respect to ϕ at $r = R$ in the manner indicated in Fig. 21.5.

$$H_\phi = \sum_{-\infty}^{+\infty} b_n \exp(jn\phi) \tag{21.13}$$

b_n is obtained by integrating each contribution about the port in question.

$$b_n = \left(\frac{\sin n\psi_{\rm p}}{n\pi}\right)(H_1\alpha_n^0 + H_2\alpha_n^2 + H_3\alpha_n^4) + \left(\frac{\sin n\psi_{\rm s}}{n\pi}\right) \times \left(H_4\alpha_n^1 + H_5\alpha_n^3 + H_6\alpha_n^5\right) \tag{21.14}$$

where

$$\alpha_n = \exp(-jn\pi/3)$$

The unknown coefficient, A_n, is now obtained by comparing the two descriptions for H_ϕ at $r = R_{\rm i}$. The required result is

$$A_n = j\eta_{\rm e}\eta_0 b_n \left(J_n'(kR_{\rm i}) - \frac{\kappa}{\mu}\frac{nJ_n(kR_{\rm i})}{kR_{\rm i}}\right)^{-1} \tag{21.15}$$

The average value of the electric field over the connecting striplines and stubs is now evaluated. It may be obtained by having recourse to Eq. (21.6).

$$E_1 = \frac{1}{2\psi_{\rm p}}\int_{-\psi_{\rm p}}^{\psi_{\rm p}} E_z\,d\phi = \eta_{11}H_1 + \eta_{12}H_2 + \eta_{13}H_3 + \eta_{14}H_4 + \eta_{15}H_5 + \eta_{16}H_6 \tag{21.16a}$$

$$E_2 = \frac{1}{2\psi_{\rm p}}\int_{2\pi/3-\psi_{\rm p}}^{2\pi/3+\psi_{\rm p}} E_z\,d\phi = \eta_{13}H_1 + \eta_{11}H_2 + \eta_{12}H_3 + \eta_{16}H_4 + \eta_{14}H_5 + \eta_{15}H_6 \tag{21.16b}$$

$$E_3 = \frac{1}{2\psi_{\rm p}}\int_{4\pi/3-\psi_{\rm p}}^{4\pi/3+\psi_{\rm p}} E_z\,d\phi = \eta_{12}H_1 + \eta_{13}H_2 + \eta_{11}H_3 + \eta_{15}H_4 + \eta_{16}H_5 + \eta_{14}H_6 \tag{21.16c}$$

$$E_4 = \frac{1}{2\psi_{\rm s}}\int_{2\pi/3-\psi_{\rm s}}^{2\pi/3+\psi_{\rm s}} E_z\,d\phi = \eta_{41}H_1 + \eta_{42}H_2 + \eta_{43}H_3 + \eta_{44}H_4 + \eta_{45}H_5 + \eta_{46}H_6 \tag{21.16d}$$

$$E_5 = \frac{1}{2\psi_{\rm s}}\int_{\pi-\psi_{\rm s}}^{\pi+\psi_{\rm s}} E_z\,d\phi = \eta_{43}H_1 + \eta_{41}H_2 + \eta_{42}H_3 + \eta_{46}H_4 + \eta_{44}H_5 + \eta_{45}H_6 \tag{21.16e}$$

$$
\begin{aligned}
E_6 &= \frac{1}{2\psi_s} \int_{5\pi/3-\psi_s}^{5\pi/3+\psi_s} E_z \, d\phi \\
&= \eta_{42} H_1 + \eta_{43} H_2 + \eta_{41} H_3 + \eta_{45} H_4 + \eta_{46} H_5 + \eta_{44} H_6 \qquad (21.16\text{f})
\end{aligned}
$$

A careful scrutiny of the preceding equations indicates that the open-circuit parameters may be written in terms of linear combinations of the impedance poles of two constituent symmetrical structures. These are obtained by decomposing the original six-port structure with twofold symmetry into two symmetrical constituent six-port geometries. Since the nature of these impedance poles is well understood, the calculations may proceed in terms of known functions.

$$
\eta_{11} = \sum_{n=-\infty}^{\infty} \frac{\eta_{pn} \alpha_n^0}{m}, \quad \eta_{12} = \sum_{n=-\infty}^{\infty} \frac{\eta_{pn} \alpha_n^2}{m}, \quad \eta_{13} = \sum_{n=-\infty}^{\infty} \frac{\eta_{pn} \alpha_n^4}{m} \qquad (21.17\text{a})
$$

$$
\eta_{14} = \sum_{n=-\infty}^{\infty} \left(\frac{\sin n\psi_s}{\sin n\psi_p} \right) \frac{\eta_{pn} \alpha_n^1}{m}, \quad \eta_{15} = \sum_{n=-\infty}^{\infty} \left(\frac{\sin n\psi_s}{\sin n\psi_p} \right) \frac{\eta_{pn} \alpha_n^3}{m},
$$

$$
\eta_{16} = \sum_{n=-\infty}^{\infty} \left(\frac{\sin n\psi_s}{\sin n\psi_p} \right) \frac{\eta_{pn} \alpha_n^5}{m} \qquad (21.17\text{b})
$$

$$
\eta_{41} = \sum_{n=-\infty}^{\infty} \left(\frac{\sin n\psi_p}{\sin n\psi_s} \right) \frac{\eta_{sn} \alpha_n^5}{m}, \quad \eta_{42} = \sum_{n=-\infty}^{\infty} \left(\frac{\sin n\psi_p}{\sin n\psi_s} \right) \frac{\eta_{sn} \alpha_n^1}{m},
$$

$$
\eta_{43} = \sum_{n=-\infty}^{\infty} \left(\frac{\sin n\psi_p}{\sin n\psi_s} \right) \frac{\eta_{sn} \alpha_n^3}{m} \qquad (21.17\text{c})
$$

$$
\eta_{44} = \sum_{n=-\infty}^{\infty} \frac{\eta_{sn} \alpha_n^0}{m}, \quad \eta_{45} = \sum_{n=-\infty}^{\infty} \frac{\eta_{sn} \alpha_n^2}{m}, \quad \eta_{46} = \sum_{n=-\infty}^{\infty} \frac{\eta_{sn} \alpha_n^4}{m} \qquad (21.17\text{d})
$$

η_{pn} and η_{sn} are the poles of the constituent problem regions of the overall arrangement considered here.

The impedance poles in the description of the open-circuited parameters of a symmetrical m-port gyromagnetic disk are a standard result in the literature. The relationship is given by

$$
\eta_n(m) = \frac{jm\eta_e\eta_0\psi}{\pi} \left(\frac{\sin n\psi}{n\psi} \right)^2 \left[\frac{J_n'(kR)}{J_n(kR)} - \left(\frac{\kappa}{\mu} \right) \left(\frac{n}{kR} \right) \right]^{-1} \qquad (21.18)
$$

ψ is ψ_s over the ports defined by the stubs and ψ_p over the coupling ports.

For design purposes, it is preferable to work with the characteristic impedance of the problem region rather than its wave impedance. This may be done by forming V

and I at a typical port instead of E and H. This amounts to replacing the wave impedance in the description of a typical pole by the characteristic impedance of the strip under consideration. Writing a typical pole of the problem region under consideration in terms of Z_0 instead of η_0 gives

$$Z_n(m) = \frac{jm\eta_e Z_0 \psi}{\pi} \left(\frac{\sin n\psi}{n\psi}\right)^2 \left[\frac{J_n'(kR)}{J_n(kR)} - \left(\frac{\kappa}{\mu}\right)\left(\frac{n}{kR}\right)\right]^{-1} \tag{21.19}$$

where $m = 6$ for a six-port junction.

Z_0 is the free space characteristic impedance of a typical port or stub:

$$Z_0 = 30\pi \ln\left(\frac{W + t + 2H}{W + t}\right) \tag{21.20}$$

W is the width of the center conductor (in meters), H is the thickness of each garnet resonator (meters) of radius R (meters), and t is the thickness of the center conductor (meters). The other quantities have the usual meanings.

21.5 CONSTITUENT PROBLEM REGIONS OF A SIX-PORT JUNCTION HAVING TWO DEGREES OF THREEFOLD SYMMETRY

A scrutiny of the open-circuit parameters of the six-port network with two degrees of threefold symmetry dealt with here indicates that these only involve the poles of each constituent problem region and the ratio of the coupling angles of the two geometries. The open-circuit parameters under consideration reduce to those of a symmetrical six-port network when the coupling angles are equal. The classic three-port topology is obtained when the coupling angles of the stubs are zero. Figure 21.6 indicates the decomposition of the problem region in question. Figure 21.7 illustrates the situation met in connection with a regular hexagonal resonator. This latter problem may readily be solved by having recourse to any number of numerical techniques.

The description of the open-circuit parameters may be clarified for the purpose of calculation by collating the impedance poles of the two families of poles associated with each sextet of ports into in-phase and counterrotating series.

$$\eta_{11} = \frac{(\eta_{p,0} + \eta_{p,3}) + (\eta_{p,1} + \eta_{p,-2}) + (\eta_{p,2} + \eta_{p,-1})}{6} \tag{21.21a}$$

$$\eta_{12} = \frac{(\eta_{p,0} + \eta_{p,3}) + (\eta_{p,1} + \eta_{p,-2})\,\alpha_1^2 + (\eta_{p,2} + \eta_{p,-1})\,\alpha_1^4}{6} \tag{21.21b}$$

$$\eta_{13} = \frac{(\eta_{p,0} + \eta_{p,3}) + (\eta_{p,1} + \eta_{p,-2})\,\alpha_1^4 + (\eta_{p,2} + \eta_{p,-1})\,\alpha_1^2}{6} \tag{21.21c}$$

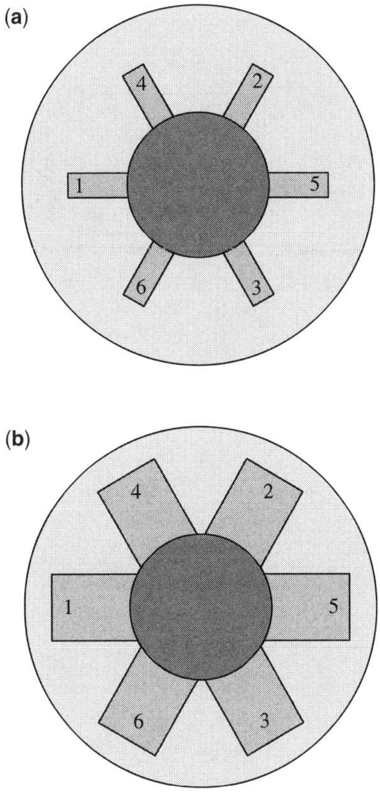

FIGURE 21.6 Topology of a six-port constituent disk resonator with (a) coupling ports of six-port geometry and (b) stubs of six-port geometry.

$$\eta_{14} = \frac{(g_0\,\eta_{p,0} - g_3\,\eta_{p,3}) + (g_2\,\eta_{p,-2} - g_1\,\eta_{p,1})\,\alpha_1^4 + (g_2\,\eta_{p,2} - g_1\,\eta_{p,-1})\,\alpha_1^2}{6} \tag{21.21d}$$

$$\eta_{15} = \frac{(g_0\,\eta_{p,0} - g_3\,\eta_{p,3}) + (g_2\,\eta_{p,-2} - g_1\,\eta_{p,1}) + (g_2\,\eta_{p,2} - g_1\,\eta_{p,-1})}{6} \tag{21.21e}$$

$$\eta_{16} = \frac{(g_0\,\eta_{p,0} - g_3\,\eta_{p,3}) + (g_2\,\eta_{p,-2} - g_1\,\eta_{p,1})\,\alpha_1^2 + (g_2\,\eta_{p,2} - g_1\,\eta_{p,-1})\,\alpha_1^4}{6} \tag{21.21 f}$$

$$\eta_{41} = \frac{(f_0\,\eta_{s,0} - f_3\,\eta_{s,3}) + (f_2\,\eta_{s,-2} - f_1\,\eta_{s,1})\,\alpha_1^2 + (f_2\,\eta_{s,2} - f_1\,\eta_{s,-1})\,\alpha_1^4}{6} \tag{21.21g}$$

$$\eta_{42} = \frac{(f_0\,\eta_{s,0} - f_3\,\eta_{s,3}) + (f_2\,\eta_{s,-2} - f_1\,\eta_{s,1})\,\alpha_1^4 + (f_2\,\eta_{s,2} - f_1\,\eta_{p,-1})\,\alpha_1^2}{6} \tag{21.21h}$$

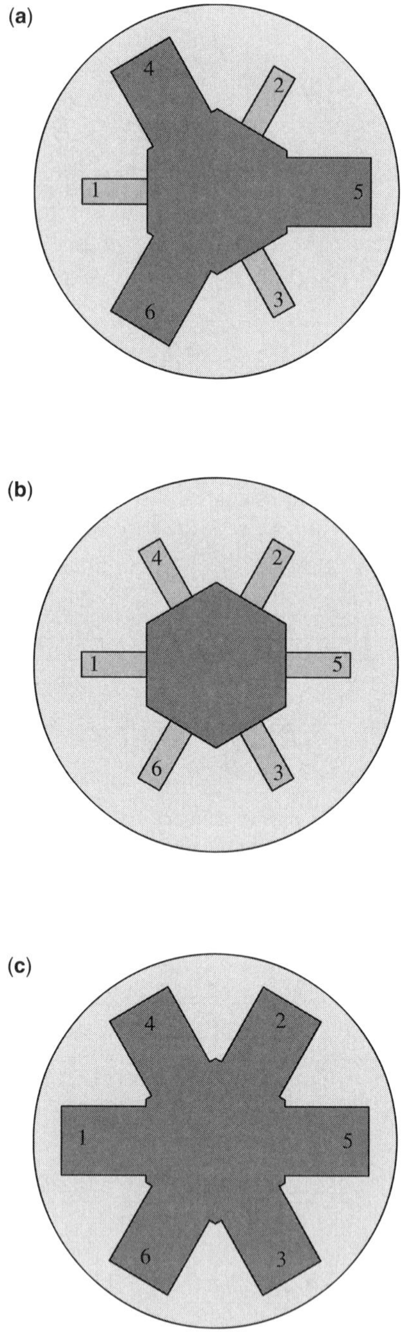

FIGURE 21.7 (a) Topology of a six-port irregular hexagonal resonator with two degrees of threefold symmetry. Topology of a six-port constituent hexagonal resonator with (b) coupling ports of six-port geometry and (c) stubs of six-port geometry.

$$\eta_{43} = \frac{(f_0\,\eta_{s,0} - f_3\,\eta_{s,3}) + (f_2\,\eta_{s,-2} - f_1\,\eta_{s,1}) + (f_2\,\eta_{s,2} - f_1\,\eta_{s,-1})}{6} \tag{21.21i}$$

$$\eta_{44} = \frac{(\eta_{s,0} + \eta_{s,3}) + (\eta_{s,1} + \eta_{s,-2}) + (\eta_{s,2} + \eta_{s,-1})}{6} \tag{21.21j}$$

$$\eta_{45} = \frac{(\eta_{s,0} + \eta_{s,3}) + (\eta_{s,1} + \eta_{s,-2})\,\alpha_1^2 + (\eta_{s,2} + \eta_{s,-1})\,\alpha_1^4}{6} \tag{21.21k}$$

$$\eta_{46} = \frac{(\eta_{s,0} + \eta_{s,3}) + (\eta_{s,1} + \eta_{s,-2})\,\alpha_1^4 + (\eta_{s,2} + \eta_{s,-1})\,\alpha_1^2}{6} \tag{21.21l}$$

where

$$f_n = \sin(n\psi_p)/\sin(n\psi_s) \tag{21.21m}$$

$$g_n = \sin(n\psi_s)/\sin(n\psi_p) \tag{21.21n}$$

If the junction has sixfold symmetry then

$$f_n = g_n = 1 \tag{21.21o}$$

$$\eta_p = \eta_s \tag{21.21p}$$

The open-circuit parameters reduce to those met in connection with the six-port solution.

A minimum of six impedance poles is necessary in the description of this sort of junction. Not all of the entries appearing in the impedance matrix are positive real functions.

21.6 3 × 3 IMPEDANCE MATRIX OF CIRCULATORS USING WYE RESONATORS

The 3 × 3 impedance matrix for the circulator is obtained from the 6 × 6 one by meeting the boundary conditions on the open-circuited transmission lines comprising the resonator.

$$\begin{bmatrix} V_4 \\ V_5 \\ V_6 \end{bmatrix} = -Z_L \begin{bmatrix} 1 & 0 & 0 \\ 0 & 1 & 0 \\ 0 & 0 & 1 \end{bmatrix} \begin{bmatrix} I_4 \\ I_5 \\ I_6 \end{bmatrix} \tag{21.22}$$

$V_{4,5,6}$ and $I_{4,5,6}$ are the voltages and currents at ports 4, 5, and 6 and Z_L is the input impedance of a typical line at $r = R_i$. This latter quantity is related to the geometry of the line and the medium in the region $R_i \leq r \leq R_0$. The situation treated here is one where the lines are uniform and filled with a magnetized ferrite material. In this case Z_L has the form

$$Z_L = -j\eta_e Z_s \cot\theta_s \tag{21.23}$$

Z_s is the free space characteristic impedance of the stubs and θ_s is its electrical length.

$$\theta_s = \frac{2\pi}{\lambda_0}\sqrt{\mu_e \varepsilon_f}(R_0 - R_i) \tag{21.24}$$

The possibility of an edge mode effect in the striplines is identified but neglected. The development outlined here is therefore restricted to a moderately magnetized arrangement.

Making use of matrix partitioning, the voltage–current relationship for the 6 × 6 impedance matrix at the ports of the gyromagnetic circular region may be written as

$$\begin{bmatrix} V_1 \\ V_2 \\ V_3 \\ - \\ V_4 \\ V_5 \\ V_6 \end{bmatrix} = \left[\begin{array}{ccc|ccc} Z_{11} & Z_{12} & Z_{13} & Z_{14} & Z_{15} & Z_{16} \\ Z_{13} & Z_{11} & Z_{12} & Z_{16} & Z_{14} & Z_{15} \\ Z_{12} & Z_{13} & Z_{11} & Z_{15} & Z_{16} & Z_{14} \\ \hline Z_{41} & Z_{42} & Z_{43} & Z_{44} & Z_{45} & Z_{46} \\ Z_{43} & Z_{41} & Z_{42} & Z_{46} & Z_{44} & Z_{45} \\ Z_{42} & Z_{43} & Z_{41} & Z_{45} & Z_{46} & Z_{44} \end{array}\right] = \begin{bmatrix} I_1 \\ I_2 \\ I_3 \\ - \\ I_4 \\ I_5 \\ I_6 \end{bmatrix} \tag{21.25}$$

This may be rewritten in terms of the following submatrices:

$$\begin{bmatrix} \mathbf{V}_1 \\ \mathbf{V}_2 \end{bmatrix} = \begin{bmatrix} \mathbf{Z}_1 & \mathbf{Z}_2 \\ \mathbf{Z}_3 & \mathbf{Z}_4 \end{bmatrix} \begin{bmatrix} \mathbf{I}_1 \\ \mathbf{I}_2 \end{bmatrix} \tag{21.26}$$

$\mathbf{Z}_{1,2,3,4}$ are defined by inspection. $\mathbf{V}_1$ and $\mathbf{I}_1$ are the voltages and currents on the triplet of circulator terminals and $\mathbf{V}_2$ and $\mathbf{I}_2$ are the voltages and currents on the triplet of open-circuited transmission lines.

The 3 × 3 impedance matrix for the circulator is now defined by solving the preceding equation for $\mathbf{V}_1\mathbf{I}_1^{-1}$. This may be done by having recourse to the relationship between the submatrices $\mathbf{V}_2$ and $\mathbf{I}_2$ defined by the boundary condition. The result is

$$\mathbf{Z}' = \mathbf{V}_1\mathbf{I}_1^{-1} = \mathbf{Z}_1 - \mathbf{Z}_2(Z_L\mathbf{I} + \mathbf{Z}_4)^{-1}\mathbf{Z}_3 \tag{21.27}$$

I is the unit matrix and **Z**$'$ is the required 3×3 impedance matrix having the form

$$\mathbf{Z}' = \begin{bmatrix} Z'_{11} & Z'_{12} & Z'_{13} \\ Z'_{13} & Z'_{11} & Z'_{12} \\ Z'_{12} & Z'_{13} & Z'_{11} \end{bmatrix} \tag{21.28}$$

The matrix obtained in this way is symmetric. It therefore embodies an eigenvalue problem. These are given in terms of the open-circuit parameters by

$$Z^0 = Z'_{11} + Z'_{12} + Z'_{13} \tag{21.29a}$$

$$Z^+ = Z'_{11} + \alpha^2 Z'_{12} + \alpha Z'_{13} \tag{21.29b}$$

$$Z^- = Z'_{11} + \alpha Z'_{12} + \alpha^2 Z'_{13} \tag{21.29c}$$

$\mathbf{Z}^0$, $\mathbf{Z}^{\pm}$ are pure imaginary quantities.

21.7 SHORT-CIRCUIT PARAMETERS OF QUARTER-WAVE COUPLED CIRCULATOR

While a knowledge of the open-circuit parameters at the terminals of the gyromagnetic resonator is sufficient for synthesis purposes, those at the terminals of the impedance transformer are necessary for analysis or optimization. The purpose of this section is to deal with this situation.

One way to proceed is on an eigenvalue basis. This amounts to mapping each of the eigenvalues at the terminals of the resonator to those of the impedance transformer. Note that the impedance matrix at the terminals of a conventional quarter-wave coupled circulator is ill behaved. It is therefore preferable to work with short-circuit parameters at these terminals. Figure 21.8 indicates the required transformations. The necessary bilinear transformations between the immittance eigenvalues at the two planes are given by

$$Y^0 = \frac{D + jCZ^0}{jB + AZ^0} \tag{21.30a}$$

$$Y^+ = \frac{D + jCZ^+}{jB + AZ^+} \tag{21.30b}$$

$$Y^- = \frac{D + jCZ^-}{jB + AZ^-} \tag{21.30c}$$

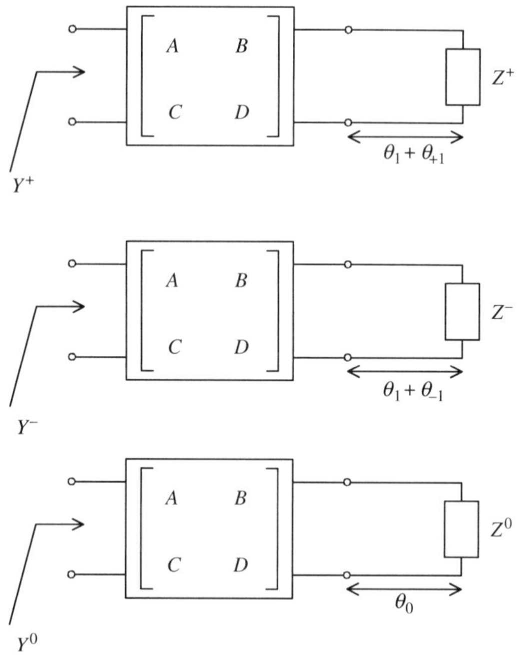

FIGURE 21.8 Eigen-networks of quarter-wave coupled junction circulator.

The required open-circuit parameters are obtained in terms of the eigenvalues at the terminals of the impedance transformer by

$$Y_{11} = \frac{Y^0 + Y^+ + Y^-}{3} \tag{21.31a}$$

$$Y_{12} = \frac{Y^0 + \alpha^2 Y^+ + \alpha^4 Y^-}{3} \tag{21.31b}$$

$$Y_{13} = \frac{Y^0 + \alpha^4 Y^+ + \alpha^2 Y^-}{3} \tag{21.31c}$$

While the immittance matrices are appropriate for synthesis purposes, the scattering parameters are necessary for the purpose of analysis. The relationship between immittance and scattering parameters is a standard matrix operation.

CHAPTER TWENTY-TWO

The Four-Port Single Junction Stripline Circulator

22.1 INTRODUCTION

The three-port single junction circulator is the best understood arrangement but a four-port one may also be realized without too much difficulty. The purpose of this chapter is to catalog some solutions. In common with three-port devices, these junctions have some of the properties of a transmission line cavity resonator between ports 1 and 2, and a definite standing wave pattern exists within the junction with nulls at ports 3 and 4 also. An important difference between the two, however, is that the four-port device cannot be adjusted with external tuning elements only. This remark may be understood by recognizing that a four-port device can be matched without being a circulator. The adjustment of this junction may be done in a systematic way by perturbing each of its reflection eigenvalues at a time, on the unit circle, until these coincide with those of an ideal circulator. Each eigenvalue corresponds to one of the four possible generator settings or eigenvectors at the ports of the junction. The two rotating eigen-networks contain the odd poles of the problem region, the in-phase eigen-network contains the even poles with the symmetry of the problem region, and the fourth counterrotating eigen-network contains the remaining even poles. One possible synthesis of the four-port junction circulator involves the realization of one impedance pole, one pair of split impedance poles, and one impedance zero. The split impedance poles are fixed by the two rotating eigen-networks; the single impedance pole is chosen from either the in-phase or counterrotating eigen-network. The impedance zero is assumed satisfied by the remaining eigen-network without ado provided it is not in the neighborhood of a pole at the frequency under consideration. The synthesis procedure is facilitated provided the nature of the field patterns used to construct the device are known. The chapter describes

The Stripline Circulator: Theory and Practice. By J. Helszajn

two possible geometries based on various combinations of single and degenerate pairs of $TM_{m,n,0}$ modes of the isotropic resonator and one that relies on the use of a suitable degeneracy of the gyromagnetic resonator for its operation.

The chapter also includes the standing wave patterns and cutoff numbers of the first six modes in a cross planar resonator formed by the junction of four open-circuited stubs. This sort of planar resonator is encountered in the construction of the four-port single junction circulator.

22.2 SYMMETRY PROPERTIES OF CIRCULATORS

The total number of independent entries in the scattering or immittance matrices of the four-port junction under consideration may be deduced by having recourse to its symmetry. This may be done by satisfying the commutation relation between the scattering or immittance matrices and the symmetry operators of the junction. The symmetry operators are matrices that indicate how the terminal fields transform under reflections and rotations, which map the junction on itself. The symmetry operator for a clockwise junction rotation of 90° is

$$\mathbf{F}_1 = \begin{bmatrix} 0 & 1 & 0 & 0 \\ 0 & 0 & 1 & 0 \\ 0 & 0 & 0 & 1 \\ 1 & 0 & 0 & 0 \end{bmatrix} \tag{22.1}$$

The other two rotational operators are specified by

$$\mathbf{F}_2 = \mathbf{F}_1^2 \tag{22.2}$$

$$\mathbf{F}_3 = \mathbf{F}_1^3 \tag{22.3}$$

$\mathbf{F}_2$ and $\mathbf{F}_3$ represent junction rotations of 180° and 270°, respectively.

The general form of the scattering matrix of a four-port network is

$$\mathbf{S} = \begin{bmatrix} S_{11} & S_{12} & S_{13} & S_{14} \\ S_{21} & S_{22} & S_{23} & S_{24} \\ S_{31} & S_{32} & S_{33} & S_{34} \\ S_{41} & S_{42} & S_{43} & S_{44} \end{bmatrix} \tag{22.4}$$

Making use of the fact that the symmetry and scattering matrices commute gives

$$\mathbf{F}_1\mathbf{S} = \mathbf{S}\mathbf{F}_1 \tag{22.5}$$

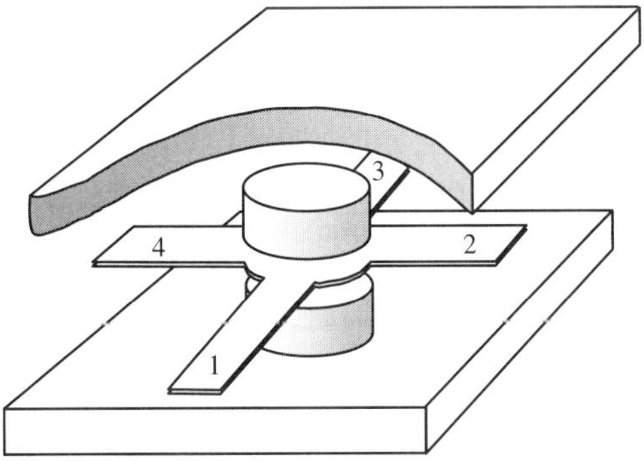

FIGURE 22.1 Schematic of single junction four-port stripline circulator.

The general expression for the scattering matrix of a symmetrical four-port junction is therefore described by

$$\mathbf{S} = \begin{bmatrix} S_{11} & S_{41} & S_{31} & S_{21} \\ S_{21} & S_{11} & S_{41} & S_{31} \\ S_{31} & S_{21} & S_{11} & S_{41} \\ S_{41} & S_{31} & S_{21} & S_{11} \end{bmatrix} \tag{22.6}$$

Meeting the commutation relations for the other symmetry operators produces a similar result.

Figure 22.1 indicates the topology of the four-port stripline circulator. The convention with regard to the port numbering in the earlier literature is to label the ports in an anticlockwise sense. This notation, however, has not been strictly adhered to in the later engineering literature. The one adopted in this work corresponds to the former convention.

22.3 SIMILARITY TRANSFORMATION

An important property of any symmetrical network is that its immittance or scattering parameters can be decomposed into linear combinations of its so-called eigenvalues. These quantities are one-port scalar variables, which have the nature of the main diagonal of the matrix under consideration. The connection between the two sets of variables may be deduced by having recourse to the standard similarity transformation in terms of the eigenvectors of the problem region introduced in Chapter 4. An ideal four-port circulator has no impedance or admittance matrix so the development proceeds in terms of its scattering matrix.

$$\overline{\mathbf{S}} = \overline{\mathbf{U}}\,\overline{\boldsymbol{\lambda}}\,\overline{\mathbf{U}}^{-1} \tag{22.7}$$

The columns of the matrix $\overline{\mathbf{U}}$ are constructed in terms of the eigenvectors $\overline{\mathbf{U}}_i$ of the problem region, and $\overline{\mathbf{U}}^{-1}$ is the inverse of $\overline{\mathbf{U}}$. The matrix $\overline{\boldsymbol{\lambda}}$ is a diagonal matrix with the eigenvalues s_i of the scattering matrix along its main diagonal. The relationships between the eigenvalues and the coefficients of the scattering matrix are then obtained by multiplying out the similarity identity. The eigenvalue equation of the four-port single junction circulator is illustrated in Fig. 22.2. The port nomenclature is in keeping with a trigonometric coordinate system.

The eigenvectors may be derived from symmetry conditions only. The result is

$$(U_n)_\mathrm{p} = \frac{1 \exp[-j2\pi n(q-1)/m]}{\sqrt{m}} \tag{22.8}$$

One set of eigenvectors that satisfy the symmetry of the four-port network with $n = 0, +1, -1, +2$ is

$$\overline{\mathbf{U}}_0 = \frac{1}{\sqrt{4}}\begin{bmatrix}1\\1\\1\\1\end{bmatrix}, \quad \overline{\mathbf{U}}_{-1} = \frac{1}{\sqrt{4}}\begin{bmatrix}-1\\j\\-1\\-j\end{bmatrix}, \quad \overline{\mathbf{U}}_{+1} = \frac{1}{\sqrt{4}}\begin{bmatrix}1\\-j\\-1\\j\end{bmatrix}, \quad \overline{\mathbf{U}}_2 = \frac{1}{\sqrt{4}}\begin{bmatrix}1\\-1\\1\\-1\end{bmatrix} \tag{22.9}$$

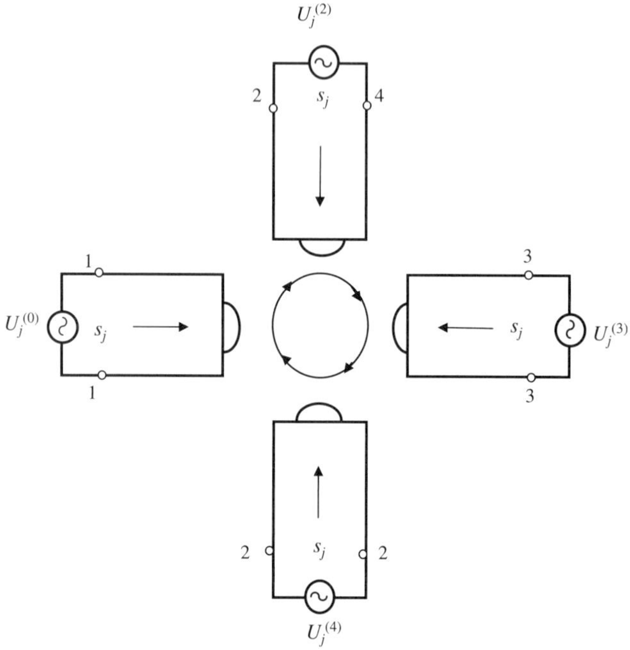

FIGURE 22.2 Eigenvalue equation of four-port single junction circulator.

These eigenvectors again have the property that a single input at any one port may be decomposed into a linear combination of the individual eigenvectors with equal amplitude.

The eigenvectors under consideration are indicated in Fig. 22.3. A scrutiny of each of the counterrotating eigenvectors, one at a time, indicates that one produces a clockwise rotating magnetic field on the axis of the junction and the other produces a counterrotating wave there. In each case the electric field is zero on the axis of the junction. The eigenvalues appearing in the description of the scattering coefficients may again be revealed by having recourse to the diagonalization procedure utilized in the

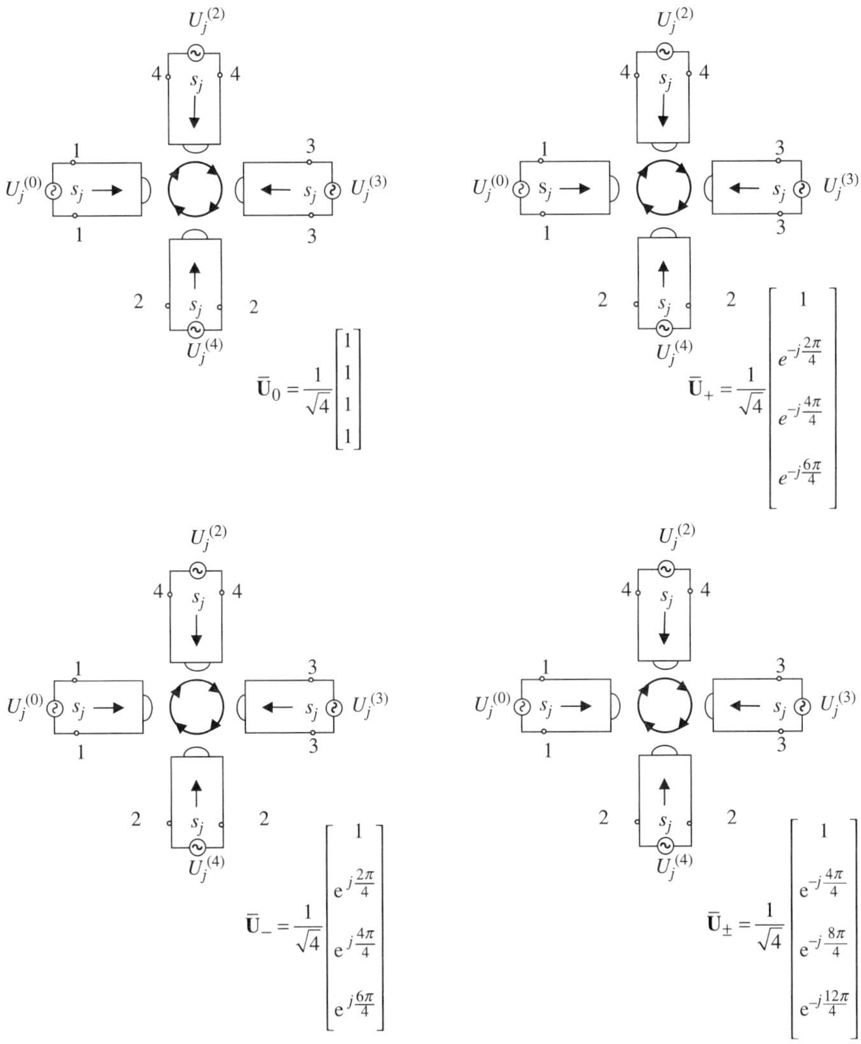

FIGURE 22.3 Eigenvectors of four-port junction circulator.

description of the three-port circulator. In this instance the matrix $\overline{\mathbf{U}}$ is given by

$$\overline{\mathbf{U}} = \frac{1}{\sqrt{4}}\begin{bmatrix} 1 & 1 & 1 & 1 \\ 1 & -j & -1 & j \\ 1 & -1 & 1 & -1 \\ 1 & j & -1 & -j \end{bmatrix} \tag{22.10}$$

The diagonal matrix $\overline{\boldsymbol{\lambda}}$ has one degenerate pair of eigenvalues and two nondegenerate ones:

$$\overline{\boldsymbol{\lambda}} = \begin{bmatrix} s_0 & 0 & 0 & 0 \\ 0 & s_{+1} & 0 & 0 \\ 0 & 0 & s_{-1} & 0 \\ 0 & 0 & 0 & s_2 \end{bmatrix} \tag{22.11}$$

The required result is

$$S_{11} = S_{22} = S_{33} = S_{44} = \frac{s_0 + s_{+1} + s_{-1} + s_2}{4} \tag{22.12a}$$

$$S_{12} = S_{23} = S_{34} = S_{41} = \frac{s_0 + js_{+1} - js_{-1} - s_2}{4} \tag{22.12b}$$

$$S_{13} = S_{24} = S_{31} = S_{42} = \frac{s_0 - s_{+1} - s_{-1} + s_2}{4} \tag{22.12c}$$

$$S_{14} = S_{43} = S_{32} = S_{21} = \frac{s_0 - js_{+1} + js_{-1} - s_2}{4} \tag{22.12d}$$

A scrutiny of the first of these four equations indicates that the spur or trace of the scattering matrix is equal to the sum of the eigenvalues. This is a general result.

22.4 EIGENVALUE ADJUSTMENT

One property of a single four-port junction circulator is that S_{11} is equal to zero. Scrutiny of this condition suggests that the eigenvalues of an ideal four-port single junction circulator lie equally spaced on a unit circle. One possible synthesis procedure is indicated in Fig. 22.4. Each eigenvalue diagram may be visualized as a separate problem. The first one is met provided

$$S_{11} = S_{21} = S_{41} = 0 \tag{22.13a}$$

$$S_{31} = -1 \tag{22.13b}$$

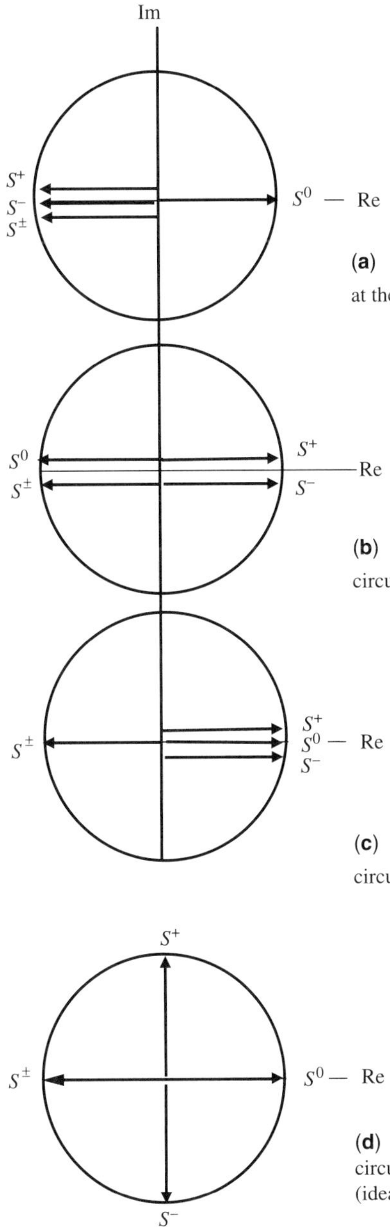

(a) Scattering eigenvalues for four-port junction at the origin

(b) Scattering eigenvalues after first circulation adjustment of four-port junction

(c) Scattering eigenvalues after second circulation adjustment of four-port junction

(d) Scattering eigenvalues after third circulation adjustment of four-port junction (ideal circulator)

FIGURE 22.4 Eigenvalue adjustment of four-port single junction circulator.

The second adjustment coincides with

$$S_{11} = S_{21} = S_{41} = \tfrac{1}{2} \tag{22.14a}$$

$$S_{31} = -\tfrac{1}{2} \tag{22.14b}$$

The third adjustment satisfies the condition of the ideal circulator without ado:

$$S_{11} = S_{31} = S_{41} = 0 \tag{22.15a}$$

$$S_{21} = 1 \tag{22.15b}$$

The reflection eigenvalue of a typical eigen-network has a value of $+1$ at a typical impedance pole and a value of -1 at a zero.

A feature of the four-port circulator is that while the isolation at port 3 is dependent on the return loss at port 1, that at port 4 is not. This property may be demonstrated by idealizing the reflection eigenvalues s_0 and s_2. This gives

$$S_{11} = S_{22} = S_{33} = S_{44} = \frac{s_{+1} + s_{-1}}{4} \tag{22.16a}$$

$$S_{12} = S_{23} = S_{34} = S_{41} = \frac{2 + j(s_{+1} - s_{-1})}{4} \tag{22.16b}$$

$$S_{13} = S_{24} = S_{31} = S_{42} = \frac{-(s_{+1} + s_{-1})}{4} \tag{22.16c}$$

$$S_{14} = S_{43} = S_{32} = S_{21} = \frac{2 - j(s_{+1} - s_{-1})}{4} \tag{22.16d}$$

A scrutiny of these relationships indicates that minimizing S_{11} is compatible with minimizing S_{31} as asserted.

22.5 EIGENVECTORS, EIGENVALUES, AND EIGEN-NETWORKS

In order to proceed with a physical adjustment of the junction, a connection between the reflection eigenvalues and the physical circuit is necessary. The required connection is that between reflection and impedance.

A first Foster expansion of the impedance eigenvalues in terms of the poles of the problem region is given by

$$Z^0 = Z_0 + Z_{+4} + Z_{-4} + \cdots \tag{22.17a}$$

$$Z^+ = Z_{+1} + Z_{-3} + Z_{+5} + \cdots \tag{22.17b}$$

$$Z^- = Z_{-1} + Z_{+3} + Z_{-5} + \cdots \tag{22.17c}$$

$$Z^{\pm} = Z_{+2} + Z_{-2} + Z_{+6} + Z_{-6} + \cdots \tag{22.17d}$$

The two rotating eigen-networks contain the odd poles of the problem region, the in-phase one contains the even poles with the symmetry of the problem region, and the fourth counterrotating eigen-network contains the remaining even poles. The four eigen-networks of the problem region under consideration are summarized in Fig. 22.5.

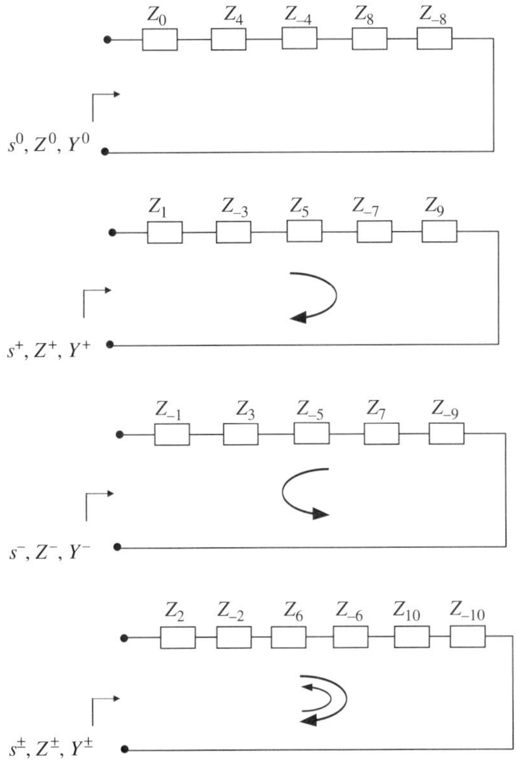

FIGURE 22.5 Eigen-networks of four-port junction circulator.

The corresponding admittance eigenvalues are

$$Y^0 = 1/Z^0 \tag{22.18a}$$

$$Y^+ = 1/Z^+ \tag{22.18b}$$

$$Y^- = 1/Z^- \tag{22.18c}$$

$$Y^\pm = 1/Z^\pm \tag{22.18d}$$

The reflection eigenvalues are

$$s^0 = \frac{Z_0 + Z^0}{Z_0 - Z^0} \tag{22.19a}$$

$$s^+ = \frac{Z_0 + Z^+}{Z_0 - Z^+} \tag{22.19b}$$

$$s^- = \frac{Z_0 + Z^-}{Z_0 - Z^-} \tag{22.19c}$$

$$s^\pm = \frac{Z_0 + Z^\pm}{Z_0 - Z^\pm} \tag{22.19d}$$

The ideal eigenvalue diagram involves the realization of one impedance pole, one pair of split counterrotating poles, and one impedance zero.

The nomenclature used to itemize the eigen-networks of the four-port junction breaks down for junctions with more than four ports. The lowest impedance pole in the first Foster expansion of the impedance eigenvalues may be used to describe the different eigenvalues in the general case.

22.6 PHENOMENOLOGICAL ADJUSTMENT

The synthesis of the four-port single junction circulator involves the realization of a degenerate pole from each counterrotating eigen-network, a single pole from one of the other two eigen-networks, and a zero from the remaining one. The actual pole selection is not unique and more than one realization has been described. In any practical situation, the physical adjustment is simplified provided the cutoff numbers and field patterns associated with the various poles are known. Figure 22.6 indicates the field patterns of the first four modes in a simple disk resonator with top and bottom electric walls and a magnetic side wall. Its mode chart is given in Chapter 13. A scrutiny of this chart indicates that some way of perturbing the poles of the problem region is necessary in order to realize a four-port single junction circulator. One means of doing so is to introduce suitable radial or circumferential magnetic walls into the resonator plate. Another is to introduce a capacitance post on the axis of the structure.

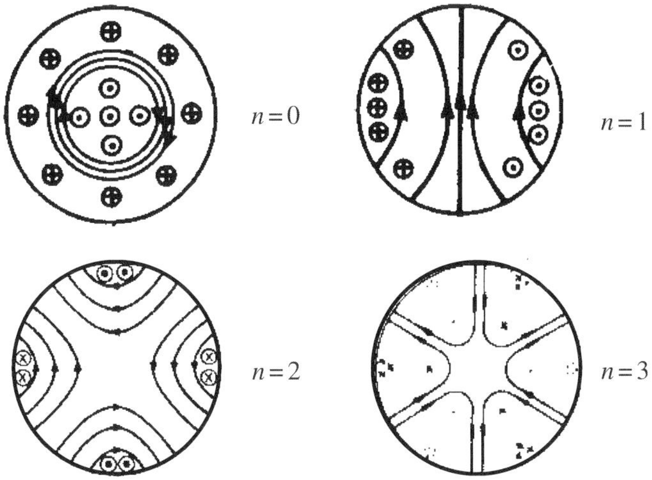

FIGURE 22.6 Modes in circular resonator with top and bottom electric walls and magnetic side walls. (Reproduced with permission from H. Bosma, On stripline circulation at U.H.F., *Trans. IEEE Microwave Theory Tech.*, Vol. MTT-12, pp. 61–72, Jan. 1964.)

22.7 FOUR-PORT SINGLE JUNCTION CIRCULATOR USING $TM_{\pm 1,1,0}$ AND $TM_{0,1,0}$ MODES

The most straightforward phenomenological adjustment of a four-port single junction circulator is deduced by constructing a linear combination of radial $TM_{\pm 1,1,0}$ and $TM_{0,1,0}$ modes in a disk resonator. The poles of the two modes, however, do not coincide. Some means of tuning one or the other is therefore necessary. One possibility is to tune the $TM_{0,1,0}$ pole to the frequency of the degenerate $TM_{\pm 1,1,0}$ poles with the help of a thin nonresonant capacitive post at the center of the junction. Once these poles are established within the junction, the $TM_{\pm 1,1,0}$ one is split by the application of a direct magnetic field until two output ports are decoupled from one input port. The electric fields employed in this arrangement are given in Fig. 22.7. The illustrations in Figs. 22.7a and 22.7b depict the standing wave solutions of the $TM_{\pm 1,1,0}$ and $TM_{0,1,0}$ modes of the unmagnetized junction. Figures 22.7c and 22.7d give the same field patterns with the $TM_{\pm 1,1,0}$ mode rotated through 45°. Adding the amplitudes of the electric fields at the four ports indicates that circulation takes place between ports 1 and 2 and that ports 3 and 4 are decoupled.

The angle through which the $TM_{\pm 1,1,0}$ hybrid mode is rotated is established by taking a linear combination of the electric fields around the periphery of the ferrite disk.

$$E(R, \phi) = a_{01} + a_{11} \cos(\tau_{11} + \phi) \tag{22.20}$$

where a_{01} and a_{11} are arbitrary constants and τ_{11} is the phase angle through which the $TM_{\pm 1,1,0}$ standing wave pattern is rotated.

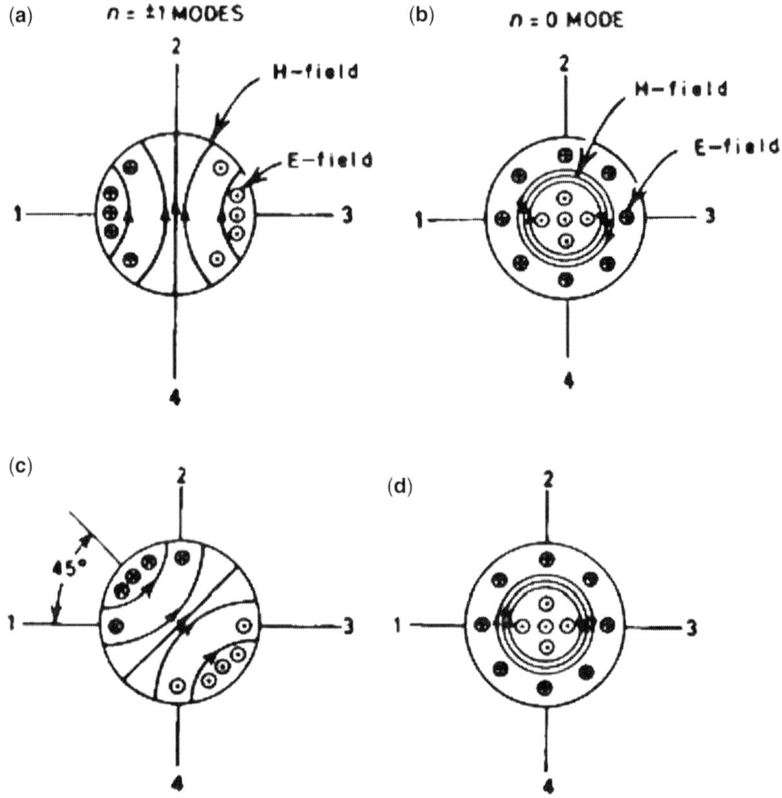

FIGURE 22.7 (a) $TM_{\pm 1,1,0}$ field patterns for unmagnetized ferrite post. (b) $TM_{0,1,0}$ field pattern in ferrite disk. (c) $TM_{\pm 1,1,0}$ modes in magnetized ferrite post. (d) $TM_{0,1,0}$ field pattern in ferrite disk. (Reproduced with permission from C. E. Fay and R. L. Comstock, Operation of the ferrite junction circulator, *IEEE Trans. Microwave Theory Tech.*, Vol. MTT-13, pp. 15–27, Jan. 1965.)

Applying the boundary conditions of an ideal circulator at the four ports gives

$$E(R, 0) = a_{01} + a_{11} \cos \tau_{11} = +1 \tag{22.21a}$$

$$E(R, \pi/2) = a_{01} + a_{11} \cos (\tau_{11} + \pi/2) = +1 \tag{22.21b}$$

$$E(R, \pi) = a_{01} + a_{11} \cos (\tau_{11} + \pi) = 0 \tag{22.21c}$$

$$E(R, 3\pi/2) = a_{01} + a_{11} \cos (\tau_{11} + 3\pi/2) = 0 \tag{22.21d}$$

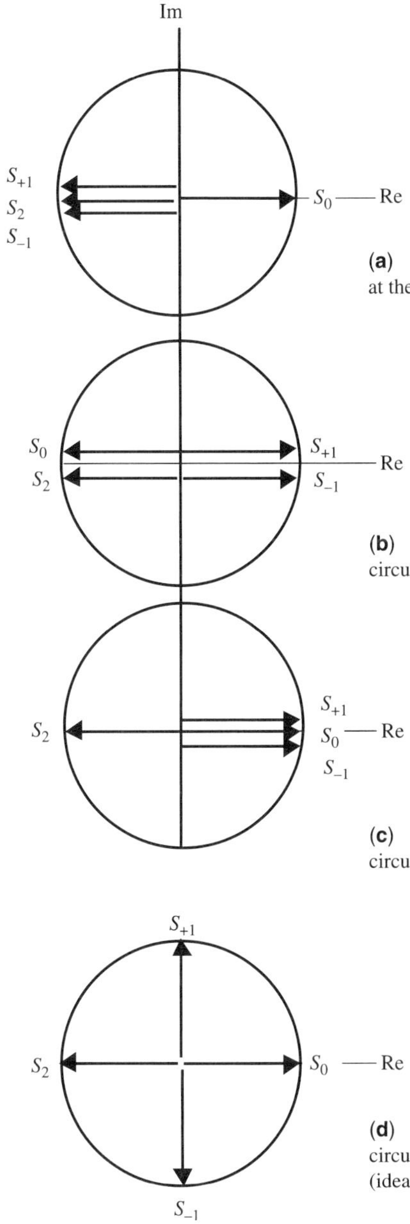

(a) Scattering eigenvalues for four-port junction at the origin

(b) Scattering eigenvalues after first circulation adjustment of four-port junction

(c) Scattering eigenvalues after second circulation adjustment of four-port junction

(d) Scattering eigenvalues after third circulation adjustment of four-port junction (ideal circulator)

FIGURE 22.8 Eigenvalue adjustment of four-port junction circulator using split $TM_{\pm1,1,0}$ modes.

The result is

$$\tau_{11} = 45^\circ \tag{22.22a}$$

$$a_{01} = \frac{1}{2} \tag{22.22b}$$

$$a_{11} = 1/\sqrt{2} \tag{22.22c}$$

The solution obtained in this way is consistent with the field patterns illustrated in Figs. 22.7c and 22.7d. The eigenvalue diagram in connection with this arrangement is indicated in Fig. 22.8.

22.8 FOUR-PORT SINGLE JUNCTION CIRCULATOR USING $TM_{\pm 3,1,0}$ AND $TM_{0,1,0}$ MODES

The adjustment of one four-port single junction circulator, which relies for its operation on the use of a degenerate pair of higher order impedance poles, is summarized in this section. It again consists of a simple gyromagnetic disk at the junction of four striplines. The direct magnetic field is applied perpendicular to the plane of the resonator in the usual way. The symmetric $TM_{0,1,0}$ and the radial $TM_{\pm 3,1,0}$ modes in a simple planar ferrite disk are used in this arrangement with top and bottom electric walls and a magnetic side wall. Its experimental adjustment relies on the fact that the impedance pole associated with the $TM_{0,1,0}$ mode already lies approximately midway between those of the split $TM_{\pm 3,1,0}$ ones.

Figures 22.9a and 22.9b depict the field patterns of the modes in question. The radial wavenumbers for such a demagnetized ferrite disk are $kR = 3.83$ for $TM_{0,1,0}$ and $kR = 4.20$ for $TM_{\pm 3,1,0}$. In this instance, the eigenvalue diagram is identical to that for the previous case except that the ± 1 poles are replaced by the ± 3 poles.

An understanding of the phenomenological solution of this arrangement starts with the distribution of the electric field around the periphery of the ferrite disk of the four-port circulator

$$E_z = a_0 + a_3 \cos 3(\tau_3 + \phi) \tag{22.23}$$

a_0 and a_3 are arbitrary constants and τ_3 is the phase angle through which the standing wave solution formed by the degenerate modes is rotated. The electric fields at the ports are

$$E_z(\phi = 0) = a_0 + a_3 \cos\, 3\tau_3 = +1 \tag{22.24a}$$

$$E_z(\phi = \pi/2) = a_0 + a_3 \cos 3(\tau_3 + \pi/2) = +1 \tag{22.24b}$$

$$E_z(\phi = \pi) = a_0 + a_3 \cos 3(\tau_3 + \pi) = 0 \tag{22.24c}$$

$$E_z(\phi = 3\pi/2) = a_0 + a_3 \cos 3(\tau_3 + 3\pi/2) = 0 \tag{22.24d}$$

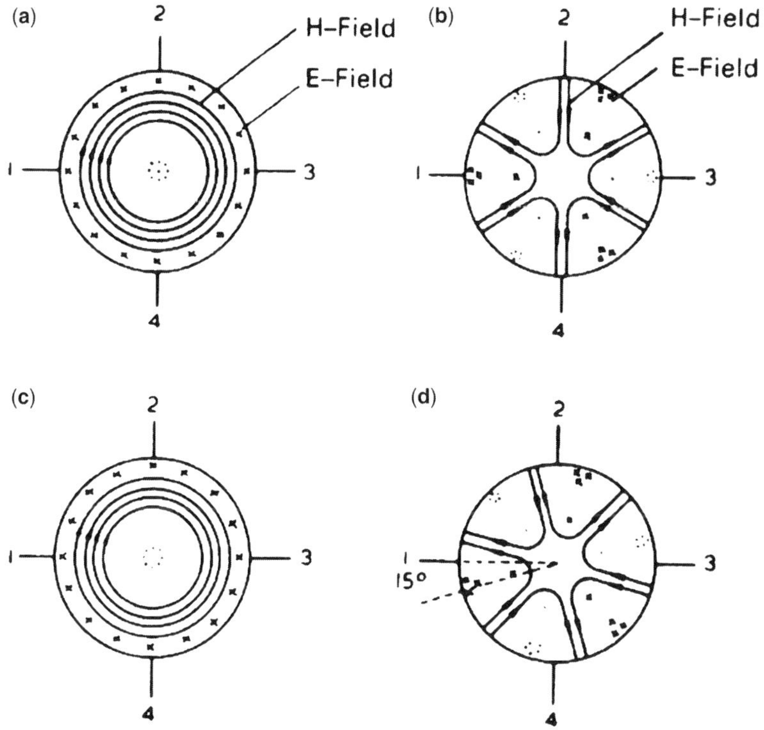

FIGURE 22.9 (a) $TM_{0,1,0}$ field patterns for unmagnetized disk. (b) $TM_{\pm3,1,0}$ field patterns for unmagnetized disk. (c) $TM_{0,1,0}$ field patterns for magnetized disk. (d) $TM_{\pm3,1,0}$ field patterns for magnetized disk. (Reproduced with permission from J. Helszajn, Waveguide and stripline four-port single junction circulators, *IEEE Trans. Microwave Theory Tech.*, Vol. MTT-20, pp. 630–633, 1973.)

The result is

$$\tau_3 = 15^\circ \tag{22.25a}$$

$$a_0 = \frac{1}{2} \tag{22.25b}$$

$$a_3 = 1/\sqrt{2} \tag{22.25c}$$

Figures 22.9c and 22.9d indicate the field patterns of the modes with the $TM_{\pm3,1,0}$ ones rotated by 15° by magnetizing the junction with a direct magnetic field. If the magnitudes of the electric fields for the two patterns are equal (after rotation) at the four ports, transmission occurs between ports 1 and 2 and ports 3 and 4 are isolated.

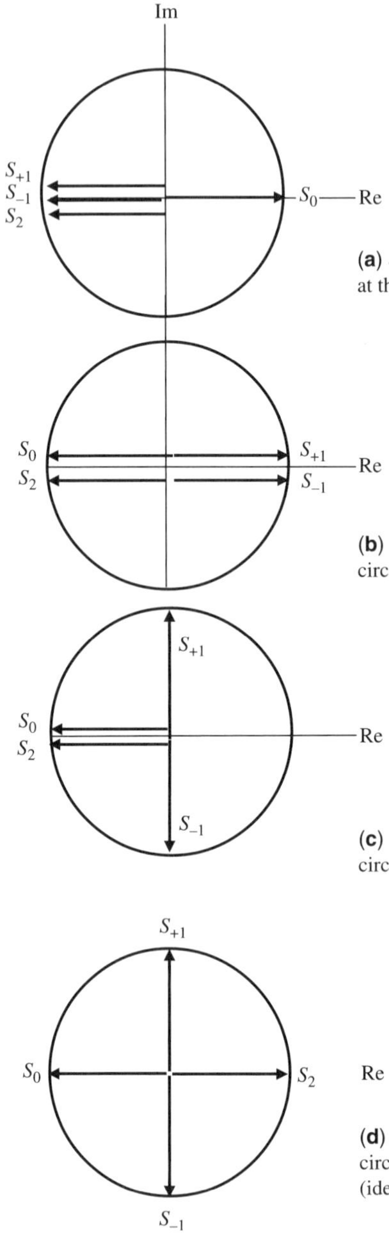

(**a**) Scattering eigenvalues for four-port junction at the origin

(**b**) Scattering eigenvalues after first circulation adjustment of four-port junction

(**c**) Scattering eigenvalues after second circulation adjustment of four-port junction

(**d**) Scattering eigenvalues after third circulation adjustment of four-port junction (ideal circulator)

FIGURE 22.10 Eigenvalue adjustment of four-port junction circulator using $TM_{1,2,0}$ and $TM_{1,\pm1,0}$ modes.

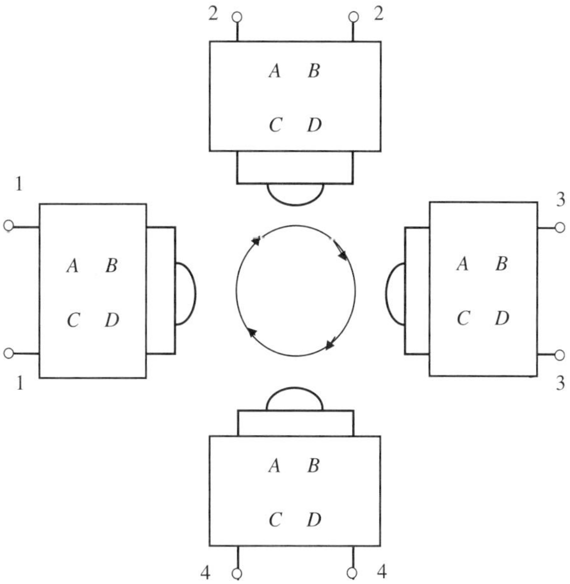

FIGURE 22.11 Four-port single junction circulator using quarter-wave long impedance transformers.

22.9 FOUR-PORT SINGLE JUNCTION CIRCULATOR USING $TM_{1,1,0}$ AND $TM_{1,2,0}$ MODES

The eigenvalue adjustment of the four-port single junction circulator in Fig. 22.4 is actually not unique. Another possibility is indicated in Fig. 22.10. The degree of

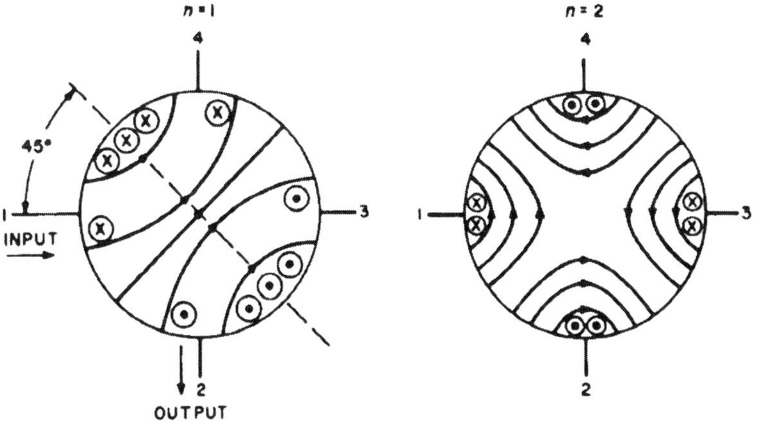

FIGURE 22.12 Adjustment of four-port junction circulator using $TM_{\pm 1,1,0}$ and $TM_{2,1,0}$ modes. (Reproduced with permission from C. E. Fay and W. A. Dean, The four-port single junction circulator in stripline, PG-MTT, *Digest Int. Symp. IEEE*, 1966.)

splitting between the $TM_{1,\pm1,0}$ modes in this arrangement is fixed by ensuring that the $TM_{1,2,0}$ eigenvalue intercepts the split $TM_{1,\pm1,0}$ ones rather than by the need to satisfy the gyrator conductance. This condition may be understood by scrutinizing the mode chart of the disk resonator in Chapter 13. The gyrator conductance is then separately satisfied by introducing quarter-wave transformers at each port of the device in the manner indicated in Fig. 22.11. The phenomenological adjustment is indicated in Fig. 22.12.

22.10 STANDING WAVE SOLUTIONS OF PLANAR X RESONATORS

Symmetrical planar resonators may also be constructed by the symmetrical connection of any number of quarter-wave stubs. The purpose of this section is to give the equipotential lines and cutoff numbers of an X resonator formed by the junction of four open-circuited stubs. This resonator has the symmetry required for the construction of the four-port single junction circulator. Its toplogy is indicated in Fig. 22.13. Figure 22.14 indicates the equipotential lines of first four-typical modes of the resonator. The corresponding cutoff numbers are

$$kR = 1.68$$

$$kR = 1.68$$

$$kR = 1.86$$

$$kR = 3.48$$

The standing wave solutions of the first two nonsymmetrical modes are constructed by taking linear combinations of the intersection of two half-wave long stubs with appropriate open-circuited terminals. These may be interpreted

FIGURE 22.13 Schematic diagram of cross resonator.

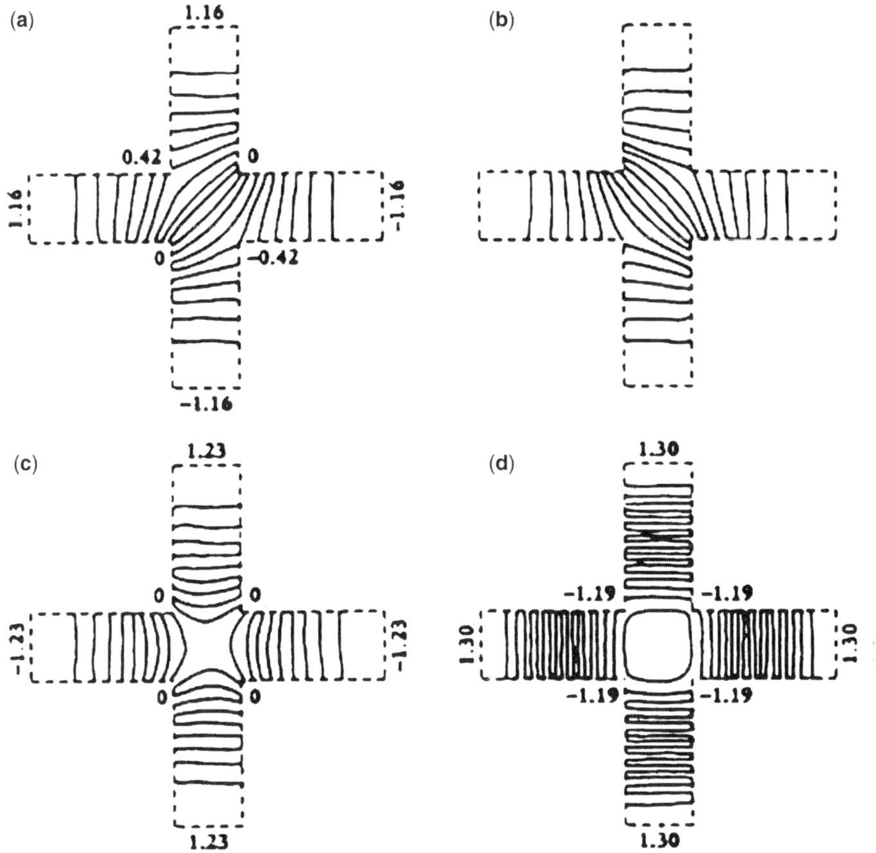

FIGURE 22.14 Equipotential lines for (a, b) dominant modes in X resonator; (c) first symmetric mode in X resonator; (d) second higher order mode in X resonator. (Reproduced with permission from J. Helszajn, Standing wave solution of planar irregular hexagonal and wye Resonators, *IEEE Trans. Microwave Theory Tech.*, Vol. MTT-29, No. 6, pp. 562–567, June 1981.)

as $\pm 45^\circ$ rotations of the half-wave long resonators. This is also the case for the first higher order mode except that now the intersecting stubs are $1\frac{1}{2}$ wave long resonances.

Figure 22.15 depicts one possible linear combination of X resonator modes that is suitable for the construction of a four-port single junction circulator. This adjustment may be realized experimentally by tuning the cutoff number of the symmetric mode $kR = 3.48$ to that of the first nonsymmetric one with $kR = 1.86$, with the aid of a capacitive post through the center of the resonator.

The modes of the X resonator are degenerate with those of a planar square resonator with similar magnetic side walls in the limit as the shape angle ϕ subtended at the origin by the width of the outside terminals approaches 90°. The modes of

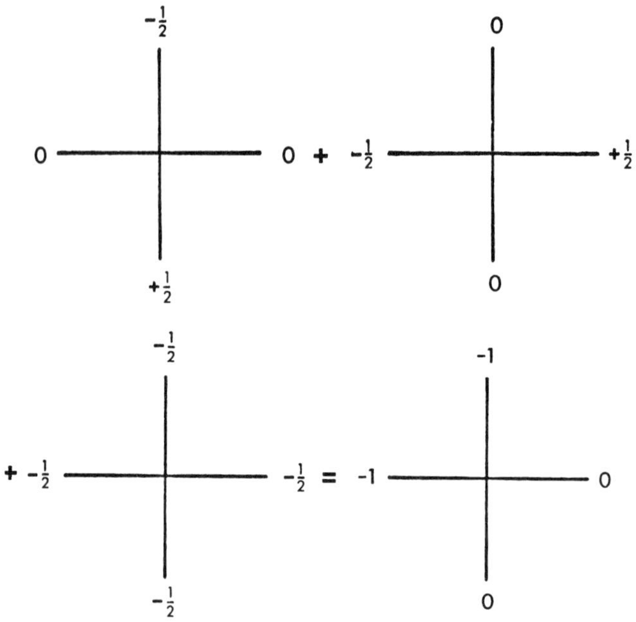

FIGURE 22.15 Circulation solution using modes in cross resonator.

the X resonator may therefore be labeled $TM_{m,n,0}$ limit modes. The two lowest order degenerate modes are obtained by a superposition of the $TM_{1,1,0}$ and $TM_{0,1,0}$ modes in a planar square resonator.

22.11 FREQUENCIES OF FOUR-PORT UE LOADED DISK MAGNETIZED RESONATORS

One possible variation of the cross resonator is the disk–stub gyromagnetic arrangement consisting of a circular plate to which are connected four short UEs. In this geometry, the coupling angle of a typical stub is not restricted to $\pi/2$ rad. Its frequency may be deduced by visualizing it as a symmetric four-port region loaded by suitable UEs or by a eight-port arrangement with four of its ports closed by magnetic walls and the other four terminated by suitable stubs. The equivalence between the two models suggests that the first eight impedance poles of the problem region, strictly speaking, are necessary to reproduce the boundary conditions of the resonator. The topology under consideration is indicated in Fig. 22.16. It is fixed by a coupling or shape angle (ψ) and the ratio of the radii R_i and R_0. Its degenerate or split resonance may be deduced by resonating the stubs with the eigen-networks of the circular gyromagnetic region. The description of a typical impedance pole of an m-port symmetrical region is a standard result in the text.

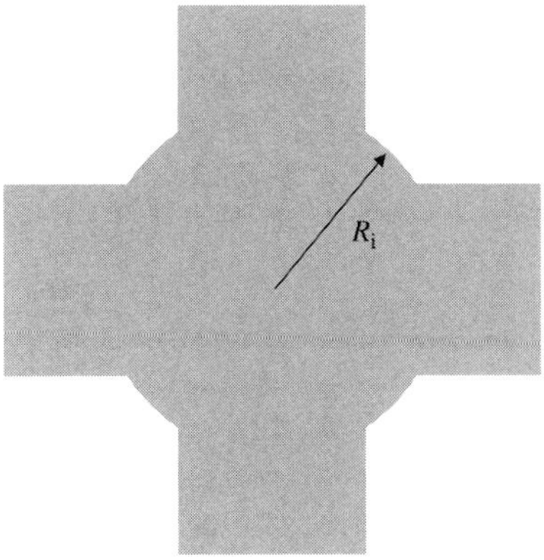

FIGURE 22.16 Cross resonator formed by circular gyromagnetic disk symmetrically loaded by a quartet of UEs.

If the loosely coupled resonator is visualized as a symmetrical eight-port network with alternate ports open-circuited, then the eigenvalues at the reactive walls of the four-port network are given in terms of the poles of the eight-port one by

$$Z^0 \approx \frac{Z_0 + Z_4}{2} \tag{22.26a}$$

$$Z^+ \approx \frac{Z_{+1} + Z_{-3}}{2} \tag{22.26b}$$

$$Z^- \approx \frac{Z_{-1} + Z_{+3}}{2} \tag{22.26c}$$

$$Z^{\pm} \approx \frac{Z_{-2} + Z_{+2}}{2} \tag{22.26d}$$

If the eigenvalues are written in terms of the impedance poles of the four-port network the result is

$$Z^0 \approx Z_0 + Z_4 \tag{22.27a}$$

$$Z^+ \approx Z_{+1} + Z_{-3} \tag{22.27b}$$

$$Z^- \approx Z_{-1} + Z_{+3} \tag{22.27c}$$

$$Z^{\pm} \approx Z_{-2} + Z_{+2} \tag{22.27d}$$

A typical pole of a four-port symmetrical isotropic disk junction is given by

$$Z_n = \frac{j4\psi\eta_r Z_r}{\pi}\left(\frac{\sin n\psi}{n\psi}\right)^2\left(\frac{J'_n(kR_i)}{J_n(kR_i)}\right)^{-1} \tag{22.28}$$

The shape or coupling angle of a typical strip (ψ) is defined by

$$\sin\psi = (W/2R_i) \tag{22.29}$$

η_r is the constituent wave impedance of the substrate:

$$\eta_r = \sqrt{\mu_{eff}/\varepsilon_f} \tag{22.30}$$

Z_r is the characteristic impedance of a typical stripline:

$$Z_r = 30\pi\ \ln\left(\frac{W+t+2H}{W+t}\right) \tag{22.31}$$

The phase constant (k) is

$$k = \omega\sqrt{\mu_0\mu_{eff}\varepsilon_0\varepsilon_f}$$

μ_{eff} and ε_f are the relative permeability and relative dielectric constant of the ferrite or garnet material; W, t, and H are the linear dimensions of the uniform striplines.

The characteristic equation for the frequencies of the resonator is obtained by resonating the degenerate impedance eigenvalues using suitable open-circuited uniform transmission lines. The characteristic equation for the first two pair of degenerate resonances is now established by forming a transverse resonance condition between the radial and uniform lines:

$$Z^{\pm} = j\eta_r Z_r \cot(kL) \tag{22.32}$$

L is the length of a typical open-circuited stub:

$$L = R_0 - R_i \tag{22.33}$$

The split cutoff space of a gyromagnetic resonator is also readily established. The description of a typical pole of this sort of problem region is also a classic result in the literature. It is given by

$$Z_n \approx \frac{j4\psi\eta_r Z_r}{\pi}\left(\frac{\sin n\psi}{n\psi}\right)^2\left[\frac{J_{n-1}(kR_i)}{J_n(kR_i)} - n\left(\frac{1+\kappa/\mu}{kR_i}\right)\right]^{-1} \tag{22.34}$$

μ and κ are the usual diagonal and off-diagonal entries of the permeability tensor. The ratio of κ and μ is known as the gyrotropy of the problem region.

CHAPTER TWENTY-THREE

Frequency Responses of Quarter-Wave Coupled Reciprocal Stripline Junctions

23.1 INTRODUCTION

The adjustment of the classic three-port junction circulator involves two so-called circulation conditions. The first coincides with the maximum power transfer condition of the reciprocal arrangement; the second merely involves removal of the degeneracy of its counterrotating eigen-networks. A scrutiny of these two conditions indicates that the frequency responses of the two arrangements are related so that the problem of constructing wideband circulators is essentially that of making wideband reciprocal junctions. The required relationship may be demonstrated once the equivalent circuit of the degree-1 reciprocal junction is derived in terms of the degenerate admittance eigen-networks or eigenvalues of the three-port network. These eigenvalues coincide with those in the description of the susceptance slope parameter of the magnetized junction or circulator. The development in this chapter is restricted to the classic H-plane junction for which the equivalent circuit of the reciprocal junction is represented as a shunt connection of the three lines. Once the equivalent network of the reciprocal junction is postulated, it is possible to connect matching networks at each port and calculate the overall response of the return loss at one typical port. This is done for degree-2 and degree-3 circuits by taking a suitable linear combination of the eigenvalues of the junction. The frequency responses are also deduced at the terminals of each network. This allows the precise experimental adjustment of each matching network entering into the description of the desired overall response. The element values for the matching networks are chosen to

The Stripline Circulator: Theory and Practice. By J. Helszajn

coincide with those in connection with the realization of typical circulators with overall Chebyshev responses.

23.2 EQUIVALENT CIRCUIT OF RECIPROCAL THREE-PORT JUNCTION

The equivalent circuit of any symmetrical junction may be deduced by starting with the standard relationship between the reflection coefficient S_{11} of the scattering matrix and its eigenvalues. The one-port immittance of the circuit is then obtained by having recourse to the bilinear relationship between the immittance and scattering variables. The reflection coefficient at port 1 with ports 2 and 3 terminated in their characteristic impedances is the trace of the eigenvalues.

$$S_{11} = \frac{s_0 + s_{+1} + s_{-1}}{3} \tag{23.1}$$

The frequency dependence of the scattering matrix eigenvalues is obtained by making use of the bilinear relationship between that of the scattering and admittance eigenvalues:

$$s_0 = \frac{1 - y_0}{1 + y_0} \tag{23.2}$$

$$s_{+1} = \frac{1 - y_{+1}}{1 + y_{+1}} \tag{23.3}$$

$$s_{-1} = \frac{1 - y_{-1}}{1 + y_{-1}} \tag{23.4}$$

For a reciprocal junction

$$s_{+1} = s_{-1} = s_1 \tag{23.5}$$

and

$$y_{+1} = y_{-1} = y_1 \tag{23.6}$$

In the approximation employed here the in-phase eigen-network is idealized by an electric wall. This gives

$$s_0 = -1 \tag{23.7}$$

This condition corresponds to $S_{21} = -1$ when the junction is magnetized to form a circulator. The nature of the normalized degenerate admittance eigenvalue y_1 is taken as that of a $\lambda/4$ short-circuited transmission line.

$$y_1 = -j[(4b_1' \cot \theta)/\pi] \tag{23.8}$$

$$\theta = (\pi/2)(1 + \delta) \tag{23.9}$$

δ is a normalized frequency variable

$$\delta = (-\omega + \omega_0)/\omega_0 \tag{23.10}$$

b_1' is the normalized susceptance slope parameter of the degenerate eigen-network, and θ is its radian length. Combining the above relationships gives

$$S_{11} = \frac{1 - 3y_1}{3 + 3y_1} \tag{23.11}$$

The corresponding input admittance is

$$Y_{\text{in}} = Y_0[(1 + 3y_1)/2] \tag{23.12}$$

The equivalent network at $\theta = \pi/2$ is shown in Fig. 23.1. The general circuit is depicted in Fig. 23.2.

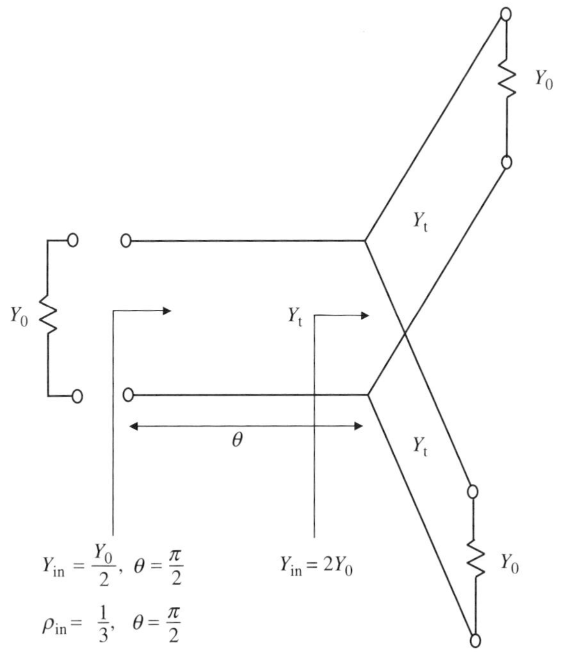

FIGURE 23.1 Equivalent circuit of reciprocal H-plane junction in terms of 90° UEs.

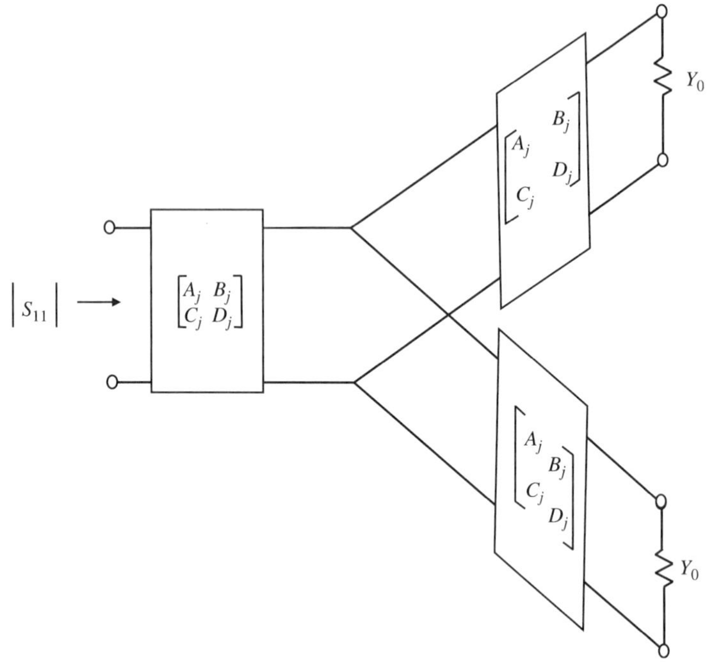

FIGURE 23.2 Equivalent circuit of reciprocal *H*-plane junction in terms of *ABCD* parameters.

An approximate relation between the VSWR at one port and the susceptance slope parameter of the input admittance at the same port is readily derived. This may be done in the neighborhood of the resonance frequency by replacing the distributed network by a lumped element one. The required result is obtained by making use of the bilinear relationship between admittance and reflection coefficient and that between the latter quantity and the VSWR.

$$b_1' = \frac{\frac{2}{3}\left[\frac{\mathrm{VSWR}^2 - 2.5\,\mathrm{VSWR} + 1}{2\mathrm{VSWR}}\right]^{1/2}}{2\delta} \tag{23.13}$$

In obtaining this result y_1 in Eq. (23.8) has been replaced by

$$y_1 \approx -j2\delta b_1' \tag{23.14}$$

23.3 EIGEN-NETWORKS OF RECIPROCAL JUNCTION

The nature of the eigen-networks at the terminals of the three-port network from which the eigenvalues may be deduced are indicated in Fig. 23.3. The eigenvalues

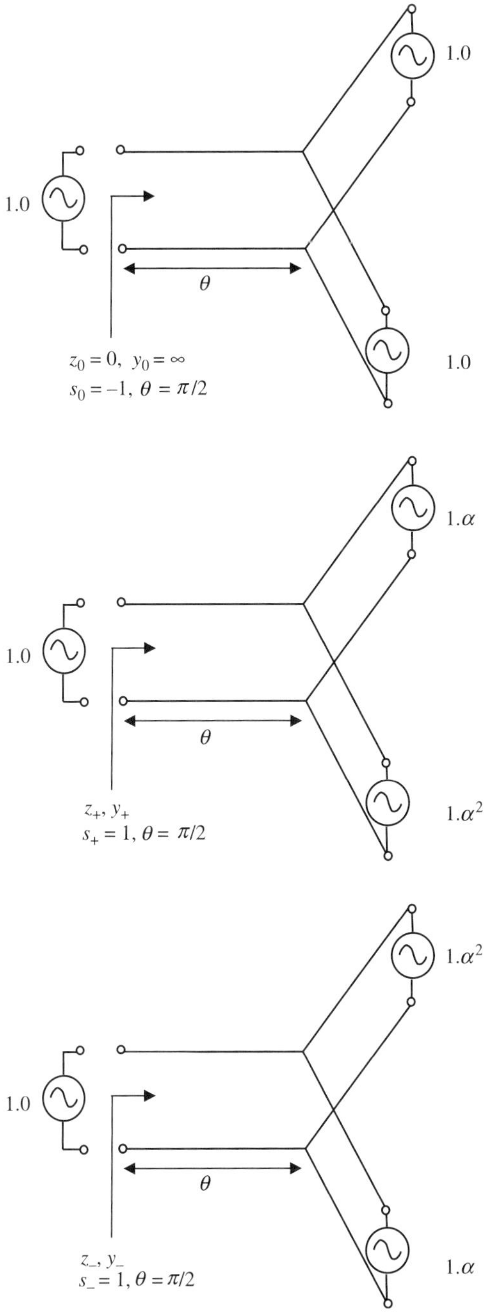

FIGURE 23.3 Eigen-networks of reciprocal *H*-plane junction.

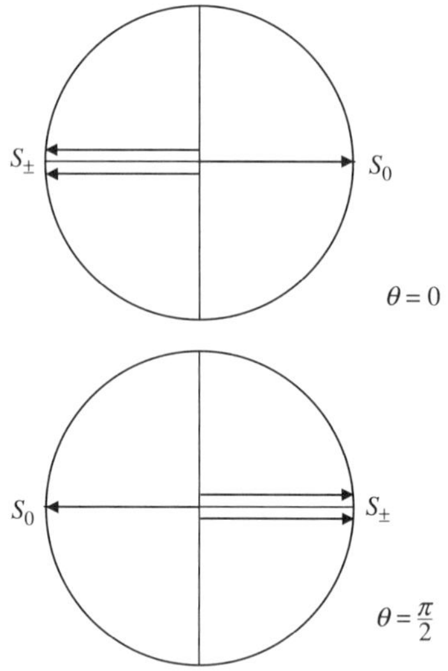

FIGURE 23.4 Eigenvalue diagrams of reciprocal three-port junction.

are the usual one-port transmission line variables of the circuit. These are obtained by applying each eigenvector of the junction one at a time at the ports of the circuit. This suggests that the in-phase eigen-network has a magnetic wall on the axis, whereas the counterrotating ones have electric walls there. This sort of circuit has no impedance or admittance matrices at $\theta = \pi/2$. The scattering matrix, of course, always exists.

A valuable insight into the adjustment of a symmetrical junction may readily be established by displaying its reflection eigenvalues on a unit circle. Figure 23.4 indicates the eigenvalue diagrams of a three-port reciprocal *H*-plane junction for both $\theta = 0$ and $\theta = \pi/2$ radians. The latter diagram coincides with the first passband of the junction at finite frequency.

23.4 REFLECTION COEFFICIENT OF RECIPROCAL JUNCTION

The equivalent circuit of the overall network is shown in Fig. 23.5. Each matching network is represented in terms of an overall *ABCD* matrix. The ones studied here consist of quarter-wave long impedance transformers. The derivation of the reflection coefficient of the degree-1 reciprocal junction proceeds by taking a linear combination of its eigenreflection coefficients. The result is

$$\Gamma_{11} = (\gamma_0 + 2\gamma_1)/3 \tag{23.15}$$

where

$$\gamma_+ = \gamma_- = \gamma_1 \tag{23.16}$$

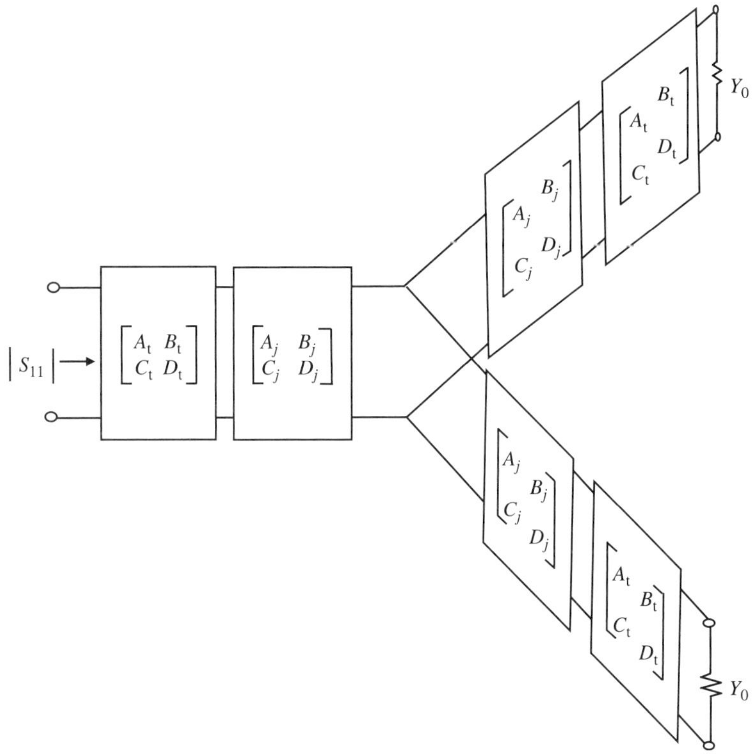

FIGURE 23.5 Equivalent circuit of wideband reciprocal junction in terms of *ABCD* networks.

and

$$\gamma_0 = \exp(-j2\phi_0) \tag{23.17}$$

$$\gamma_1 = \exp[-j2(\phi_1 + \pi/2)] \tag{23.18}$$

ϕ_0 and ϕ_1 are reflection angles.

In the problem under consideration γ_0 is associated with a one-port open-circuited transmission line and γ_1 with a short-circuit one. The reflection eigenvalues are related separately to the admittance ones by the usual bilinear mapping between the two:

$$\gamma_0 = \frac{1 - y_0}{1 + y_0} \tag{23.19}$$

$$\gamma_1 = \frac{1 - y_1}{1 + y_1} \tag{23.20}$$

The reflection angles ϕ_0 and ϕ_1 in the description of the reflection coefficients are now obtained without ado by introducing the following substitutions into the above transformations:

$$y_0 = j\tan(\phi_0) \tag{23.21}$$

$$y_1 = j\tan(\phi_1 + \pi/2) \tag{23.22}$$

The convention adopted here is to represent the various building blocks by *ABCD* parameters. The connection between the two representations is then specified by

$$j\tan(\phi_0) = \frac{C + Dy'_0}{A + By'_0} \tag{23.23}$$

$$j\tan\left(\phi_1 + \frac{\pi}{2}\right) = \frac{C + Dy'_1}{A + By'_1} \tag{23.24}$$

A and *D* are real numbers and *B* and *C* are pure imaginary quantities.

The bilinear transformations between the origins of the eigen-networks considered here and the input terminals can be simplified by introducing the boundary conditions there:

$$y'_0 = 0 \tag{23.25}$$

$$y'_1 = \infty \tag{23.26}$$

This gives

$$j\tan(\phi_0) = C/A \tag{23.27}$$

$$j\tan(\phi_1 + \pi/2) = D/B \tag{23.28}$$

For the degree-1 topology

$$\begin{bmatrix} A_j & B_j \\ C_j & D_j \end{bmatrix} = \begin{bmatrix} \cos\theta & j(\sin\theta)/y_j \\ jy_j\sin\theta & \cos\theta \end{bmatrix} \tag{23.29}$$

y_j is the characteristic admittance of the region.

$$y_j = 4b'_j/\pi \tag{23.30}$$

The magnitude squared of the reflection coefficient can now be formed as

$$9|\Gamma_{11}|^2 = \{\cos(2\phi_0) + 2\cos[2(\phi_1 + \pi/2)]\}^2 + \{\sin(2\phi_0) + 2\sin[2(\phi_1 + \pi/2)]\}^2 \tag{23.31}$$

The minimum value for the reflection coefficient is obtained with $\theta = 90^\circ$ as

$$|\rho| = \frac{1}{3} \tag{23.32}$$

This condition corresponds with the maximum power transfer condition through the junction.

23.5 FREQUENCY RESPONSE OF DEGREE-*n* NETWORK

The derivation of the frequency response of the degree-2 circuit proceeds by replacing the $A_j\ B_j\ C_j\ D_j$ parameters of the degree-1 by an overall $ABCD$ description. It is also necessary to respect the order of the multiplication in forming this matrix. The overall $ABCD$ matrix is given by

$$\begin{bmatrix} A & B \\ C & D \end{bmatrix} = \begin{bmatrix} A_t & B_t \\ C_t & D_t \end{bmatrix} \begin{bmatrix} A_j & B_j \\ C_j & D_j \end{bmatrix} \tag{23.33}$$

A_j, B_j, C_j, and D_j have the meaning met in connection with the degree-1 topology and A_t, B_t, C_t, and D_t are those of the overall matching network. For a degree-2 circuit

$$A_t = \cos\ \theta \tag{23.34a}$$

$$B_t = (j\ \sin\ \theta)/y_t \tag{23.34b}$$

$$C_t = jy_t \sin\ \theta \tag{23.34c}$$

$$D_t = \cos\ \theta \tag{23.34d}$$

where y_t is the normalized admittance of the UE. The arrangement under consideration is illustrated in Fig. 23.5.

A degree-3 network is obtained by replacing the single UE by two UEs in cascade. A_t, B_t, C_t, and D_t are given, for the degree-2 network, in Chapter 19.

The eigen-networks of the general problem are depicted in Fig. 23.6.

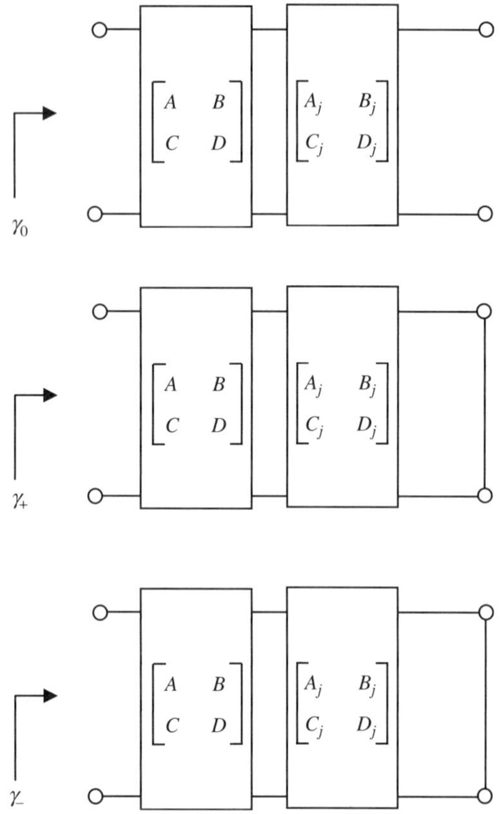

FIGURE 23.6 Eigen-networks of quarter-wave coupled reciprocal three-port junction.

23.6 DEGREE-1 THREE-PORT JUNCTION CIRCULATOR

The purpose of this section is to deduce a correspondence between the frequency responses of a three-port reciprocal junction and that of a junction circulator embedded in the same matching networks. In order to do so it is necessary to relate the two circuits. The derivation of the equivalent circuit of the three-port junction circulator starts with the definition of its eigenvalues in terms of the admittance y_1 encountered in the equivalent circuit of the reciprocal junction.

The split counterrotating admittance eigenvalues of an ideal circulator in the vicinity of the circulation frequency are

$$y_{+1} = y_1 - j/\sqrt{3} \tag{23.35a}$$

$$y_{-1} = y_1 + j/\sqrt{3} \tag{23.35b}$$

where y_1 is the degenerate counterrotating admittance eigenvalue of the reciprocal junction. The reflection coefficient s_{+1} of an ideal circulator in the vicinity of the circulation frequency is now given by

$$s_{+1} = \frac{1 - (y_1 - j/\sqrt{3})}{1 + (y_1 - j/\sqrt{3})} \tag{23.36a}$$

Similarly, one has for the eigenvalue s_{-1}

$$s_{-1} = \frac{1 - (y_1 + j/\sqrt{3})}{1 + (y_1 + j/\sqrt{3})} \tag{23.36b}$$

The eigenvalue for which the frequency variation is omitted is again

$$s_0 = -1 \tag{23.36c}$$

The derivation of the required scattering parameters is facilitated by factoring the numerator and denominator polynomials of the reflection coefficient in such a way as to reveal the midband relationships.

$$s_+ = \left(\frac{1 - j\tan(\phi_+ + \pi/2)}{1 + j\tan(\phi_+ + \pi/2)}\right)\frac{(1 + j/\sqrt{3})}{(1 - j/\sqrt{3})} \tag{23.37a}$$

$$s_- = \left(\frac{1 - j\tan(\phi_- + \pi/2)}{1 + j\tan(\phi_- + \pi/2)}\right)\frac{(1 - j/\sqrt{3})}{(1 + j/\sqrt{3})} \tag{23.37b}$$

where

$$\frac{(1 \pm j/\sqrt{3})}{(1 \mp j/\sqrt{3})} = \exp\left(\pm j\frac{\pi}{3}\right)$$

and

$$\cot\ \phi_\pm = \frac{3\zeta_e Y_r \cot\ \theta}{4Y_0 \pm 3\zeta_e Y_r \cot\ \theta}$$

The preceding equations give the connection between the reflection and transmission line angles.

s_+ and s_- may also be written as

$$s_+ \approx \exp\left[-j2(\phi_1 - \pi/6 + \pi/2)\right] \tag{23.38a}$$

$$s_- \approx \exp\left[-j2(\phi_1 + \pi/6 + \pi/2)\right] \tag{23.38b}$$

The reflection coefficient is now obtained by taking a linear combination of s_0, s_{+1}, and s_{-1}.

$$S_{11} = \frac{-1 + \exp[-j2(\phi_1 - \pi/6 + \pi/2)] + \exp[-j2(\phi_1 + \pi/6 + \pi/2)]}{3} \tag{23.39}$$

Neglecting second order terms in y_1, the result is given by

$$S_{11} \approx y_1/2 \tag{23.40}$$

The normalized input admittance of the circuit is given in terms of the original variables by

$$y_{\text{in}} = 1 - j\,[(4b_1' \cot\theta)/\pi] \tag{23.41}$$

In obtaining this result it has been assumed that

$$\cot\ \phi_+ \approx \cot\ \phi_- \approx (3\zeta_e Y_r/4Y_0)\cot\ \theta \tag{23.42}$$

An approximate equivalent network for this equation is an ideal circulator available at any frequency with an admittance y_1 connected at each port. This is depicted in Fig. 23.7.

Writing S_{11} in terms of the VSWR now gives

$$b_1' = \frac{2(\text{VSWR} - 1)}{2\delta(\text{VSWR} + 1)} \tag{23.43}$$

A scrutiny of the two circuits under consideration indicates that each involves the same variables. The frequency variation of either circuit may therefore be used to infer that of the other.

Some experimental results in the literature on the frequency responses of such circuits afford some correlation between theory and experiment. One family of solutions, in the open literature, for the case of a demagnetized junction, produces values of susceptance slope parameter given by $b_1' = 5.65$, 3.93, 3.33, 1.84. The corresponding quantities in the magnetized case are $b_1' = 6.6$, 3.89, 3.51, 1.62.

23.7 DEGREE-2 AND DEGREE-3 CIRCUITS

Some computations on the frequency responses of degree-2 and degree-3 networks are summarized in this section. The impedance levels employed here coincide with the design of equal ripple circulators with VSWR = 1.15 and $2\delta_0 = 0.35$, 0.50, 0.66. The element values for the networks are given in Chapter 11. The results are shown in Figs. 23.8 and 23.9 in the case of a degree-2 topology and in

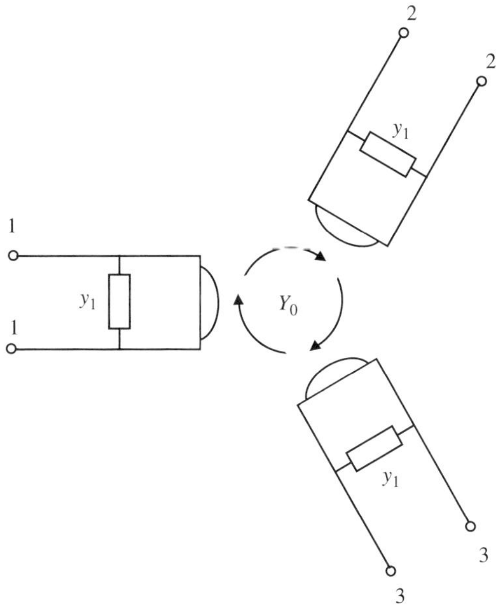

FIGURE 23.7 Equivalent circuit of ideal circulator.

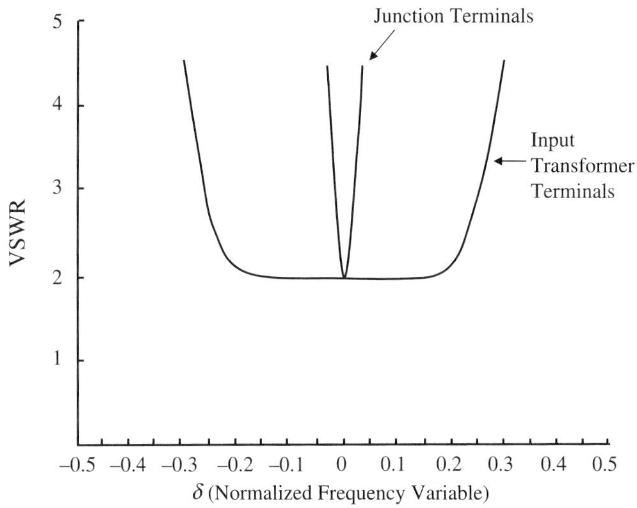

FIGURE 23.8 Frequency response for degree-2 junction circulator with element values of $r = 1.15$ and $2\delta_0 = 0.35$. (Reproduced with permission from J. Helszajn, Frequency response of quarter-wave coupled reciprocal stripline junctions, *IEEE Trans. Microwave Theory Tech.*, Vol. MTT-21, pp. 533–537, Aug. 1973.)

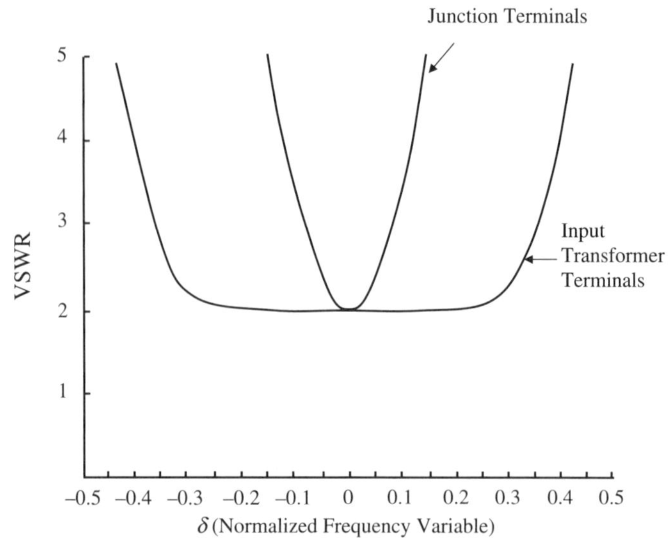

FIGURE 23.9 Frequency response for degree-2 junction circulator with element values of $r = 1.15$ and $2\delta_0 = 0.50$. (Reproduced with permission from J. Helszajn, Frequency response of quarter-wave coupled reciprocal stripline junctions, *IEEE Trans. Microwave Theory Tech.*, Vol. MTT-21, pp. 533–537, Aug. 1973.)

Figs. 23.10–23.12 in the case of degree-3. The frequency responses are given in each case at the terminals of each network. The frequency behavior at the intermediate transformer terminals in the case of degree-3 is obtained using the material with y_t replaced by y_{02}. This allows the precise experimental adjustment of each matching network to obtain the overall response. These illustrations suggest that the bandwidth of the overall reciprocal network is comparable with that of the junction magnetized to form a circulator. It is also observed that the frequency responses of the degree-2 and degree-3 reciprocal junctions associated with identical circulator characteristics are similar. The only difference is that the slope at the bandedges for the degree-2 network is less than that for the degree-3 network. Here, the bandwidth of the reciprocal junction refers to that which coincides with maximum power transfer through the junction. The construction of wideband circulators is therefore closely related to that of wideband reciprocal three-port junctions.

23.8 DEGREE-2 CIRCULATOR

The experimental frequency responses of three-port reciprocal and nonreciprocal stripline junctions with degree-1 and degree-2 networks are reproduced in this section. The degree-1 network consists of a garnet disk with a magnetization of $\mu_0 M_0 = 0.0500$ T and a dielectric constant of $\varepsilon_r = 14.3$ at the junction of three 50 Ω striplines.

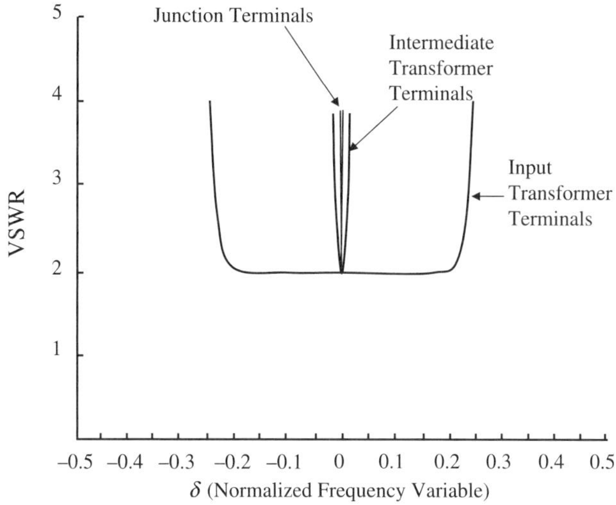

FIGURE 23.10 Frequency response for degree-3 junction circulator with element values of $r = 1.15$ and $2\delta_0 = 0.35$. (Reproduced with permission from J. Helszajn, Frequency response of quarter-wave coupled reciprocal stripline junctions, *IEEE Trans. Microwave Theory Tech.*, Vol. MTT-21, pp. 533–537, Aug. 1973.)

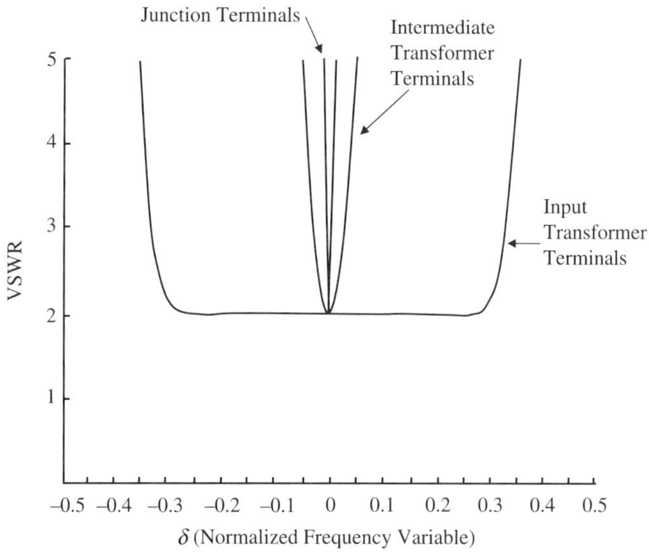

FIGURE 23.11 Frequency response for degree-3 junction circulator with element values of $r = 1.15$ and $2\delta_0 = 0.50$. (Reproduced with permission from J. Helszajn, Frequency response of quarter-wave coupled reciprocal stripline junctions, *IEEE Trans. Microwave Theory Tech.*, Vol. MTT-21, pp. 533–537, Aug. 1973.)

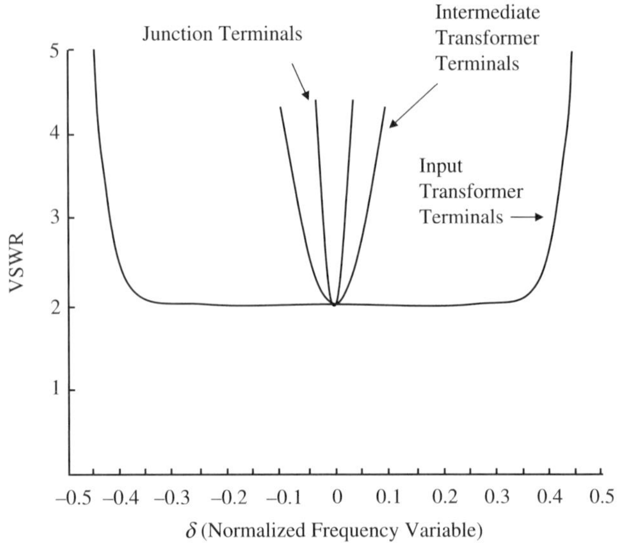

FIGURE 23.12 Frequency response for degree-3 junction circulator with element values of $r = 1.15$ and $2\delta_0 = 0.65$. (Reproduced with permission from J. Helszajn, Frequency response of quarter-wave coupled reciprocal stripline junctions, *IEEE Trans. Microwave Theory Tech.*, Vol. MTT-21, pp. 533–537, Aug. 1973.)

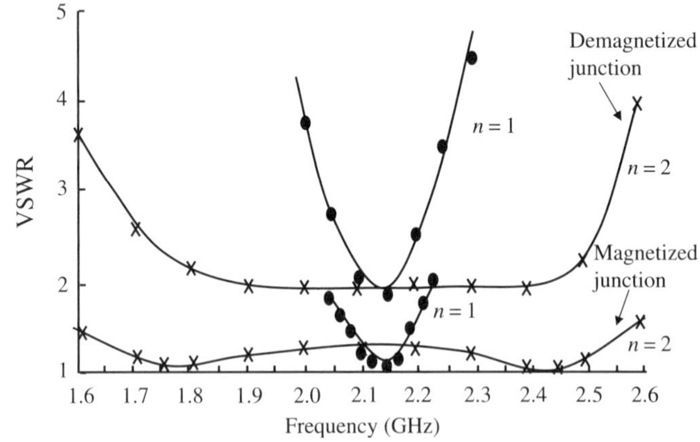

FIGURE 23.13 Experimental frequency response of demagnetized and magnetized degree-1 and degree-2 stripline circuits. (Reproduced with permission from J. Helszajn, Frequency response of quarter-wave coupled reciprocal stripline junctions, *IEEE Trans. Microwave Theory Tech.*, Vol. MTT-21, pp. 533–537, Aug. 1973.)

Its demagnetized and magnetized frequency responses are superimposed in Fig. 23.13. The susceptance slope parameters of the degree-1 networks are given by $b_1' = 4.65$ and 5.7, respectively. The frequency responses of the degree-2 networks are also shown in this illustration. The dielectric constant ε_r of the quarter-wave impedance transformer used in this arrangement is $\varepsilon_r = 3.2$. This value corresponds to $y_t = 1.79$. Such a degree-2 arrangement coincides with a circulator characteristic with $r = 1.25$ and $2\delta_0 = 0.40$. The bandwidth of the experimental magnetized junction is $2\delta_0 = 43\%$ at the same voltage standing wave ratio. The bandwidth between the minimum VSWR points of the magnetized junction is approximately 0.707 that of the full bandwidth in keeping with theory. It is of note that the frequency responses of reciprocal and nonreciprocal solutions in Fig. 23.13 are of the same order. The frequency behavior of the reciprocal degree-2 network is slightly asymmetrical due to the onset of low-field loss.

CHAPTER TWENTY-FOUR

Scattering Matrices of Junction Circulators with Chebyshev Characteristics

24.1 INTRODUCTION

The reflection coefficient at port 1 of a junction circulator with a degree-2 Chebyshev characteristic is often established in terms of its one-port complex gyrator circuit. However, it may also be obtained in terms of its eigenvalues and this is the approach employed in this chapter. It provides not only the reflection coefficient at a typical input port but also the transmission coefficients between the other ports. The method introduced here is quite general and applies to the m-port junction also. It starts by representing the matching network at each port by its *ABCD* matrix. The overall reflection eigenvalues at the input terminals of the junction are then obtained one at a time in terms of the *ABCD* parameters and the eigenvalues at the junction terminals as a preamble to constructing the scattering parameters of the junction. This chapter deals both with the case where the frequency variation of the in-phase eigen-network at the gyrator terminals is neglected compared with those of the counterrotating ones, and with that where it is included. The former approach is in excellent agreement with that obtained by assuming a one-port model for the circulator. The influence of the in-phase eigen-network on the overall frequency response is studied separately in the case of the stripline circulator.

The Stripline Circulator: Theory and Practice. By J. Helszajn

24.2 EIGENVALUES OF THE SCATTERING MATRIX

The three-port junction under consideration is depicted in Fig. 24.1, in terms of ideal two-port gyrators of characteristic admittance Y_0. The relationships between its scattering parameters and its eigenvalues are constructed in the usual way:

$$S_{11} = \frac{s_0 + s_+ + s_-}{3} \tag{24.1a}$$

$$S_{21} = \frac{s_0 + \alpha s_+ + \alpha^2 s_-}{3} \tag{24.1b}$$

$$S_{31} = \frac{s_0 + \alpha^2 s_+ + \alpha s_-}{3} \tag{24.1c}$$

where $\alpha = \exp(j2\pi/3)$.

The eigenvalues of the scattering matrix are the reflection coefficients associated with each possible way of exciting the junction. These eigenvalues have unit amplitude and only differ from each other by the nature of reflection angles ϕ_0, $\phi_\pm$. The nature of these quantities is specified in Chapter 4 and will not be repeated here. The admittance eigenvalues, in the reflection plane, are defined in terms of

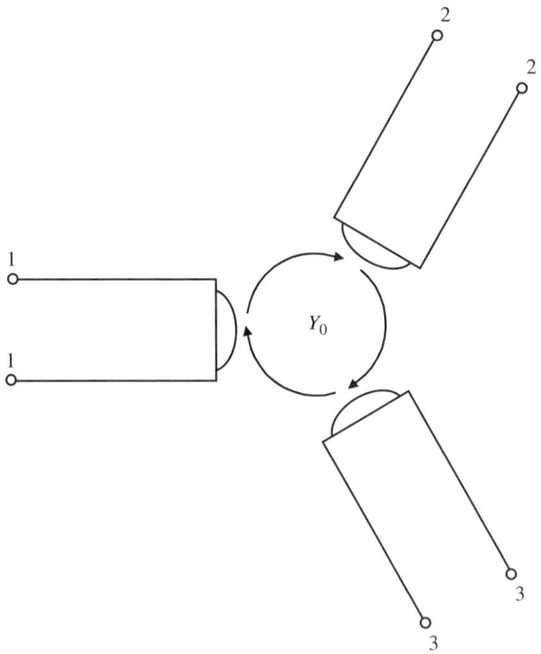

FIGURE 24.1 Schematic of ideal three-port junction circulator in terms of two-port gyrators.

reflection angles by

$$Y_0 = jY_0 \tan(\phi_0) \tag{24.2a}$$

$$Y_+ = jY_0 \tan(\phi_0 + \phi_+ + \pi/2) \tag{24.2b}$$

$$Y_- = jY_0 \tan(\phi_0 + \phi_- + \pi/2) \tag{24.2c}$$

It is also helpful, in what follows, to recall the relationships between the 1-port immittances of the 1-port eigen-networks in terms of the electrical lengths θ and admittances Y' instead of the reflection ones. This gives

$$Y_0 = jY'_0 \tan(\theta_0) \tag{24.3a}$$

$$Y_+ = -jY'_+ \cot(\theta_1 + \theta_+) \tag{24.3b}$$

$$Y_- = jY'_- \cot(\theta_1 + \theta_-) \tag{24.3c}$$

The eigen-networks associated with the admittance eigenvalues Y_+ and Y_- are short-circuited transmission lines of electrical length $\theta_1 + \theta_\pm$. These are shown in Fig. 24.2. For a three-port circulator for which $S_{21} = -1$, the equivalent circuit for the admittance eigenvalue Y_0 is a quarter-wave long open-circuited transmission line of length θ_0.

The required relationships between the reflection and transmission angles ϕ_i and θ_i are given without ado by introducing the following equalites:

$$Y_0 \tan(\phi_0) = Y'_0 \tan(\theta_0) \tag{24.4a}$$

$$Y_0 \tan(\phi_1 + \phi_+ + \pi/2) = -Y'_+ \cot(\theta_1 + \theta_+) \tag{24.4b}$$

$$Y_0 \tan(\phi_1 + \phi_- + \pi/2) = -Y'_- \cot(\theta_1 + \theta_-) \tag{24.4c}$$

A one-to-one correspondence between the admittance eigenvalues in the θ-plane defined here and those met in the Bosma description of a junction employing the classic disk resonator is obtained by approximating the circuit ones by

$$y_0 = jy_0 \tan(\theta_0) \tag{24.5a}$$

$$y_+ \approx -jy_1 \cot(\theta_1) + jy_+ \tan(\theta_+) \tag{24.5b}$$

$$y_- \approx -jy_1 \cot(\theta_1) + jy_- \tan(\theta_-) \approx -jy_1 \cot(\theta_1) - jy_+ \tan(\theta_+) \tag{24.5c}$$

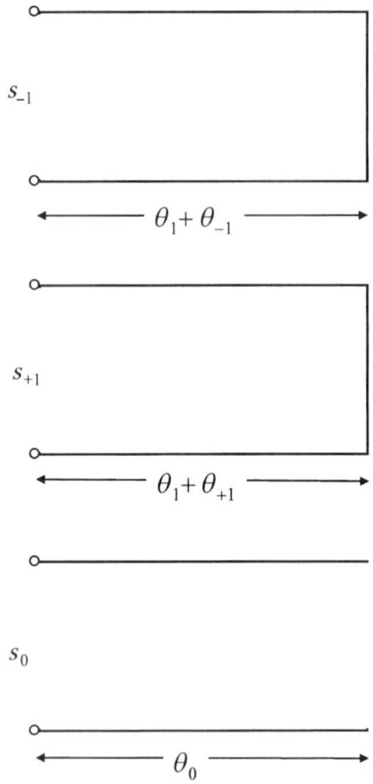

FIGURE 24.2 Eigen-networks of three-port junction circulator.

The equivalent circuit for the admittance y_+ is a transmission line of admittance y_+ and electrical length θ_+ in shunt with a short-circuited transmission line of admittance y_1 of electrical length θ_1. A similar statement applies to the admittance y_-.

The in-phase eigen-network is unperturbed by the gyrotropy. The approximate eigen-network obtained in this way is shown in Fig. 24.2.

24.3 EIGENVALUES OF AUGMENTED SCATTERING MATRIX

The equivalent circuit of the junction under consideration with matching networks at each port is illustrated in Fig. 24.3. Its eigen-networks are shown in Fig. 24.4. The entries of the scattering matrix at the new terminals are now given by

$$\Gamma_{11} = \frac{\gamma_0 + \gamma_+ + \gamma_-}{3} \tag{24.6a}$$

$$\Gamma_{21} = \frac{\gamma_0 + \alpha\gamma_+ + \alpha^2\gamma_-}{3} \tag{24.6b}$$

$$\Gamma_{31} = \frac{\gamma_0 + \alpha^2\gamma_+ + \alpha\gamma_-}{3} \tag{24.6c}$$

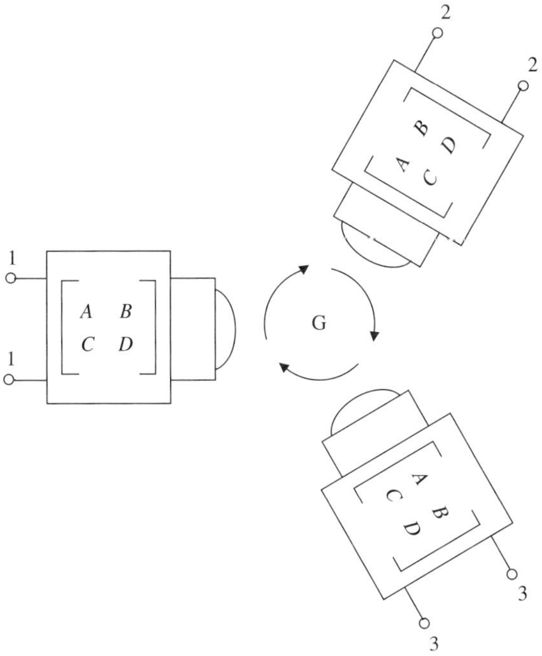

FIGURE 24.3 Schematic of three-port junction circulator with matching networks.

The reflection eigenvalues at the input terminals of the *ABCD* networks are given by straightforward calculation:

$$\gamma_0 = \exp(-j2\psi_0) \tag{24.7a}$$

$$\gamma_+ = \exp[-j2(\psi_+ + \pi/2)] \tag{24.7b}$$

$$\gamma_- = \exp[-j2(\psi_- + \pi/2)] \tag{24.7c}$$

where

$$j\tan(\psi_0) = \frac{jC + Dy_0}{A + jBy_0} \tag{24.8a}$$

$$j\tan(\psi_+ + \pi/2) = \frac{jC + Dy_+}{A + jBy_+} \tag{24.8b}$$

$$j\tan(\psi_- + \pi/2) = \frac{jC + Dy_-}{A + jBy_-} \tag{24.8c}$$

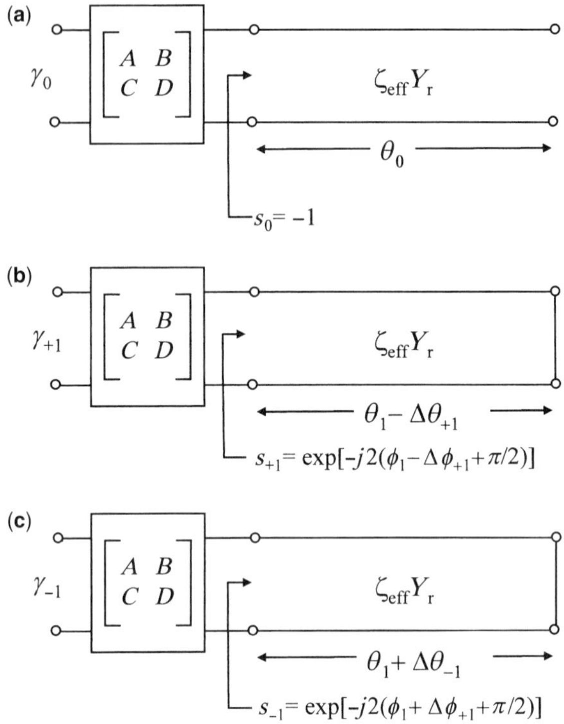

FIGURE 24.4 Eigen-networks of three-port circulator with matching networks.

It is separately observed that each reflection eigenvalue satisfies $\gamma\gamma^* = 1$. This means that the new eigenvalues reside on the unit circle also.

The *ABCD* parameters, for a single quarter-wave transformer, are

$$A = \cos\theta \tag{24.9a}$$

$$B = (\sin\theta)/y_{\mathrm{t}} \tag{24.9b}$$

$$C = y_{\mathrm{t}} \sin\theta \tag{24.9c}$$

$$D = \cos\theta \tag{24.9d}$$

The frequency variable θ is defined by

$$\theta = (\pi/2)(1 + \delta) \tag{24.10}$$

where

$$\delta = (\omega - \omega_0)/\omega_0 \tag{24.11}$$

At the bandedges the frequency variable becomes

$$\delta_0 = (\omega_1 - \omega_0)/\omega_0 \tag{24.12}$$

The Chebyshev polynomial for $n = 2$ passes through zero when

$$\sqrt{2}\cos\,\theta = \cos\,\theta_0 \tag{24.13}$$

The frequency response for a degree-2 circuit is defined separately in Chapter 25.

24.4 CIRCULATION ADJUSTMENT

The adjustment under consideration proceeds by having the frequency at which the reflection coefficient passes through its zeros and maxima coincide with those of a Chebyshev polynomial.

When the reflection coefficient passes through zero, the eigenvalues lie equally spaced on the unit circle.

$$\psi_+ + \pi/2 = \psi_0 + \pi/3 \tag{24.14a}$$

$$\psi_- + \pi/2 = \psi_0 - \pi/3 \tag{24.14b}$$

Substituting these two conditions into Eq. (24.8) gives

$$j\tan(\psi_0) = -j\frac{D}{B} \tag{24.15a}$$

$$j\tan(\psi_0 + \pi/3) = \frac{jC + Dy_+}{A + jBy_+} \tag{24.15b}$$

$$j\tan(\psi_0 - \pi/3) = \frac{jC + Dy_-}{A + jBy_-} \tag{24.15c}$$

The simplified form for the admittance eigenvalue, y_0, comes about because $y_0 = \infty$ for a circulator for which $s_0 = -1$.

The derivation of the required result continues by writing $t = \tan(\psi_0)$ and putting

$$y_\pm = j(\lambda \mp \mu) \tag{24.16}$$

This gives

$$t = -D/B \tag{24.17}$$

$$\frac{t-\sqrt{3}}{1+t\sqrt{3}} = \frac{C-\lambda D-\mu D}{A+\lambda B+\mu B} = \frac{t-(C-D\lambda)/\mu B}{1+t(A+\lambda B)/\mu D} \tag{24.18a}$$

and

$$\frac{t+\sqrt{3}}{1-t\sqrt{3}} = \frac{C-\lambda D-\mu D}{A+\lambda B+\mu B} = \frac{t+(C-D\lambda)/\mu B}{1-t(A+\lambda B)/\mu D} \tag{24.18b}$$

These equations are consistent provided

$$y_1 \cot(\theta) = \frac{CD-AB}{B^2+D^2} \tag{24.19a}$$

and

$$\sqrt{3}y_+ \tan(\theta_+) = \frac{1}{B^2+D^2} \tag{24.19b}$$

which must be evaluated at $\sqrt{2}\cos\theta = \cos\theta_0$ in keeping with Eq. (24.13).

To obtain the transformer admittance y_t it is necessary to directly evaluate Γ_{11} at the center frequency in terms of the original variables. For $g < y_t^2$ the result is

$$\sqrt{3}y_+ \tan(\theta_+) = y_t^2\sqrt{(2-r)/r} \tag{24.20}$$

Writing the preceding equation in terms of the original variables gives

$$g = \frac{\sqrt{r/(2-r)} - \sin^2(\theta_0)}{\sqrt{r/(2-r)}\cos^2(\theta_0)} \tag{24.21}$$

$$y_1 = \left(g\sqrt{(2-r)/r}\right)^{1/2}\left(g\sqrt{r/(2-r)}-1\right)\sin^2(\theta_0) \tag{24.22}$$

$$y_t^2 = g\sqrt{r/(2-r)} \tag{24.23}$$

The present results are identical with those based on the one-port gyrator circuit model provided

$$r \approx \sqrt{r/(2-r)} \tag{24.24}$$

This condition is readily satisfied for the sort of values of r usually encountered in circulator design.

24.5 QUARTER-WAVE COUPLED-BELOW-RESONANCE STRIPLINE CIRCULATOR

The theory developed so far will now be combined with the electromagnetic problem in the case of the stripline circulator. The admittance eigenvalues of this problem region are

$$y_+ = \frac{-j\pi\zeta_e Y_r}{3Y_0 \sin(\psi)}\left(\frac{J_1'(kR)}{J_1(kR)} - \frac{\kappa\mu}{kR}\right) \tag{24.25a}$$

$$y_- = \frac{-j\pi\zeta_e Y_r}{3Y_0 \sin(\psi)}\left(\frac{J_1'(kR)}{J_1(kR)} + \frac{\kappa/\mu}{kR}\right) \tag{24.25b}$$

$$y_0 = \frac{j\pi\zeta_e Y_r}{3Y_0\psi}\left(\frac{J_0'(kR)}{J_0(kR)}\right) \tag{24.25c}$$

Figure 24.5 depicts the frequency behavior of the preceding eigenadmittances as a function of kR for a circulator which when magnetized gives $r = 1.07$. The reflection eigenvalues of the corresponding phase angles are indicated in Fig. 24.6. These results show that the equivalent circuits for y_{+1} and y_{-1} are short-circuited radial transmission lines, while that of the eigen-network y_0 is an open-circuited transmission line as asserted. A one-to-one correspondence between the electromagnetic and network problems may be undertaken provided the frequency variation of the in-phase eigen-network may be neglected compared with that of the split counter-rotating ones. This gives

$$g = \sqrt{3}y_+ \tan(\theta_+) = \frac{\pi\zeta_e Y_r}{\sqrt{3}Y_0 \sin(\psi) kR}\left(\frac{\kappa}{\mu}\right) \tag{24.26a}$$

$$b' = \frac{\pi y_1}{4} = \frac{\pi\zeta_e Y_r}{3Y_0 \sin(\psi)}\left(\frac{(kR)^2 - 1}{2kR}\right) \tag{24.26b}$$

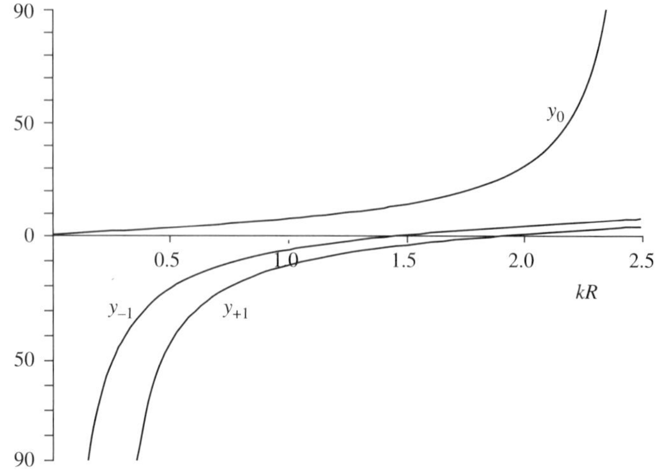

FIGURE 24.5 Eigenadmittances of stripline circulator with $r = 1.07$ and $2\delta_0 = 20\%$.

In obtaining this result it has been assumed that the susceptance slope parameters of the two equivalent reciprocal circuits are the same. Here g is the normalized gyrator conductance of the junction, and b' is the normalized susceptance slope parameter of the reciprocal counterrotating eigen-networks. The electromagnetic problem that

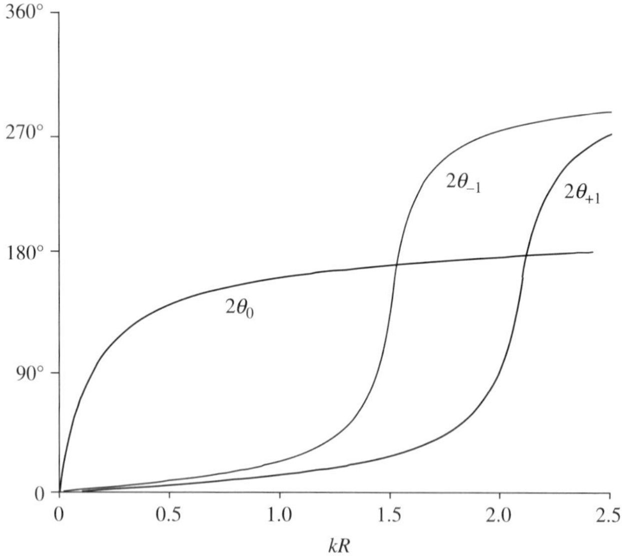

FIGURE 24.6 Electrical lengths of eigen-networks for stripline circulator with $r = 1.07$ and $2\delta_0 = 20\%$.

must be satisfied when the frequency variation of the s_0 eigenvalue is neglected compared to that of $s_\pm$ is therefore, in the case of a quarter-wave coupled circulator,

$$\frac{\pi \zeta_e Y_r}{3Y_0 \sin(\psi)}\left(\frac{(kR)^2-1}{2kR}\right)=\frac{\pi}{4}\left(\frac{r-\sin^2(\theta_0)}{r^2\cos^2(\theta_0)}\right)(r-1)\tan^2(\theta_0) \tag{24.27a}$$

$$\frac{\pi \zeta_e Y_r}{\sqrt{3}Y_0 \sin(\psi)kR}\left(\frac{\kappa}{\mu}\right)=\frac{r-\sin^2(\theta_0)}{r\cos^2(\theta_0)} \tag{24.27b}$$

The first equation determines the coupling angle and the second the gyrotropy. The wave admittance in the preceding equations is defined by

$$\zeta_e = \sqrt{\varepsilon_r/\mu_e} \tag{24.28}$$

and

$$Y_r = \sqrt{\varepsilon_d}Y_0 \tag{24.29}$$

where

$$Y_0 = 30\pi \ln\left(\frac{W+2H+t}{W+t}\right)$$

and

$$\sin(\psi) = W/2R$$

W is the width of the stripline, H is half the ground-plane spacing, t is the thickness of the center conductor, and the other quantities have the usual meanings.

24.6 FREQUENCY VARIATION OF QUARTER-WAVE COUPLED CIRCULATOR

The assumption used throughout this development is that the frequency variation of the in-phase eigenvalue may be omitted compared to that of the counterrotating eigenvalues. This assumption will now be tested in the case of stripline circulators with $2\delta = 20\%$ and $r = 1.10$ and 1.22. The physical variables $\sin(\psi)$ and κ/μ in Eqs. (24.27a) and (24.27b) are given once the bandwidth and VSWR are specified.

Figures 24.7 and 24.8 indicate the frequency behavior of the overall quarter-wave coupled circulator for the ideal and actual cases. These results have been obtained by

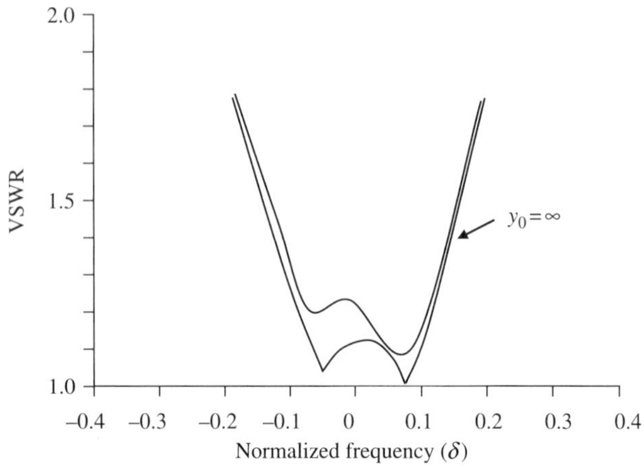

FIGURE 24.7 Frequency response of quarter-wave coupled circulator with and without third eigen-network for $r = 1.07$ and $2\delta_0 = 20\%$. [Reproduced with permission from J. Helszajn, The synthesis of quarter-wave coupled circulators with Chebyshev characteristics, *IEEE Trans. Microwave Theory Tech.* (Short Papers), Vol. MTT-20, pp. 764–769, Nov. 1972.]

assuming that the center frequency lies midway between the two split frequencies rather than at $kR = 1.84$.

The above results suggest that although the frequency variation of the in-phase reflection coefficient is small compared to that of the counterrotating ones, it cannot be ignored for devices with relatively wide bands and low VSWRs.

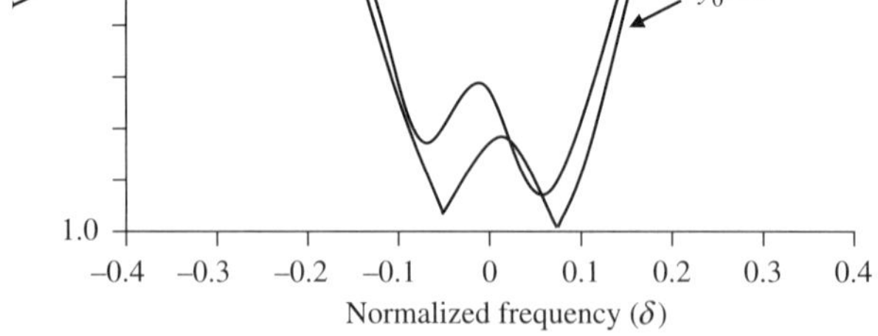

FIGURE 24.8 Frequency response of quarter-wave coupled circulator with and without third eigen-network for $r = 1.22$ and $2\delta_0 = 20\%$. [Reproduced with permission from J. Helszajn, The synthesis of quarter-wave coupled circulators with Chebyshev characteristics, *IEEE Trans. Microwave Theory Tech.* (Short Papers), Vol. MTT-20, pp. 764–769, Nov. 1972.]

24.7 FREQUENCY RESPONSE OF QUARTER-WAVE COUPLED CIRCULATOR WITH CAPACITIVE TUNING

One way in which the theoretical results can be obtained in practice is to introduce a triplet of circumferential magnetic walls or a thin metal post in the center of the junction. Equivalently, thin metal posts may be introduced at the input terminals of the transformer, which is the normal approach in stripline devices. This last statement comes about because the in-phase eigen-network is an open-circuited half-wave resonator, which can be tuned at either input or output terminals.

The *ABCD* matrix for a single shunt capacitor is

$$A = 1 \tag{24.30a}$$

$$B = 0 \tag{24.30b}$$

$$C = \omega C / Y_0 \tag{24.30c}$$

$$D = 1 \tag{24.30d}$$

Cascading this *ABCD* network with that for the single quarter-wave transformer gives the following *ABCD* parameters for the overall matching network:

$$A = \cos(\theta) \tag{24.31a}$$

$$B = \frac{\sin(\theta)}{y_t} \tag{24.31b}$$

$$C = y_t \sin(\theta) + \left(\frac{\omega C}{Y_0}\right) \frac{\sin(\theta)}{y_t} \tag{24.31c}$$

$$D = \cos(\theta) - \left(\frac{\omega C}{Y_0}\right) \frac{\sin(\theta)}{y_t} \tag{24.31d}$$

The arrangement in question is shown in Fig. 24.9.

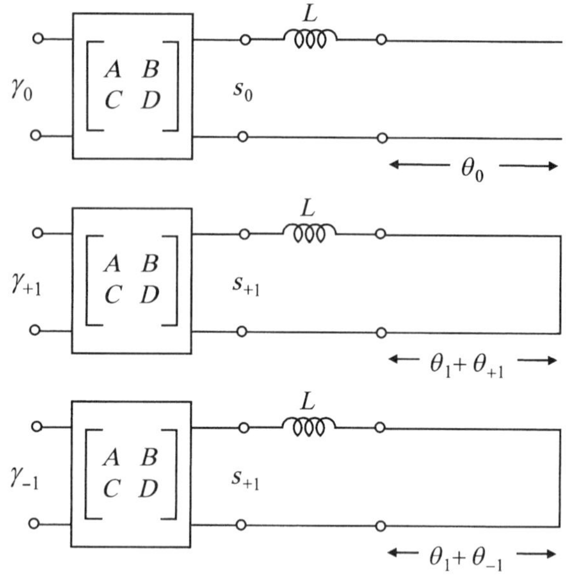

FIGURE 24.9 Eigen-networks of three-port circulator with capacitive tuning. [Reproduced with permission from J. Helszajn, The synthesis of quarter-wave coupled circulators with Chebyshev characteristics, *IEEE Trans. Microwave Theory Tech.* (Short Papers), Vol. MTT-20, pp. 764–769, Nov. 1972.]

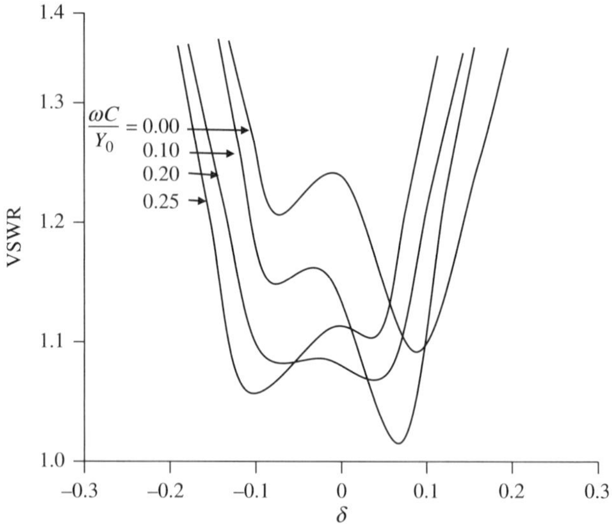

FIGURE 24.10 Frequency response of quarter-wave coupled circulator with capacitive tuning for $r = 1.1$ and $2\delta_0 = 20\%$. [Reproduced with permission from J. Helszajn, The synthesis of quarter-wave coupled circulators with Chebyshev characteristics, *IEEE Trans. Microwave Theory Tech.* (Short Papers), Vol. MTT-20, pp. 764–769, Nov. 1972.]

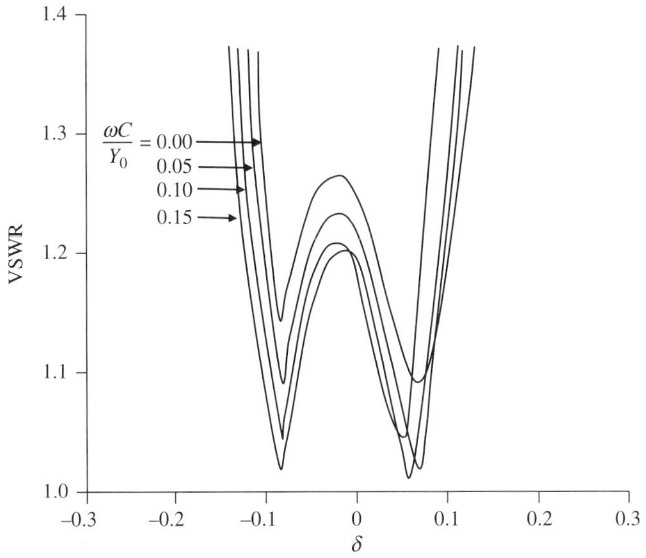

FIGURE 24.11 Frequency response of quarter-wave coupled circulator with capacitive tuning for $r = 1.22$ and $2\delta_0 = 20\%$. [Reproduced with permission from J. Helszajn, The synthesis of quarter-wave coupled circulators with Chebyshev characteristics, *IEEE Trans. Microwave Theory Tech.* (Short Papers), Vol. MTT-20, pp. 764–769, Nov. 1972.]

Figures 24.10 and 24.11 indicate the influence of the capacitor C on the overall frequency response of the device for $r = 1.10$ and 1.22, and $2\delta = 20\%$. These suggest that such a capacitor can indeed be used to improve the correlation between the two and three eigen-network models of the stripline circulator.

CHAPTER TWENTY-FIVE

Synthesis of Stepped Impedance Transducers

25.1 INTRODUCTION

A property of any three-port junction circulator is that the isolation between one input port and one output port is infinite and the insertion loss between the same input port and the other output port is zero when the return loss at each port is infinite. The classic equivalent circuit of this type of junction is a complex shunt STUB-G load consisting of a short-circuited quarter-wave long UE in shunt with a real conductance. One possible solution met in connection with this topology is the quarter-wave impedance transformer. A second solution based on short-line transformers is obtained by replacing the 90° stub by short, typically 30°, open- and short-circuited ones. One property of either network is that its gain-bandwidth product is uniquely fixed by the quality factor of the load circuit. The purpose of this chapter is to develop the exact t-plane synthesis of each topology for a degree-2 circuit. This is done both in closed and in tabular forms. The classic solution to any of these types of problems, as in the related filter ones, is based on an insertion loss specification in the θ-plane, the use of the unitary condition to deduce a squared amplitude reflection coefficient from a knowledge of that of the transmission one, the mapping of the θ-plane into the t-plane, the construction of a bounded real reflection coefficient, the use of the bilinear transformation between immittance and reflection coefficient, and the synthesis of a one-port immittance function in terms of UEs and t-plane inductors and capacitors that have the topology of the required two-port network.

A semitutorial approach to the actual t-plane synthesis problem has been adopted in this chapter to familiarize the nonspecialist reader with this type of method.

25.2 θ-PLANE INSERTION LOSS FUNCTION FOR QUARTER-WAVE LONG STEPPED IMPEDANCE TRANSDUCERS

The most important circuit topology met in the design of commercial junction circulators is that consisting of one or more transformers (UEs) and a complex G-STUB load in terms of a microwave specification. Since the solution of this problem is crucial in the design of practical circulators, it will be developed from first principles in closed form.

The first step in the synthesis of any commensurate line network is to establish an appropriate insertion loss specification. One topology that displays a suitable equi-ripple bandpass response centered about 90°, which is applicable to a complex G-STUB load and n UEs, is indicated in Fig. 25.1. The required amplitude squared frequency response is illustrated in Fig. 25.2. It is defined by

$$L(\theta) = 1 + K^2 + \varepsilon^2\left(\frac{(1+\sin\theta_c)T_{n+1}(\cos\theta/\cos\theta_c) - (1-\sin\theta_c)T_{n-1}(\cos\theta/\cos\theta_c)}{2\sin\theta}\right)^2 \tag{25.1}$$

K and ε are related by the minimum and maximum values of the voltage standing wave ratio within the passband

$$S(\max) = \left(\sqrt{1+K^2+\varepsilon^2} + \sqrt{1+K^2}\right)^2 \tag{25.2}$$

$$S(\min) = \left(\sqrt{1+K^2} + K\right)^2 \tag{25.3}$$

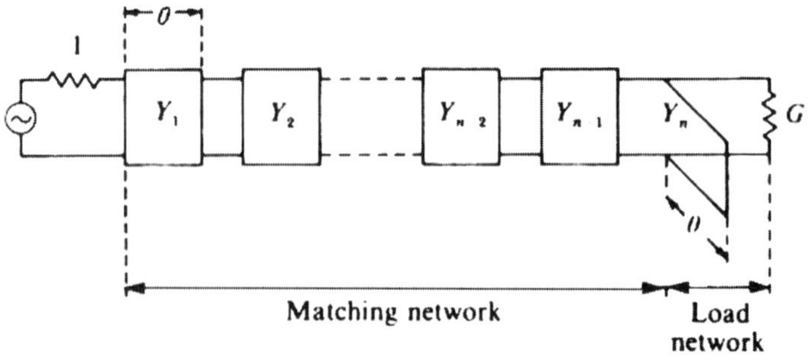

FIGURE 25.1 Schematic diagram of n UEs terminated in G-STUB load.

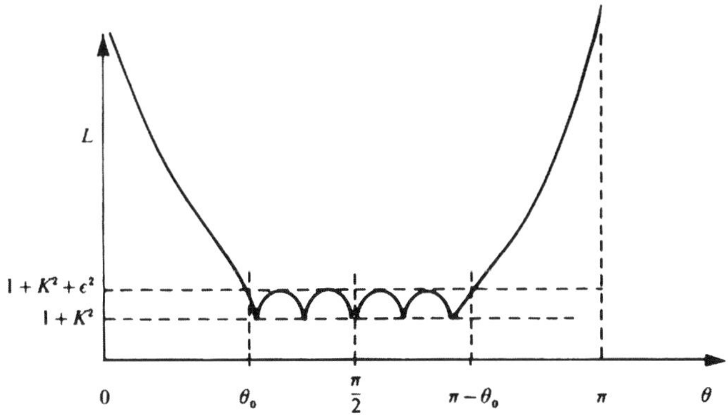

FIGURE 25.2 Frequency response of n UEs loaded by shunt G-STUB load.

or

$$K = \frac{S(\min) - 1}{2\sqrt{S(\min)}} \tag{25.4}$$

$$\sqrt{K^2 + \varepsilon^2} = \frac{S(\max) - 1}{2\sqrt{S(\max)}} \tag{25.5}$$

θ and θ_c are the usual radian frequency and the lower bandedge one:

$$\theta = (\pi/2)(\omega/\omega_0) \tag{25.6}$$

$$\theta_c = (\pi/2)(\omega_c/\omega_0) \tag{25.7}$$

respectively. θ is also sometimes written in terms of a normalized bandwidth parameter as

$$\theta = \pi/2(1 + \delta) \tag{25.8}$$

where

$$\delta = \frac{\omega - \omega_0}{\omega_0} \tag{25.9}$$

ω_0 is the center frequency of the transformer, and ω is the normal frequency variable.

θ_c is related to the normalized bandwidth W by

$$W = 2 - 4\theta_c/\pi \tag{25.10}$$

$T_n(x)$ is the Chebyshev polynomial of order n and argument x.

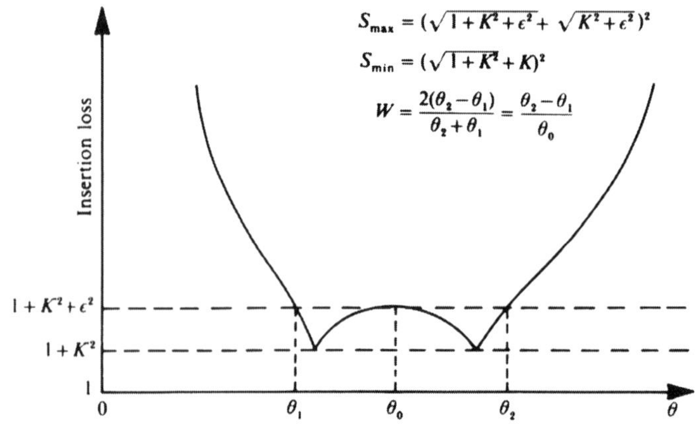

FIGURE 25.3 Frequency response of degree-2 network terminated in G-STUB load.

The amplitude square of the reflection coefficient is related to the insertion loss function by the unitary condition in the usual way:

$$|\Gamma(\theta)|^2 = \frac{L(\theta) - 1}{L(\theta)} \tag{25.11}$$

Once the reflection coefficient at either the input or output port in the t-plane is specified, $Z(t)$ at the corresponding port is obtained by making use of the classic bilinear transformation between the two.

$$\frac{Y(\theta)}{G} = \frac{1 - \Gamma(\theta)}{1 + \Gamma(\theta)} \tag{25.12}$$

For a degree-2 network the insertion loss function becomes

$$L(\theta) = 1 + K^2 + \varepsilon^2 \left(\frac{(1 + \sin \theta_c)[T_2(\cos \theta / \cos \theta_c) - (1 - \sin \theta_c)]}{2 \sin \theta} \right)^2 \tag{25.13}$$

The frequency response of this insertion loss function is indicated in Fig. 25.3.

25.3 *t*-PLANE SYNTHESIS OF QUARTER-WAVE LONG STEPPED IMPEDANCE TRANSDUCERS

The derivation of a realizable immittance proceeds by forming the amplitude square of the reflection coefficient in the t-plane by introducing the t-plane variable throughout:

$$t = j \tan \theta \tag{25.14}$$

If the network is synthesized from the load terminals, as is done here, then it is necessary to construct $S_{22}(t)S_{22}(-t)$ instead of $S_{11}(t)S_{11}(-t)$. It continues by forming $S_{22}(t)$ from a knowledge of $S_{22}(t)S_{22}(-t)$. The numerator polynomial of $S_{22}(t)$ need not be Hurwitz, which means that its zeros may be selected from either its left or right half-plane roots. The denominator polynomial is Hurwitz so that its poles are restricted to its L.H.P. ones. The normalized admittance at the output terminals is then given from a knowledge of $S_{22}(t)$. The result in terms of the original variables is

$$\frac{Y(t)}{G} = \frac{d_2 t^2 + dt + d_0}{n_2 t^2 + n_1 t} \tag{25.15}$$

The coefficients in the numerator and denominator polynomials are given by

$$n_2 = \sqrt{a+1} - \sqrt{a} \tag{25.16a}$$

$$d_2 = \sqrt{a-1} + \sqrt{a} \tag{25.16b}$$

$$n_1 = [2\sqrt{(a-1)}\,c - b + 1]^{1/2} - (2\sqrt{ac} - b)^{1/2} \tag{25.16c}$$

$$d_1 = [2\sqrt{(a-1)}\,c - b + 1]^{1/2} + (2\sqrt{ac} - b)^{1/2} \tag{25.16d}$$

$$n_0 = 0 \tag{25.16e}$$

$$d_0 = 2\sqrt{c} \tag{25.16f}$$

where

$$a = K^2 + \varepsilon^2 \tag{25.17a}$$

$$b = 2\beta\varepsilon^2 - K^2 \tag{25.17b}$$

$$c = \beta^2\varepsilon^2 \tag{25.17c}$$

and

$$\beta = \tan^2\theta_c + (\tan\theta_c)/(\cos\theta_c) \tag{25.18}$$

Making use of the relationships between the numerator and denominator coefficients $n_{1,2}$ and $d_{0,1,2}$ also indicates that

$$d_1 n_1 - d_0 n_2 = 1 \tag{25.19a}$$

$$n_2 d_2 = 1 \tag{25.19b}$$

The synthesis of this immittance commences by extracting a shunt t-plane inductor in a second Cauer form. This step produces a t-plane inductor,

$$L = Gd_0/n_1$$

and a remainder admittance,

$$Y'(t) = G\frac{n_1 d_2 t + n_1 d_1 - n_2 d_0}{n_1(n_2 t + n_1)}$$

The synthesis of the required circuit continues by extracting a UE by replacing t by unity in $Y'(t)$. The result is

$$Y'(1) = G\frac{n_1 d_2 + n_1 d_1 - n_2 d_0}{n_1(n_2 t + n_1)}$$

Making use of the relationships between the various coefficients indicates that $Y'(1)$ may also be written as

$$Y'(1) = G/n_1 n_2$$

The remainder admittance $Y''(t)$ is then given by Richards Theorem by

$$Y''(t) = Y'(1)\frac{Y'(t) - tY'(1)}{Y'(1) - tY'(t)}$$

The remainder admittance after canceling a common term $(t^2 - 1)$ is

$$Y''(t) = G/n_1^2$$

This admittance fixes the load at the generator terminals without ado:

$$\frac{G}{n_1^2} = 1$$

The t-plane inductor, the UE, and the load conductance are now specified in terms of the original variables for the purpose of calculations by

$$G = n_1^2 \tag{25.20}$$

$$Y_1' = n_1/n_2 \tag{25.21}$$

$$L = n_1 d_0 \tag{25.22}$$

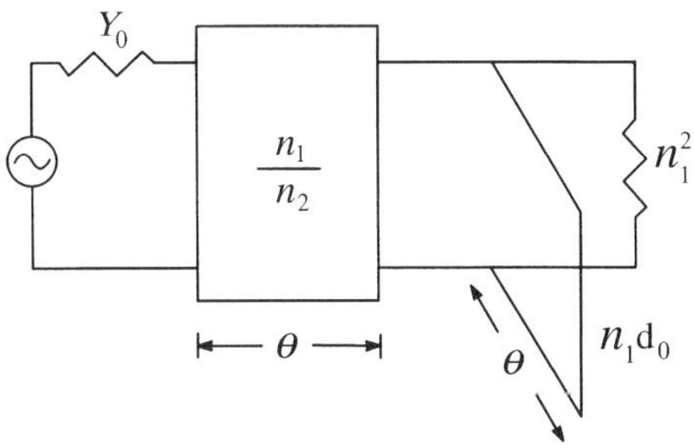

FIGURE 25.4 Topology of degree-2 G-STUB network

The topology produced by this procedure is now established by realizing the t-plane inductor by a short-circuited transmission line with a characteristic impedance equal to that of the t-plane inductor in shunt with the load conductance G, by realizing the UE by a series transmission line with a characteristic admittance equal to that of the UE, and by fixing the G by the condition at the generator terminals. The required topology is depicted in Fig. 25.4.

25.4 NETWORK PARAMETERS OF QUARTER-WAVE LONG IMPEDANCE TRANSDUCERS

It is usual, in tabulating solutions to this type of problem, to replace the stub admittance L by its equivalent susceptance slope parameter B':

$$B' = \pi L/4 \tag{25.23}$$

This notation has the merit that it permits the Q-factor of the load to also be defined without difficulty:

$$Q_{\text{L}} = B'/G \tag{25.24}$$

A typical family of solutions for this sort of degree-2 network is tabulated in Table 25.1. One feature of these tables is that the quality factor is uniquely fixed once the bandwidth and the maximum value of the VSWR in the passband are specified. Another important aspect of this solution is that the immittance level of the complex gyrator circuit can be adjusted by varying the minimum value of the VSWR in the passband. Since the latter parameter is of little interest in the design of any circulator, it may be used to absorb any uncertainty in the absolute immittance level of the gyrator circuit.

The synthesis of a degree-3 topology proceeds in a like manner. Table 25.2 summarizes some results.

TABLE 25.1 Degree-2 Network Variables of 90° UE

	Degree $N = 2$ $S(\max) = 1.2$ $S(\min) = 1$					Degree $N = 2$ $S(\max) = 1.2$ $S(\min) = 1.06$			
W	G	B'	Q	Y_f	W	G	B'	Q	Y_f
0.100	54.899	343.593	6.259	8.117	0.100	45.124	295.209	6.542	7.359
0.150	24.878	102.468	4.119	5.464	0.150	20.502	88.154	4.300	4.960
0.200	14.371	43.608	3.035	4.153	0.200	11.885	37.583	3.162	3.776
0.250	9.507	22.568	2.374	3.378	0.250	7.896	19.492	2.468	3.078
0.300	6.865	13.222	1.926	2.870	0.300	5.730	11.448	1.998	2.622
0.350	5.272	8.440	1.601	2.515	0.350	4.425	7.328	1.656	2.304
0.400	4.239	5.737	1.353	2.255	0.400	3.577	4.994	1.396	2.072
0.450	3.530	4.089	1.158	2.058	0.450	2.997	3.571	1.192	1.896
0.500	3.023	3.026	1.001	1.905	0.500	2.582	2.650	1.026	1.760
0.550	2.648	2.307	0.871	1.782	0.550	2.275	2.026	0.891	1.652
0.600	2.362	1.802	0.763	1.684	0.600	2.042	1.587	0.777	1.563
0.667	2.077	1.336	0.643	1.579	0.667	1.809	1.182	0.653	1.473
	Degree $N = 2$ $S(\max) = 1.2$ $S(\min) = 1.02$					Degree $N = 2$ $S(\max) = 1.2$ $S(\min) = 1.08$			
0.100	52.013	332.465	6.392	7.900	0.100	41.195	269.883	6.551	7.031
0.150	23.584	99.178	4.205	5.320	0.150	18.747	80.656	4.302	4.743
0.200	13.634	42.225	3.097	4.045	0.200	10.890	34.423	3.161	3.615
0.250	9.028	21.863	2.422	3.292	0.250	7.255	17.876	2.464	2.951
0.300	6.527	12.816	1.964	2.799	0.300	5.280	10.515	1.991	2.517
0.350	5.019	8.186	1.631	2.454	0.350	4.090	6.741	1.648	2.215
0.400	4.040	5.567	1.378	2.202	0.400	3.318	4.602	1.387	1.995
0.450	3.369	3.971	1.179	2.011	0.450	2.789	3.296	1.182	1.829

0.500	2.889	2.941	1.0.18	1.862	0.500	2.411	2.450	1.016	1.701
0.550	2.534	2.243	0.885	1.744	0.550	2.132	1.877	0.880	1.599
0.600	2.264	1.754	0.774	1.648	0.600	1.920	1.473	0.767	1.518
0.667	1.995	1.301	0.652	1.547	0.667	1.708	1.098	0.643	1.432
	Degree $N = 2$	S(max) = 1.2	S(min) = 1.04			Degree $N = 2$	S(max) = 1.2	S(min) = 1.1	
0.100	48.738	316.180	6.487	7.648	0.100	36.949	240.386	6.506	6.659
0.150	22.118	94.360	4.266	5.152	0.150	16.851	71.919	4.268	4.497
0.200	12.801	40.197	3.140	3.919	0.200	9.818	30.739	3.131	3.432
0.250	8.489	20.827	2.453	3.192	0.250	6.563	15.991	2.437	2.806
0.300	6.147	12.219	1.988	2.716	0.300	4.795	9.424	1.965	2.399
0.350	4.735	7.812	1.650	2.384	0.350	3.730	6.055	1.623	2.116
0.400	3.819	5.318	1.393	2.141	0.400	3.039	4.143	1.363	1.910
0.450	3.191	3.797	1.190	1.937	0.450	2.566	2.974	1.159	1.755
0.500	2.742	2.814	1.027	1.814	0.500	2.228	2.216	0.994	1.635
0.550	2.409	2.149	0.892	1.700	0.550	1.979	1.701	0.859	1.541
0.600	2.157	1.681	0.779	1.609	0.600	1.790	1.337	0.747	1.465
0.667	1.905	1.249	0.656	1.512	0.667	1.601	1.000	0.625	1.386

Source: R. Levy and J. Helszajn, Specific equations for one and two section quarter wave matching network for stub-resistor loads, *IEEE Trans. Microwave Theory Tech.*, Vol. MTT-30, pp. 55–62, Jan. 1982. Reproduced with permission.

TABLE 25.2 Degree-3 Network Variables of 90° UE

W	G	B'	G	Y_1	Y_2
Degree $N = 3$ $S(\max) = 1.15$ $S(\min) = 1$					
0.200	969.349	3486.515	3.597	4.732	147.312
0.250	399.553	1138.049	2.843	3.818	76.309
0.300	194.299	455.324	2.343	3.214	44.805
0.350	106.011	209.635	1.977	2.789	28.712
0.400	62.989	106.970	1.698	2.474	19.635
0.450	39.988	59.049	1.477	2.233	14.123
0.500	26.777	34.687	1.295	2.044	10.579
0.550	18.741	21.430	1.143	1.893	8.196
0.600	13.620	13.805	1.014	1.770	6.533
0.667	9.356	8.106	0.866	1.639	5.013
0.700	7.902	6.336	0.802	1.584	4.454
0.750	6.265	4.472	0.714	1.514	3.789
0.800	5.083	3.229	0.635	1.454	3.277
0.850	4.211	2.378	0.565	1.403	2.878
0.900	3.555	1.782	0.501	1.359	2.562
0.950	3.054	1.356	0.444	1.321	2.309
1.000	2.665	1.045	0.392	1.289	2.104
Degree $N = 3$ $S(\max) = 1.15$ $S(\min) = 1.02$					
0.200	893.332	3313.304	3.709	4.559	137.630
0.250	368.379	1081.745	2.936	3.681	71.355
0.300	179.240	432.918	2.415	3.102	41.940
0.350	97.862	199.388	2.037	2.693	26.909
0.400	58.196	101.785	1.749	2.392	18.428
0.450	36.983	56.215	1.520	2.161	13.275
0.500	24.794	33.042	1.333	1.981	9.961
0.550	17.378	20.428	1.176	1.836	7.731
0.600	12.649	13.170	1.041	1.719	6.175
0.667	8.709	7.742	0.889	1.594	4.752
0.700	7.366	6.055	0.822	1.542	4.228
0.750	5.852	4.279	0.731	1.475	3.604
0.800	4.758	3.093	0.650	1.418	3.125
0.850	3.951	2.280	0.577	1.370	2.751
0.900	3.344	1.711	0.512	1.329	2.455
0.950	2.880	1.303	0.453	1.294	2.217
1.000	2.520	1.006	0.399	1.263	2.025
Degree $N = 3$ $S(\max) = 1.15$ $S(\min) = 1.04$					
0.200	801.796	3043.685	3.796	4.363	125.987
0.250	330.850	994.045	3.005	3.525	65.394
0.300	161.116	397.989	2.470	2.974	38.491
0.350	88.060	183.398	2.083	2.585	24.738
0.400	52.436	93.683	1.787	2.298	16.973

(Continued)

TABLE 25.2 *Continued*

Degree $N = 3$ $S(\max) = 1.15$ $S(\min) = 1.04$					
W	G	B'	G	Y_1	Y_2
0.450	33.374	51.781	1.552	2.080	12.253
0.500	22.416	30.463	1.359	1.909	9.216
0.550	15.744	18.854	1.198	1.772	7.171
0.600	11.488	12.170	1.059	1.662	5.743
0.667	7.939	7.168	0.903	1.544	4.437
0.700	6.728	5.612	0.834	1.495	3.956
0.750	5.362	3.972	0.741	1.432	3.383
0.800	4.375	2.876	0.657	1.379	2.942
0.850	3.646	2.124	0.583	1.334	2.598
0.900	3.098	1.597	0.515	1.296	2.326
0.950	2.678	1.219	0.455	1.263	2.108
1.000	2.353	0.943	0.401	1.235	1.932
Degree $N = 3$ $S(\max) = 1.15$ $S(\min) = 1.06$					
0.200	697.205	2688.010	3.855	4.139	112.522
0.250	287.967	878.302	3.050	3.348	58.498
0.300	140.406	351.866	2.506	2.828	34.498
0.350	76.861	162.270	2.111	2.462	22.222
0.400	45.854	82.970	1.809	2.193	15.286
0.450	29.252	45.912	1.570	1.988	11.067
0.500	19.700	27.047	1.373	1.827	8.351
0.550	13.880	16.766	1.208	1.700	6.521
0.600	10.163	10.841	1.067	1.597	5.242
0.667	7.062	6.402	0.907	1.488	4.071
0.700	6.002	5.020	0.836	1.443	3.640
0.750	4.806	3.561	0.741	1.385	3.126
0.800	3.941	2.585	0.656	1.336	2.731
0.850	3.302	1.914	0.580	1.295	2.423
0.900	2.820	1.443	0.512	1.260	2.178
0.950	2.451	1.104	0.450	1.230	1.983
1.000	2.165	0.856	0.396	1.205	1.825
Degree $N = 3$ $S(\max) = 1.15$ $S(\min) = 1.08$					
0.200	580.786	2253.269	3.880	3.881	97.189
0.250	240.220	736.770	3.067	3.144	50.639
0.300	117.341	295.436	2.518	2.660	29.945
0.350	64.382	136.404	2.119	2.321	19.351
0.400	38.519	69.843	1.813	2.071	13.360
0.450	24.657	38.714	1.570	1.882	9.713
0.500	16.672	22.852	1.371	1.735	7.362
0.550	11.801	14.198	1.203	1.618	5.777
0.600	8.687	9.205	1.060	1.524	4.668
0.667	6.083	5.458	0.897	1.425	3.653
0.700	5.193	4.288	0.826	1.384	3.279
0.750	4.187	3.052	0.729	1.332	2.832

(*Continued*)

TABLE 25.2 *Continued*

Degree $N = 3$ $S(\max) = 1.15$ $S(\min) = 1.08$					
W	G	B'	G	Y_1	Y_2
0.800	3.458	2.224	0.643	1.288	2.489
0.850	2.918	1.653	0.566	1.251	2.222
0.900	2.511	1.250	0.498	1.220	2.010
0.950	2.200	0.961	0.437	1.194	1.840
1.000	1.957	0.748	0.382	1.171	1.703
Degree $N = 3$ $S(\max) = 1.15$ $S(\min) = 1.1$					
0.200	452.332	1742.610	3.853	3.572	79.668
0.250	187.508	570.432	3.042	2.900	41.652
0.300	91.859	229.069	2.494	2.460	24.733
0.350	50.587	105.957	2.095	2.153	16.060
0.400	30.402	54.376	1.789	1.928	11.150
0.450	19.567	30.223	1.545	1.758	8.157
0.500	13.315	17.897	1.344	1.627	6.225
0.550	9.494	11.160	1.175	1.523	4.4921
0.600	7.047	7.266	1.031	1.440	4.008
0.667	4.996	4.334	0.868	1.353	3.171
0.700	4.293	3.417	0.796	1.317	2.862
0.750	3.497	2.445	0.699	1.272	2.494
0.800	2.920	1.791	0.613	1.234	2.212
0.850	2.492	1.338	0.537	1.203	1.991
0.900	2.168	1.018	0.470	1.176	1.816
0.950	1.920	0.787	0.410	1.154	1.677
1.000	1.727	0.616	0.356	1.135	1.564
Degree $N = 3$ $S(\max) = 1.15$ $S(\min) = 1.12$					
0.200	309.011	1152.644	3.730	3.173	59.027
0.250	128.630	378.096	2.939	2.587	31.049
0.300	63.360	152.247	2.403	2.205	18.573
0.350	35.134	70.666	2.011	1.939	12.165
0.400	21.296	36.419	1.710	1.746	8.529
0.450	13.846	20.345	1.469	1.602	6.308
0.500	9.535	12.120	1.271	1.491	4.872
0.550	6.391	7.609	1.104	1.404	3.901
0.600	5.192	4.991	0.961	1.335	3.220
0.667	3.763	3.010	0.800	1.264	2.596
0.700	3.272	2.387	0.730	1.236	2.366
0.750	2.714	1.724	0.635	1.199	2.091
0.800	2.308	1.274	0.552	1.170	1.881
0.850	2.006	0.961	0.479	1.145	1.717
0.900	1.778	0.738	0.415	1.125	1.587
0.950	1.602	0.575	0.359	1.108	1.484
1.000	1.466	0.454	0.310	1.093	1.401

(Continued)

TABLE 25.2 *Continued*

Degree $N = 3$ $S(\max) = 1.15$ $S(\min) = 1.14$					
W	G	B'	G	Y_1	Y_2
0.200	138.408	458.742	3.314	2.528	31.758
0.250	58.354	151.441	2.595	2.084	16.998
0.300	29.231	61.496	2.104	1.799	10.384
0.350	16.559	28.850	1.742	1.604	6.968
0.400	10.302	15.064	1.462	1.465	5.020
0.450	6.908	8.546	1.237	1.363	3.825
0.500	4.928	5.182	1.051	1.287	3.050
0.550	3.704	3.317	0.896	1.229	2.525
0.600	2.911	2.222	0.763	1.184	2.157
0.667	2.238	1.381	0.617	1.139	1.820
0.700	2.005	1.111	0.554	1.122	1.696
0.750	1.739	0.820	0.472	1.100	1.549
0.800	1.545	0.620	0.401	1.083	1.437
0.850	1.400	0.477	0.341	1.069	1.351
0.900	1.290	0.374	0.290	1.058	1.283
0.950	1.523	0.410	0.270	1.084	1.428
1.000	1.738	0.446	0.257	1.105	1.556

Source: R. Levy and J. Helszajn, Specific equations for one and two section quarter-wave matching networks for stub-resistor loads, *IEEE Trans. Microwave Theory Tech.*, Vol. MTT-30, pp. 55–62, Jan. 1982. Reproduced with permission.

25.5 *t*-PLANE SYNTHESIS OF SHORT-LINE MATCHING NETWORK

It is also possible to employ short-line impedance transformers to match a complex load such as met in connection with the complex gyrator circuit of a typical junction circulator. In order to ensure, for the purpose of synthesis, that all line lengths are commensurate, it is necessary to replace the quarter-wave long stub in the gyrator circuit by two stubs in parallel—one short-circuited and the other open-circuited. The equivalent circuit obtained in this way is indicated in Fig. 25.5. The admittance of the load is then described by

$$Y_{in} = G + j(Y_C \tan\theta - Y_L/\tan\theta) \tag{25.25}$$

Y_C and Y_L are the characteristic admittance of its open- and short-circuited stubs, respectively, and θ_t is the commensurate electrical length

$$\theta_t = \tan^{-1}\sqrt{Y_L/Y_C} \tag{25.26}$$

It is observed that the commensurate length at resonance is now an arbitrary parameter, not limited to 90°, and this is the key to a precise synthesis technique.

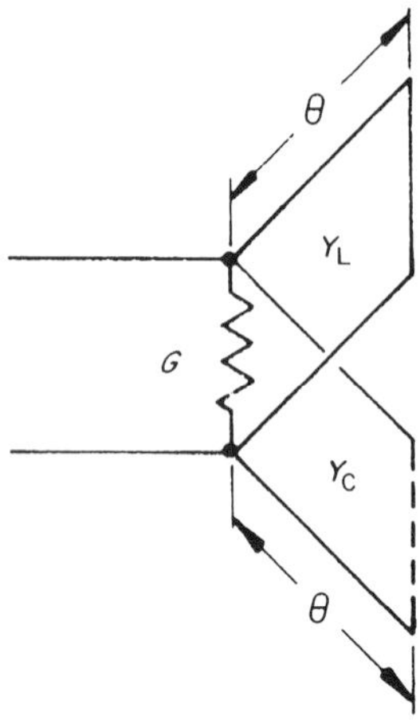

FIGURE 25.5 Equivalence between degree-1 and degree-2 complex loads.

An equivalence between this load network and the usual single-stub representation is obtained by equating the susceptance slope parameters of the two arrangements. One definition of the susceptance slope parameter of a load in terms of its susceptance is

$$B' = \frac{\omega_t}{2} \frac{dB}{d\omega}\bigg|_{\theta=\theta_t} \tag{25.27}$$

Evaluating the preceding equation in terms of the original variables gives

$$B' = \frac{\omega_t}{2}(Y_C \sec^2 \theta + Y_L \operatorname{cosec}^2 \theta) \frac{d\theta}{d\omega}\bigg|_{\theta=\theta_t} \tag{25.28}$$

or

$$B' = (Y_C + Y_L) \tan^{-1} \sqrt{Y_L/Y_C} \tag{25.29}$$

Expressions for Y_C and Y_L in terms of θ_t and B' may be obtained but are of no value here.

Figure 25.6 indicates the network circuit of such an $(n-2)$ matching technology.

The alternative representation of the load network shown by the equivalence of Fig. 25.7 is also instructive. The admittance values of the compound short-circuited stub are

$$Y_1' = Y_C + Y_L \tag{25.30}$$

$$Y_2' = (Y_C + Y_L)Y_L/Y_C \tag{25.31}$$

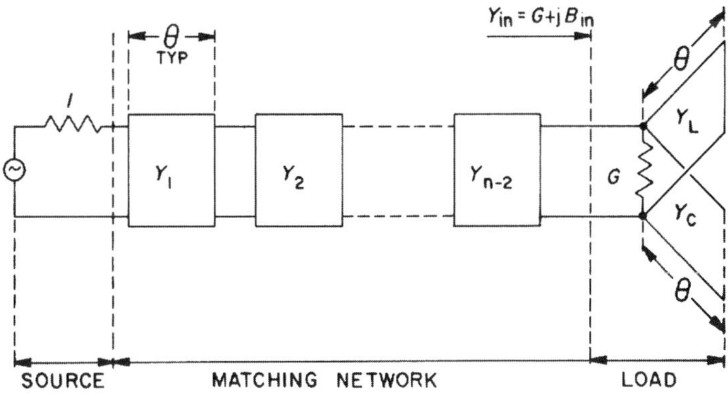

FIGURE 25.6 Topology of $(n - 2)$ UEs in cascade with degree-2 complex load.

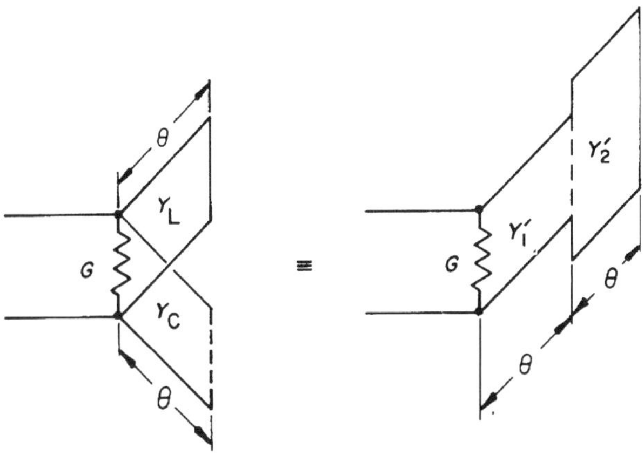

FIGURE 25.7 Equivalence between degree-2 complex loads.

and the resonant length is the same as given in Eq. (25.26). When $\theta_t = 45°$ then $Y_C = Y_L$, and the compound stub degenerates to a $90°$ stub of uniform admittance $2Y_L$, as expected.

25.6 NETWORK PARAMETERS OF SHORT-LINE IMPEDANCE TRANSDUCERS

Sample results for these short-line matching networks of order $n = 4$ (two unit elements matching) are presented in Tables 25.3–25.5. These are given to illustrate the general features of this type of network. There are even more possibilities than the quarter-wave network presented previously because the electrical length of the unit elements θ_t represents yet another variable in addition to $S(\min)$ (for given bandwidth and $S(\max)$). Both θ_t and $S(\min)$ may be varied to control the load admittance

TABLE 25.3 Short-line Matching Transformers where Theta (Midband) $\theta\_ = 27°$

$S_{max} = 1.1$
$S_{min} = 1$

W	Y_1	Y_2	G	Y_C	Y_L	B'	Q-factor	θ_t
0.150	6.427	1.746	10.331	51.850	12.810	29.827	2.887	26.43
0.200	4.878	1.368	6.200	23.319	5.558	13.114	2.115	26.02
0.250	3.960	1.155	4.290	12.891	2.942	7.056	1.645	25.53
0.300	3.357	1.023	3.254	8.132	1.767	4.317	1.327	24.99
0.350	2.93	0.937	2.630	5.620	1.157	2.887	1.098	24.41
0.400	2.621	0.879	2.227	4.150	0.807	2.059	0.925	23.80
0.450	2.383	0.840	1.951	3.218	0.590	1.541	0.790	23.19
0.500	2.196	0.813	1.755	2.592	0.448	1.198	0.683	25.57
0.550	2.046	0.796	1.610	2.149	0.350	0.958	0.595	21.97
0.600	1.924	0.785	1.501	1.823	0.279	0.784	0.523	21.37
0.667	1.793	0.778	1.392	1.506	0.213	0.618	0.444	20.60
0.700	1.737	0.776	1.350	1.381	0.188	0.554	0.410	20.23
$S_{max} = 1.1$								
$S_{min} = 1.04$								
0.150	5.755	1.575	8.368	42.848	10.508	24.534	2.932	26.35
0.200	4.375	1.243	5.073	19.464	4.582	10.862	2.141	25.88
0.250	3.558	1.057	3.550	10.881	2.439	5.890	1.659	25.33
0.300	3.023	0.943	2.725	6.946	1.474	3.634	1.334	24.73
0.350	2.648	0.871	2.229	4.856	0.971	2.450	1.099	24.09
0.400	2.373	0.824	1.909	3.626	0.681	1.762	0.923	23.44
0.450	2.163	0.793	1.691	2.841	0.501	1.329	0.786	22.78
0.500	1.999	0.773	1.537	2.309	0.382	1.040	0.677	22.13
0.550	1.868	0.762	1.423	1.931	0.300	0.837	0.588	21.50
0.600	1.761	0.756	1.338	1.650	0.240	0.689	0.515	20.89
0.667	1.647	0.754	1.254	1.374	0.184	0.547	0.436	20.10
0.700	1.599	0.755	1.221	1.265	0.163	0.491	0.403	19.72
$S_{max} = 1.1$								
$S_{min} = 1.08$								
0.150	4.339	1.227	4.969	22.087	5.242	12.389	2.493	25.97
0.200	3.325	0.996	3.141	10.468	2.334	5.646	1.797	25.29
0.250	2.730	0.873	2.300	6.110	1.271	3.158	1.373	24.52
0.300	2.343	0.804	1.846	4.072	0.785	2.010	1.089	23.71
0.350	2.075	0.765	1.575	2.963	0.528	1.395	0.886	22.89
0.400	1.880	0.743	1.402	2.292	0.378	1.030	0.735	22.10
0.450	1.733	0.734	1.286	1.852	0.283	0.795	0.619	21.34
0.500	1.619	0.731	1.204	1.545	0.219	0.635	0.527	20.63
0.550	1.529	0.734	1.145	1.319	0.174	0.520	0.454	19.96
0.600	1.456	0.740	1.101	1.147	0.141	0.434	0.395	19.33
0.667	1.380	0.751	1.058	0.972	0.109	0.350	0.331	18.55
0.700	1.348	0.757	1.042	0.901	0.097	0.317	0.304	18.19

TABLE 25.4 Short-line Matching Transformers where Theta (Midband) $\theta_{-} = 36°$

$S_{max} = 1.1$
$S_{min} = 1$

W	Y_1	Y_2	G	Y_C	Y_L	B'	Q-factor	θ_t
0.150	4.856	2.642	11.718	35.925	18.342	33.668	2.873	35.55
0.200	3.706	2.060	6.992	15.917	7.934	14.663	2.097	35.22
0.250	3.028	1.727	4.805	8.640	4.185	7.798	1.623	34.84
0.300	2.587	1.518	3.618	5.338	2.504	4.710	1.302	34.41
0.350	2.280	1.378	2.903	3.605	1.634	3.105	1.069	33.95
0.400	2.055	1.281	2.439	2.597	1.136	2.181	0.894	33.48
0.450	1.886	1.212	2.122	1.963	0.828	1.607	0.757	33.00
0.500	1.755	1.161	1.895	1.539	0.625	1.228	0.648	32.52
0.550	1.651	1.124	1.728	1.241	0.487	0.967	0.560	32.05
0.600	1.567	1.096	1.600	1.024	0.387	0.773	0.486	31.59
0.667	1.478	1.070	1.474	0.815	0.294	0.600	0.407	30.99
0.700	1.442	1.060	1.424	0.734	0.259	0.532	0.374	30.70
$S_{max} = 1.1$								
$S_{min} = 1.04$								
0.150	4.355	2.381	9.479	29.597	15.036	27.639	2.916	35.48
0.200	3.332	1.868	5.709	13.217	6.535	12.105	2.120	35.11
0.250	2.732	1.577	3.966	7.238	3.466	6.479	1.634	34.68
0.300	2.342	1.395	3.020	4.513	2.085	3.939	1.304	34.21
0.350	2.072	1.276	2.451	3.076	1.369	2.614	1.067	33.71
0.400	1.876	1.194	2.082	2.234	0.956	1.849	0.888	33.20
0.450	1.729	1.137	1.830	1.702	0.701	1.370	0.749	32.69
0.500	1.615	1.097	1.651	1.343	0.532	1.054	0.638	32.19
0.550	1.625	1.067	1.519	1.090	0.416	0.833	0.549	31.70
0.600	1.453	1.046	1.419	0.905	0.332	0.674	0.475	31.23
0.667	1.378	1.026	1.320	0.724	0.254	0.522	0.396	30.62
0.700	1.346	1.029	1.281	0.654	0.254	0.464	0.362	30.33
$S_{max} = 1.1$								
$S_{min} = 1.08$								
0.150	3.305	1.847	5.599	15.048	7.481	13.835	2.471	35.19
0.200	2.560	1.486	3.506	6.943	3.315	6.203	1.769	34.65
0.250	2.128	1.288	2.541	3.935	1.796	3.405	1.340	34.04
0.300	1.853	1.172	2.019	2.536	1.104	2.123	1.051	33.41
0.350	1.664	1.099	1.706	1.781	0.739	1.442	0.845	32.79
0.400	1.530	1.053	1.505	1.329	0.526	1.042	0.692	32.19
0.450	1.431	1.023	1.369	1.035	0.392	0.787	0.575	31.62
0.500	1.356	1.003	1.273	0.832	0.302	0.616	0.484	31.08
0.550	1.297	0.991	1.202	0.686	0.240	0.494	0.411	30.58
0.600	1.251	0.983	1.150	0.575	0.194	0.404	0.352	30.12
0.667	1.204	0.978	1.098	0.466	0.150	0.318	0.289	29.56
0.700	1.184	0.976	1.078	0.423	0.133	0.284	0.263	29.28

TABLE 25.5 Short-line Matching Transformers where Theta (Midband) $\theta_{-} = 45°$

$S_{max} = 1.1$
$S_{min} = 1.08$

W	Y_1	Y_2	G	Y_C	Y_L	B'	Q-factor	θ_t
0.150	3.928	3.928	14.024	25.578	25.578	40.178	2.865	45.00
0.200	3.020	3.020	8.293	11.014	11.014	17.300	2.086	45.00
0.250	2.491	2.491	5.640	5.779	5.779	9.078	1.609	45.00
0.300	2.149	2.149	4.199	3.438	3.438	5.400	1.286	45.00
0.350	1.914	1.914	3.330	2.230	2.230	3.503	1.052	45.00
0.400	1.744	1.744	2.767	1.541	1.541	2.420	0.875	45.00
0.450	1.618	1.618	2.380	1.116	1.116	1.754	0.737	45.00
0.500	1.521	1.521	2.103	0.839	0.839	1.318	0.627	45.00
0.550	1.445	1.445	1.899	0.649	0.649	1.020	0.537	45.00
0.600	1.385	1.385	1.743	0.515	0.515	0.808	0.464	45.00
0.667	1.322	1.322	1.588	0.388	0.388	0.610	0.384	45.00
0.700	1.296	1.296	1.526	0.341	0.341	0.535	0.351	45.00
$S_{max} = 1.1$								
$S_{min} = 1.04$								
0.150	3.529	3.529	11.323	20.948	20.948	32.905	2.906	45.00
0.200	2.725	2.725	6.750	9.057	9.057	14.226	2.108	45.00
0.250	2.258	2.258	4.634	4.775	4.775	7.500	1.618	45.00
0.300	1.958	1.958	3.486	2.855	2.855	4.485	1.287	45.00
0.350	1.753	1.753	2.793	1.862	1.862	2.924	1.047	45.00
0.400	1.606	1.606	2.345	1.293	1.293	2.031	0.866	45.00
0.450	1.497	1.497	2.037	0.941	0.941	1.479	0.726	45.00
0.500	1.414	1.414	1.818	0.711	0.711	1.117	0.614	45.00
0.550	1.350	1.350	1.656	0.553	0.553	0.868	0.524	45.00
0.600	1.299	1.299	1.533	0.440	0.440	0.691	0.451	45.00
0.667	1.246	1.246	1.411	0.334	0.334	0.524	0.371	45.00
0.700	1.224	1.224	1.363	0.293	0.293	0.461	0.338	45.00
$S_{max} = 1.1$								
$S_{min} = 1.08$								
0.150	2.701	2.701	6.631	10.378	10.378	16.301	2.458	45.00
0.200	2.122	2.122	4.092	4.565	4.565	7.170	1.752	45.00
0.250	1.792	1.792	2.918	2.453	2.453	3.853	1.320	45.00
0.300	1.584	1.584	2.282	1.495	1.495	2.349	1.029	45.00
0.350	1.446	1.446	1.900	0.994	0.994	1.561	0.822	45.00
0.400	1.349	1.349	1.654	0.703	0.703	1.104	0.668	45.00
0.450	1.278	1.278	1.486	0.520	0.520	0.817	0.550	45.00
0.500	1.226	1.226	1.366	0.399	0.399	0.627	0.459	45.00
0.550	1.186	1.186	1.279	0.315	0.315	0.494	0.386	45.00
0.600	1.155	1.155	1.213	0.253	0.253	0.398	0.328	45.00
0.667	1.124	1.124	1.148	0.195	0.195	0.306	0.267	45.00
0.700	1.111	1.111	1.123	0.172	0.172	0.271	0.241	45.00

level G and susceptance slope parameter B' while leaving the Q-factor essentially unchanged. Actually, as θ_t is reduced, the bandwidth increases somewhat, but a disadvantage of carrying this too far is that the admittance levels may become impractical.

The cases where $Y_2 = 1$ allow the characteristics of loads obtained using measured results in 50 Ω lines to be retained. The matching networks may commence at the load itself, and it is not necessary to use reference planes some distance away from the load to avoid perturbing the junction.

It is interesting to observe that $Y_1 = Y_2$, when $\theta_t = 45^\circ$, and the double unit element degenerates to the classic single quarter-wave transformer. We observe also that here $Y_L = Y_C$, and the susceptance slope parameter (Eq. (25.29)) degenerates to give

$$B' = \pi Y_L/2$$

This is related to the susceptance slope parameter of the single-stub-resistor load network by a factor of 2, which is due to the fact that the stubs in our present case are of half the electrical length and hence have twice the effective bandwidth. Incidentally, these identities for the degenerate cases represent a good test of the validity of the synthesis developed in this chapter.

CHAPTER TWENTY-SIX

Fabrication of UHF Circulators Using Irregular Hexagonal Gyromagnetic Resonators

26.1 INTRODUCTION

Gyromagnetic materials for use in the design of ferrite phase shifters and junction circulators may be biased below the main Kittel resonance or above it. In the former case the origin of the direct magnetic field intensity coincides with that at which it is zero, whereas in the latter case it corresponds with that at which it is infinite. UHF devices are normally biased above the main resonance. An important classic difference between the two situations is that the normalized magnetization and effective permeability are both less than unity below the main resonance but are both usually larger than unity above it. Another distinguishing feature between the two is that the frequency behavior and magnitudes of the elements of the permeability tensor are different. A further important difference between the two regimes is that while the effective permeability varies very little with the direct magnetic field below the Kittel resonance, it varies rapidly above it. The main emphasis of this chapter is the study of an irregular gyromagnetic resonator biased above the Kittel line as a preamble to the design of one commercial circulator. The exact connections between the effective permeability and gyrotropy and the direct magnetic parameters of the magnetic insulator introduced in Chapter 2 are often approximated in the description of above-resonance devices and this is the approach adopted here. Ideally, the frequency interval defined by the split frequencies of the resonator must separately exclude the possibility of the split higher order modes intruding into the passband defined by the dominant split ± 1 modes, or by the specification,

The Stripline Circulator: Theory and Practice. By J. Helszajn

and must also avoid the intrusion of the main resonance skirt in the passband. The possible onset of low-field loss must also be respected. One difficulty with assembling experimental data with this sort of arrangement at a single frequency is that the midband frequency varies with the direct magnetic field. One way to deal with this problem is to reset the size of the resonator at each and every field setting. Another way is to introduce suitably located magnetic walls. An important phenomenon, not dealt with specifically in this text, is that of third order intermodulation distortion (IMD) in magnetic insulators. Some results on this feature are included in this chapter for completion sake.

26.2 WAVE IMPEDANCE AND WAVENUMBER IN UHF GYROMAGNETIC CIRCUITS

A distinguishing difference between the constitutive effective permeabilities above and below the Kittel line is that it is larger than unity above it, whereas it *is* approximately unity below it. These features have some importance in the design of practical ferrite devices. The constitutive parameters in a ferrite substrate magnetized above the uniform mode resonance and perpendicular to its plane are typically described by

$$\varepsilon_f = 15.0$$

and

$$\mu_{eff} = 3.5$$

so that

$$\mu_{eff}\varepsilon_f = 52.5$$

$$\varepsilon_f/\mu_{eff} = 4.29$$

The value of the equivalent dielectric constant perpendicular to the plane of the substrate is therefore quite different from that in the plane. The situation below the Kittel line is quite different. In this instance,

$$\mu_{eff} \approx 1$$

so that

$$\varepsilon_f\mu_{eff} = \varepsilon_f$$

and

$$\varepsilon_f / \mu_{eff} \approx \varepsilon_f$$

The material, in this case, displays the same effective constitutive parameters in both the direction of propagation and perpendicular to it. Both the electrical length of any transformer region and the size of any resonator are therefore less above the Kittel line than below it. Another consequence of this result is that whereas the wave admittance of the transformer circuit in the design of the matching network has a similar value to that met in below-resonance devices with dielectric materials, that of the resonator is quite different. The latter property is the explanation for the need of thin assemblies in the design of UHF devices.

26.3 GYROMAGNETIC SPACE OF ABOVE-RESONANCE CIRCULATORS

The magnetic circuit of the magnetic insulator is determined by both the gyrotropy and the effective permeability. The gyrotropy is set by the weakly magnetized theory of the circulator. The upper bound of this quantity is approximately restricted at the upper frequency of the specification to

$$\kappa / \mu = 0.35 \tag{26.1}$$

This value of gyrotropy actually ensures that the intersection between the upper branch of the dominant splitting mode and the lower branch of the first higher order one is always outside the upper bandedge of the specification. The situation may be understood by recognizing that the gyrotropy, unlike the case with below-resonance devices, increases rather than decreases with frequency. One consequence of this feature is that the frequency variation of the gyrator conductance is different below and above the Kittel line.

The upper bound on the gyrotropy places a lower bound on the loaded Q-factor of the junction or gyrator circuit, which is approximately given by

$$Q_L \approx 2.5$$

The effective permeability, however, is set by one of three possible considerations. One possible design procedure in the case of a side-coupled triangular resonator (say) is to fix the effective permeability (μ_{eff}) in terms of the normalized gyrator conductance (g). A second is to fix it in terms of the cutoff number of the resonator. The third possibility is to set it in terms of the susceptance slope parameter of the junction.

The three relationships are

$$\mu_{\text{eff}} = \varepsilon_{\text{f}}/rg \tag{26.2}$$

$$k_0\sqrt{\varepsilon_{\text{f}}\mu_{\text{eff}}}A = 4\pi/3 \tag{26.3}$$

$$b' = B'/Y_0 = \frac{0.855(kA)^2}{Y_0(k_0H)\mu_{\text{eff}}}\sqrt{\frac{\varepsilon_0}{\mu_0}} \tag{26.4}$$

r is the midband VSWR of the device, A is the side dimension of the resonator, and H is the half-space thickness of the device.

The purpose of this chapter is to outline the design of a degree-2 circulator by fixing the direct magnetic parameters of the device in the manner indicated here.

26.4 APPROXIMATE RELATIONSHIPS OF PERMEABILITY TENSOR

The exact descriptions of the gyrotropy (κ/μ) and the effective permeability (μ_{eff}) entering into the description of gyromagnetic resonators are tabulated in Chapter 2 in terms of the parameters of a magnetized magnetic insulator. Some useful approximations that are applicable above the uniform line resonance are summarized here.

$$\frac{\kappa}{\mu} \approx \frac{1}{\sigma}\left(\frac{p/\sigma}{1+p/\sigma}\right) \tag{26.5}$$

or

$$\sigma \approx \frac{-p+\sqrt{p^2+4\mu/\kappa}}{2} \tag{26.6}$$

and

$$\mu_{\text{eff}} \approx 1 + p/\sigma \tag{26.7}$$

or

$$\sigma \approx p/(\mu_{\text{eff}} - 1) \tag{26.8}$$

provided $\sigma(p+\sigma) >> 1$.

The ties between the two variables and the two direct magnetic field quantities may again be connected in one of six ways. Table 26.1 summarizes the different approximate relationships under consideration.

TABLE 26.1 Approximate Equations (Above Kittel Line)

p, σ	$\dfrac{\kappa}{\mu} \approx \left(\dfrac{1}{\sigma}\right)\dfrac{p/\sigma}{(1+p/\sigma)}$
	$\mu_{\text{eff}} \approx 1 + p/\sigma$
p, κ/μ	$\sigma \approx \dfrac{-p + \sqrt{p^2 + 4[1 + p(\mu/\kappa)]}}{2}$
	$\mu_{\text{eff}} \approx 1 + p/\sigma$
p, μ_{eff}	$\sigma \approx \dfrac{p}{\mu_{\text{eff}} - 1}$
	$\dfrac{\kappa}{\mu} \approx \dfrac{(\mu_{\text{eff}} - 1)^2}{p\mu_{\text{eff}}}$
σ, κ/μ	$p \approx \dfrac{\sigma(\kappa/\mu)}{\kappa/\mu - 1/\sigma}$
	$\mu_{\text{eff}} \approx \dfrac{-\mu/\kappa}{\sigma - (\mu/\kappa)}$
σ, μ_{eff}	$p \approx \sigma(\mu_{\text{eff}} - 1)$
	$\dfrac{\kappa}{\mu} = \dfrac{1}{\sigma}\dfrac{(\mu_{\text{eff}} - 1)}{\mu_{\text{eff}}}$
κ/μ, μ_{eff}	$p \approx \left(\dfrac{\mu}{\kappa}\right)\dfrac{(\mu_{\text{eff}} - 1)^2}{\mu_{\text{eff}}}$
	$\sigma \approx \left(\dfrac{\mu}{\kappa}\right)\dfrac{\mu_{\text{eff}} - 1}{\mu_{\text{eff}}}$

The normalized magnetization (p) and the normalized internal magnetic field intensity (σ) have the meaning introduced in Chapter 2:

$$p = \gamma M_0/\omega$$

$$\sigma = \gamma(H_0 - N(r)M_0)/\omega$$

$N_z(r)$ is the shape demagnetizing factor along the axis of the magnetic insulator. In a nonellipsoidal body, such as a disk, this quantity is in principle dependent on the radial coordinate of the geometry, as discussed in Chapter 3. One consequence of this feature is that the magnetic field intensity inside a disk is nonuniform even in the presence of a uniform direct magnetic field. This effect, however, is usually disregarded in everyday engineering. Strictly speaking the value of H_0 entering in the calculation of the internal field is that in the absence of the resonator.

TABLE 26.2 Approximate Equations (Above Kittel Line)

p, σ	$\mu = 1 + p/\sigma$ $\kappa = p/\sigma^2$
κ, μ	$p = (\mu - 1)^2/\kappa$ $\sigma = (\mu - 1)/\kappa$
p, κ	$\mu = 1 + \sqrt{p\kappa}$ $\sigma = \sqrt{p/\kappa}$
p, μ	$\sigma = p/(\mu - 1)$ $\kappa = (\mu - 1)^2/p$
σ, κ	$p = \kappa\sigma^2$ $\mu = 1 + \kappa\sigma$
σ, μ	$p = \sigma(\mu - 1)$ $\kappa = (\mu - 1)/\sigma$

The condition $\sigma(p + \sigma) > 1$ introduced in connection with the approximate magnetic variables reduces to

$$\sigma(p + \sigma) > p^2 \left(\frac{H_0}{M_0}\right)\left(\frac{H_0}{M_0} - 1\right) \tag{26.9}$$

The interval over which this approximation holds is bracketed by

$$5.1 \geq p \geq 2.2 \tag{26.10}$$

provided

$$1.3 \leq H_0/M_0 \leq 2$$

Approximate relationships between μ, κ, p, and σ may also be constructed by taking linear combinations of μ_{eff} and κ/μ. Table 26.2 gives the results obtained in this way.

A feature of devices biased above the Kittel line, unlike the situation that prevails below it, is that the normalized magnetization (p) is not restricted to unity by the onset of low-field loss. The relaxation in this requirement comes about because the direct field necessary to bias the material there is in most circumstances sufficient to saturate most materials.

A scrutiny of the relationship for μ_{eff} suggests that its value is preserved between any two frequencies provided p/σ is kept constant. This condition is satisfied provided $B_0/\mu_0 M_0$ is preserved. This last remark may readily be understood by writing p/σ in terms of the original variables:

$$\frac{p}{\sigma} = \frac{(M_0/H_0)}{1 - N(r)(M_0/H_0)} \tag{26.11}$$

The absolute values of B_0 and $\mu_0 M_0$ are fixed by the required gyrotropy. One design procedure is therefore to fix the ratio of $B_0/\mu_0 M_0$ or p/σ from a statement of μ_{eff}, and then to obtain σ in terms of the gyrotropy (κ/μ). p or $\mu_0 M_0$ is thereafter calculated from p/σ or $B_0/\mu_0 M_0$. If p or $\mu_0 M_0$ obtained in this way is realistic, then the design may proceed. Recall that while the magnetic field intensifies inside and outside a magnetic insulator are different, the magnetic flux density is the same in each region.

26.5 H_0/M_0 SPACE

The direct magnetic field entering into the description of the effective permeability and gyrotropy may be defined in a number of different ways. The purpose of this section is to do so in terms of H_0/M_0 in the case for which $N_z = 1$. Adopting the approximate form for μ_{eff} by way of example gives

$$\mu_{eff} \approx 1 + \frac{1}{(H_0/M_0 - 1)} \tag{26.12}$$

Figure 26.1 indicates the connection between μ_{eff} and H_0/M_0. It suggests that μ_{eff} is solely fixed by the ratio H_0/M_0.

Similarly,

$$\left(\frac{\kappa}{\mu}\right) \approx \frac{1}{p}\left(\frac{1}{(H_0/M_0) - 1}\right)\left(\frac{M_0}{H_0}\right) \tag{26.13}$$

Figure 26.2 gives the normalized magnetization versus the direct magnetic field for parametric values of gyrotropy. It suggests that there is more than one combination of p and H_0/M_0 for a given value of gyrotropy.

The normalized magnetization versus the direct magnetic field for parametric values of gyrotropy is shown in Fig. 26.3.

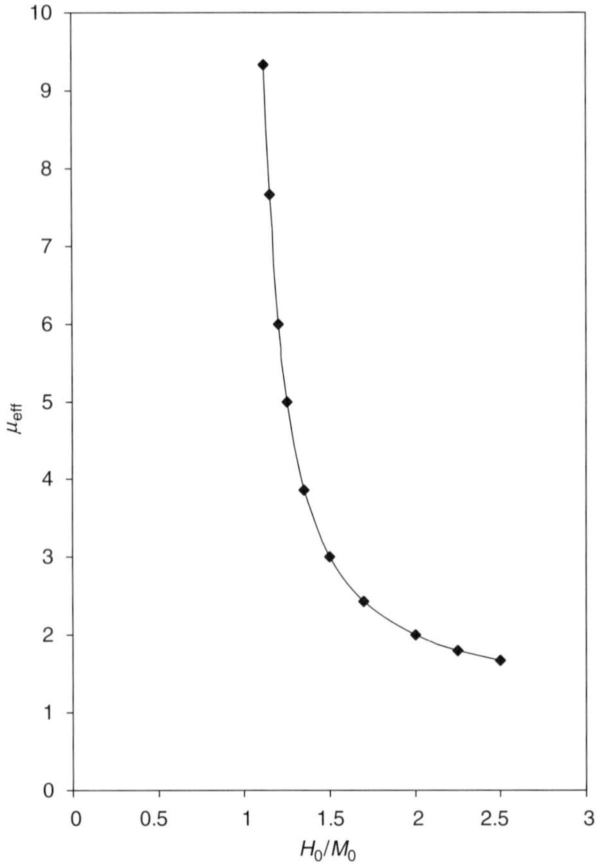

FIGURE 26.1 Effective permeability above the Kittel line versus direct magnetic field for parametric values of normalized magnetization.

26.6 THE KITTEL LINE

In practice the insertion loss of any gyromagnetic device is largely determined by the proximity of the Kittel line. This quantity fixes the magnetic field intensity of the magnetic insulator. In the frequency domain, the line is fixed by the radian frequencies of the specification ω_0 and the Kittel line (ω_r) by the bandwidth of the circuit ($\Delta\omega$) and the uniform or effective linewidth (ΔH_{eff}).

The required condition is obtained by recognizing that, in an above-resonance device, the lower skirt of the Kittel line coincides with the upper bandedge frequency of the microwave specification.

$$\omega_r - (n\gamma\Delta H_{eff}/2) = \omega_0 + (\Delta\omega_0/2) \tag{26.14}$$

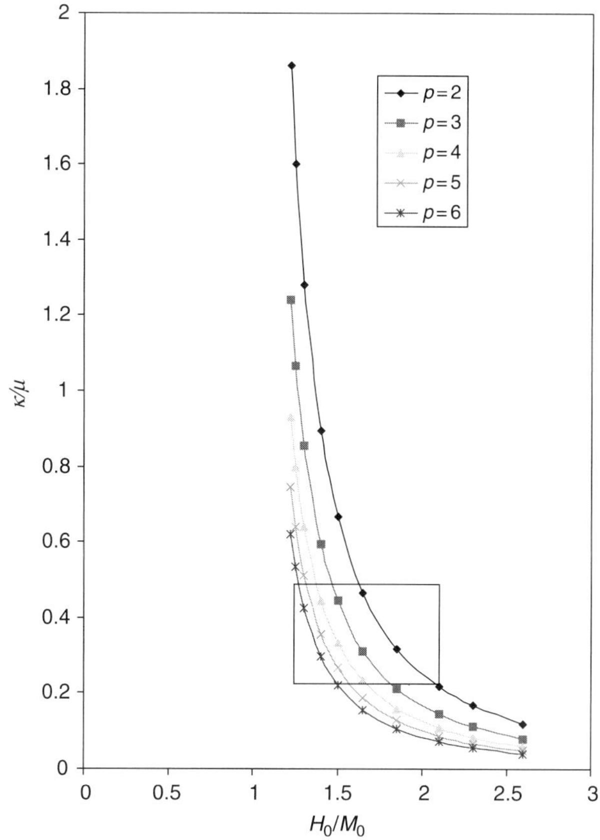

FIGURE 26.2 Gyrotropy above the Kittel line versus direct magnetic field for parametric values of normalized magnetization.

n is the number of half-linewidths at the skirt of the Kittel line and is the unknown of the problem. Rearranging the preceding condition gives

$$\frac{\omega_\mathrm{r}}{\omega_0} = 1 + \left(\frac{\Delta\omega_0}{2\omega_0}\right) + \left(\frac{n\gamma\,\Delta H_\mathrm{eff}}{2\omega_\mathrm{r}}\right)\left(\frac{\omega_\mathrm{r}}{\omega_0}\right) \tag{26.15}$$

A second relationship for $\omega_\mathrm{r}/\omega_0$ is obtained by having recourse to the Kittel frequency in Chapter 2:

$$\frac{\omega_\mathrm{r}}{\omega_0} = \left(\frac{\gamma M_0}{\omega_0}\right)\left(\frac{H_0}{M_0} - 1\right) \tag{26.16}$$

with $N_z = 1$ and $N_t = 0$.

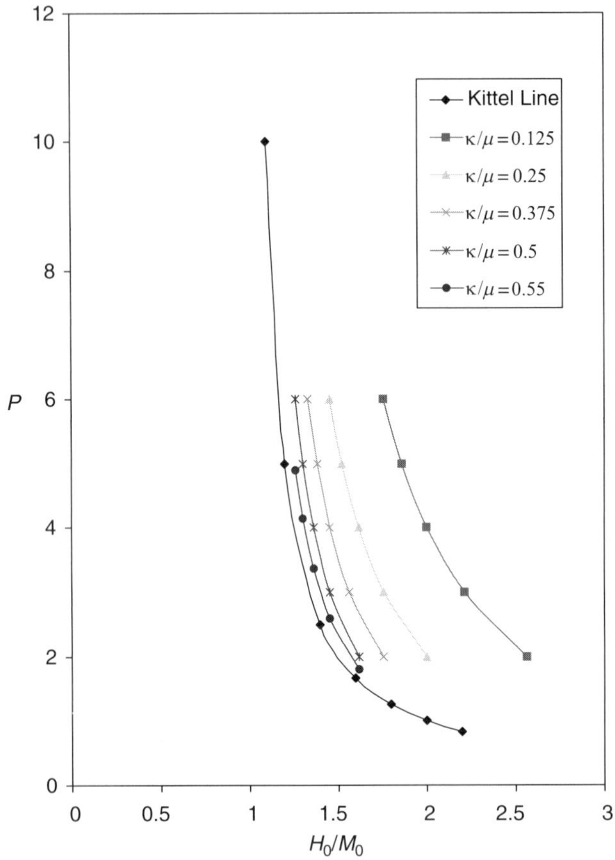

FIGURE 26.3 Normalized magnetization above the Kittel line versus direct magnetic field for parametric values of gyrotropy.

The required result is now established by combining the previous two conditions

$$\left(\frac{\gamma M_0}{\omega_0}\right)\left(\frac{H_0}{M_0}-1\right) \approx 1+\left(\frac{\Delta\omega_0}{2\omega_0}\right)+\left(\frac{n\gamma\,\Delta H_{\mathrm{eff}}}{2\omega_{\mathrm{r}}}\right) \tag{26.17}$$

A more accurate condition is obtained by using the exact definition for ω_{r} in Chapter 2 given by

$$\omega_{\mathrm{r}} = \gamma M_0[H_0/M_0 + (N_t - N_z)] \tag{26.18}$$

Also recall that the demagnetizing factors for a disk satisfy

$$N_z + 2N_t = 1 \tag{26.19}$$

The condition in Eq. (26.18) indicates that the minimum value of H_0/M_0, other things being equal, is fixed by the product $n\ \Delta H_{\text{eff}}/2$. Of course, the exact value is determined by both the insertion loss and bandwidth of the specification. One difficulty with the description of the linewidth, however, is that the lineshape is not usually symmetric about the main Kittel line. One way of dealing with this difficulty is to measure the split unloaded quality factors of the resonator as a preamble to design in the manner indicated in Chapter 11. A trial value of H_0/M_0 between 1.20 and 1.30 has been found satisfactroy in many instances. The exact condition is experimental.

In order to avoid so-called low-field loss, it is also necessary to ensure that the material is saturated. This requirement is satisfied provided the internal direct magnetic field required to establish the Kittel line is zero

$$H_\text{i} = 0 \tag{26.20}$$

or

$$1/p \geq N_t \tag{26.21}$$

This condition dictates the maximum magnetization that may be utilized in any design. The nature of this type of loss is pictorially illustrated in Chapter 1.

26.7 TEMPERATURE STABILITY OF KITTEL LINE

An important aspect entering into the description of the permeability tensor is the dependence of its entries on temperature. This is attributed primarily to the variation of the saturation magnetization of the material with temperature. While some materials may be temperature compensated by doping, this is not always the case. The situation in the case of a pure YIG material is depicted in Chapter 28.

The temperature at which the magnetization vanishes is known as the Curie temperature. The change in magnetization with temperature is specified in ferrite brochures by

$$\frac{\Delta M_0}{M_0\,\Delta T} \tag{26.22}$$

The value in the case of a YIG material is

$$\frac{\Delta M_0}{M_0\,\Delta T} = 1.8 \times 10^{-3} (^\circ\text{C})^{-1} \tag{26.23}$$

The frequency shift of the Kittel line for an infinite sheet is given without ado by

$$\left(\frac{\omega_r - \Delta\omega_r}{\gamma}\right) = (H_0 - \Delta H_0) - (M_0 - \Delta M_0) \tag{26.24}$$

$\Delta\omega_r = 0$ provided

$$\Delta H_0 = \Delta M_0 \tag{26.25}$$

Similar relationships are readily deduced for μ, $\kappa\mu_{\text{eff}}$, and κ/μ. Those are left as an exercise for the reader.

26.8 MODE CHARTS OF UHF GYROMAGNETIC IRREGULAR HEXAGONAL PLANAR RESONATOR

The resonator shape under consideration in this chapter is an irregular hexagonal geometry. Its details are specified in Chapter 16 by one radius and one

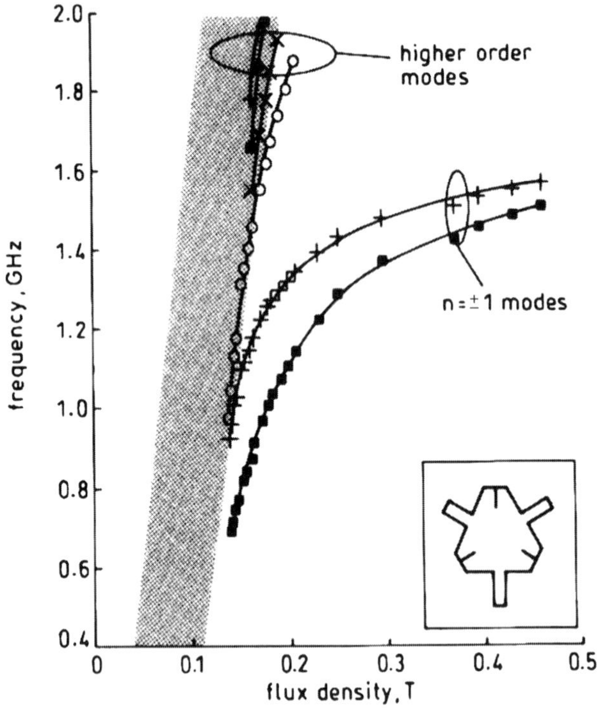

FIGURE 26.4 Mode chart of irregular hexagonal resonator ($\mu_0 M_0 = 0.1100$ T). (Reproduced with permission from J. Helszajn, C. S. Cheng, and D. D'Orazio, UHF irregular hexagonal gyromagnetic resonators symmetrically Loaded with magnetic walls, *IEE Proc. Microwave Antennas Propag.*, Vol. 146, No. 6, Dec. 1999.)

internal angle. The side dimensions are related to the circumscribed radius and shape angle by

$$A = 2r \sin(\phi/2) \tag{26.26a}$$

$$B = 2r \sin(60^\circ - \phi/2) \tag{26.26b}$$

Some experimental results on a three-port junction obtained using a side-coupled irregular gyromagnetic resonator with ports 2 and 3 terminated in matched loads for three different values of magnetization are illustrated in Figs. 26.4–26.6. One feature of these charts is that the midband frequency varies significantly with the direct magnetic field intensity. Another is that the splitting between the dominant pair of degenerate modes increases rapidly as the magnetic field approaches the main resonance line.

The lower bound on the direct magnetic field is fixed by the Kittel line in practice. This relationship is superimposed separately on the data in Figs. 26.4–26.6. It is fixed

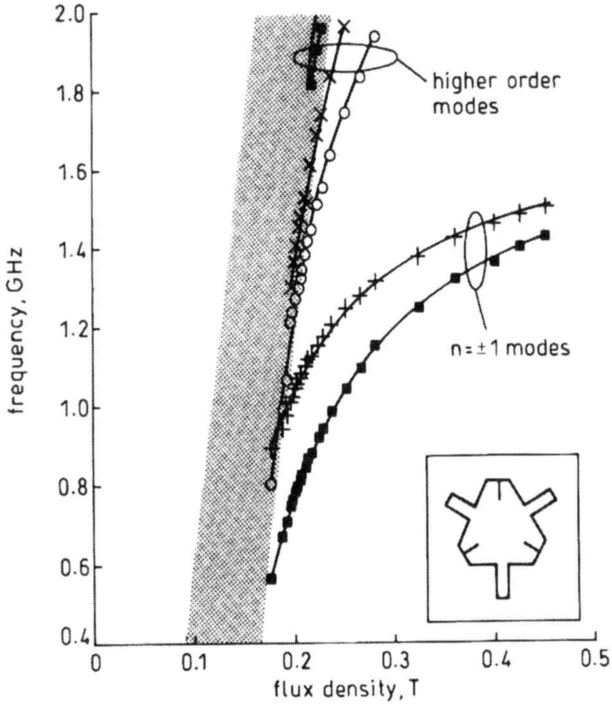

FIGURE 26.5 Mode chart of irregular hexagonal resonator ($\mu_0 M_0 = 0.1600$ T). (Reproduced with permission from J. Helszajn, C. S. Cheng, and D. D'Orazio, UHF irregular hexagonal gyromagnetic resonators symmetrically Loaded with magnetic walls, *IEE Proc. Microwave Antennas Propag.*, Vol. 146, No. 6, Dec. 1999.)

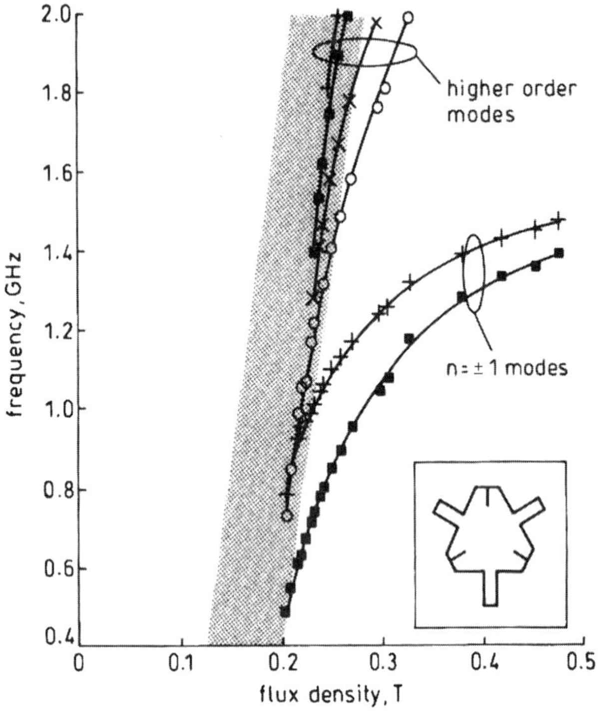

FIGURE 26.6 Mode chart of irregular hexagonal resonator ($\mu_0 M_0 = 0.1850$ T). (Reproduced with permission from J. Helszajn, C. S. Cheng, and D. D'Orazio, UHF irregular hexagonal gyromagnetic resonators symmetrically Loaded with magnetic walls, *IEE Proc. Microwave Antennas Propag.*, Vol. 146, No. 6, Dec. 1999.)

in Eq. (26.12). The value employed for N_z is 0.768. It is empirically defined here by replacing the irregular hexagonal plate by an equivalent circular geometry. B_0 is obtained here by measuring the magnetic flux density between the ferrites. In doing so, the small difference between the thicknesses of the center conductor and the probe is disregarded. Some measurements elsewhere suggest this assumption is permissible for engineering purposes provided the reluctance of the overall magnetic circuit is large. The definition employed here, however, is compatible with engineering practice.

A finite element plot of the magnetic flux density of a typical magnet arrangement is available in Chapter 3.

A scrutiny of the magnetized mode charts of the three materials employed in this work in Figs. 26.4–26.6 indicates that the ratios of $B_0/\mu_0 M_0$ at, say, 1 GHz are identical. The value obtained at 1 GHz in the case of the resonator geometry utilized here is about 1.37. The corresponding effective permeability in this instance is equal to 3.18. The spacing between the upper face of a typical ferrite and a typical pole piece is 1.62 mm; the thickness of the center conductor is 0.15 mm and that of a typical ferrite is 1.62 mm.

One polynomial approximation for the split frequencies of the dominant counter-rotating modes of this sort of weakly magnetized resonator based on a finite element solution of six different shape angles (ϕ) is given in Chapter 16. A polynomial approximation for the circumscribed radius r is also given in the same chapter. The resonator shape adopted in that work is defined by its shape angle (ϕ) and circumscribed radii (r) by

$$\phi = 21.79^\circ$$

$$r = 10.07\,\text{mm}$$

The values of the polynomials obtained under these conditions are 0.84 and 2.10, respectively. The values of the magnetization of the ferrite substrates employed here are $\mu_0 M_0 = 0.1000$ T, 0.1600 T, and 0.1850 T, respectively.

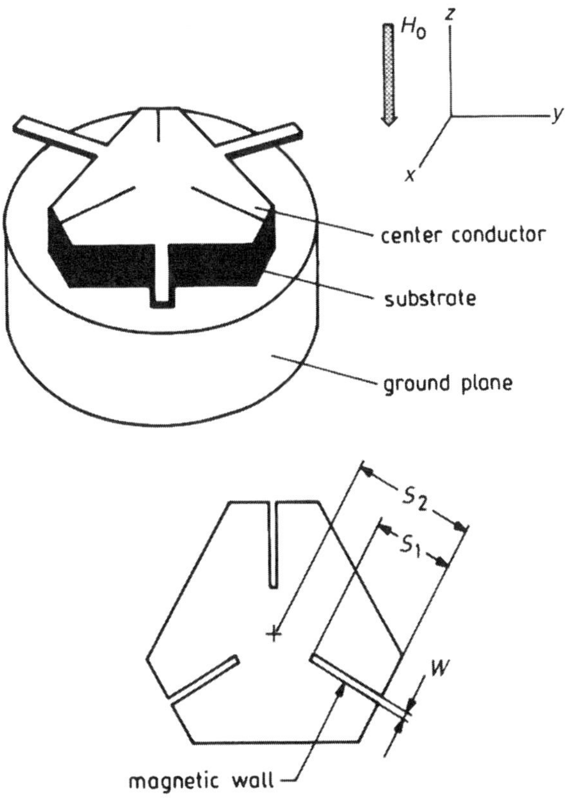

FIGURE 26.7 Schematic diagram of irregular hexagonal planar resonator loaded by radial magnetic walls.

26.9 RADIAL MAGNETIC WALLS

A classic means of tuning a gyromagnetic resonator is to symmetrically load it with magnetic walls in the manner described in Chapter 10. The schematic diagram of this arrangement is illustrated in Fig. 26.7. This section summarizes two typical mode charts for this type of resonator for a substrate with $\mu_0 M_0 = 0.1600$ T. Figure 26.8 indicates one mode chart for an irregular hexagonal resonator using this material. It suggests, in keeping with some prior art, that the resonant frequency of such resonators can be varied by such means. Figure 26.9 depicts a similar result for another geometry. Changes in the radial wavenumber due to changes in μ_{eff} with magnetization can be separated from that due to the introduction of the magnetic slots by constructing constant $B_0/\mu_0 M_0$ lines. This feature is clearly shown in Fig. 26.10. Polynomial approximations applicable to the resonator shape investigated here are

$$k_0 r = -23.698\left(\frac{s_1}{s_2}\right)^3 + 11.11\left(\frac{s_1}{s_2}\right)^2 - 1.3926\left(\frac{s_1}{s_2}\right) + 28.2, \quad \frac{B_0}{\mu_0 M_0} = 2.1 \tag{26.27a}$$

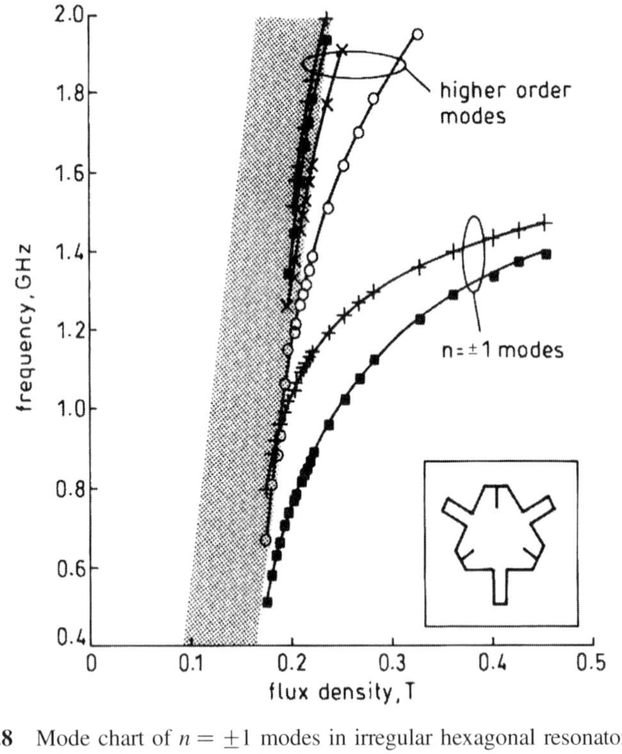

FIGURE 26.8 Mode chart of $n = \pm 1$ modes in irregular hexagonal resonator loaded with radial magnetic walls ($\mu_0 M_0 = 0.1600$ T, $s_1/s_2 = 0.50$). (Reproduced with permission from J. Helszajn, C. S. Cheng, and D. D'Orazio, UHF irregular hexagonal gyromagnetic resonators symmetrically with magnetic walls, *IEE Proc. Microwave Antennas Propag.*, Vol. 146, No. 6, Dec. 1999.)

$$k_0 r = -17.319\left(\frac{s_1}{s_2}\right)^3 + 5.0866\left(\frac{s_1}{s_2}\right)^2 - 0.4752\left(\frac{s_1}{s_2}\right) + 26.3, \quad \frac{B_0}{\mu_0 M_0} = 1.8 \tag{26.27b}$$

$$k_0 r = -14.892\left(\frac{s_1}{s_2}\right)^3 + 4.4964\left(\frac{s_1}{s_2}\right)^2 - 0.0133\left(\frac{s_1}{s_2}\right) + 22.8, \quad \frac{B_0}{\mu_0 M_0} = 1.5 \tag{26.27c}$$

$$k_0 r = -9.0065\left(\frac{s_1}{s_2}\right)^3 + 0.5805\left(\frac{s_1}{s_2}\right)^2 + 0.1452\left(\frac{s_1}{s_2}\right) + 20.0, \quad \frac{B_0}{\mu_0 M_0} = 1.3 \tag{26.27d}$$

with $\phi = 21.79^\circ$.

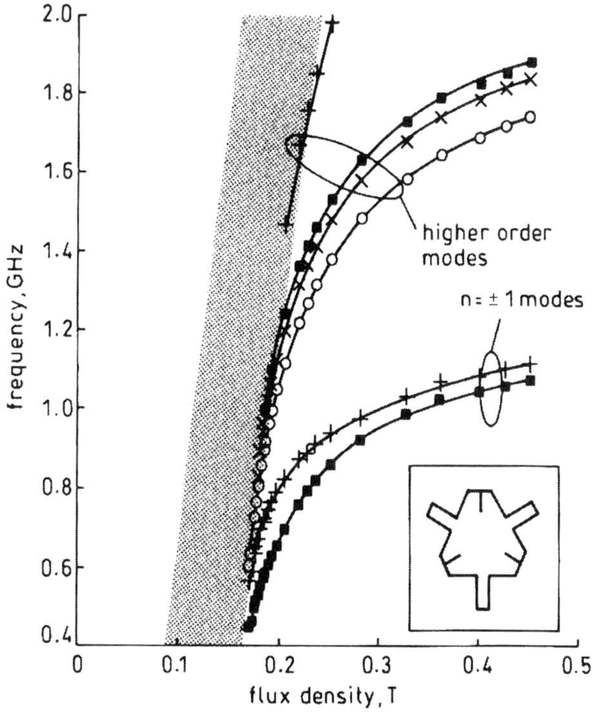

FIGURE 26.9 Mode chart of $n = \pm 1$ modes in irregular hexagonal resonator loaded with radial magnetic walls ($\mu_0 M_0 = 0.1600$ T, $s_1/s_2 = 0.70$). (Reproduced with permission from J. Helszajn, C. S. Cheng, and D. D'Orazio, UHF irregular hexagonal gyromagnetic resonators symmetrically with magnetic walls, *IEE Proc. Microwave Antennas Propag.*, Vol. 146, No. 6, Dec. 1999.)

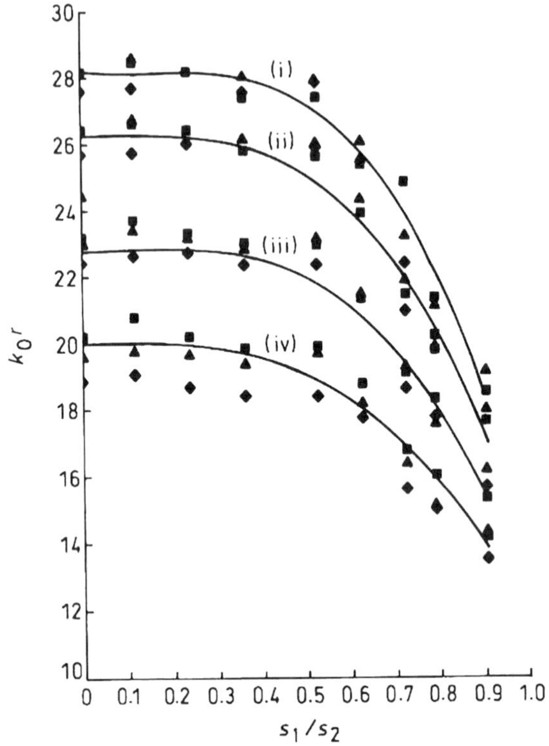

FIGURE 26.10 Relationship between k_0r and s_1/s_2 for parametric values of B_0/μ_0M_0 and μ_0M_0. (Reproduced with permission from J. Helszajn, C. S. Cheng, and D. D'Orazio, UHF irregular hexagonal gyromagnetic resonators symmetrically with magnetic walls, *IEE Proc. Microwave Antennas Propag*., Vol. 146, No. 6, Dec. 1999.)

The overlap of the different magnetization curves in the $k_0r/(s_1/s_2)$ space may be taken as one indication of the quality of the magnetic flux density calibration.

26.10 MAGNETIC VARIABLES OF GYROMAGNETIC RESONATORS

The diagonal and off-diagonal elements of the gyromagnetic substrate may be verified by taking linear combinations of μ_{eff} and κ/μ. Figures 26.11 and 26.12 summarize relationships between the normalized internal direct magnetic field and the effective permeability and gyrotropy, respectively.

The relationship between σ and μ_{eff} in Fig. 26.11 is established by experimentally determining the radial wavenumber at the midband frequency and solving it for μ_{eff}

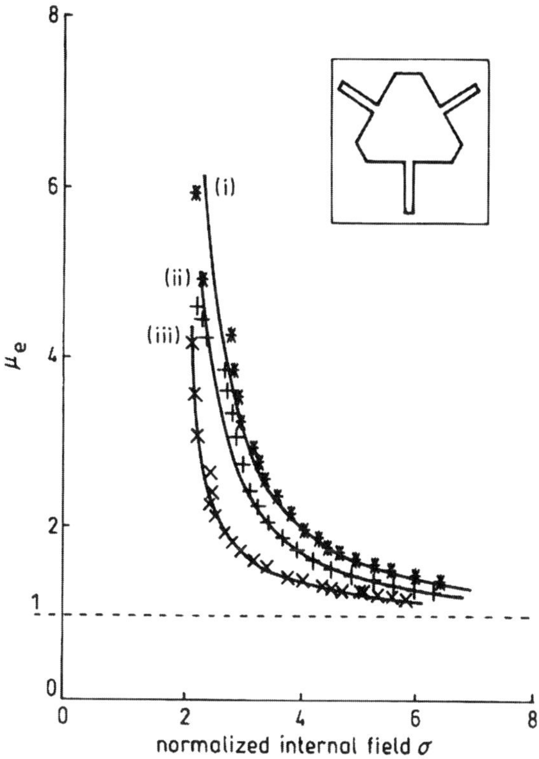

FIGURE 26.11 Relationship between normalized internal direct magnetic field intensity and effective permeability.

as a preamble of making use of that between p, σ, and μ_{eff}:

$$k_0\sqrt{\mu_{\text{eff}}\varepsilon_{\text{f}}} = 2.10/r \tag{26.28}$$

The relationship between σ and κ/μ shown in Fig. 26.12 is deduced separately by writing κ/μ in terms of σ and μ_{eff}:

$$\frac{\kappa}{\mu} \approx \frac{(\mu_{\text{eff}} - 1)}{\sigma\mu_{\text{eff}}} \tag{26.29}$$

The only meaningful solutions in the present effort are those that correspond to 1GHz. This gives $p = 3.08$, $\sigma = 11$; $p = 4.48$, $\sigma = 11$; $p = 5.18$, $\sigma = 11$.

A scrutiny of Fig. 26.12 suggests that there is more than one combination of p and σ for a given gyrotropy. One way to move σ or the direct magnetic field intensity away from the Kittel line is to use a large value of p or magnetization.

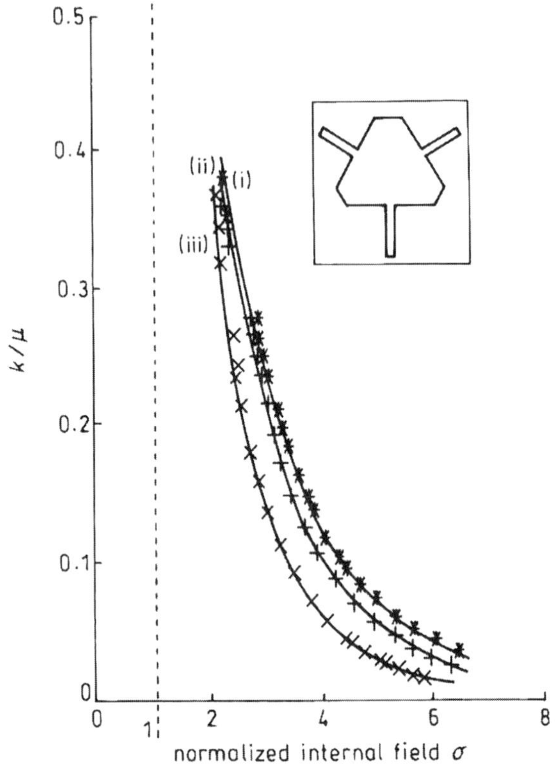

FIGURE 26.12 Relationship between normalized internal direct magnetic field intensity and gyrotropy.

26.11 COMPLEX GYRATOR CIRCUIT OF UHF CIRCULATORS

The quantities entering into the description of the one-port complex STUB-G gyrator circuit of a UHF junction using a weakly magnetized irregular hexagonal resonator are its quality factor (Q_L), its susceptance slope parameter (b′), and its gyrator conductance (g). The experimental evaluation of these quantities has been dealt with elsewhere in the text and is reproduced here for completeness only.

$$\frac{1}{Q_L} = \sqrt{3}\left(\frac{\omega_+ - \omega_-}{\omega_0}\right) \tag{26.30}$$

$$g = \sqrt{2(\text{VSWR}) - 1}, \quad g \geq 1 \tag{26.31}$$

$$g = \sqrt{\frac{2(\text{VSWR}) - 1}{(\text{VSWR})}}, \quad g \leq 1 \tag{26.32}$$

and

$$b' = gQ_L \tag{26.33}$$

The above relationships are also related by

$$g = \sqrt{3}b'\left(\frac{\omega_+ - \omega_-}{\omega_0}\right) \tag{26.34}$$

Q_L is obtained by measuring ω_0, ω_+, and ω_- versus direct field intensity; g is obtained by measuring the return loss or VSWR of the junction at one port at the midband frequency with the other two ports terminated in 50 Ω loads.

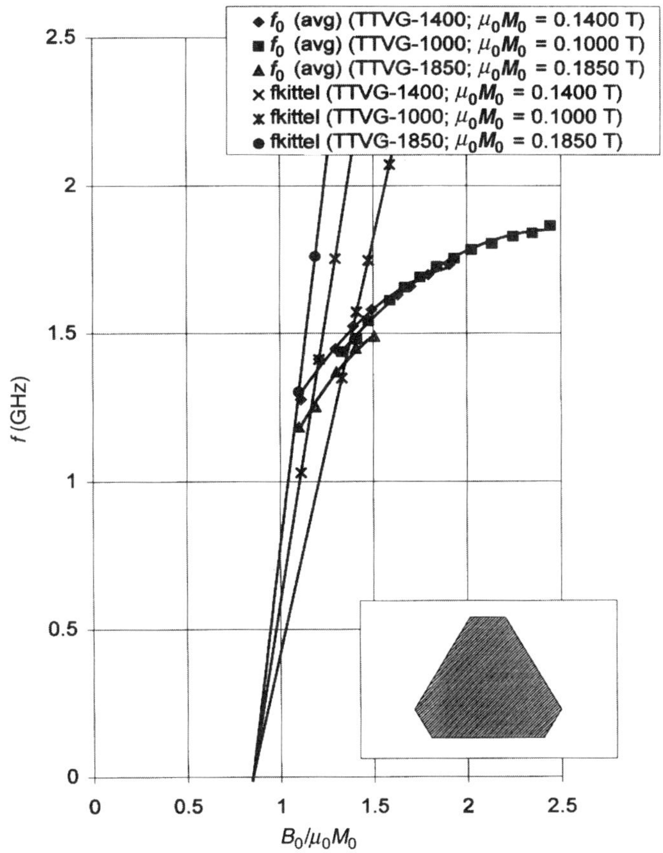

FIGURE 26.13 Frequency of UHF circulator using irregular gyromagnetic resonator for parametric values of magnetisation. (Reproduced with permission from J. Helszajn, Apollo Microwave Ltd. Montreal.)

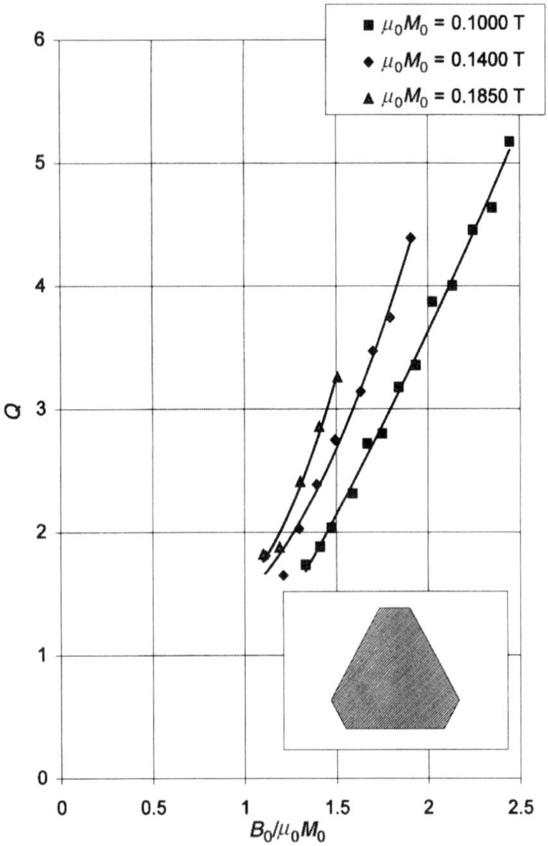

FIGURE 26.14 Quality factor of UHF circulator using irregular gyromagnetic resonator for parametric values of magnetisation. (Reproduced with permission from J. Helszajn, Apollo Microwave Ltd. Montreal.)

The data obtained in this way are summarized in Figs. 26.13–26.16. This has been done for three different garnet materials. The quality factor is experimentally deduced by measuring the split frequencies of the junction and the gyrator conductance by measuring its VSWR at the midband frequency. The susceptance slope parameter is then calculated in terms of the quality factor and the gyrator conductance.

26.12 REAL PART CONDITION

An important distinction between a stripline circulator using a gyromagnetic resonator biased above and below the Kittel line is the nature of the real part of the complex gyrator circuit. While it is essentially flat with frequency below the Kittel

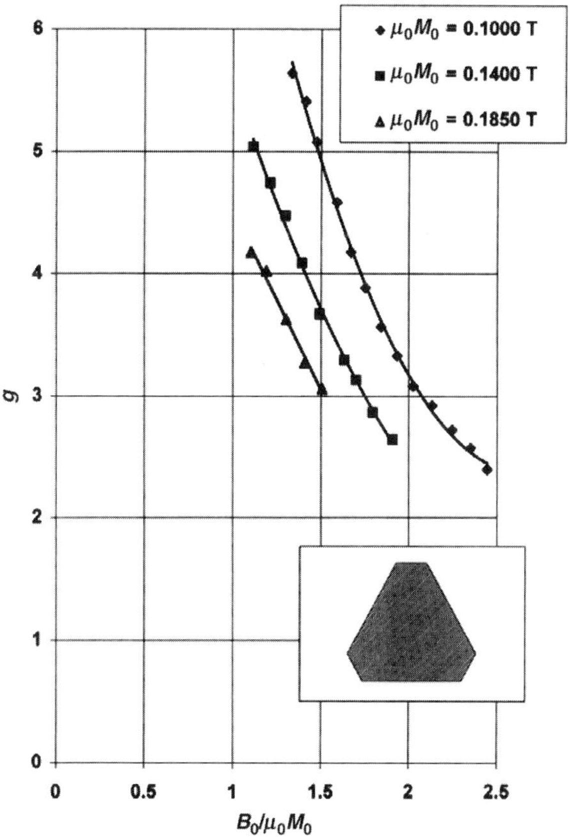

FIGURE 26.15 Gyrator conductance of UHF circulator using irregular gyromagnetic resonator for parametric values of magnetisation. (Reproduced with permission from J. Helszajn, Apollo Microwave Ltd. Montreal.)

line, it is not above it. One means of tackling this problem is to introduce a short UE between the terminals of the resonator and the external network, in order to wrap the real part about a constant conductance circle. In practice, the center conductor of the UE is often the same as that of the coupling angle so that experimentally it has no identity. The net effect, however, is to lower the midband frequency of the circulator.

In practice the ensuing shift in frequency is somewhat experimentally masked by resetting the direct magnetic field on the junction. It does produce a discrepancy, however, between the calculated and experimental field settings.

26.13 QUALITY FACTOR OF UHF CIRCULATORS

The quality factor of a junction circulator, as is now understood, may be calculated by measuring the split frequencies of the junction. It may also be estimated, however, by

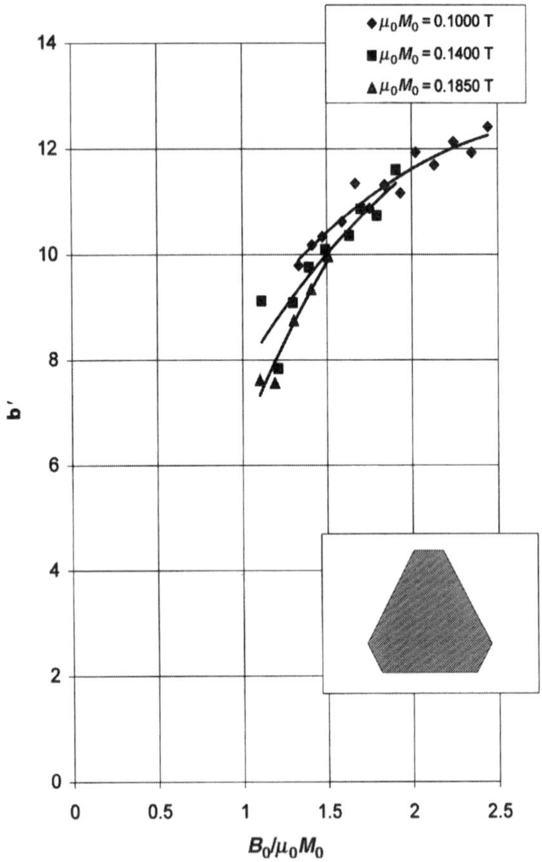

FIGURE 26.16 Susceptance slope parameter of UHF circulator using irregular gyromagnetic resonator for parametric values of magnetisation. (Reproduced with permission from J. Helszajn, Apollo Microwave Ltd. Montreal.)

analyzing the frequency response of the circulator by having recourse to the gain-bandwidth of a degree-2 arrangement. Figure 26.17 indicates one typical UHF response. A value of 2.30 is compatible with this data. In keeping with practice, this work suggests, that such values are realizable with the materials employed here without encroaching on the Kittel resonance line. The overall insertion loss of the circulator is 0.20 dB maximum. The size of the resonator employed in obtaining this result is somewhat smaller than that utilized in obtaining the raw mode charts. The ratio $B_0/\mu_0 M_0 = 1.24$.

Some estimate of the effective unloaded quality factor may also be deduced by having recourse to the insertion loss function between any two typical ports in Chapter 11.

$$L(\mathrm{dB}) = 20 \log_{10}(1 + Q_\mathrm{L}/Q_\mathrm{eff}) \tag{26.35}$$

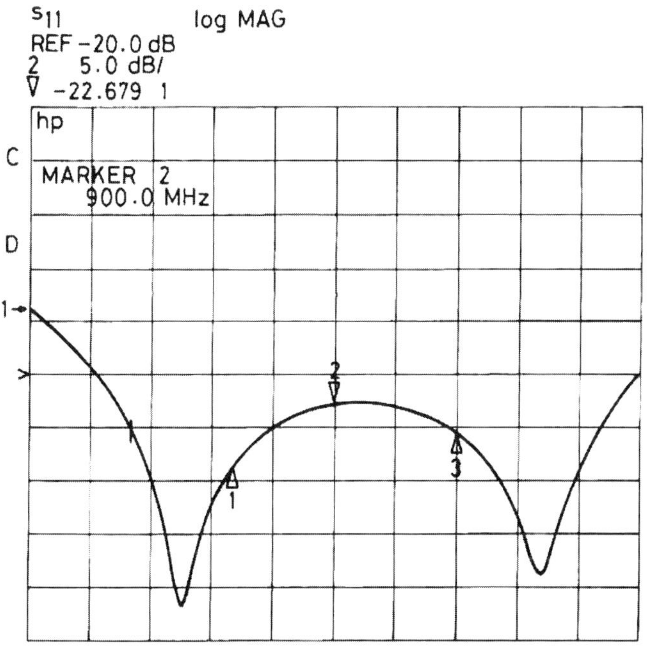

FIGURE 26.17 Frequency response of degree-2 junction circulator. (Reproduced with permission from J. Helszajn, C. S. Cheng, and W. D'Orazio, UHF irregular hexagonal gyromagnetic resonators symmetrically with magnetic walls, *IEE Proc. Microwave Antennas Propag*, Vol. 146, No. 6, Dec. 1999.)

where

$$\frac{1}{Q_{\text{eff}}} = \frac{1}{Q_+} + \frac{1}{Q_-} \tag{26.36}$$

$Q_{\pm}$ are the unloaded quality factors of the split eigen-networks of the problem region.

Taking Q_L as 2.5 (say) and the insertion loss as 0.10 dB, and dividing the overall loss equally between the impedance transformers in a typical transmission path and the resonator, gives Q_{eff} as 250.

CHAPTER TWENTY-SEVEN

Fabrication of Very Weakly and Weakly Magnetized Microstrip Circulators

27.1 INTRODUCTION

One difficulty with the design of a microstrip circulator is that the substrate thickness is often specified by equipment requirements rather than by the circulator design. One family of solutions that is fairly attractive for use in the construction of such devices is based on a weakly magnetized resonator in which the resonator shape is used in the synthesis problem. Some possible resonator shapes in the design of three-port circulators are indicated in Fig. 27.1. The realization of this class of circulator offers some useful insights into trade-offs between the resonator geometry, the substrate thickness, the maximum and minimum ripple levels of the network problem, and the gyrotropy of this circulator. The use of the alternate line transformer deserves a special mention in its design. The family of solutions outlined here is the only one possible at frequencies above about 50 GHz. One resonator for which a very weakly magnetized solution is available in closed form is the classic disk resonator. In this arrangement the susceptance slope parameter of the gyrator circuit is dependent on the substrate thickness, the quality factor, or gain bandwidth on the gyrotropy. In this sort of problem the gyrator conductance is the dependent quantity and is met once the other two parameters are specified. Another is the triangular geometry coupled at each of its symmetrical triplets of ports. While the quality factor of each resonator is of the same order, each has a quite different value of susceptance slope parameter. Specifically, the susceptance slope parameter of the side-coupled triangular resonator is three times that of a simple disk resonator, whereas that of

The Stripline Circulator: Theory and Practice. By J. Helszajn

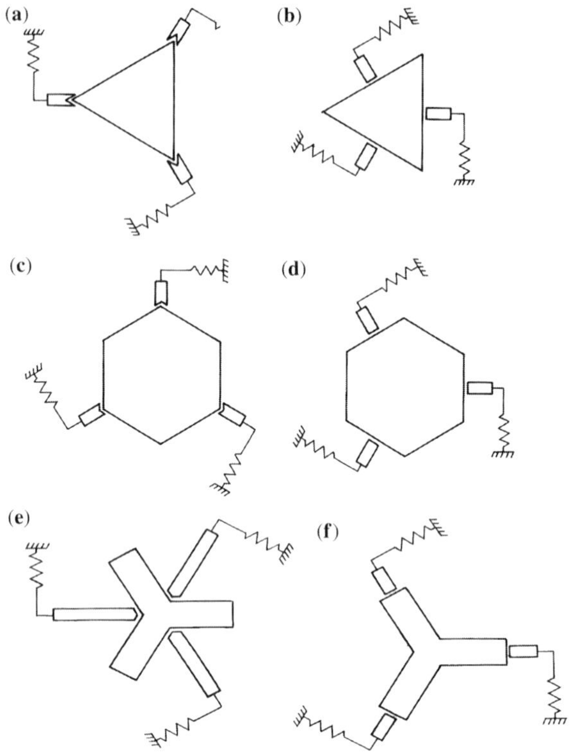

FIGURE 27.1 Schematic diagram of planar resonators for use in planar junction circulators.

the apex fed one is one-third that of the disk geometry. Each circuit is therefore associated with a quite different microwave solution. A complete description of a circulator using a very weakly magnetized regular hexagonal resonator also exists. Suffice it to note that its gyrator circuit is not very different from that of a simple disk resonator. The sort of specifications that can be realized with this class of circulation solutions depends, of course, on what is meant by a very weakly magnetized resonator. A resonator is either very weakly magnetized, weakly magnetized, moderately magnetized, or strongly magnetized. The meanings of these regimes have been dealt with in Chapters 13 and 28. While the gain-bandwidth product of the class of solutions dealt with in this chapter is restricted by the gyrotropy of the problem region, it still provides practical commercial solutions. Of course, different conclusions are possible if a moderately or a strongly magnetized resonator is utilized.

27.2 PARALLEL PLATE WAVEGUIDE MODEL OF MICROSTRIP CIRCULATORS

The usual approach to the design of microstrip passive circuits and circulators using weakly magnetized resonators is to replace the problem region with imperfect magnetic side walls by an equivalent waveguide model, and this is the method adopted here. Figure 27.2 indicates possible equivalences in the cases of microstrip circulators using circular and triangular planar resonators. Once any design is complete, in terms of the equivalent parallel plate waveguide approximation, it is necessary to invoke the relationship between the actual and effective parameters of the problem region.

The effective quantities entering into the description of a planar triangular resonator with open walls are defined in terms of an effective inscribed radius (r). No justification is sought for this approach other than that it permits some initial engineering parts to be laid out on the basis of an equivalent waveguide model.

A separate matter of concern in the design of such circuits is that if the fringing fields on a typical resonator contour are excessive, then it becomes difficult to preserve the definitions of its coupling angle and its shape. The effect on the resonator shape is particularly worrying in that it may no longer have any resemblance to the actual assumed one. This situation may be understood by recognizing that the effective side walls of a high impedance strip are displaced further

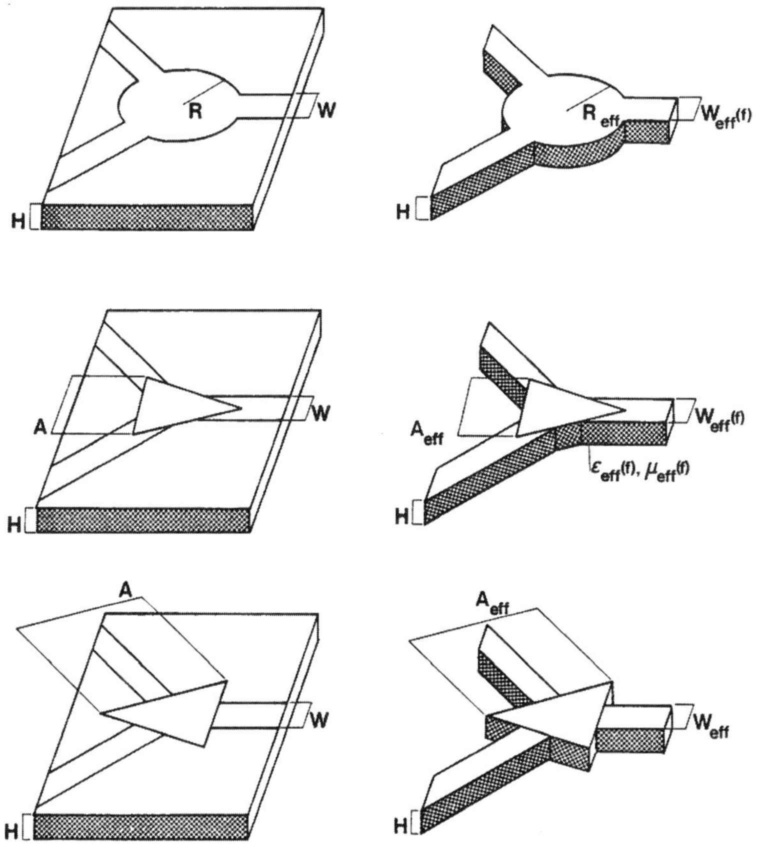

FIGURE 27.2 Equivalence between microstrip resonators and parallel waveguide circuits.

out from the physical ones than is the case with those of a low impedance strip. This sort of effect would readily distort the equilateral resonator in Fig. 27.1a (say) into something more like the irregular hexagonal resonator with concave side walls. It is also necessary to ensure that fringing effects do not corrupt the complex gyrator description of the circuit. One way to partially avoid both difficulties is to impose a lower bound on the aspect ratio (r/H) of the resonator. The influence of the aspect ratio of the resonator region upon the uniformity of the gyrotropy has also been mentioned in connection with this sort of device. One arbitrary bound on this quantity is

$$r/H \geq 5 \tag{27.1}$$

r either represents the inscribed radius defined by the triangular resonator,

$$r = A/2\sqrt{3} \tag{27.2a}$$

or the actual radius of the disk,

$$r = R \tag{27.2b}$$

The effective quantities entering into the descriptions of disk and triangular arrangements are

$$\begin{aligned} \varepsilon_{\text{eff}} &= 1 + q_{\text{e}}(\varepsilon_{\text{f}} - 1) \\ 1/\mu_{\text{eff}} &= 1 + q_{\text{m}}(1/\mu_e - 1) \\ A_{\text{eff}} &= A + (H \ln 2)/\pi \\ R_{\text{eff}} &= R + (H \ln 2)/\pi \end{aligned}$$

One possible definition of the filling factors (q_{e}, q_{m}) entering into the descriptions of the effective quantities is

$$q_{\text{e}} \approx q_{\text{m}} \approx \tfrac{1}{2}[1 + (1 + 5H/\text{r})^{-0.55}] \tag{27.3}$$

q_{e} or q_{m} is the filling factor of a strip of thickness H whose width (W) equals the diameter of a segmented resonator $2r$.

Strictly speaking, it is also necessary to introduce the notion of an effective gyrotropy in the description of the quality factor of the resonator. One semiempirical formulation is

$$(\kappa/\mu)_{\text{eff}} \approx q_m(\kappa/\mu) \tag{27.4}$$

The effective permeability μ_{eff} is given separately:

$$\mu_{\text{eff}} = \frac{\mu^2 - (q_{\text{m}}\kappa)^2}{\mu} \tag{27.5}$$

where q_{m} is the filling factor introduced in Eq. (27.3). It is assumed here that $\mu_{\text{eff}} = \mu$ and $\kappa_{\text{eff}} = q_{\text{m}}\kappa$.

27.3 VERY WEAKLY MAGNETIZED PROBLEM REGION

A unique feature of a junction circulator using a very weakly or weakly magnetized resonator (on which the design procedure of this chapter rests) is that its complex gyrator circuit is essentially independent of the coupling angle at its ports. This means that the design of the transformer circuit does not perturb the junction in any way and that its adjustment may be dealt with by the appropriate parallel plate model of a microstrip line on a demagnetized ferrite substrate.

The design of a very weakly magnetized microstrip circulator using a disk or a triangular resonator proceeds with the aid of its equivalent waveguide model and a knowledge of the corresponding design procedure for stripline circulators. If the in-phase eigen-network of the device may be idealized by a frequency independent short-circuit boundary condition at its terminals, then its complex gyrator circuit may be realized in the form indicated in Fig. 27.3. In this circuit, G is the absolute conductance at the terminals of the junction and also at the output microstrip line, B' is the absolute susceptance slope parameter, and Q_{L} is the loaded Q-factor. Tables 27.1–27.3 summarize some classic design equations for the realization of microstrip circulators using weakly magnetized disk and triangular resonators.

The first column in these tables gives the susceptance slope parameter, gyrator conductance, and loaded Q-factor at the terminals of the device. The second column defines the ratio of the entries of the permeability tensor κ/μ of the magnetized resonator in terms of the loaded Q-factor (or in terms of the gyrator conductance) and susceptance slope parameter of the device.

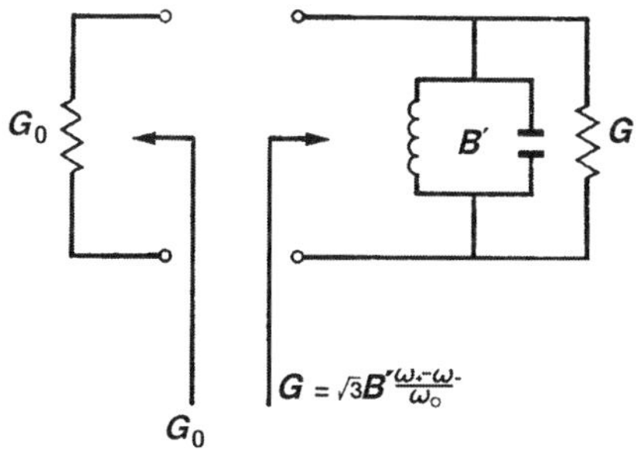

FIGURE 27.3 1-port complex gyrator circuit of 3-port junction circulator.

TABLE 27.1 Design Data for Very Weakly Magnetized Disk Resonator ($k_{\mathrm{eff}}(f)R_{\mathrm{eff}}(f) = 1.84$)

Network Variables	Magnetic Variables	Physical Variables	Frequency Variables
B'	–	$\dfrac{0.740\zeta_0(k_{\mathrm{eff}}(f)R_{\mathrm{eff}}(f))^2}{\mu_{\mathrm{e,eff}}(f)k_0H}$	–
G	$\left[\dfrac{2\sqrt{3}B'}{(k_{\mathrm{eff}}(f)R_{\mathrm{eff}}(f))^2 - 1}\right]\left(\dfrac{\kappa}{\mu}\right)_{\mathrm{eff}}$	$\zeta_{\mathrm{eff}}\zeta_0\dfrac{W_{\mathrm{eff}}(f)}{H}$	$\sqrt{3}B'\left(\dfrac{\omega_+ - \omega_-}{\omega_0}\right)$
$Q_{\mathrm{L}}(\mathrm{eff})$	$\dfrac{(k_{\mathrm{eff}}(f)R_{\mathrm{eff}}(f))^2 - 1}{2\sqrt{3}}\left(\dfrac{\mu}{\kappa}\right)_{\mathrm{eff}}$	$\dfrac{0.740\zeta_0(k_{\mathrm{eff}}(f_{\mathrm{eff}}(f))^2}{G\mu_{\mathrm{e,eff}}(f)k_0H}$	$\left[\sqrt{3}\left(\dfrac{\omega_+ - \omega_-}{\omega_0}\right)\right]^{-1}$

Once the magnetic variables of the resonator are fixed, its linear dimensions may be calculated using the entries in the third column in Tables 27.1–27.3. The radius of the circular disk resonator or the side dimension of the triangular one is obtained from the respective cutoff numbers. The substrate thickness H is then obtained from the specification of the susceptance slope parameter. The calculation of $W_{\mathrm{eff}}(f)$ completes the design and ensures that the network, magnetic, and physical variables in Tables 27.1–27.3 are compatible. If the specifications of the device are incompatible with the substrate thickness H, or the width $W_{\mathrm{eff}}(f)$ of the connecting lines is incompatible with the resonator area, the design must be repeated with a less severe specification.

The last column in these tables gives the description of the gyrator conductance and loaded Q-factor in terms of the split frequencies of the two counterrotating

TABLE 27.2 Design Data for Very Weakly Magnetized Triangular Resonator Coupled at Corners ($k_{\mathrm{eff}}(f)A_{\mathrm{eff}}(f) = 4\pi/3$)

Network Variables	Magnetic Variables	Physical Variables	Frequency Variables
B'	–	$\dfrac{0.0475\zeta_0(k_{\mathrm{eff}}(f)A_{\mathrm{eff}}(f))^2}{\mu_{\mathrm{e,eff}}(f)k_0H}$	–
G	$\left[\dfrac{3B'}{\pi}\right]\left(\dfrac{\kappa}{\mu}\right)_{\mathrm{eff}}$	$\zeta_{\mathrm{eff}}\zeta_0\dfrac{W_{\mathrm{eff}}(f)}{H}$	$\sqrt{3}B'\left(\dfrac{\omega_+ - \omega_-}{\omega_0}\right)$
$Q_L(\mathrm{eff})$	$\dfrac{\pi}{3}\left(\dfrac{\mu}{\kappa}\right)_{\mathrm{eff}}$	$\dfrac{0.0475\zeta_0(k_{\mathrm{eff}}(f)A_{\mathrm{eff}}(f))^2}{G\mu_{\mathrm{e,eff}}(f)k_0H}$	$\left[\sqrt{3}(\dfrac{\omega_+ - \omega_-}{\omega_0})\right]^{-1}$

TABLE 27.3 Design Data for Very Weakly Magnetized Triangular Resonator Coupled Midway Along Side Dimensions ($k_{\mathrm{eff}}(f)A_{\mathrm{eff}}(f) = 4\pi/3$)

Network Variables	Magnetic Variables	Physical Variables	Frequency Variables
B'	–	$\dfrac{0.4275\zeta_0(k_{\mathrm{eff}}(f)A_{\mathrm{eff}}(f))^2}{\mu_{\mathrm{eff}}(f)k_0H}$	–
G	$\left[\dfrac{3B'}{\pi}\right]\left(\dfrac{\kappa}{\mu}\right)_{\mathrm{eff}}$	$\zeta_{\mathrm{eff}}\zeta_0\dfrac{W_{\mathrm{eff}}(f)}{H}$	$\sqrt{3}B'\left(\dfrac{\omega_+ - \omega_-}{\omega_0}\right)$
$Q_L(\mathrm{eff})$	$\dfrac{\pi}{3}\left(\dfrac{\mu}{\kappa}\right)_{\mathrm{eff}}$	$\dfrac{0.4275\zeta_0(k_{\mathrm{eff}}(f)A_{\mathrm{eff}}(f))^2}{G\mu_{\mathrm{eff}}(f)k_0H}$	$\left[\sqrt{3}\left(\dfrac{\omega_+ - \omega_-}{\omega_0}\right)\right]^{-1}$

modes of the magnetized resonator. The quantities in Tables 27.1–27.3 are related through the following relationship that has already appeared in a number of places in this text:

$$G = \sqrt{3}B'\left(\frac{\omega_+ - \omega_-}{\omega_0}\right) \tag{27.6}$$

This equation indicates that G is proportional to the product of the difference between the two split frequencies of the counterrotating modes of the magnetized junction and the susceptance slope parameter of the junction. This is a general result for any very weakly magnetized junction circulator.

In order to have a trustworthy synthesis procedure for the design of these devices, it is necessary to place an upper bound on the gyrotropy (splitting) over which the entries in Tables 27.1–27.3 may be used with confidence. A very weakly magnetized junction is defined for the purpose of this chapter as one for which the split counterrotating eigen-networks or eigenvalues may be described by single split poles and for which the in-phase eigenvalue may be represented by a frequency independent electric wall at the terminals of the resonator. One definition of a weakly magnetized junction is therefore the verification of this condition. Another measure that is in keeping with its very weakly magnetized model is to ensure that its quality factor is independent of the coupling angle at the resonator terminals. Still another measure is to ensure that the susceptance slope parameter is independent of both its gyrotropy and its coupling angle. The very weakly magnetized model in the case of a disk resonator already displays significant deterioration in the closed form description of its complex gyrator circuit when the value of its gyrotropy (κ) is as little as 0.25. The gyrotropy in the case of a junction using a side-coupled triangle is likewise restricted to about 0.25. While such values of gyrotropy place some restriction on the gain-bandwidth product of this sort of solution, it is nevertheless adequate in the design of practical specifications. The moderately magnetized region actually produces a more attractive gain-bandwidth product but at the cost of a more difficult adjustment procedure.

27.4 WEAKLY MAGNETIZED PROBLEM REGION

One approach that may be used to extend the description of the very weakly magnetized description of a junction circulator using a disk resonator has been to introduce suitable correction terms based on some exact calculations. A scrutiny of this solution indicates that its quality factory and susceptance slope parameter deteriorate more rapidly than its gyrator conductance. This feature has been observed elsewhere. One means of dealing with this problem is to write the actual absolute susceptance slope parameter $B'(\kappa/\mu)$ in terms of a modified quality factor $Q_L(\kappa/\mu)$ and the weakly magnetized description of the gyrator conductance G as

$$B'(\kappa/\mu) = Q_L(\kappa/\mu)G, \ 0 \le \kappa \le 0.35 \tag{27.7}$$

or

$$B'(\kappa/\mu) = Q_L(\kappa/\mu)\left(\frac{B'(0)}{Q_L}\right), \ 0 \le \kappa \le 0.35 \tag{27.8}$$

G is the absolute value of the gyrator conductance in Tables 27.1–27.3 and Q_L is reproduced from Chapter 28:

$$Q_L \approx 0.689\left(\frac{\mu}{\kappa}\right) + \left[0.046 - 2.632\left(\frac{\kappa}{\mu}\right) + 3.551\left(\frac{\kappa}{\mu}\right)^2\right]$$
$$0 \le \kappa/\mu \le 0.30, \ 0.10 \le \psi \le 0.50 \tag{27.9}$$

One suitable polynomial expansion for the quality factor $Q_L(\text{eff})$ of a weakly magnetized microstrip circulator using a disk resonator is obtained by introducing the filling factor q_m in the stripline relationship:

$$Q_L(\text{eff}) \approx 0.689\left(\frac{1}{q_m}\right)\left(\frac{\mu}{\kappa}\right) + \left[0.046 - 2.632\, q_m\left(\frac{\kappa}{\mu}\right) + 3.551 q_m^2\left(\frac{\kappa}{\mu}\right)^2\right]$$
$$0 \le q_m(\kappa/\mu) \le 0.30, \ 0.10 \le \psi \le 0.50 \tag{27.10}$$

The description of the quality factor of a junction using a weakly magnetized triangular resonator based on some calculations on a side-coupled triangular resonator based on the contour integral method is

$$Q_L(\text{eff}) = \frac{\pi}{3}\left(\frac{1}{q_m}\right)\left(\frac{\mu}{\kappa}\right) + \left[1.486 - 12.71\,q_m\left(\frac{\kappa}{\mu}\right) + 19.74\,q_m^2\left(\frac{\kappa}{\mu}\right)^2\right]$$

$$0 \le (\kappa/\mu) \le 0.35,\ 0.52 \le \psi \le 1.20 \qquad (27.11)$$

where

$$\sin\psi = W/2R \qquad (27.12)$$

and

$$\sin\psi = W/2r \qquad (27.13)$$

respectively.

The values of the quality factors bracketed by the above two relationships are valid for both the very weakly and weakly magnetized resonators adopted here. Since it is usually desirable, in any design, to maximize the gain-bandwidth product of the load, the only values of gyrotropy that are of any interest are those associated with the upper bounds defined by these intervals. The corresponding lower bounds on the quality factors of the two arrangements are 1.86 and 2.42, respectively. These results suggest, other factors being equal, that the gain-bandwidth product of a junction circulator employing a circular disk resonator is more attractive for the design of commercial devices than one using a triangular one. One feature of either solution is that the quality factor is independent of the coupling angle.

A separate scrutiny of the first circulation condition suggests that it also needs some modification. One possible explanation of this sort of discrepancy is that, strictly speaking, the resonant frequency displayed by a junction circulator coincides with that at which the reflection eigenvalues of the in-phase and degenerate counterrotating eigen-networks are out of phase rather than with that of the counterrotating ones.

This means that the actual boundary condition that needs to be satisfied is

$$X_{in} = 0 \qquad (27.14)$$

instead of

$$X_1 = 0 \qquad (27.15)$$

X_{in} represents the imaginary part of the impedance of the complex gyrator circuit for the case for which the reactance of the in-phase eigen-network is different from zero; X_1 represents the same quantity when it is equal to zero.

The second circulation condition is met by satisfying the appropriate entries specified by Eqs. (27.7) and (27.8).

27.5 EXPERIMENTAL EVALUATION OF COMPLEX GYRATOR CIRCUITS

The split frequencies of one microstrip junction using a side coupled triangular resonator is indicated in Fig. 27.4 for a number of different values of magnetization. The quality factor for two of the materials is indicated in Fig. 27.5. The relationship between the quality factor of this sort of circuit and the gyrotropy of the resonator is self-evident. Figure 27.6 shows some results for a disk resonator for three different materials, which may be used to verify the relationships between the magnetic and frequency variables tabulated in columns 2 and 4 in Tables 27.1–27.3. Figure 27.7 indicates that, for a given gyrotropy, the quality factor of a disk resonator is lower than that of the corresponding triangular geometry, in keeping with theory.

In order to cater for saturation effects in a magnetized substrate, an initial value of loaded Q-factor is also sometimes defined. This quantity is constructed by extending

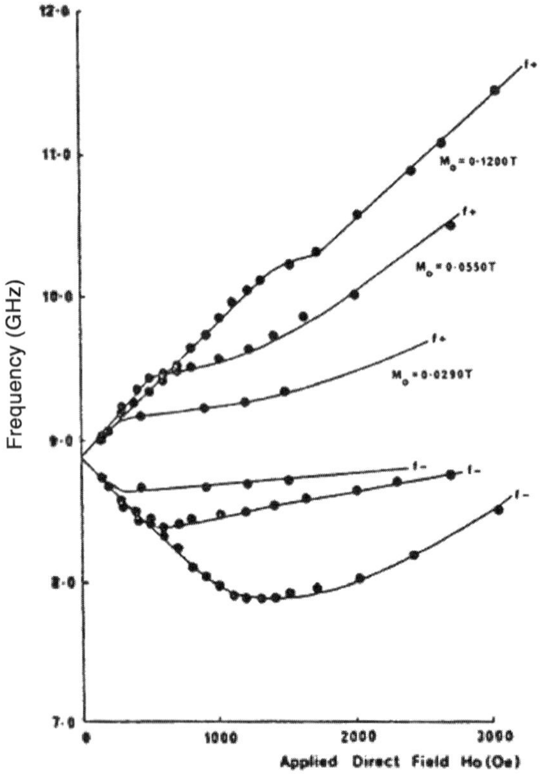

FIGURE 27.4 Split frequencies versus applied direct magnetic field for loosely coupled side-coupled triangular resonator on garnet substrate ($\mu_0 M_0 = 0.0290$T, 0.0550T and 0.1200T). (Reproduced with permission from J. Helszajn, Fabrication of very weakly and weakly magnetised microstrip circulators, *IEEE Trans. on Microwave Theory Tech.*, MTT-46, pp. 439–449, May 1998.)

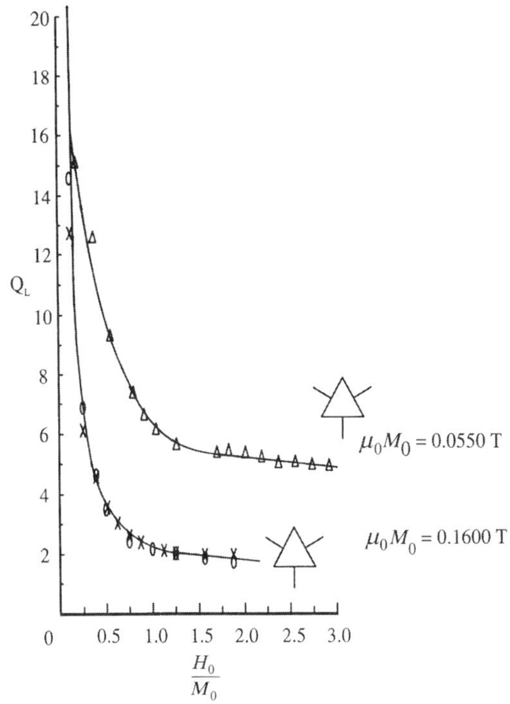

FIGURE 27.5 Quality factor versus applied direct magnetic field for loosely coupled triangular resonator on different garnet substrates ($p = 0.190$ and 0.533). (Reproduced with permission from J. Helszajn, Fabrication of very weakly and weakly magnetised microstrip circulators, *IEEE Trans. on Microwave Theory Tech.*, MTT-46, pp. 439–449, May 1998.)

the angle suspended by the split frequencies at the origin to the value of applied direct magnetic field at saturation. This field coincides with that for which the internal direct magnetic field (H_i) is equal to zero:

$$H_i = 0 \tag{27.16}$$

where

$$H_i = H_0 - N_z M_0 \tag{27.17}$$

M_0 is the saturation magnetization (A/m), N_z is the demagnetizing factor along the axis of the resonator, and H_0 is the direct magnetic field intensity (A/m).

If the material is not saturated, then it is necessary to replace q_m in the description of Q_L(eff) by $q_m(M/M_0)$, where M is the actual magnetization.

The relationship between the gyrotropy and effective quality factor Q_L(eff) of a microstrip resonator using a triangular resonator is illustrated in Fig. 27.8 for

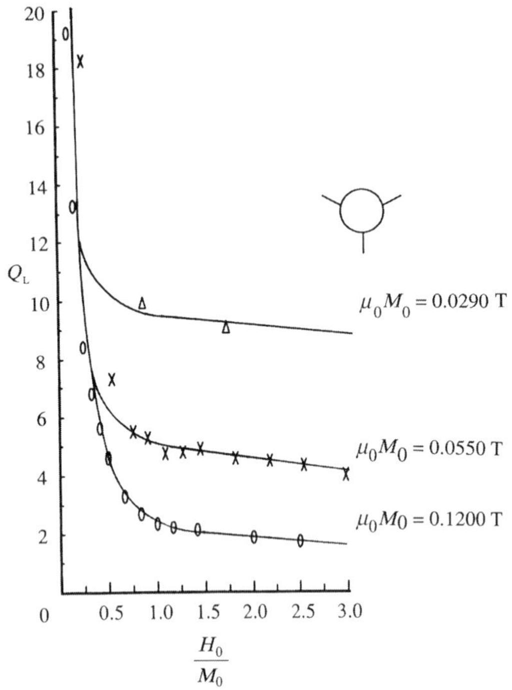

FIGURE 27.6 Quality factor versus applied direct magnetic field for loosely coupled disk resonator on different garnet substrates ($p = 0.090$, 0.174 and 0.388). (Reproduced with permission from J. Helszajn, Fabrication of very weakly and weakly magnetised microstrip circulators, *IEEE Trans. on Microwave Theory Tech.*, MTT-46, pp. 439–449, May 1998.)

$q_m = 0.80$ and 1.0. Some experimental results that rely on a definition of Q_L^i (eff) are superimposed separately on this illustration. While these quality factor curves appear at first sight well behaved, this is not in itself sufficient since it is also necessary to ensure that the susceptance slope parameter is independent of the gyrotropy over the same interval.

Figure 27.9 depicts some experimental data at about 9 GHz on the gyrator conductance of junction circulators using disk, side, and apex coupled triangular resonators printed on a 0.635 mm substrate with a saturation magnetization ($\mu_0 M_0$) equal to 0.1200 T. This sort of data is obtained by making use of the relationship between the midband return loss of a magnetized junction and its gyrator conductance with the two output ports terminated by 50 Ω loads. It is evident from this result that the gyrator conductance of the three arrangements under consideration are not in the ratio 1:3:9 of the respective values of the susceptance slope parameters in Tables 27.1–27.3. One obvious reason for these discrepancies is that the values of the quality factor of disk and triangular resonators are slightly different. Another is that the values of the aspect ratios of the

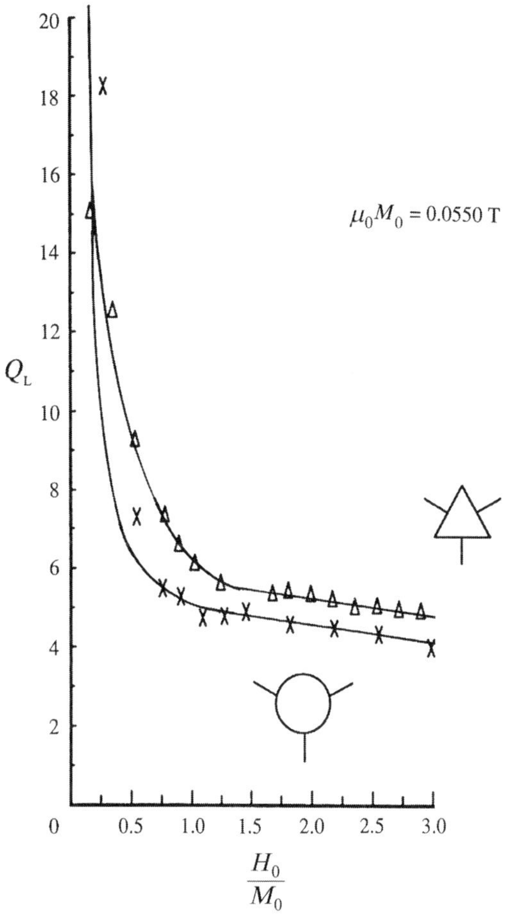

FIGURE 27.7 Comparison between quality factors of disk and triangular resonators. (Reproduced with permission from J. Helszajn, Fabrication of very weakly and weakly magnetised microstrip circulators, *IEEE Trans. on Microwave Theory Tech.*, MTT-46, pp. 439–449, May 1998.)

resonators employed to obtain this data ($r/H \approx 2.73$) are somewhat lower than that stipulated earlier in order to adequately reproduce the resonator shape of the problem region.

A scrutiny of the network problem readily indicates that none of the complex gyrator circuits displayed by these geometries are (at first sight) appropriate at 9 GHz for the design of quarter-wave coupled devices on a 0.635 mm substrate. Some experimental hardware is illustrated in Fig. 27.10.

A feature of the general network problem is that while its maximum ripple level in the passband and its bandwidth specification are essentially fixed by the quality factor of the load, the value of its minimum ripple level may be used to control the level of

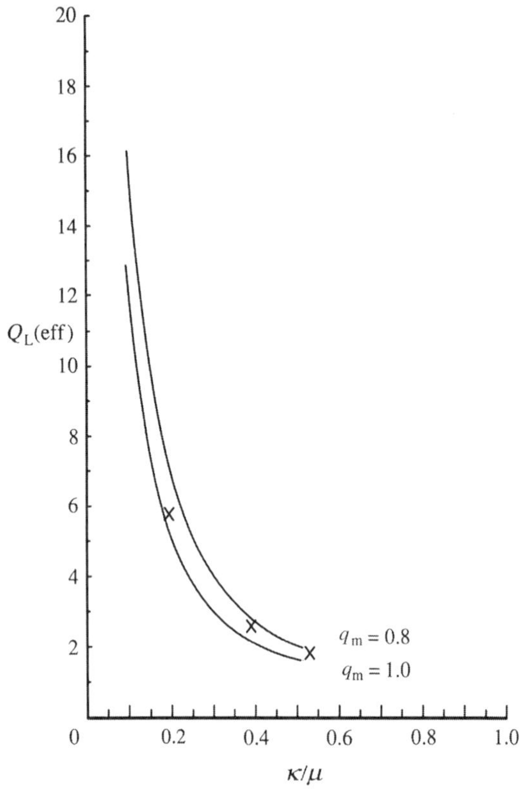

FIGURE 27.8 Comparisons between quality factors based on closed form, numerical method and experimental values of a triangular resonator versus gyrotropy ($r/H = 2.73$). (Reproduced with permission from J. Helszajn, Fabrication of very weakly and weakly magnetised microstrip circulators, *IEEE Trans. on Microwave Theory Tech.*, MTT-46, pp. 439–449, May 1998.)

its susceptance slope parameter. The equipment maker is seldom interested in this latter quantity so that it may in practice be used to absorb any uncertainty in the precise characterization of the absolute value of the susceptance slope parameter of the gyrator network.

Another feature that explains the sometimes unexpected success of commercial workers in this area is that while the frequency response of this class of circuit is akin to that of a degree-2 filter circuit, no precise lower bound in this instance is usually imposed on its quality factor. An assurance that its upper bound is not violated is therefore sufficient for design.

If a normalized value of 8 is taken for the susceptance slope parameter by way of example and if the substrate thickness is taken as 0.635 mm, then an end fed triangular resonator is best at about 1.5 GHz, a disk one at 4.5 GHz, and a side-coupled triangular resonator at 13.5 GHz.

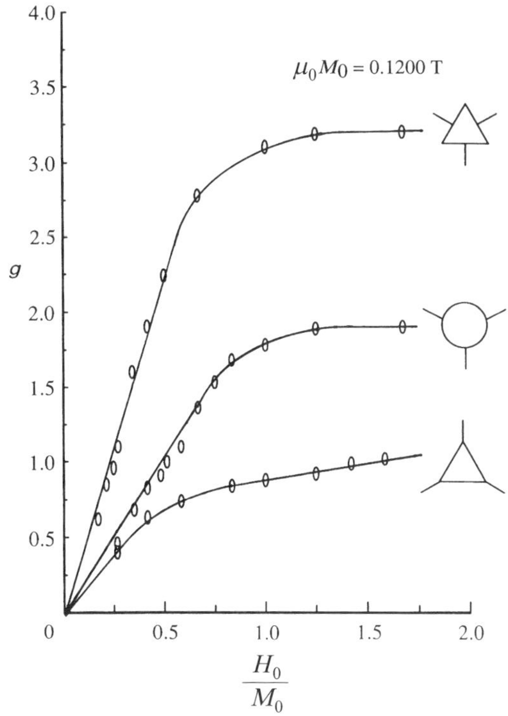

FIGURE 27.9 Gyrator conductance versus applied direct magnetic field for disk, apex and side-wall coupled triangular resonators on garnet substrate ($p = 0.388$). (Reproduced with permission from J. Helszajn, Fabrication of very weakly and weakly magnetised microstrip circulators, *IEEE Trans. Microwave Theory Tech.*, 01. MTT-46, pp. 439–449, May1998.)

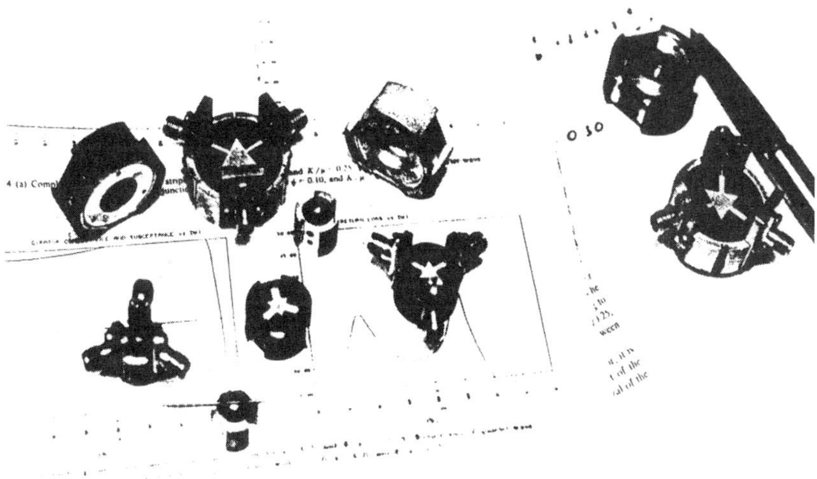

FIGURE 27.10 Photograph of microstrip circuits.

27.6 SYNTHESIS PROCEDURE

If the substrate thickness forms part of the specification then the design procedure starts by evaluating the susceptance slope parameter of disk and triangular resonators based on the modified weakly magnetized model from a statement of the frequency and substrate thickness and its dielectric constant. In the example studied here,

$$f_0 = 3950\,\text{MHz}$$
$$H = 0.635\,\text{mm}$$

Both the susceptance slope parameter and the quality factor of the complex gyrator circuit are in the modified weakly magnetized model of the junction dependent on the effective gyrotropy. The latter quantity must therefore be fixed before proceeding with design. As the gain-bandwidth product of the specification is the paramount quantity, it determines the gyrotropy in practice. The value for the triangular geometry is adopted here:

$$\kappa_{\text{eff}} = 0.35$$

Recall that the gyrotropy necessary to establish a given gain-bandwidth product is somewhat lower in the case of a junction using a disk geometry than in that employing a triangular one.

The frequency dependent effective permeabilities associated with the effective gyrotropy of the circuit are

$$\mu_{\text{e}}(f) = 0.88$$
$$\mu_{\text{e,eff}}(f) = 0.90$$

The effective constitutive parameters cannot be calculated until r/H or R/H is available. A trial calculation based on the ideal magnetic wall model of the problem region must therefore be employed, in the first instance, to obtain this quantity.

$$q_{\text{m}} = 0.87$$

In order to be able to compare the values of the normalized susceptance slope parameter of the three weakly magnetized resonator shapes under consideration, it is assumed in the first place that all three designs are based on the same gyrotropy. The required values are then obtained by scaling the susceptance slope parameters obtained in Tables 27.1–27.3 by having recourse to Eqs. (27.7) and (27.8):

$$\mu_{e,eff}(f)b' = 1.90, \quad \text{apex-coupled triangular resonator}$$
$$\mu_{e,eff}(f)b' = 5.71, \quad \text{disk resonator}$$
$$\mu_{e,eff}(f)b' = 17.11, \quad \text{side-coupled triangular resonator}$$

A realizable specification that coincides with each resonator shape may now be obtained once the quality factor of each junction is evaluated in terms of its gyrotropy. The effective calculated value for the triangular resonator is

$$Q_L(\text{eff}) = 2.42$$

The network problem now indicates that the side-coupled triangular resonator is perhaps the most practical solution at this frequency. One possibility that is compatible with the calculated value of b' is

$$S(\text{min}) = 1.075$$
$$S(\text{max}) = 1.15$$
$$2\delta_0 = 0.22$$

The equivalent gyrator circuit for this specification is defined by

$$b' = 17.22$$
$$g = 7.12$$
$$y_t = 2.86$$

If there is no restriction on the thickness of the substrate, then the design can proceed on the basis of a specification. This may be done as long as the gain-bandwidth product can be realized by the gyrator model under consideration. One possible specification (say) is given by

$$S(\text{min}) = 1.05$$
$$S(\text{max}) = 1.15$$
$$2\delta_0 = 0.26$$

The equivalent gyrator circuit is described by

$$b' = 12.0$$
$$g = 6.0$$
$$Q_L(\text{eff}) = 2.0$$
$$y_t = 2.63$$

This specification can be realized by a disk resonator on a substrate with a thickness of 0.345 mm. The required effective gyrotropy is

$$\kappa_{\text{eff}} = 0.28$$

This calculation as well as that of the previous example also suggests that small variations in the quality factor of the specifications can produce large variations in the overall frequency response and in the details of the complex gyrator circuit.

27.7 COMMERCIAL PRACTICE

The purpose of this section is to summarize the design and performance of one commercial design using a triangular resonator at 5 GHz and two designs in the open literature using disk ones at 4 GHz and 9 GHz, respectively. While the details of each adjustment (in the absence of a Smith chart display) are not sufficiently robust to draw firm conclusions about any agreement between design and practice, it does draw attention to the wide disparities met in the choices of the design parameters of practical circulators.

The frequency response of one commercial 5 GHz quarter-wave coupled device using a side-coupled triangle on a 1.90 mm substrate is indicated in Fig. 27.11. Its performance is specified by

$$S(\max) \approx 1.16$$
$$S(\min) \approx 1.04$$
$$2\delta_0 \approx 0.21$$

If an idealized template is superimposed on this data, then its gyrator circuit is defined by

$$b' = 23.98$$
$$g = 9.137$$
$$Q_L = 2.625$$

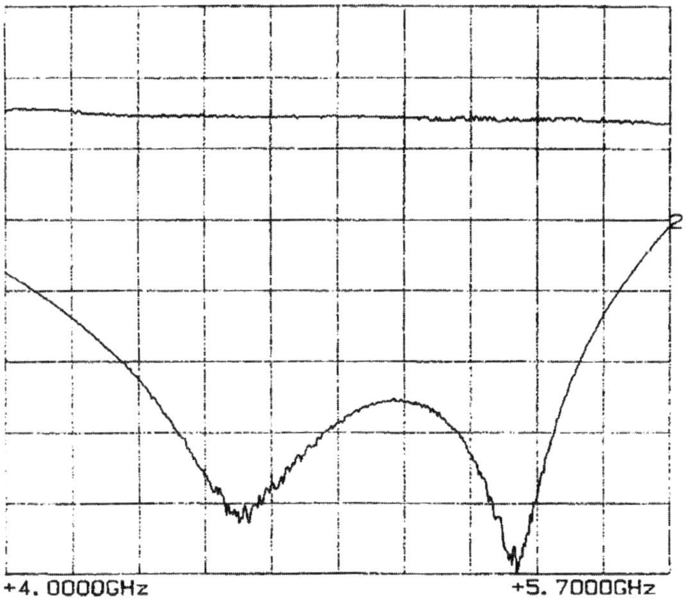

FIGURE 27.11 Frequency response of quarter-wave coupled junction circulator using side-coupled triangular resonator ($r/H = 1.44$ mm). (Courtesy of W. T. Nisbet, MESL-RACAL).

Its design is essentially specified by

$$\kappa = 0.56$$
$$q_{\mathrm{m}} = 0.73$$
$$H_0/M_0 \approx 1.20$$
$$r/H = 1.44$$

A quarter-wave coupled 4 GHz circulator using a disk gyromagnetic resonator on a 0.625 mm substrate has also been described. Its performance is defined by

$$S(\max) \approx 1.20$$
$$S(\min) \approx 1.15$$
$$2\delta_0 \approx 0.29$$

If an idealized template is superimposed on this data, then its complex gyrator circuit may be described by

$$b' = 5.97$$
$$g = 3.39$$
$$Q_{\mathrm{L}} = 1.76$$

Its design parameters are defined by

$$\kappa = 0.53$$
$$q_{\mathrm{m}} = 0.75$$
$$H_0/M_0 \approx 1.0$$
$$R/H = 6.0$$

Still another quarter-wave device using a disk resonator printed on a 0.635 mm substrate has been described in the open literature at 9 GHz. Its frequency response is defined by

$$S(\max) \approx 1.15$$
$$S(\min) \approx 1.10$$
$$2\delta_0 \approx 0.26$$

If an idealized template is also superimposed on this data, then

$$b' = 7.80$$
$$g = 4.24$$
$$Q_{\mathrm{L}} = 1.84$$

In this instance the design details are described by

$$\kappa = 0.54$$
$$q_{\mathrm{m}} = 0.811$$
$$H_0/M_0 \approx 1.43$$
$$R/H = 3.0$$

The various microstrip circulators discussed here suggest that some trade-off exists between the network specification of the device, the shape of the resonator, the substrate thickness, and indeed the gyrotropy. This observation is of significant value in the synthesis problem and goes a long way to explain the often surprising success of the development engineer. When all is said and done, an irregular hexagonal resonator is perhaps the ideal shape for the design of a 4 GHz device on a 0.635 mm thick substrate.

CHAPTER TWENTY-EIGHT

The Stripline Circulator: Theory and Practice

The principles and difficulties entering into the design of the stripline circulator are outlined in this chapter. The circulator has its origin in the unitary condition. It states that matching any symmetrical nonreciprocal three-port junction is both necessary and sufficient for the realization of an ideal circulator. The origin of the nonreciprocity has its root in the different interactions between an electron spin in a magnetized magnetic insulator and radiofrequency counterrotating magnetic fields. A typical structure is the irregular hexagonal gyromagnetic resonator coupled by one of two possible triplet of ports treated in Chapter 16. The Green's procedure, the finite element method, and the contour integral one are three numerical procedures that will always succeed in describing any junction with threefold symmetry. A surprising result is that its so-called complex gyrator circuit at any port is, under simplifying conditions, a one-port degree-1 G-STUB load with entries that are simple linear combinations of its counterrotating eigen-networks. Such a circuit is completely characterized by its gyrator conductance, susceptance slope parameter, and quality factor. In practice the gain bandwidth of the device is restricted by the quality factor of the complex gyrator circuit. The frequency response of this degree-1 network may be extended to a degree-n arrangement by having recourse to classic filter theory. An important and attractive feature of the network problem is that the absolute values of the elements of the complex gyrator circuit, for a given gain bandwidth, are controlled by the minimum ripple level of the passband response. More generally, the complex gyrator circuit consists of a series element that is equal to the reactance of the in-phase eigen-network within a factor of $4/3$ in cascade with the G-STUB load. The dissipation between any two ports is described separately by the unloaded quality factors of its eigen-networks and by the loaded quality factor of the one-port gyrator circuit in a simple way. A difficulty entering

The Stripline Circulator: Theory and Practice. By J. Helszajn

into the design of this class of device is the need to cater for the effects of the open walls on the character of the complex gyrator circuit. This effect is particularly serious in resonators with small ferrite aspect ratios. The meaning of very weakly, weakly, moderately, and strongly magnetized gyromagnetic resonators is given special attention. Another concern is the effect of small air gaps in the assembly of commercial products. Nonlinear loss or spinwave instability at large signal level and third order intermodulation products are other matters of concern. Edge mode effects and the radial variation of the shape demagnetizing factor perpendicular to a ferrite disk are two other features that can cause discrepancies between theory and practice.

28.1 COMPLEX GYRATOR CIRCUIT OF WEAKLY MAGNETIZED JUNCTION CIRCULATOR

A description of the complex gyrator circuit of any junction circulator is mandatory for its design. It consists of a one-port STUB-R load provided the frequency variation of its in-phase eigen-network may be neglected compared to those of its split counter-rotating ones and provided that its resonator is weakly magnetized. The elements of the STUB-R load obtained under these assumptions are then described in terms of the sum and differences of the counterrotating admittance eigenvalues of the junction. The gyrator circuit under consideration is reproduced in Fig. 28.1.

$$Y_{\text{in}} = G_{\text{in}} + jB_{\text{in}} \tag{28.1}$$

where

$$G_{\text{in}} = \sqrt{3}B'\left(\frac{\omega_+ - \omega_-}{\omega_0}\right) \tag{28.2}$$

$$B_{\text{in}} = -(4/\pi)B' \cot\theta \tag{28.3}$$

and

$$\theta = (\pi/2)(\omega/\omega_0) \tag{28.4}$$

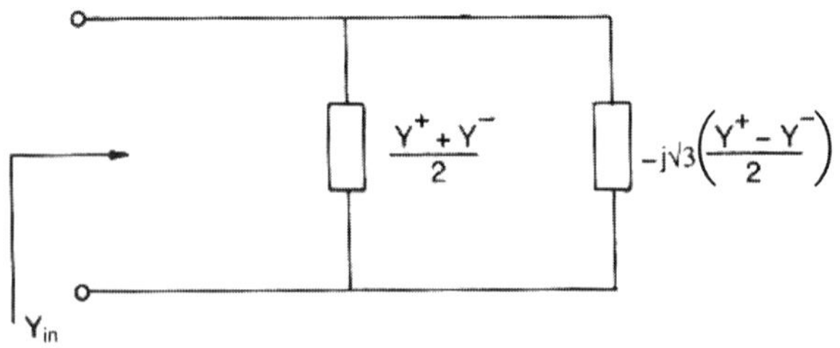

FIGURE 28.1 Degree-1 complex gyrator circuit of junction circulator.

At $\omega = \omega_0$

$$\theta = \theta_0 = \pi/2 \tag{28.5}$$

The susceptance slope parameter (B') is defined as the slope of the imaginary part of the complex gyrator circuit in the vicinity of the circulator midband frequency in the usual way by

$$B' = \frac{\theta_0}{2} \frac{\partial B_{\text{in}}}{\partial \theta}\bigg|_{\theta_0 = \frac{\pi}{2}} \tag{28.6}$$

Figure 28.2 depicts the ideal situation. The real part condition is also often written

$$\frac{1}{Q_L} = \sqrt{3}\left(\frac{\omega_+ - \omega_-}{\omega_0}\right) \tag{28.7}$$

Q_L defines the gain-bandwidth product of the network problem.

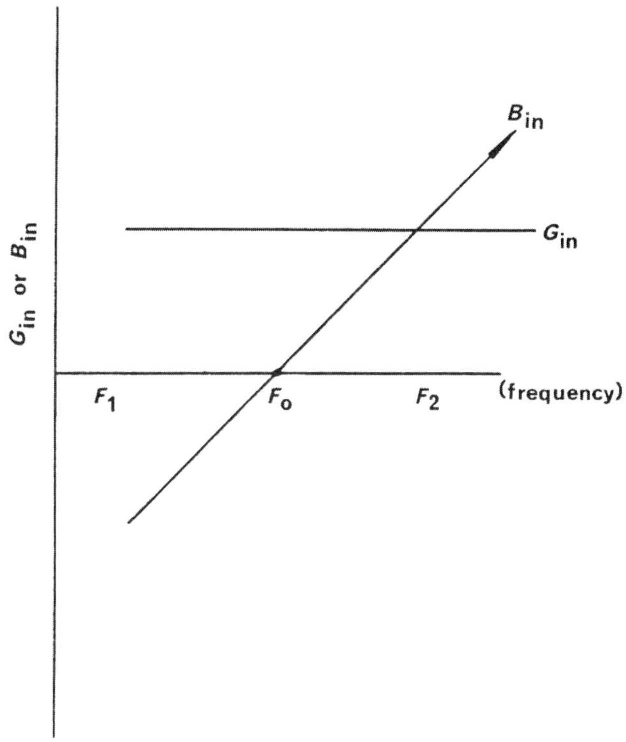

FIGURE 28.2 Ideal frequency response of complex gyrator circuit.

TABLE 28.1 Quality Factor Q_L of Weakly Magnetized Circulator

ψ \ κ/μ	0.05	0.10	0.20	0.25	0.30	0.35	0.40
0.1	13.72	6.728	3.139	2.444	2.150	2.538	4.494
0.2	13.72	6.721	3.107	2.372	1.969	2.085	3.551
0.3	13.71	6.714	3.077	2.302	1.788	1.548	2.155
0.4	13.55	6.713	3.066	2.273	1.708	1.277	0.9912
0.5	13.71	6.689	3.077	2.291	1.733	1.305	0.9539
0.6	13.72	6.723	3.100	2.330	1.796	1.403	1.106
0.7	13.72	6.728	3.118	2.360	1.843	1.472	1.197

In this sort of problem, the absolute value of the susceptance slope parameter (B') and the difference between the split frequencies of the gyromagnetic resonator $(\omega_+ - \omega_-)/\omega_0$ are the independent variables and the absolute gyrator conductance (G_{in}) is the dependent one. If any two of the three quantities entering into the description of the gyrator circuit are known, then the third one may readily be evaluated from the relationship between the three. Various means of experimentally evaluating these quantities are available throughout the text.

28.2 VERY WEAKLY MAGNETIZED GYROMAGNETIC RESONATOR

The validity of the closed form description of a junction circulator employing a very weakly magnetized disk resonator is the topic of this section. Its formulation may be verified by comparing it with some numerical calculations using the first seven poles of the problem region. Table 28.1 summarizes some data on the quality factor of the junction with the gyrotropy of the junction bracketed between 0 and 0.40 for parametric values of the coupling angle of the strips at the ports of the resonator. It indicates that the quality factor is independent of the coupling angle as long as the gyrotropy is bracketed between 0 and 0.25. It also suggests that the closed form description shows significant deterioration when the gyrotropy equals 0.35. The quality factor displayed by such a junction with a gyrotropy equal to 0.25 is about 2.30. This value is compatible with the synthesis of quarter-wave coupled junction circulators with 20 dB ripple levels over some 25% or with 26 dB ripple levels over 18%. Figure 28.3 illustrates the frequency response of one typical gyrator circuit.

28.3 WEAKLY MAGNETIZED GYROMAGNETIC RESONATOR

Table 28.1 indicates that, for κ/μ between 0 and 0.30, Q_L is nearly independent of ψ. In this interval it may be described by a polynomial approximation in terms of the gyrotropy.

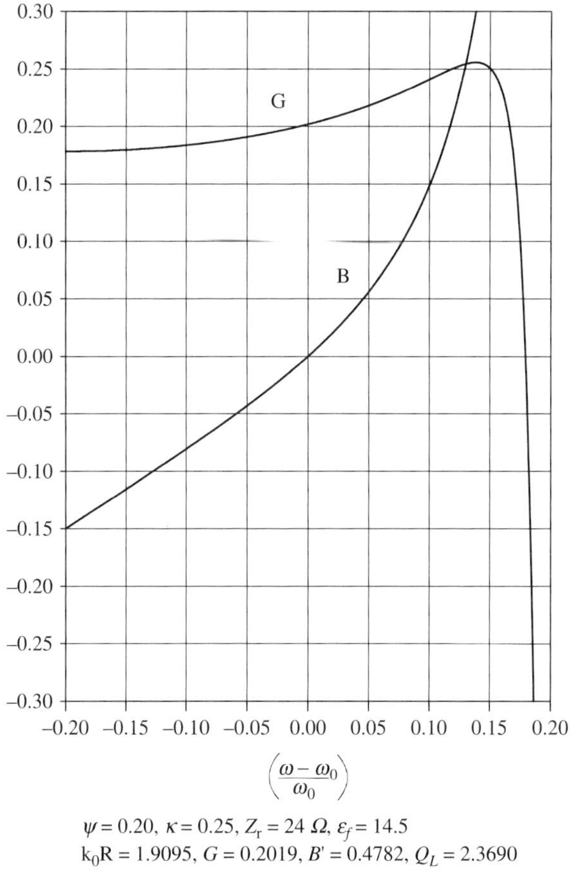

FIGURE 28.3 Frequency response of complex gyrator circuit ($\psi = 0.20$, $\kappa/\mu = 0.25$). (Reproduced with permission from J. Helszajn. Quarter-wave coupled junction circulators using weakly magnetised disk resonators, *IEEE Trans. Microwave Theory Tech.*, Vol. MTT-30, pp. 800–806, May 1982.)

$$Q_L \approx 0.689\,\frac{\mu}{\kappa} + \left(0.0463 - 2.6318\left(\frac{\kappa}{\mu}\right) + 3.5513\left(\frac{\kappa}{\mu}\right)^2\right),$$
$$0 \le \frac{\kappa}{\mu} \le 0.30 \qquad (28.8)$$

This polynomial is obtained by adding a correction polynomial of second order to the classic approximation for Q_L. It is derived by the method of least squares for the average value of Q_L over the variable ψ for the interval 0.10–0.70. The error in the classic term is about 2% at $\kappa/\mu = 0.1$, increasing to 10%, 16%, and 23% at

$\kappa/\mu = 0.20$, 0.25, and 0.30. The correction factor is derived for κ/μ between 0.05 and 0.30 using five points.

In addition to the description of the loaded Q-factor, it is also necessary to have descriptions of the gyrator conductance and susceptance slope parameter of the junction. In the approximation outlined here, it is found that the gyrator conductance is sufficiently well described by its closed form approximation. One suitable approximation for B' is therefore obtained by combining Q_L with G:

$$B' \approx \frac{\pi \zeta_e Y_r}{\sqrt{3}\,(kR)\sin\psi} \times \left[\frac{(kR)^2 - 1}{2\sqrt{3}} + 0.0463\left(\frac{\kappa}{\mu}\right) - 2.6318\left(\frac{\kappa}{\mu}\right)^2 + 3.5513\left(\frac{\kappa}{\mu}\right)^3\right] \tag{28.9}$$

It is also necessary to ensure that the equivalent circuit of the junction is well behaved over the frequency interval of the specification. Although this condition is not generally met, it is in fact satisfied in the design of junction circulators bounded by the values of loaded Q-factor in Table 28.1 for κ/μ bracketed between 0 and 0.30.

28.4 MODERATELY MAGNETIZED GYROMAGNETIC RESONATOR

A property of a very weakly or weakly magnetized junction using a gyromagnetic disk resonator is that its quality factor is independent of the coupling angle. This, however, is not the case in general. The gyrotropy of the former regime is bracketed between $0 \leq \kappa/\mu \leq 0.25$ and the latter by $0.25 \leq \kappa/\mu \leq 0.30$. The gyrotropy of the strongly magnetized or tracking solution dealt with in Chapter 17 resides between 0.50 and 1.0. It is associated with a unique coupling angle. The purpose of this section is to deal with the so-called moderately magnetized interval defined by $0.30 \leq \kappa/\mu \leq 0.50$.

A scrutiny of the entries in Table 28.1 suggests that the quality factor of the solution under consideration passes through a minimum at a unique coupling angle. Furthermore, its complex gyrator circuit is particularly attractive for the design of quarter-wave coupled devices. A detailed scrutiny of this solution indicates that it coincides with the condition for which the in-phase eigen-network has an electric wall at its terminals and the magnetized counterrotating ones display complex conjugate immittances at the same terminals. This condition is satisfied, in terms of the impedance eigenvalues of the junction, provided

$$Z^0 \approx 0 \tag{28.10a}$$

$$Z^+ + Z^- = 0 \tag{28.10b}$$

The gyrator conductance is then given in terms of the corresponding admittance eigenvalues by

$$G - j\sqrt{3}\left(\frac{Y^{+} - Y^{-}}{2}\right) \tag{28.11}$$

The impedance eigenvalues are separately expanded in terms of the impedance poles of the problem region as

$$Z^{0} \approx Z_{0} + Z_{3} \tag{28.12a}$$

$$Z^{+} \approx Z_{+1} + Z_{-2} \tag{12.12b}$$

$$Z^{-} \approx Z_{-1} + Z_{+2} \tag{28.12c}$$

and

$$Y^{+} = 1/Z^{+} \tag{28.13a}$$

$$Y^{-} = 1/Z^{-} \tag{28.13b}$$

Table 11.2 in Chapter 11 gives the sizes of the poles and eigenvalues entering into this solution.

28.5 THE DEGREE-2 CIRCULATOR

A degree-2 circulator may be realized by having recourse to either an alternate-line or quarter-wave long transformer. The synthesis of these types of networks is the topic of Chapter 25. The topology under consideration is reproduced from the same chapter in Fig. 28.4. Figure 28.5 illustrates some practical geometries met in practice. Other configurations readily come to mind. A detailed scrutiny of the gain-bandwidth product of each topology suggests that there is not much to choose between the two types of transformer. One benefit of the alternate-line transformer, other than its compactness, is of course the fact that the impedance adjacent to the terminal of the junction is usually not very different from that of the termination. The impedance of the 90° UE is approximately equal to the square root of the gyrator conductance. The adjustment of a weakly or very weakly magnetized circulator differs, however, from that of a moderately or strongly magnetized one in that the former adjustment is independent of the coupling angle. This means that either the stripwidth or the relative dielectric constant of the transformer region may, in this instance, be taken as the independent variable. In the latter case the dielectric constant is uniquely specified by the specification. The actual resonator shape in any design is usually fixed by the ground plane spacing of the microstrip or stripline at the ports

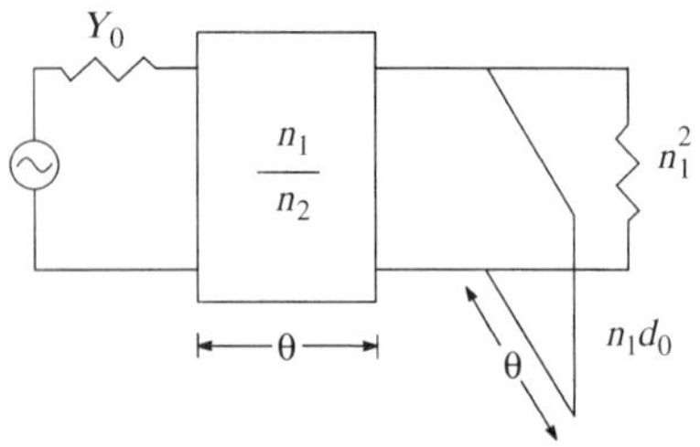

FIGURE 28.4 Topology of UE coupled complex gyrator circuit.

of the device. The sort of circulator specifications that can be realized with either a quarter-wave or alternate-line transformer is essentially fixed by the quality factor of the complex gyrator circuit. Standard tables for the design of either type of circuit are given in Chapter 25 in the case where $S(\min)$ is different from unity. Once the loaded Q-factor of the specification is specified, its value is used to check whether it is compatible with the very weakly or weakly magnetized models employed in this work. If it is, the gyrator conductance and susceptance slope parameter may be evaluated. The design procedure then continues by selecting the

FIGURE 28.5 Schematic diagrams of degree-2 junction circulators.

resonator shape and evaluating the substrate thickness and the width of the connecting lines. In this design problem the quality factor and the susceptance slope parameter are the independent variables and the gyrator conductance is the dependent one. The quality factor is fixed by the gyrotropy of the gyromagnetic resonator. The susceptance slope parameter is determined separately by the resonator shape and substrate thickness and to some extent the gyrotropy. A feature of this sort of problem is that varying the value of the minimum ripple level or VSWR in the passband leaves the maximum ripple level or VSWR and the passband specification unchanged but produces relative large variations in the absolute value of the susceptance slope parameter of the load. Figure 28.6 shows a typical situation. A detailed scrutiny of the network problem indicates that a wide family of practical specifications may in fact be realized by staying within these arbitrary bounds. One engineering decision is to suppose that

$$\mathrm{VSWR(min)} = \sqrt{\mathrm{VSWR(max)}} \tag{28.14}$$

The exact values of the elements entering into the description of the complex gyrator circuit, in any particular situation, must be determined by the actual gain-bandwidth specification.

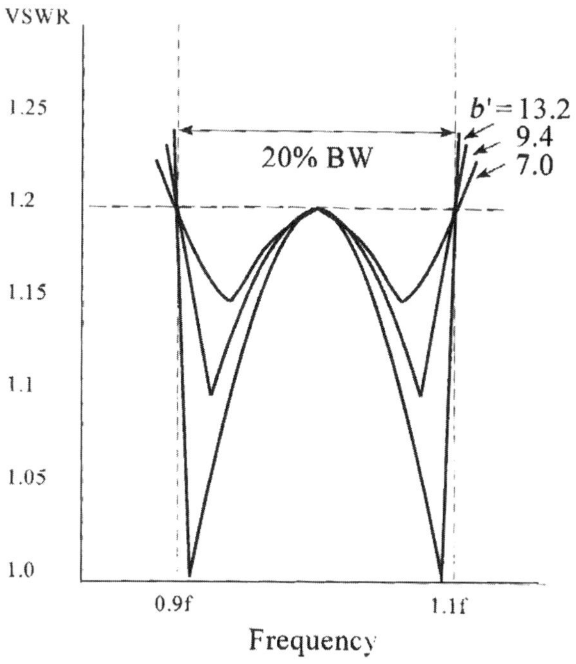

FIGURE 28.6 Connection between VSWR(min) and magnetized susceptance slope parameter (b').

It is assumed separately that the network variables apply provided

$$(\omega_+ - \omega_-)/\omega_0 \geq (\omega_2 - \omega_1)/\omega_0 \tag{28.15}$$

$\omega_{1,2}$ and ω_0 specify the bandedges and center frequencies.

One possible adjustment technique for the 90° UE arrangement may be obtained by recognizing that the gyrator conductance at the midband frequency in a quarter-wave coupled device can be written

$$\frac{y_t^2}{S(\max)} = \sqrt{3}b'\left(\frac{\omega_+ - \omega_-}{\omega_0}\right) \tag{28.16}$$

Figure 28.7 illustrates the frequency response of one degree-2 solution. It corresponds to the complex gyrator circuit specified in Fig. 28.4.

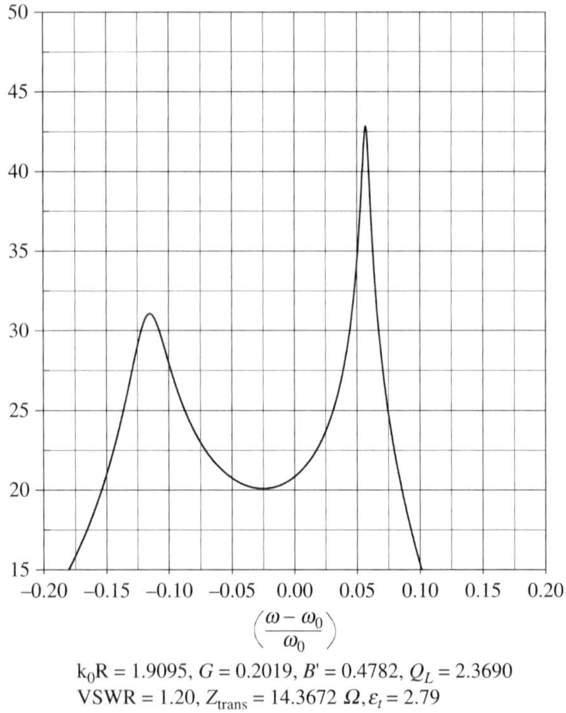

FIGURE 28.7 Frequency response of degree-2 circulator ($\psi = 0.20$ $\kappa/\mu = 0.25$). (Reproduced with permission from J. Helszajn, Quarter-wave coupled junction circulators using weakly magnetised disk resonators, *IEEE Trans. Microwave Theory Tech.*, Vol. MTT-30, pp. 800–806, May 1982.)

Scrutiny of this relationship suggests that in order to establish the required gyrotropy of the problem region care must be taken to accurately realize both the admittance (y_t) of the transformer and the susceptance slope parameter (b') of the complex gyrator circuit.

28.6 GAP EFFECTS IN CIRCULATOR ASSEMBLIES

One practical problem encountered in the assembly of a circulator is the difficulty of avoiding small gaps between the inner and outer conductors of the stripline and the ferrite surfaces. Figure 28.8 depicts the geometry under consideration. Such gaps always alter the center frequency of the device and quite often produce spurious modes in the frequency response. The origin of this difficulty may be appreciated by forming the effective dielectric constant of a partially dielectric-filled circular capacitor. It may also be calculated by constructing a transverse resonance condition along the axis of the geometry. The required result, using either approach, is given in terms of a filling factor q by

$$\frac{1}{\varepsilon_{\text{eff}}} = \frac{(1-q)}{\varepsilon_d} + \frac{q}{\varepsilon_f} \tag{28.17}$$

where

$$q = L/H \tag{2.18}$$

The above relationship satisfies the conservation of capacitance in the assembly

$$\frac{H}{\varepsilon_0 \varepsilon_{\text{eff}} A} = \frac{L}{\varepsilon_0 \varepsilon_f A} + \frac{(H-L)}{\varepsilon_0 \varepsilon_d A} \tag{28.19}$$

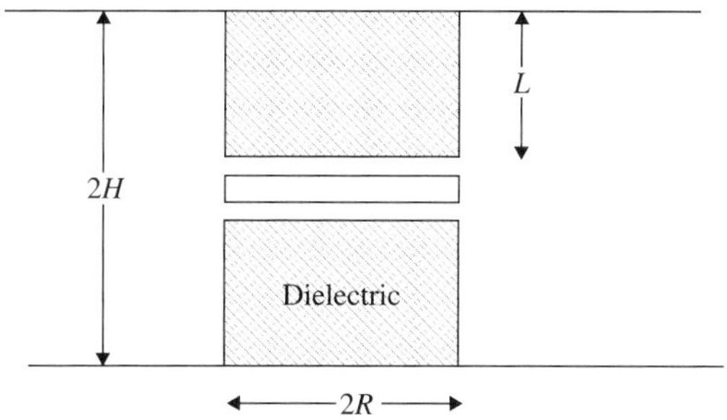

FIGURE 28.8 Ferrite resonator with gap.

Figure 28.9 depicts the connection between the effective dielectric constant of the arrangement and the filling factor q. This illustration indicates, quite clearly, the effect of a small gap on the effective relative dielectric constant of the arrangement.

The effect of a very small gap on the effective dielectric constant is readily understood by way of an example. Taking q as 0.99, ε_f, as 15.0, and ε_d as 1.0 gives ε_{eff} as 13.15. The effect of a gap on the permeability of a magnetic insulator is also described in terms of an effective quantity

$$\mu'_{eff} = (1 - q) + q\mu_{eff} \tag{28.20}$$

The connection between μ'_{eff} and μ_{eff} is described in Fig. 28.10.

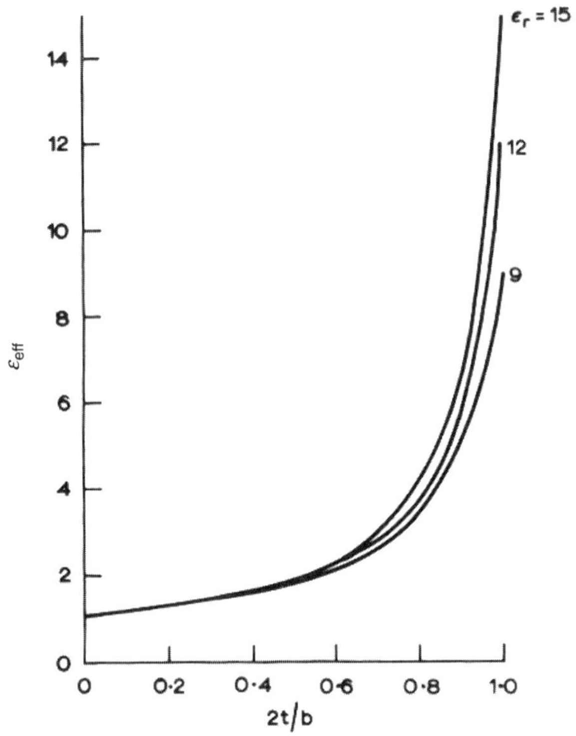

FIGURE 28.9 Effective dielectric constant of ferrite resonator with gap. High-power suspended resonator stripline circulator. (Reproduced with permission from J. Helszajn, Frequency and bandwidth of *H* plane TEM junction circulator, *Proc. IEE*, Vol. 117, pp. 1235–1238, July 1970.)

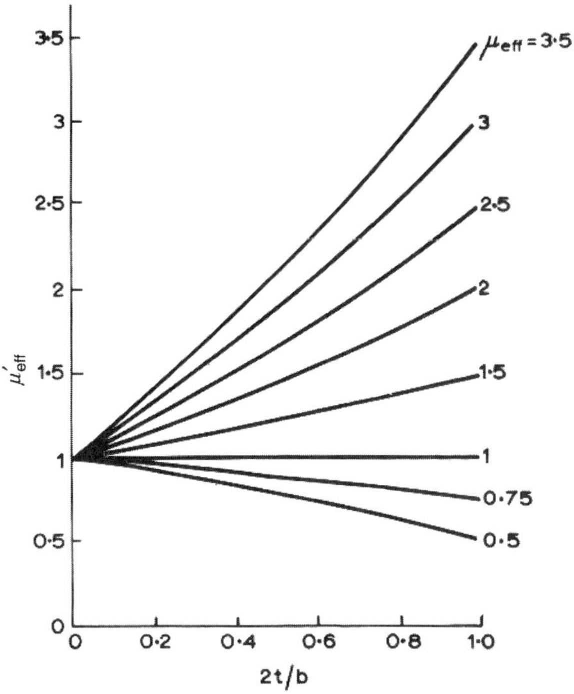

FIGURE 28.10 Effective permeability of ferrite resonator with gap (effective dielectric constant of ferrite resonator with gap). High-power suspended resonator stripline circulator (Reproduced with permission from J. Helszajn, Frequency and bandwidth of *H* plane TEM junction circulator, *Proc. IEE*, Vol. 117, pp. 1235–1238, July 1970.)

28.7 SUSPENDED PLANAR RESONATOR

One way to increase the power rating of a stripline circulator is to introduce a free space or dielectric gap at the symmetry plane of each ferrite disk. Figure 28.11 illustrates one geometry employing four ferrite disks. Figure 28.12 indicates the geometry in more detail. The quasi-suspended planar resonator obtained in this way is particularly attractive in the design of high mean power devices. This comes about in that its surface area is in this instance larger than is otherwise the case and that the temperature drop across its thickness is reduced. It also has the merit that its insertion loss is reduced in comparison with that of a conventional arrangement. The operating frequency and gyrotropy of this sort of resonator have been formulated in terms of the filling factor (q) met in connection with a gap in a conventional resonator. This is done in terms of effective constitutive parameters. The descriptions of the effective diagonal and off-diagonal elements of the tensor permeability are given by

$$\mu_{eff} = (1 - q) + q\mu \tag{28.21}$$

$$\kappa_{eff} = q\kappa \tag{28.22}$$

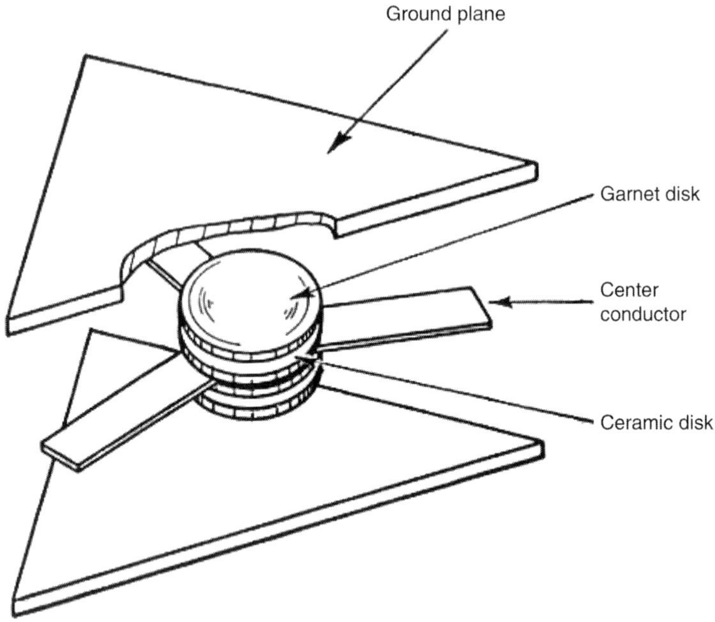

FIGURE 28.11 Stripline junction using four ferrite or garnet disks.

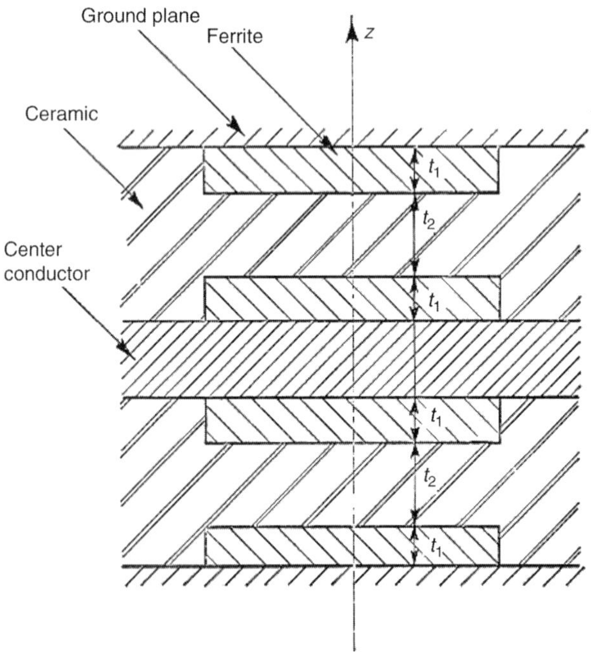

FIGURE 28.12 Resonator using four ferrite or garnet disks.

FIGURE 28.13 Commercial stripline circulator. (Reproduced with permission from J. Helszajn, An *H*-plane high power TEM ferrite circulator, *Radio Electron. Engr.*, Vol. 33, pp. 257–262, 1967.)

These two relationships are readily deduced by taking linear combinations of the scalar counterrotating circular constitutive parameters:

$$(\mu_{\text{eff}} + \kappa_{\text{eff}}) = 1 + q(\mu + \kappa - 1) \tag{28.23a}$$

$$(\mu_{\text{eff}} - \kappa_{\text{eff}}) = 1 + q(\mu - \kappa - 1) \tag{28.23b}$$

The natures of the effective dielectric and permeability constants are those introduced in connection with the gap effect of the geometry.

Figure 28.13 depicts one 1.3 GHz commercial device. The matching network consists of two quarter-wave long shunt UEs spaced by a quarter-wave UE. The mean power rating of the device is 2.5 kW and the peak power rating is 150 kW. Its bandwidth is 7.5%.

28.8 PASSBAND FREQUENCIES OF A THREE-PORT JUNCTION IN IMMITTANCE PLANE

The frequencies of the passbands of any three-port junction circulator may be calculated by having recourse to a full wave analysis, to a finite element solver, or to measurements. While it is sometimes tempting to associate the center frequency of any circulator with that of the degenerate frequencies of the counterrotating eigen-networks, it is in general incorrect to do so. This situation only holds when the in-phase eigen-network, displays an electric wall at a pair of terminals at the frequency at which the counterrotating ones exhibit a magnetic wall at the same plane. The actual passband frequency of a reciprocal three-port junction is either smaller or larger than that of the degenerate counterrotating eigen-networks according to whether the in-phase eigen-network is smaller or larger than 90°. The center frequency of any circulator is an eigenvalue problem. Figure 28.14 illustrates the three possibilities. It is preferable, however, for the purpose of calculation to replace the condition in the reflection plane by one in the immittance one. The required statement may be understood by recalling that the condition for which two unit reflection vectors are out of phase on the classic Smith chart corresponds to that for which the corresponding normalized impedances are the reciprocal of each other.

$$Z^0 - 1/Z^{\pm} = 0 \tag{28.24}$$

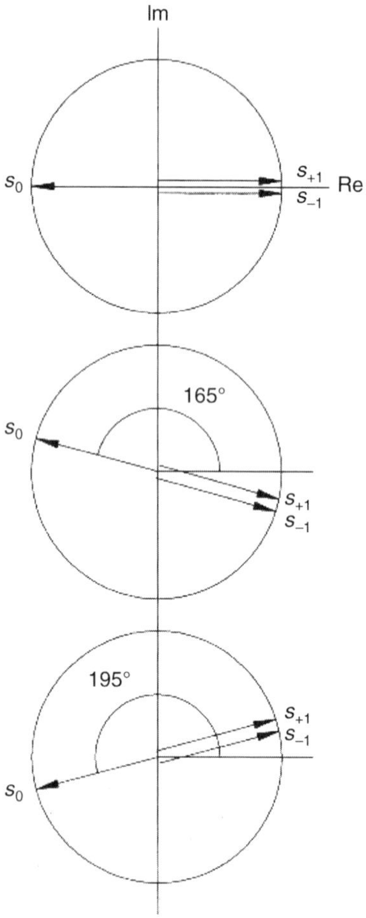

FIGURE 28.14 Passband conditions in symmetrical reciprocal junction.

The impedance eigenvalues in the preceding equation are related to the reflection ones by the standard bilinear transformation between the two

$$Z^0 = \frac{1 - s_0}{1 + s_0} \tag{28.25a}$$

and

$$Z^{\pm} = \frac{1 - s_{\pm}}{1 + s_{\pm}} \tag{28.25b}$$

A scrutiny of the characteriztic equation indicates that it is in keeping with the ideal solution for which the reflection angles are commensurate:

$$Z^0 = 0 \tag{28.26a}$$

$$Z^{\pm} = \infty \tag{28.26b}$$

In this sort of problem, the in-phase impedance eigenvalue may be expanded in terms of the impedance poles of the geometry in question:

$$Z^0 = Z_0 + Z_3 + \cdots \tag{28.27}$$

The degenerate counterrotating impedance eigenvalues are likewise expanded in terms of the corresponding impedance poles:

$$Z^+ = Z_{+1} + Z_{-2} + \cdots \tag{28.28a}$$

$$Z^- = Z_{-1} + Z_{+2} + \cdots \tag{28.28b}$$

In a reciprocal junction

$$Z^+ = Z^- \tag{28.29}$$

The relationship introduced here is exact so that it is likely to have wide applicability in the design of this sort of problem.

The failure to establish dual walls at the terminals of the junction for the in-phase and counterrotating eigen-networks is to reveal the in-phase eigen-network explicitly in its complex gyrator circuit. Its topology, based on a Cauer form procedure, is reproduced in Fig. 28.15. It consists of a series reactance with a value $4Z_0/3$ in cascade with the classic shunt UE in parallel with the gyrator conductance. In practice, the series element is absorbed in the matching network.

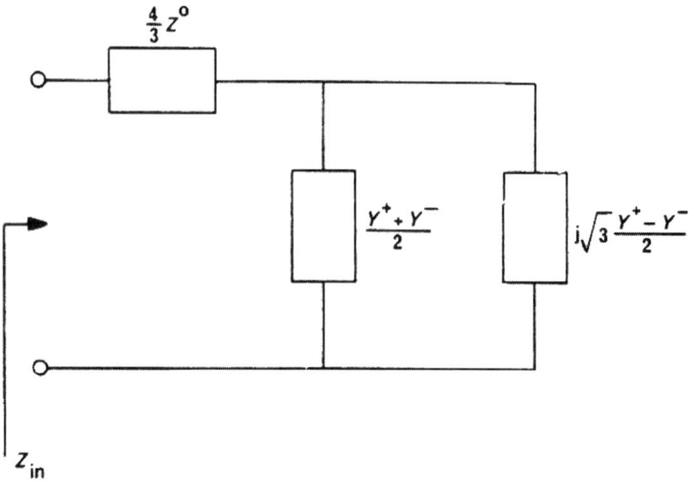

FIGURE 28.15 Degree-2 complex gyrator circuit of junction circulator.

28.9 OPEN WALLS

The classic boundary conditions of a junction circulator assume that the magnetic field is a constant over the strips and zero elsewhere. In practice, however, the walls between the strips are open rather than closed. One obvious consequence of this discrepancy is that the operating frequency of the junction is perturbed. A more serious problem is that calculations on the complex gyrator circuit do not always agree with measurements. These discrepancies are particularly severe whenever the dielectric constant of the region outside the resonator is comparable with that of the ferrite material and whenever the aspect ratio (R/H) of the resonator is excessively small.

The effect of an open wall on the frequency of the circulator may be represented either in terms of effective constitutive parameters and dimensions or by a lumped element susceptance. The latter representation, however, is preferable in connection with the description of the susceptance slope parameter of the junction.

One means of representing the effect of the open wall on the frequency is to introduce a correction term α_n in the description of a typical pole.

$$-\frac{j\pi\zeta_e Y_r}{3\psi}\left(\frac{n\psi}{\sin n\psi}\right)^2\left[\frac{J_n'(kR)}{J_n(kR)}\right]+\alpha_n(kR)=0 \tag{28.30}$$

Strictly speaking, the effects of open walls are determined, in a gyromagnetic circuit, by the wave impedances in the regions rather than by the dielectric constants. This means that the effects will be quite different above and below the Kittel line.

Figure 28.16 depicts the effect of an open wall on the cutoff space of a typical circular dielectric resonator in a dielectric medium. The frequency shift in this sort of arrangement is clearly dependent on the aspect ratio of the resonator and the ratio of the dielectric constants of the two regions. One consequence of the shifts in the degenerate cutoff numbers of the isotropic disk resonator is that the intersections between the various split branches of the gyromagnetic space are perturbed. Figure 28.17 illustrates a qualitative picture of this situation. A scrutiny of this modified mode chart suggests that the various definitions met in connection with the very weakly, weakly, moderately, and strongly magnetized circulation solutions have to be revisited. One particular solution encountered with an open instead of a closed wall between the ports of the circulator is the semitracking one with a narrow instead of wide coupling angle at the ports.

Figure 28.18 illustrates the relationships between the frequency of a triangular resonator on a dielectric substrate and its side dimension for three different substrate thicknesses. It also indicates some measurements on the magnetic wall solution.

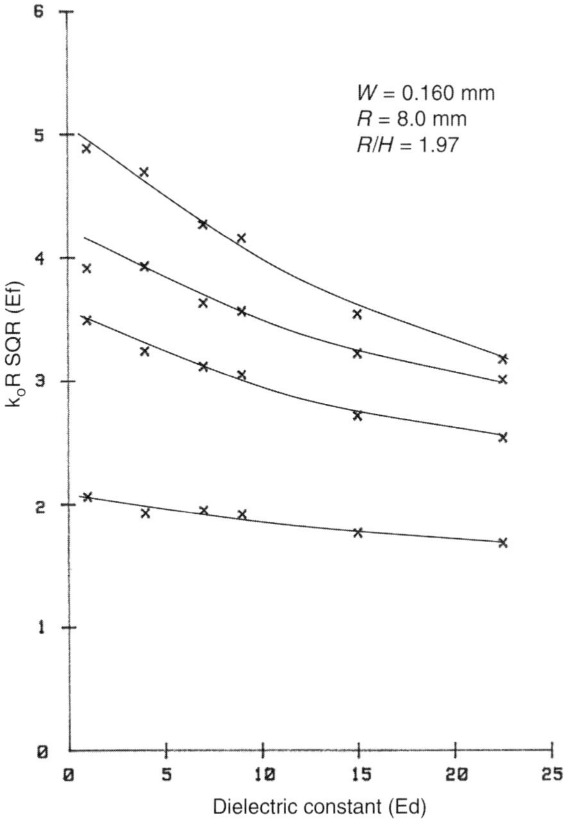

FIGURE 28.16 Relationship between dielectric constant and radial wavenumber in disk dielectric resonator.

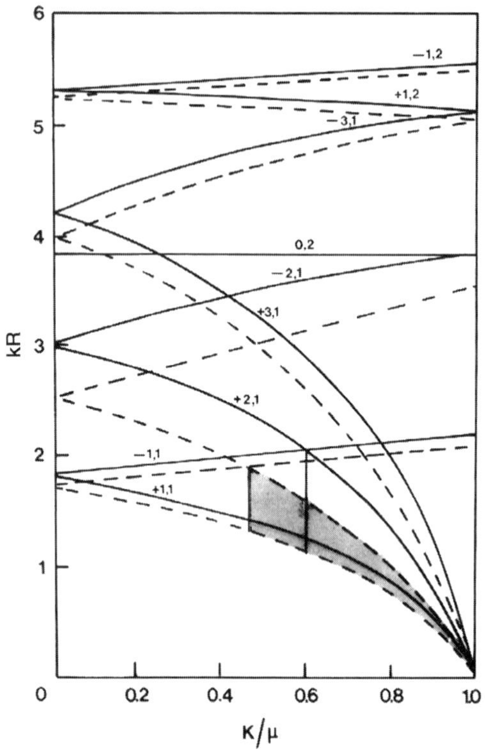

FIGURE 28.17 Cutoff space of gyromagnetic resonator with open walls.

The calculations in this illustration are obtained by segmenting the geometry into strips and replacing each strip by a parallel waveguide model as a preamble to employing a transverse resonance procedure.

28.10 SPINWAVE INSTABILITY IN MAGNETIC INSULATORS

A feature of some importance in the design of high-power devices is the appearance of a subsidiary resonance at large signal levels at a direct magnetic field below the main Kittel resonance. The subsidiary resonance is a power sensitive phenomenon that manifests itself as a nonlinear insertion loss that limits the peak power rating of ferrite devices. This feature is well understood and is related to the transfer of power from the microwave magnetic field to spinwaves at half the frequency of the microwave signal. One standard way of avoiding this instability relies on material technology to widen the spinwave linewidth at the expense of the overall small-signal insertion loss. A second way to suppress this difficulty is to bias the material below the peak of the subsidiary resonance. Still a third solution is to ensure that the

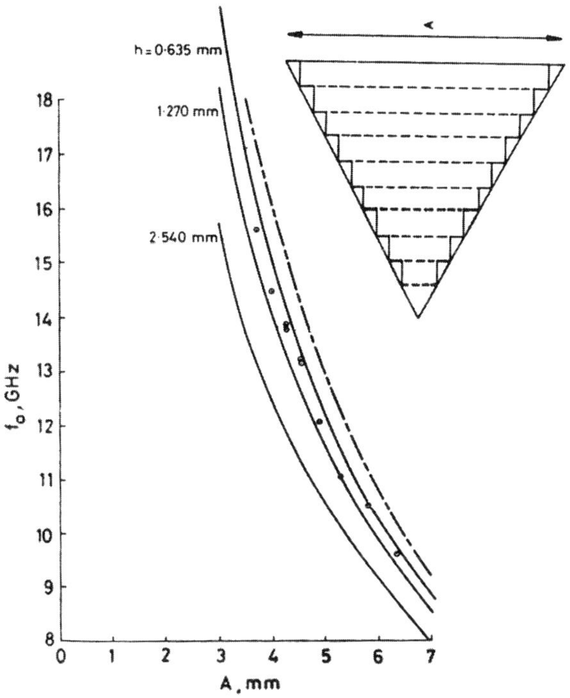

FIGURE 28.18 Relationship between side dimension of triangular planar resonator and frequency.

frequency relation between the microwave signal and the spinwaves at half its frequency cannot be satisfied. The region in question is defined in Fig. 28.19.

One additional way to suppress spinwave instability is to bias the ferrite material between the subsidiary and main resonances. Its width may be established without difficulty but is outside the scope of this work. This may be done by forming the difference between the two fields defined by the spinwave dispersion relation and the main resonance. The essential considerations entering into its definition are the direct magnetic field, the separation between the skirts of the subsidiary and main resonances, the bandwidth of this region, and the relation between magnetization and linewidth. It is also necessary to ensure that the effective or scalar permeabilities do not take on negative values over the operating band of the device.

One important advantage of biasing the ferrite material between the subsidiary and main resonances is that the overall magnetic loss is less under this condition than it is below the subsidiary resonance. This is due to the fact that the small signal interaction between the microwave signal and the spinwave manifold is also not permissible in this region. Since spinwave instability at half the pump frequency is completely suppressed in this situation, the uniform mode linewidth may be selected without regard to that of the spinwave one. Therefore, in this sort of design, the value of the spinwave linewidth has no bearing on the overall insertion loss of the device.

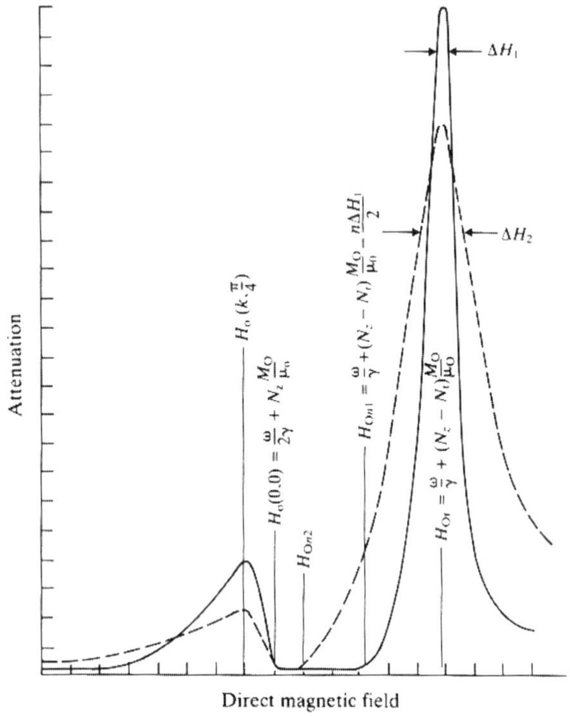

FIGURE 28.19 Subsidiary resonance at high signal level in magnetic insulator. (Reproduced with permission from R. W. Damon, Relaxation effects in ferromagnetic resonance, *Rev. Mod. Phys.*, Vol. 25, pp. 239–245, 1953.)

28.11 FREQUENCY DOUBLING IN FERRITES

A property of a magnetic insulator, at large signal level, is the generation of a harmonic component m_z at twice the frequency of the fundamental alternating magnetic field. This feature may readily be demonstrated by retaining the products $m_x h_y$ and $m_y h_x$ in the governing equation of motion of the electron spin. A simple physical interpretation of frequency doubling, however, is sufficient by way of introduction. It is obtained by making use of the conservation of magnetization in a magnetic insulator

$$M_0^2 = m_x^2 + m_y^2 + m_z^2 \tag{28.31}$$

A scrutiny of this relationship indicates that if the RF magnetization in the transverse plane is circularly polarized, the magnetization vector along the z-coordinate is a constant and therefore m_z is a constant also. This means that there is no second harmonic component in the absence of shape and anisotropy demagnetizing factors. However, if the transverse magnetization is elliptically polarized, the magnetization vector

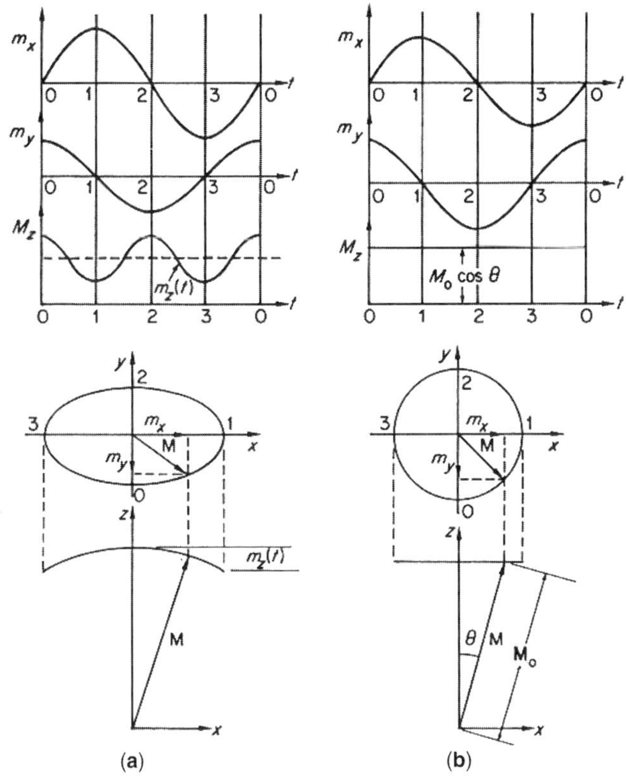

FIGURE 28.20 Magnetization vector in x-z plane for (a) $m_x \neq m_y$, (b) $m_x = m_y$.

along the z-coordinate is not a constant. Indeed, since m_z reaches its maximum value, when $m_x^2 + m_y^2$ is a minimum, twice as the tip of **M** traverses once around its elliptical orbit, its frequency is twice that of the applied RF field. The construction of this solution is illustrated in Fig. 28.20. This simple picture indicates that the second harmonic component is dependent on the ellipticity of the transverse magnetization. The most favorable geometry is a thin disk magnetized in the plane.

In the circulator arrangement the counterrotating magnetic fields are circularly polarized on the axis of the resonator and elliptically polarized elsewhere. This means that frequency doubling in this sort of geometry does not take place on the axis of the junction.

28.12 SECOND-ORDER INTERMODULATION IN MAGNETIC INSULATORS

An important intermodulation product that enters into the specification of the junction circulator in the presence of adjacent carriers is a third-order product. An exact solution of this phenomenon is outside the scope of this work. The development of second-order intermodulation, however, is not too difficult provided the two tones are circularly polarized. Such pumps at ω_1 and ω_2 in the transverse plane produce sum and difference sidebands along the z-coordinate. The governing differential equations are

$$\frac{\partial^2 m_x}{\partial t^2} + \omega_0^2 m_x = \omega_0 \omega_m h_x + \omega_m \left(\frac{\partial h_y}{\partial t} \right) \tag{28.32a}$$

$$\frac{\partial^2 m_y}{\partial t^2} + \omega_0^2 m_y = \omega_0 \omega_m h_y - \omega_m \left(\frac{\partial h_x}{\partial t} \right) \tag{28.32b}$$

$$\frac{\partial m_z}{\partial t} = \gamma (m_x h_y - m_y h_x) \tag{28.32c}$$

The alternating magnetic field intensity for a two-tone arrangement in general is specified by

$$h_x = h_{x1} \cos \omega_1 t + h_{x2} \cos (\omega_2 t + \phi) \tag{28.33a}$$

$$h_y = h_{y1} \cos(\omega_1 t - \alpha) + h_{y2} \cos(\omega_2 t + \phi - \delta) \tag{28.33b}$$

The nature of the polarization of the two tones is fixed by α and δ, and the relative phase angle between the two tones byϕ. One possible arrangement consists of two in-phase RHPW tones, another of two similar LHPW tones, and still another of counterrotating circularly polarized tones.

$$h_x = h_1 \cos \omega_1 t + h_2 \cos \omega_2 t \tag{28.34a}$$

$$h_y = \pm h_1 \sin \omega_1 t \pm h_2 \sin \omega_2 t \tag{28.34b}$$

The existence of second-order products in the arrangement considered here neglects any nonlinearity in the transverse plane. This means that the total magnetization due to the two tones may be established by having recourse to superposition. Taking two RHPW tones by way of an example gives, in the case of $m_{x1,2}$,

$$m_{x1} = \chi_1^+ h_1 \cos \omega_1 t \tag{28.35a}$$

$$m_{x2} = \chi_2^+ h_2 \cos \omega_2 t \tag{2.35b}$$

$\chi_{1,2}^+$ are the RHPW scalar susceptibilities at $\omega_{1,2}$:

$$\chi_{1,2}^+ = \frac{\omega_m}{-\omega_{1,2} + \omega_0} \tag{28.36}$$

The total magnetization along the x-coordinate is then given by

$$m_{tx} = \chi_1^+ h_1 \cos \omega_1 t + \chi_2^+ h_2 \cos \omega_2 t \tag{28.37}$$

Likewise

$$m_{ty} = \chi_1^+ h_1 \sin \omega_1 t + \chi_2^+ h_2 \sin \omega_2 t \tag{2.38}$$

It may sometimes be useful to write the total magnetization in the transverse plane in terms of circular variables. The result is

$$m_t = (m_{tx} + jm_{ty}) = \chi_1^+ h_1 \exp j\omega_1 t + \chi_2^+ h_2 \exp j\omega_2 t \tag{28.39}$$

The required intermodulation product along the z-coordinate is now established by evaluating the magnetization there in terms of the transverse variables. The result is

$$\frac{\partial m_z}{\partial t} = 2\gamma h_1 h_2 \left(\chi_1^+ - \chi_2^+\right) \sin(\omega_2 - \omega_1)t \tag{28.40}$$

This arrangement therefore produces a sideband at the difference between the frequencies of the two tones.

The corresponding result for two LHPW tones is obtained by replacing the RHPW tones by the former ones in the previous development and replacing $\chi_{1,2}^+$ by $\chi_{1,2}^-$.

$$\chi_{1,2}^- = \frac{\omega_m}{\omega_{1,2} + \omega_0} \tag{28.41}$$

This solution also reveals a sideband at the difference frequency of the two tones.

The possibility of having the two tones rotating in opposite directions is also understood. It produces a single sideband at a frequency equal to the sum of those tones. The absence of second-harmonic terms in the three different solutions developed here is compatible with the fact that there is no such contribution unless one or the other of the tones is linearly or elliptically polarized.

28.13 TEMPERATURE STABILITY OF MAGNETIC INSULATORS BELOW THE KITTEL LINE

The temperature stability of the Kittel line in a magnetic insulator has been dealt with in Chapter 26. The temperature stability of the magnetization is dealt with here. While the saturation magnetization of some ferrite and garnet materials may be stabilized by suitable doping, this solution may not always be possible. Another possibility, as described later, is to employ series or shunt compensation of the direct magnetic circuit. Still another one is to enforce the condition between the direct magnetic field and the direct magnetization of the material at which the magnetization curves at different temperatures intersect. This approach may be understood by noting that a typical hysteresis loop of a magnetic material at room temperature both shrinks and collapses as the temperature approaches its Curie value. The required condition may readily be expressed in terms of the ratio of the applied magnetic field and the saturation magnetization by starting with the definition of internal field intensity (H_{in}) in a magnetic insulator:

$$H_{\text{in}} = H_0 - N_z M \tag{28.42}$$

where N_z is the direct demagnetizing factor along the axis of the resonator and M is the actual direct magnetization.

If the internal direct magnetic field is small compared to the other quantities, then the preceding relationship may be written

$$\frac{H_0}{M_0} \approx N_z \left(\frac{M}{M_0} \right) \tag{28.43}$$

Some experimental work on a number of different ferrite materials suggests that the magnetization curves at different temperatures approximately intersect at the direct magnetic field for which

$$\left(\frac{M}{M_0} \right) \approx 0.70 \tag{28.44}$$

A temperature-stable condition for the purpose of design is therefore given by

$$\left(\frac{H_0}{M_0} \right) \approx 0.70 \tag{28.45}$$

Another quantity that enters into the description of a ferrite or garnet material is the squareness of its hysteresis loop. A typical squareness is given by

$$\left(\frac{M_r}{M_0} \right) = 0.62 \tag{28.46}$$

Writing M_0 in terms of M_r suggests that an equally good or possibly better criterion for the temperature stability of a junction circulator is, in this instance, given by

$$H_0 \approx N_z M_r \tag{28.47}$$

28.13.1 Series and Shunt Temperature Compensation

Not only is the magnetization of microwave magnetic insulators temperature dependent but so is the magnetic field intensity of magnets. While the variation of the magnetic field intensity of magnets may sometimes be used to track the magnetization of magnetic insulators, this is not always possible. One means of dealing with this problem is to introduce some form of temperature compensation in the magnetic circuit. This takes the form of special steels for which the Curie temperature is approximately 70 °C. The Curie temperature of microwave magnetic insulators is typically 180 °C and that of magnets 350 °C. Figure 28.21 indicates the relationship between magnetization and temperature for a pure YIG material. Typical series and shunt compensation arrangements are illustrated in Fig. 28.22. The shunt geometry is often used in the design of circulators biased below the Kittel line and the series geometry is used in devices biased above it.

28.13.2 Nonuniform Magnetic Field

The shape demagnetizing factor perpendicular to a flat disk is often assumed to be independent of the radial coordinate of the geometry. This assumption is only correct in a ellipsoidal body. One consequence of this feature is that the magnetic field intensity inside a disk is nonuniform even in the presence of a uniform direct magnetic field. The exact internal field is

$$H_{\text{in}}(r) = H_0 - N_z(r)M_0$$

The nature of $H_{\text{in}}(r)$ is fully discussed in Chapter 3.

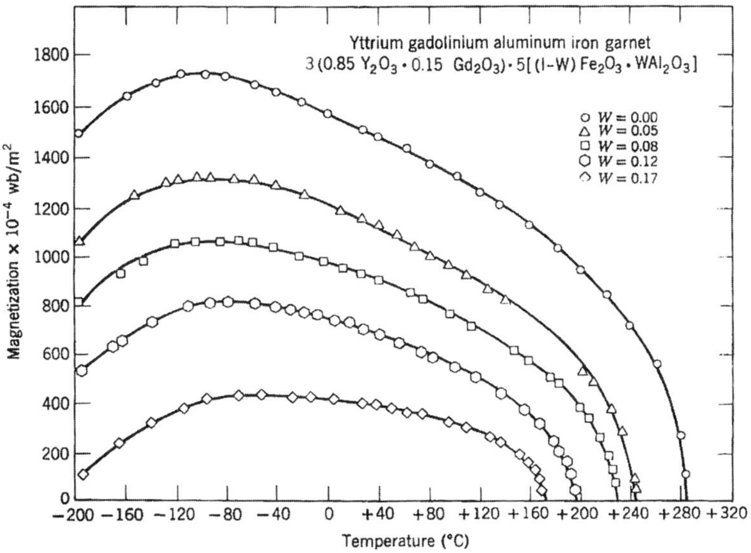

FIGURE 28.21 Magnetization versus temperature for pure YIG.

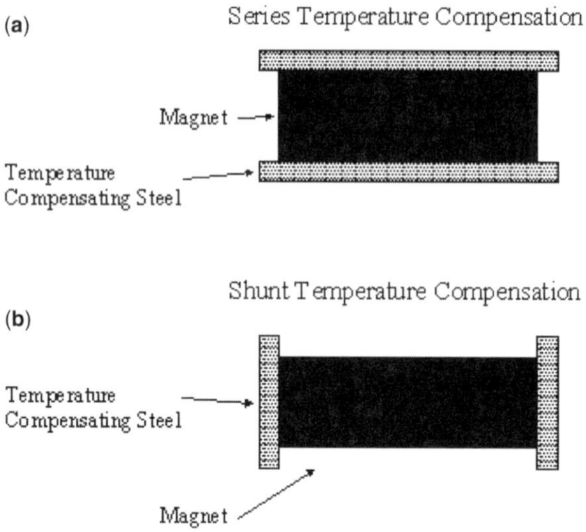

FIGURE 28.22 (a) Series temperature compensation of a gyromagnetic resonator. (b) Shunt temperature compensation of a gyromagnetic resonator.

28.13.3 Edge Mode Effect

One of the more interesting features of a suitably magnetized planar gyromagnetic circuit is a nonreciprocal edge mode or edge displacement effect in a ferrite-loaded parallel plate waveguide first explained by Hines. This feature may be demonstrated without difficulty by assuming a quasi-TE solution in the magnetized ferrite region.

$$E_x \neq 0, \quad E_y \neq 0, \quad E_z \neq 0 \tag{28.48a}$$

$$E_x = E_y = E_z = 0 \tag{28.48b}$$

It is also assumed that the spatial variations of the field patterns are described by

$$\frac{\partial}{\partial x} = -\alpha_x \tag{28.49a}$$

$$\frac{\partial}{\partial y} = 0 \tag{28.49b}$$

$$\frac{\partial}{\partial z} = -j\beta_z \tag{28.49c}$$

The tensor permeability with the direct magnetic field intensity perpendicular to the direction of propagation is described in Chapter 2 by

$$[\mu] = \begin{pmatrix} \mu & 0 & -j\kappa \\ 0 & 1 & 0 \\ j\kappa & 0 & \mu \end{pmatrix} \tag{28.50}$$

Introducing these conditions into Maxwell's first curl equation then gives

$$H_x = \frac{E_y}{\omega\mu_0\mu_{\text{eff}}}\left(-\beta_z + \alpha_x\frac{\kappa}{\mu}\right) \tag{28.51}$$

$$H_z = \frac{jE_y}{\omega\mu_0\mu_{\text{eff}}}\left(\beta_z\frac{\kappa}{\mu} - \alpha_x\right) \tag{28.52}$$

The relationship between the separation constants is given with the aid of the wave equation by

$$\alpha_x^2 - \beta_z^2 + \omega^2\mu_0\varepsilon_0\mu_{\text{eff}}\varepsilon_f = 0 \tag{28.53}$$

The required result may now be established by placing a magnetic wall boundary condition at the plane $x = 0$ and assuming a unidirectional wave confined to this edge:

$$E_y = A\exp(-\alpha_x x)\cdot\exp(-j\beta_z z) \tag{28.54}$$

$$H_z = 0 \text{ at } x = 0 \tag{28.55}$$

Introducing the latter boundary condition in Eq. (28.52) indicates that

$$\alpha_x = \frac{\kappa}{\mu}\beta_z \tag{28.56}$$

The wave equation is satisfied with μ_{eff} either positive or negative provided

$$\beta_z = \omega\sqrt{\mu_0\varepsilon_0\varepsilon_f\mu} \tag{28.57}$$

$$\alpha_x = \omega\frac{\kappa}{\mu}\sqrt{\mu_0\varepsilon_0\varepsilon_f\mu} \tag{28.58}$$

The field components in the transverse plane therefore decay exponentially with an attenuation coefficient α_x, which is proportional to the gyrotropy κ/μ and H_z is zero everywhere.

$$E_y = A\exp(-\alpha_x x)\cdot\exp(-j\beta_z z) \tag{28.59}$$

$$H_x = \eta E_y \tag{28.60}$$

$$H_z = 0 \tag{28.61}$$

and η is the wave impedance

$$\eta = \sqrt{\frac{\mu_0\mu}{\varepsilon_0\varepsilon_f}} \tag{28.62}$$

This type of line therefore supports a TEM solution and displays no low frequency cutoff number. Furthermore, since μ_{eff} does not enter directly into its description, it may smoothly straddle the two regions where μ_{eff} is either positive or negative.

If the material is just saturated then

$$\kappa = \frac{\omega_m}{\omega} \tag{28.63}$$

$$\mu = 1 \tag{28.64}$$

and β_z and α_x, in Eqs. (28.57) and (28.58) may also be written

$$\beta_z = \omega\sqrt{\mu_0\varepsilon_0\varepsilon_f} \tag{28.65}$$

$$\alpha_x = \omega_m\sqrt{\mu_0\varepsilon_0\varepsilon_f} \tag{28.66}$$

This result indicates that the edge mode effect is frequency independent and that the decoupling between the two edges is solely dependent on the relationship between ω_m and the width (l) of the ferrite section; the only frequency limitation in this class of device is thus the onset of higher-order modes. Figure 28.23 indicates the field pattern and transmitted power in the forward and backward directions of propagation for one solution in this sort of transmission line.

One application of the edge mode effect is in the fabrication of a nonreciprocal phase shifter; another is in the construction of a ferrite isolator. The former is obtained by lining one of the edges by a suitable dielectric slab; the latter device is realized by loading one of the edges by some resistive material.

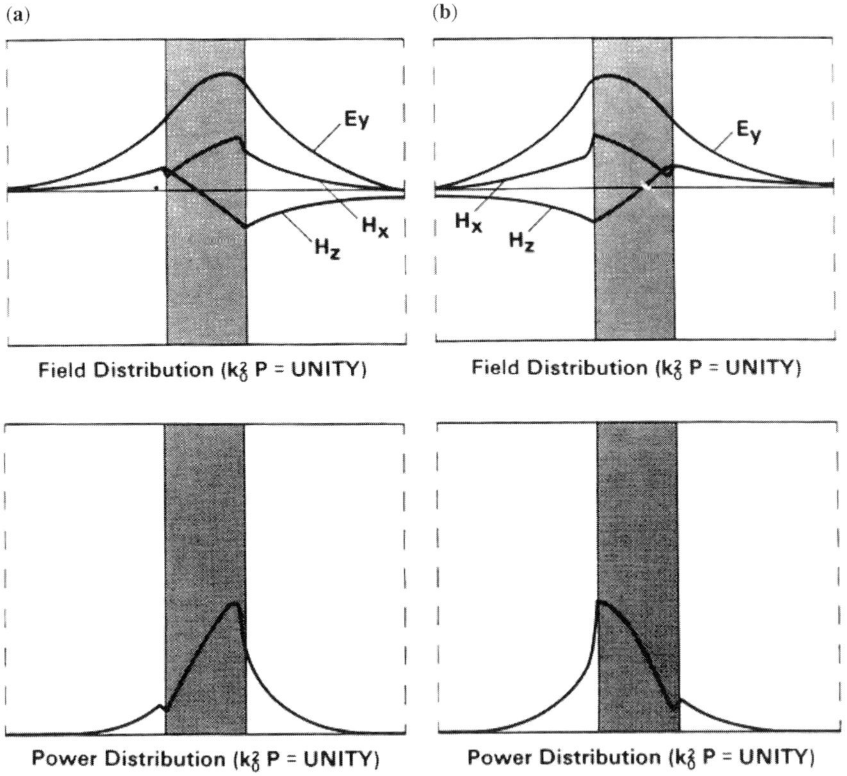

FIGURE 28.23 Edge mode effect in forward and backward directions of propagation of parallel plate gyromagnetic waveguide. (Reproduced with permission from J. Helszajn and H. Downs, Field displacement, circular polarisation, scalar permeabilities and differential phase shift in edge mode ferrite devices, *Microwellen Mag.*, Vol. 14, No. 3, pp. 269–278, April 1987.)

28.14 THIRD-ORDER INTERMODULATION PRODUCTS IN NONLINEAR DEVICES

In a linear device the output signal is strictly proportional to the input one. This, however, is not the case in a nonlinear situation. In the latter instance, new frequencies are produced by the nonlinearity. The nonlinearity has its origin in either electric effects in semiconductors or magnetic effects in magnetic insulators. The principal area of interest in this work is that of third-order intermodulation products (IMPs) in a magnetic insulator in the presence of two circularly polarized tones. This nonlinearity or IMD (intermodulation distortion) is of particular concern in the design of many UHF circulators.

The cubed term has six third-order intermodulation products: $3\omega_1$, $3\omega_2$, $2\omega_l + \omega_2$, $2\omega_2 + \omega_1$, $2\omega_1 - \omega_2$, and $2\omega_2 - \omega_1$. The first four of these are located far from ω_1 and ω_2 and will typically be outside the passband of interest. The two difference

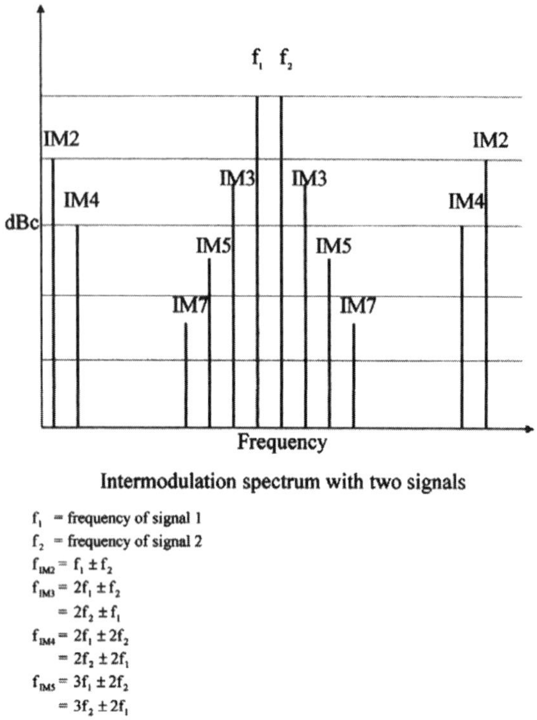

FIGURE 28.24 Two tone intermodulation spectrum.

terms produce products located near the original input signals at ω_1 and ω_2 and so cannot be easily filtered from the passband. Figure 28.24 shows a typical spectrum of second-order and third-order two-tone intermodulation products.

A typical term is

$$m\omega_1 + n\omega_2$$

with $m, n = 0, \pm 1, \pm 2, \pm 3, \ldots$. The various combinations of the two input frequencies are called intermodulation products, and the order of a given product is defined as $|m| + |n|$.

A property of the third-order intermodulation product is that if the amplitude of each carrier is increased by 1 dB, then the third-order intermodulation products increase by 3 dB. If the amplitude of the lower frequency line is increased by 1 dB, the lower-order intermodulation product increases by 2 dB whereas the upper one increases by only 1 dB. The dual situation is true for the upper frequency carrier.

The third-order IMP power is given by

$$\mathrm{IM3(dBm)} = 10 \log_{10} (P(2\omega_1 - \omega_2)) \text{ dBm}$$

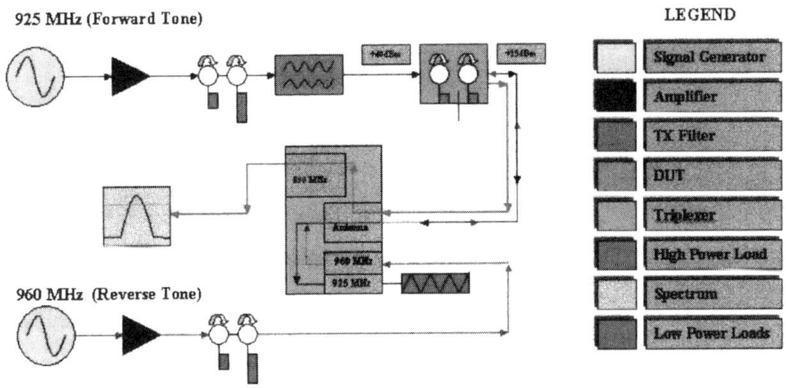

FIGURE 28.25 Block diagram of IMP test set with tones at same ports.

Some equipment manufacturers prefer to specify the IMD specification in dBc (decibels below the carrier power) instead of pure dBm results and this is the convention adopted here.

$$\mathrm{IM3(dBc)} = 10\log_{10}(P(2\omega_1 - \omega_2)) - 10\log_{10}(P(\omega_1))\ \mathrm{dBc}$$

The third-order IMP of a circulator may be measured in one of two ways depending on the system layout. In one arrangement the two tones, typically of equal magnitude, are applied at the input port of the device. This simulates a multicarrier environment where the IMP may reside in the receive band of the equipment. The other situation involves injecting one tone at port 1 and a second tone at port 2. It corresponds to a reverse signal incident on the system caused by breakthrough from a colocated antenna. The tone applied at port 2 in this instance is generally at a lower power level than that at port 1. Figures 28.25 and 28.26 show the two test sets.

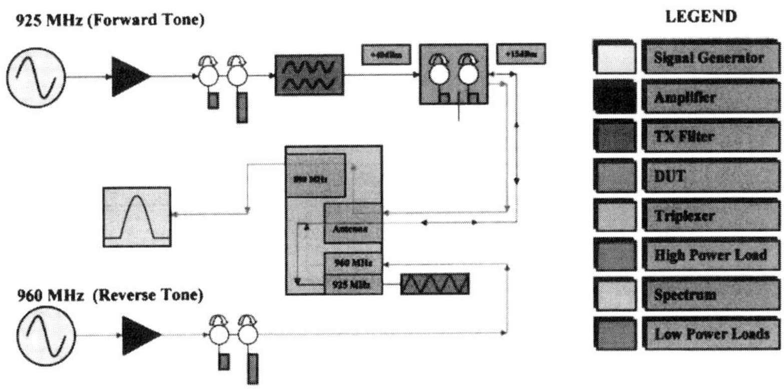

FIGURE 28.26 Block diagram of IMP test set with tones at different ports.

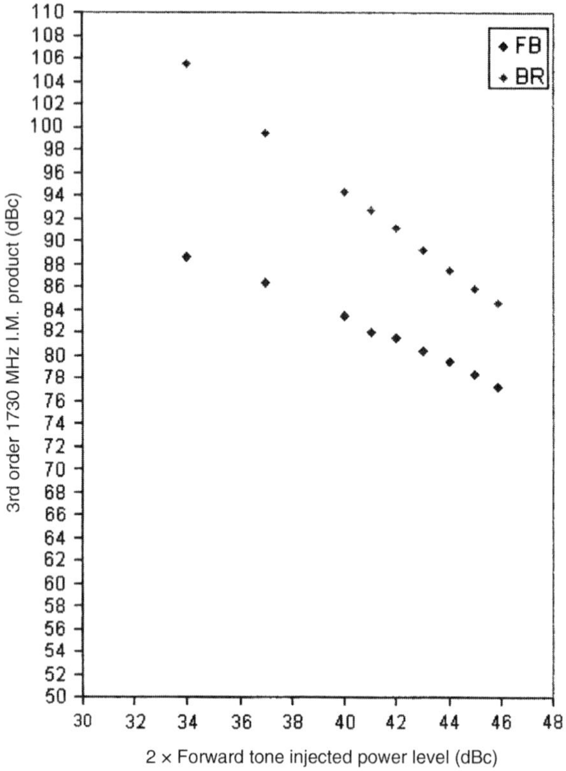

FIGURE 28.27 IMP below and above Kittel line.

Figure 28.27 compares the third-order intermodulation tone products of two 1800 MHz junction circulators biased above and below the Kittel line. The frequencies of the two input tones are in each case 1805 and 1880 MHz, producing a third-order harmonic at 1730 MHz. The material employed in the construction of the device operating below the Kittel line has a magnetization $\mu_0 M_0$ of 0.0400 T and the normalized direct magnetic field intensity of $H_0/M_0 = 1.1$. The material employed in the device biased above the Kittel line has a magnetization $\mu_0 M_0$ of 0.1100 T and its direct magnetic field intensity is typically 1.30. The two geometries also differed in that the gyromagnetic resonator of the former geometry uses a circular plate while the latter relied on an irregular hexagonal configuration for its operation. The IMP tone can be reduced, in the above Kittel devices, by moving the bias field intensity away from the line. This solution, however, is at the expense of both the size of the circulator and its gain-bandwidth product.

Bibliography

Aitken, F. M. and Mclean, R., Some properties of the waveguide Y circulator, *Proc. Inst. Elec. Eng.*, Vol. 110, No. 2, February 1963.

Akaiwa, Y., Operation modes of a waveguide Y-circulator, *IEEE Trans. Microwave Theory Tech.*, Vol. MTT-22, pp. 954–959, November 1974.

Akaiwa, Y., Bandwidth enlargement of a millimetre-wave Y circulator with half-wavelength line resonators, *IEEE. Trans. Microwave Theory Tech.*, Vol. MTT-22, pp. 1283–1286, December 1974.

Akaiwa, Y., Mode classification of a triangular ferrite post for Y-circulator, *IEEE Trans. Microwave Theory Tech.*, Vol. MTT-25, pp. 59–61, January 1977.

Allanson, J. G., Cooper, R., and Cowling, Z. G., The theory and experimental behaviour of right-angled junctions in rectangular-section waveguides, *J. Inst. Elec. Eng.*, Vol. 93, pt. III, pp. 177–187, 1949.

Anderson, L. K., An analysis of broadband circulators with external tuning elements, *IEEE Trans. Microwave Theory Tech.*, Vol. MTT-15, pp. 42–47, January 1967.

Araki, K., Kim, D., and Naito, Y., A study on circular disk resonators on a ferrite substrate, *IEEE Trans. Microwave Theory Tech.*, Vol. MTT-30, No. 2, pp. 147–154, February 1982.

Artman, J. O., Microwave resonance relations in anisotropic single-crystal ferrites, *Phys. Rev.*, Vol. 105, p. 62, 1957.

Auld, B. A., The synthesis of symmetrical waveguide circulators, *IRE Trans. Microwave Theory Tech.*, Vol. MTT-7, pp. 238–246. April 1959.

Ayashi, Y., Analysis of wideband stripline circulators for wideband operation, *IEEE Trans. Microwave Theory Tech.*, Vol. MTT-28, pp. 210–214, 1980.

Ayter, S. and Ayasli, Y., The frequency behaviour of stripline circulator junctions, *IEEE Trans. Microwave Theory Tech.*, Vol. MTT-26, pp. 197–202, 1978.

Barsony, P., Miniature strip-line circulators and isolators, *International Conference on Microwave Ferrites,* SMOLENICE, 1972.

The Stripline Circulator: Theory and Practice. By J. Helszajn

Barsony, P., Some problems of microstrip circulators, Annual of the Research Institute for Telecommunication, Budapest, 1974.

Belson, H. S. and Kreissman, C. J., Microwave resonance in hexagonal ferromagnetic single crystals, *J. Appl. Phys. Suppl.*, Vol. 30, No. 4, p. 175S, April 1959.

Bergman, J. O. and Christensen, C., Equivalent circuit for a lumped element Y circulator, *IEEE Trans. Microwave Theory Tech.*, Vol. MTT-16, pp. 308–310, May 1968.

Bex, H. and Schwartz, E., Performance limitation of lossy circulators, *IEEE Trans. Microwave Theory Tech.*, Vol. MTT-19, pp. 493–494, 1971.

Bittar, J. and Verszely, J., A general equivalent network of the input impedance of symmetric three-port circulators, *IEEE Trans. Microwave Theory Tech.*, Vol. MTT-28, p. 807–808, July 1980.

Bloembergen, N., Magnetic resonance in ferrites, *Proc. IRE*, Vol. 444, p. 1259, October 1956.

Bogdanov, A. G., Design of waveguide x-circulators, *Radio Eng. Electron. Phys.*, Vol. 14, No. 4, pp. 528–536, 1969.

Bolle, D. M. and Lewin, L., On the definition of parameters in ferrite electromagnetic wave interactions, *IEEE Trans. Microwave Theory Tech.*, Vol. MTT-21, p. 118, February 1973.

Bolle, D. M. and Talisa, S. H., The edge guide mode nonreciprocal phase shifter, *IEEE Trans. Microwave Theory Tech.*, Vol. MTT-27, p. 878, November 1979.

Bonetti, R. R. and Tissi, R., Analysis of planar disk networks, *IEEE Trans. Microwave Theory Tech.*, Vol. MTT-26, pp. 471–477, July 1978.

Bosch, F. and Mourzi, Z., Zur Bestimmung der Niedrigfeldarbeitspunktes von Streifenleitungszirkulatoren, *A.E.U.*, Vol. 22, pp. 605–607, December 1968.

Bosma, H., A general model for junction circulators; choice of magnetisation and bias field, *IEEE Trans. Magnetics*, Vol. MAG-4, No. 3, pp. 587–596, September 1968.

Bosma, H., On the principle of stripline circulation, *Proc. IEE Suppl.*, Vol. 109, pp. 137–146, 1962.

Bosma, H., Performance of lossy *H*-plane Y-circulators, *IEEE Trans. Magnetics*, Vol. MAG-2, pp. 273–277, September 1966.

Bosma, H., On stripline circulation at UHF, *IEEE Trans. Microwave Theory Tech.*, Vol. MTT-12, pp. 61–72, January 1964.

Bramham, B., A convenient transformer for matching coaxial lines, *Electron. Eng.*, Vol. 33, pp. 42–44, 1961.

Brigginshaw, P. M. and Riches, E. E., Developments of M.I.C. circulators at frequencies from 1 to 40 GHz, *IEEE Trans. Magnetics*, Vol. 11, pp. 1273–1275, September 1975.

Brug, A. A. and Wolf, W. P. Demagnetizing fields in magnetic measurements, *J. Appl. Phys.*, Vol. 57, pp. 4685–4701, 1985.

Buehler, G. V. and Eikenberg, A. F., Stripline *Y*-circulators for the 100 to 400 Mc region, *Proc. IRE*, Vol. 49, pp. 518–519, 1961.

Buffler, C. R. Resonance properties of single crystal hexagonal ferrites, *J. Appl. Phys. Suppl.* Vol. 33, No. 3, pp. 1350–1362, 1962.

Buffler, C. R. and Helszajn, J., The use of composite junctions in the design of high power circulators, *IEEE Int. Microwave Symp.*, pp. 20–22, May 1968.

Buffler, C. R., Nomogram for determining effective junction circulator VSWR, *Electron. Communicator*, Vol. 2, No. 6, p. 119, November/December 1967.

Butterweck, H. J., The *Y*-circulator, *Arch. Elek. Ubertragung*, Vol. 17, No. 4, pp. 163–176, December 1963.

Carlin, H. J., Principles of gyrator networks, in *Proceedings of Symposium on Modern Advances in Microwave Techniques*, pp. 175–204, November 1954.

Carlin, H. J. and Friedenson, R. A., Gain bandwidth properties of a distributed parameter load, *IEEE Trans. Circuit Theory*, Vol. CT-15, pp. 455–464, December 1968.

Carlin, H. J. and Giodano, Q. B., *Network theory, an Introduction to Reciprocal and Nonreciprocal Circuits*, Prentice Hall, Englewood Cliffs, NJ, 1964.

Carlin, H. J. and Kohler, W. E., Direct synthesis of band-pass transmission line structures, *IEEE Trans. Microwave Theory Tech.* Vol. MTT-13, pp. 283–297, May 1965.

Castillo, J. B. and Davis, L. E., Computer-aided design of 3-port waveguide junction circulators, *IEEE Trans. Microwave Theory Tech.* Vol. MTT-18, pp. 25–34, January 1970.

Cendes, Z. J. and Lee, J. F., The transfinite element method for modelling MMIC devices, *IEEE Trans. Microwave Theory Tech.*, Vol. MTT-36, No. 12, pp. 1639–1649, December 1988.

Chait, H. N. and Curry, T. R., *Y*-circulator, *J. Appl. Phys.*, Vol. 30, p. 152, April 1959.

Champlin, K. S. and Glover, G. H., Gap effect in measurement of large permittivities, *IEEE Trans. Microwave Theory Tech.*, Vol. MTT-14, pp. 397–398, 1966.

Chiang, K. S., Finite element method for cut-off frequencies of weakly guiding fibres of arbitrary cross-section, *Opt. Quantum Electron.*, Vol. 16, No. 6, pp. 487–493, November 1984.

Chiron, B., Forterre, G., and Rannou, C., Nouveaux dispositifs non reciproques à tres grande largueur de bande utilisant des ondes de surface electromagnetiques, *Onde Electrique*, Vol. 51, No. 9, pp. 816–818, 1971.

Chiron, B., Girard, P., and Prache, P. M., Theory and practice of lumped parameter circulators, *IEE International Conference on the Microwave Behaviour of Ferrimagnetics and Plasmas*, 1965.

Collins, J. H. and Watters, A. R., A field map for the waveguide *Y*-junction circulator, *Electron. Engr. London*, Vol. 35, pp. 540–543, August 1963.

Csendes, Z. J. and Silvester, P., Numerical solution of dielectric loaded waveguides, *IEEE Trans. Microwave Theory Tech.*, Vol. MTT-18, pp. 1124–1131, 1970.

Damon, R. W., Relaxation effects in ferromagnetic resonance, *Rev. Mod. Phys.*, Vol. 25, pp. 239–245, 1953.

Davies, J. B., An analysis of the *m*-port symmetrical *H*-plane waveguide junction with central ferrite post, *IRE Trans. Microwave Theory Tech.*, Vol. MTT-10, pp. 596–604, November 1962.

Davies, J. B. and Cohen, P., Theoretical design of symmetrical junction stripline circulators, *IEEE Trans. Microwave Theory Tech.*, Vol. MTT-11, pp. 117–123, March 1963.

Davies, J. B. and Cohen, P., Theoretical design of symmetrical junction stripline circulators, *IEEE Trans. Microwave Theory Tech.*, Vol. MTT-11, pp. 506–512, 1963.

Davis, L. E., Computed phase-shift and performance of a latching 3-port waveguide circulator, *NEREM RECORD*, 1966.

Davis, L. E., Coleman, M. D., and Cotter, J. J., Four-port crossed waveguide junction circulators, *IEEE Trans. Microwave Theory Tech.*, Vol. MTT-12, pp. 43–47, 1964.

De Koning, J. G., Hamilton, R. J. Jr., and Hierl, T. L., Full-band low-loss continuous tracking circulation in K and, *IEEE Trans. Microwave Theory Tech.*, Vol. MTT-25, pp. 152–155, February 1977.

Dessert, E., The study and production of *Y*-junction circulators in the frequency range 450–1000 MHz, *Acta Electron.*, Vol. 8, p. 175, April 1964.

Deutch, J. and Weiser, B., Resonance isolator and Y circulator with lumped elements at VHF, *IEEE Trans. Magnetics*, Vol. MAG-2, pp. 278–282, 1966.

Dunn, V. E. and Roberts, R. W., Miniature VHF and UHF circulators use lumped element design, *Microwaves*, pp. 46–47, December 1963.

Dunn, V. E. and Roberts, R. W., New design techniques for miniature VHF circulators, presented at the *G-MTT Symposium,* Clearwater, FL, 1965.

Eshbach, J. R., Spin-wave propagation and the magnetoelastic interaction in yttrium iron garnet, *J. Appl. Phys.*, Vol. 34, No. 4, Pt. 2, April 1963.

Fano, R. M., Theoretical limitations on broadband matching of arbitrary impedances, *J. Franklin Inst.*, Vol. 249, pp. 57–98, January 1950; also, pp. 139–154, February 1950.

Fay, C. E. and Comstock, R. L., Operation of the ferrite junction circulator, *IEEE Trans. Microwave Theory Tech.*, Vol. MTT-13, pp. 15–27, January 1965.

Fay, C. E. and Dean, W. A., The four-port single junction circulator in stripline, *PG-MTT, Digest of the International Symposium, IEEE,* Piscataway, NJ, 1966.

Fletcher, R. C., Lecrow, R. C., and Spencer, E. G., *Phys. Rev.*, Vol. 117, p. 955, 1960.

Feoktistov, V. G., Design of a strip *Y*-circulator, *Radio Eng. Electron. Phys.*, Vol. 13, No. 7, pp. 1111–1115, 1968.

Fliegler, E., A 350 MHz broadband lumped element circulator as a protective isolator, *IEEE Trans. Microwave Theory Tech.*, Vol. MTT-17, No. 5, pp. 275–276, 1969.

Fowler, H., Paper presented at the 1950 symposium on microwave properties and applications of ferrites, Harvard University, Cambridge, MA.

French, G. N. and Fooks, E. H., Double section matching transformers, *IEEE Trans. Microwave Theory Tech.*, Vol. MTT-17, p. 719, 1969.

Gibson, A. A. P., Dillon, B. M., and Sheikh, Applied field/frequency response of planar gyromagnetic discs, *Int. J. Electron*, Vol. 76, No. 6, pp. 1073–1081, 1994.

Goebel, U. and Schieblich, C., A unified equivalent circuit representation of *H*- and *E*-plane junction circulators, *European Microwave Conference*, pp. 803–808, 1983.

Green, J. J. and Sandy, F., Microwave characterisation of partially magnetised ferrites, *IEEE Trans. Microwave Theory Tech.*, Vol. MTT-22, pp. 641–645, June 1974.

Gurevich, A. G., *Ferrites at Microwave Frequencies*, Heywood and Co. Ltd., London, 1960.

Hagelin, S., A flow graph analysis of 3- and 4 port junction circulators, *IEEE Trans. Microwave Theory Tech.*, Vol. MTT-14 No. 5, pp. 243–249, 1966.

Hagelin, S., Analysis of lossy symmetrical three-port networks with circulator properties, *IEEE Trans. Microwave Theory Tech.*, Vol. MTT-17, No. 6, p. 328, 1969.

Hammer, G., The analysis of the waveguide Y circulator, *Periodica Polytech. Elec. Eng.*, Vol. 14, pp. 61–76, 1970.

Hauth, W., Lenz, S., and Pivit, E., Design and realization of very-high-power waveguide junction circulators, *Frequenez*, Vol. 40, No. 1, pp. 1–24, January 1986.

Helszajn, J. A ferrite ring stripline circulator, *Radio Electron. Engineer.*, Vol. 32, pp. 55–60, 1966.

Helszajn, J. A *H* plane high power TEM ferrite circulator, *Radio Electron. Engr.*, Vol. 33, pp. 257–262, 1967.

Helszajn, J., A simplified theory of the three port junction circulator, *Radio Electron. Engineer*, Vol. 33, pp. 283–288, 1967.

Helszajn, J., Frequency and bandwidth of *H* plane TEM junction circulator, *Proc. IEE*, Vol. 117, No. 7, pp. 1235–1238, July 1970.

Helszajn, J., The adjustment of the *m*-port single junction circulator, *IEEE Trans, Microwave Theory Tech.*, Vol. MTT-18, pp. 705–711, October 1970.

Helszajn, J. Wideband circulator adjustment using $n = \pm 1$ and $n = 0$ electromagnetic-field patterns, *Electron. Lett.*, Vol. 6, pp. 729–731, 12 November 1970.

Helszajn, J., Synthesis of quarter wave coupled circulators with Chebyshev characteristics, *IEEE Trans. Microwave Theory Tech.*, Vol. MTT-20, pp. 764–769, 1972.

Helszajn, J., Dissipation and scattering matrices of lossy junctions, *IEEE Trans. Microwave Theory Tech.*, Vol. MTT-20, pp. 779–782, 1972.

Helszajn, J., Three resonant mode adjustment of the waveguide circulator, *Radio Electron. Engineer*, Vol. 42, No. 4, pp. 1–4, April 1972.

Helszajn, J., Microwave measurement techniques for below resonance junction circulators, *IEEE Trans. Microwave Theory Tech.*, Vol. MTT-21, pp. 347–351, May 1973.

Helszajn, J., Frequency response of quarter-wave coupled reciprocal stripline junctions, *IEEE Trans. Microwave Theory Tech.*, Vol. MTT-21, pp. 533–537, August 1973.

Helszajn, J., Waveguide and stripline 4-port single junction circulators, *IEEE Trans. Microwave Theory Tech.* Vol. MTT-21, pp. 630–633, 1973.

Helszajn, J., Composite-junction circulators using ferrite disks and dielectric rings, *IEEE Trans. Microwave Theory Tech.*, Vol. MTT-22, pp. 400–410, April 1974.

Helszajn, J., Scattering matrices of junction circulators with Chebyshev characteristics, *IEEE Trans. Microwave Theory Tech.*, Vol. MTT-23, pp. 548–554, July 1975.

Helszajn, J., *Nonreciprocal Microwave Junctions and Microwave Circulators*, Wiley-Interscience, Hoboken, NJ, 1975.

Helszajn, J., Ferrite parameters of the junction circulator, *Electron. Components*, pp. 785–793, July 1976.

Helszajn, J., *Nonreciprocal Microwave Junctions and Circulators*, John Wiley & Sons, Hoboken, NJ, 1976.

Helszajn, J., Standing-wave solution of 3-port junction circulator with 1-port terminated in a variable short circuit, *Proc. IEE Microwaves Optics Acoustics*, Vol. 126, pp. 67–68, March 1979.

Helszajn, J., Standing wave solution of irregular hexagonal and wye resonators, *IEEE Trans. Microwave Theory Tech.* Vol. MTT-29, pp. 562–567, June 1981.

Helszajn, J., Operation of the tracking circulator, *IEEE Trans. Microwave Theory Tech.*, Vol. MTT-29, pp. 700–707, July 1981.

Helszajn, J., Quarter-wave coupled junction circulators using weakly magnetized disk resonators, *IEEE Trans. Microwave Theory Tech.*, Vol. MTT-30, pp. 800–806, May 1982.

Helszajn, J., Complex gyrator circuits of planar circulators using higher order modes in a disk resonator, *IEEE Trans. Microwave Theory Tech.* Vol. MTT-31, pp. 931–937, November 1983.

Helszajn, J., Synthesis of quarter-wave coupled junction circulators with degree 1 and 2 complex gyrator circuits, *IEEE Trans. Microwave Theory Tech.*, Vol. MTT-32, pp. 382–390, April 1985.

Helszajn, J., Contour integral equation formulation of complex gyrator admittance of junction circulators using triangular resonators, *Proc. IEEE Microwaves Antennas Propag.*, Vol. 132, pp. 255–260, July 1985.

Helszajn, J., Complex gyrator of an evanescent mode *E*-plane junction circulator using *H*-plane turnstile resonators, *IEEE Trans. Microwave Theory Tech.*, Vol. MTT-35, pp. 797–806, 1987.

Helszajn, J., Experimental evaluation of junction circulators: a review, *Proc IEE Microwave Antennas Propag.*, Vol. 141, No. 5, pp. 351–358, 1994.

Helszajn, J., Synthesis of octave-band quarterwave coupled semi-tracking junction circulators, *IEEE Trans. Microwave Theory Tech.*, Vol. MTT-43, pp. 573–578, March 1995.

Helszajn, J., Fabrication of very weakly and weakly magnetised microstrip circulators, *IEEE Trans. Microwave Theory Tech.*, Vol. MTT-46, pp. 439–449, May 1998.

Helszajn, J. and Aitken, F. M., U.H.F. techniques for lumped constant circulators, *Electron. Eng.*, pp. 53–59, November 1973.

Helszajn, J. and Baars, R. D., Synthesis of wide-band planar circulators using narrow coupling angles and undersized disk resonators, *IEEE Trans. Microwave Theory Tech.*, Vol. MTT-39, pp. 1681–1687, 1991.

Helszajn, J. and Bradley, C., Experimental characterization of reciprocal planar microstrip 3-port junctions, *Proc. IEE Microwaves Antennas Propag.*, Vol. 138, pp. 91–97, February 1991.

Helszajn, J. and Buffler, C. R., Adjustment of the 4-port single junction circulator, *Radio Electron. Engineer*, Vol. 35, pp. 357–360, June 1968.

Helszajn, J, Cheng, C. S., and D'Orazio, D., UHF irregular hexagonal gyromagnetic resonators symmetrically with magnetic walls, *Proc. IEE Microwave Antennas Propag.* Vol. 146, No. 6, pp. 427–433, December 1999.

Helszajn, J. and Downs, H., Field displacement, circular polarisation, scalar permeabilities and differential phase shift in edge mode ferrite devices, *Microwellen Mag.* Vol. 14, No. 3, pp. 269–278, April 1987.

Helszajn, J. and James, D. S., Planar triangular resonators with magnetic walls, *IEEE Trans. Microwave Theory Tech.*, Vol. MTT-26, pp. 95–100, February 1978.

Helszajn, J. and Lapointe, Y., Complex gyrator circuit of 3-port circulators using gyromagnetic resonators with six-fold symmetry, *Proc. IEE Microwaves Antennas Propag.*, Vol. 148, pp. 318–322, October 2001.

Helszajn, J. and Lynch, D. J., Cut-off space of cloverleaf resonators with electric and magnetic walls, *IEEE Trans. Microwave Theory Tech.*, Vol. MTT-40, No. 8, pp. 1620–1629, August 1992.

Helszajn, J. and Lyon, R. W., Mode charts of magnetised and demagnetised planar hexagonal resonators on ferrite substrates using finite elements, *Proc. IEE*, Vol. 131, Pt. H, No. 6, pp. 420–422, December 1984.

Helszajn, J. and McDermott, M., The inductance of a lumped constant circulator, *IEEE Trans. Microwave Theory Tech.*, Vol. MTT-18, No. 1, pp. 50–52, 1970.

Helszajn, J. and McStay, J., External susceptibility tensor of magnetised ferrite ellipsoid in terms of uniform mode ellipticity, *Proc. IEE*, Vol. 116, No. 12, pp. 2088–2092, December 1969.

Helszajn, J. and Nisbet, W. T., Circulators using planar wye resonator, *IEEE Trans. Microwave Theory Tech.*, Vol. MTT-29, pp. 689–699, July 1981.

Helszajn, J. and Powesland, M., Low-loss high power microstrip circulators, *IEEE Trans. Microwave Theory Tech.*, Vol. MTT-29, pp. 572–578, July 1981.

Helszajn, J. and Sharp, J., Frequency response of quarter-wave coupled reciprocal stripline junctions, *Microwave Eng. Europe*, pp. 29–35, March/April 2003.

Helszajn, J. and Sharp, J., Adjustment of in-phase mode in turnstile circulator, *IEEE Trans. Microwave Theory Tech.*, Vol. MTT-32, pp. 339–343, 1984.

Helszajn, J. and Walker, P. N., Operation of high peak power differential phase shift circulators, *IEEE Trans. Microwave Theory Tech.*, Vol. MTT-26, pp. 653–658, September 1978.

Helszajn, J., Baars, R. D., and Nisbet, W. T., Characteristics of circulators using planar triangular and disk resonators symmetrically loaded with magnetic walls, *IEEE Trans. Microwave Theory Tech.*, Vol. MTT-28, pp. 616–621, June 1980.

Helszajn, J., James, D. S., and Nisbet, W. T., Circulators using planar triangular resonators, *IEEE Trans. Microwave Theory Tech.*, Vol. MTT-27, pp. 188–193, February 1979.

Helszajn, J., McKay, M., and Lynch, D. J., Complex gyrator circuit of a junction circulator using weakly magnetised planar irregular hexagonal resonator, *Proc. IEE Microwave Antennas Propag.*, Vol. 143, pp. 532–538, December 1996.

Helszajn, J., Murray, R. W., Davidson, E. G. S. and Suttie, R. A., Microwave subsidiary resonance ferrite limiters, *IEEE Trans. Microwave Theory Tech.*, Vol. MTT-25, pp. 190–195, 1977.

Helszajn, J., Walker, D., and Aitken, F. M., Varactor-tuned lumped-element circulators, *IEEE Trans. Microwave Theory Tech.*, Vol. MIT-19, No. 10, pp. 825–826, 1971.

Helszajn, J., Riblet, G. P., and Mather, J. P., Insertion loss of 3-port circulator with one port terminated in variable short circuit, *IEEE Trans. Microwave Theory Tech.*, Vol. MTT-23, pp. 926–927, 1975.

Hershenov, H., All garnet substrate microstrip circulators, *Proc. IEEE*, Vol. 55, pp. 696–697, May 1967.

Hershenov, H., X-band microstrip circulator, *Proc. IEEE*, Vol. 54, pp. 2022–2023, December 1966.

Hines, M. E., Reciprocal and nonreciprocal modes of propagation in ferrite stripline and microstrip devices, *IEEE Trans Microwave Theory Tech.*, Vol. MTT-19, pp. 442–451, 1971.

Hogan, C. L., The microwave gyrator, *Bell Syst. Tech. J.*, Vol. 31, p. 1, January 1952.

How, H., Oliver, S. A., McKnight, S. W., Zaveracky, P. M., McGruer, N. E., Vittoria, C., and Schmidt, R., Influence of non-uniform magnetic field on a ferrite junction circulator, *IEEE Trans. Microwave Theory Tech.*, Vol. 4, No. 10, pp. 1982–1989, October 1999.

How, H., Fang, R.-M., Victoria, C., and Schmidt, R., Design of six-port stripline junction circulators, *IEEE Trans. Microwave Theory Tech.*, Vol. MTT-42, pp. 1272–1275, July 1994.

Humphreys, B. L. and Davies, J. B., The synthesis of *N*-port circulators, *IRE Trans. Microwave Theory Tech.*, Vol. MTT-10, pp. 551–554, 1962.

Ito, Y. and Yochouchi, H., Microwave junction circulators, *Fujitsu Scientific Tech. J.*, pp. 55–90, 1969.

Itoh, T. and Mittra, R., Analysis of a microstrip disk resonator, *Arch. Elektron Ubertragungstech*, Vol. 27, pp. 456–458, 1973.

Ivanov, S. A. and Dankov, P. I., X-band microstrip circulator on ferrite substrate, in *Proceedings of the 8th Colloquium of Microwave Communications* (8th MICROCOLL), Budapest, Hungary, pp. 353–354, August 1986.

Joseph, R. I. and Schlomann, E., Demagnetising field in non-ellipsoidal bodies, *J. Appl. Phys.*, Vol. 36, pp. 1579–1593, 1965.

Joseph, R. I. and Schloemann, E., Microwave behaviour of stripline circulator junctions, *IEEE Trans. Microwave Theory Tech.*, Vol. MTT-26, pp. 197–202, March 1978.

Kerns, D. M., Analysis of symmetrical waveguide junctions, *J. Res. Natl. Bur. Stand.*, Vol. 46, pp. 267–282, April 1951.

Khilla, A. M., Analysis of wide-band microstrip circulators using a point matching technique, *Microwave Symp. Dig.*, pp. 284–286, 1981.

Kittel, C., On the theory of ferromagnetic resonance absorption, *Phys. Rev.*, Vol. 73, pp. 155–161, January 1948.

Kiyuzaki, T. and Yada, M., An investigation on bandwidth enlargement of a *Y*-circulator, *National Conference of the Institute of Electronics and Communications Engineers of Japan*, paper 55–7, 1963.

Knerr, R. H., A thin film lumped element circulator, *IEEE Trans. Microwave Theory Tech.*, Vol. MTT-17, No. 12, pp. 1152–1154, 1969.

Knerr, R. H., Barnes, C. E., and Bosch, F., A compact broadband thin-film limped element L-band circulator, *IEEE Trans. Microwave Theory Tech.*, Vol. MTT-18, pp. 1100–1108, December 1970.

Kobayashi, M., Kawamura, K., and Suzuki, K., High *Q*-factor tri-plate resonator with an inner conductor of melted silver, *Ceramic Trans.*, Vol. 41, pp. 355–362, 1994.

Kompa, G., *S*-matrix computation of microstrip discontinuities, *AEU*, Vol. 30, pp. 58–64, 1976.

Konishi, Y., Lumped element Y circulator, *IEEE Trans. Microwave Theory Tech.*, Vol. MTT-13, pp. 852–864, November 1965.

Konishi, Y., General theory of coaxial *Y*-junction circulator and design consideration on its UHF high power application, *Tech. J. Japan Broadcast. Corp.*, Vol. 20, No. 2, p. 19, 1968.

Konishi, Y., A high power U.H.F circulator, *IEEE Trans. Microwave Theory Tech.*, Vol. MTT-15, pp. 700–708, 1967.

Konishi, Y. and Hoshino N., Design of a new broad-band isolator, *IEEE Trans. Microwave Theory Tech.*, Vol. MTT-19, No. 3, pp. 260–269, 1971.

Krause, N., Improved microstrip circulator design by resonance components, *in Proceedings EuMC,* Amsterdam, pp. 383–388, 1981.

Kretzschmar, J. G., Theory of the elliptic disk and ring resonators, *in Proceedings of the Fifth Colloquium on Microwave Communication,* Budapest, Hungary, 1974, pp. 24–30.

Ku, W. H. and Wu, Y. S., On stripline four port circulator, *1973 IEEE Microwave Theory Tech. Int. Microwave Symp. Dig.*, IEEE Cat. No. 73, Chap. 736–9 MTT, pp. 86–88.

Lagrange, A., Lahmi, H., and Vallatin, B. R., K-band high-peak-power junction circulator: influence of the static magnetic field, *IEEE Trans. Magnetics*, Vol. MAG-9, pp. 531–534, September 1973.

Landau, L. and Lifshitz, E., On the theory of the dispersion of magnetic permeability in ferromagnetic bodies, *Phys. Z. Sowjetunion*, Vol. 8, p. 153, 1935.

Landry, D. N., A single junction four-port coaxial circulator, *Wescon*, Part 5, Session 4.2, 1963.

Laura, P. A. A., Nagaya, K., and Sarmiento, G. S., Numerical experiments on the determination of cut-off frequencies of waveguides of arbitrary cross section, *IEEE Trans. Microwave Theory Tech.*, Vol. MTT-28, No. 6, pp. 568–572, June 1980.

Lax, B., Frequency and loss characteristics of microwave ferrite devices, *Proc. IRE*, Vol. 44, pp. 1368–1386, 1956.

Leetma, J. G., Ferrite circulator with conductive plate of uniform thickness having tapered angular apexes for broad banding, U.S. Patent No. 3 104 361, 17 September 1963.

Levy, R., Synthesis of mixed lumped and distributed impedance matching networks, *IEEE Trans. Microwave Theory Tech.*, Vol. MTT-20, pp. 223–233, 1972.

Levy, R. and Helszajn, J., Closed form solutions for stub R-loads with non-zero reflection minimas, *IEEE Trans. Microwave Theory Tech.*, Vol. MTT-30, pp. 55–63, January 1982.

Levy, R. and Helszajn, J., Specific equations for one and two section quarter-wave matching networks for stub resistor loads, *IEEE Trans. Microwave Theory Tech.*, Vol. MTT-30, pp. 56–62, January 1982.

Levy, R. and Helszajn, J., Synthesis of short line matching networks for resonant loads, *Proc. IEE Microwaves Antennas Propag.*, Pt. H, Vol. 130, pp. 385–390, October 1983.

Longley, S. R., Experimental 4-port *E*-plane junction circulator, *IEEE Trans. Microwave Theory Tech.*, Vol. MTT-15, pp. 378–380, 1967.

Lyon, R. W. and Helszajn, J., A finite element analysis of planar circulators using arbitrarily shaped resonators, *IEEE Trans. Microwave Theory Tech.*, Vol. MTT-30, pp. 1964–1974, 1982.

Martinson, T. M. and Kuester, E. F., Accurate analysis of arbitrary shaped patch resonators on thin substrates, *IEEE Trans. Microwave Theory Tech.*, Vol. MTT-36, No. 2, pp. 324–331, February 1988.

Masse, D., Broadband microstrip junction circulators, *Proc. IEEE*, Vol. 56, pp. 352–353, March 1968.

Matthaei, G. L., Short-step Chebyshev impedance transformers, *IEEE Trans. Microwave Theory Tech.*, Vol. MTT-14, pp. 372–383, 1966.

McChesney, G. and Dunn, V., Broadband lumped element UHF circulator, *IEEE Trans. Microwave Theory Tech.*, Vol. MTT-15, No. 3, pp. 198–199, 1967.

Medoks, A. G., Design of a waveguide *Y*-circulator, *Telecommun. Radio Eng.*, Vol. 22, Pt. 2, No. 9, September 1967.

Melchor, J. L. and Vartanian, P. H., Temperature effects in microwave ferrite devices, *IRE Trans. Microwave Theory Tech.*, Vol. MTT-7, p. 15, 1959.

Milano, U., Sanders, J. H., and Davis, L. Jr., A *Y*-junction strip-line circulator, *IRE Trans. Microwave Theory Tech.*, Vol. MTT-8, pp. 346–351, 1960.

Miura, T., Kobayashi, M., Nagata, H., and Konishi, Y., Optimization of lumped element circulator based on eigen inductance evaluation and structural improvement, *IEEE MTT-S Dig.*, pp. 2648–2654, 1996.

Miyoshi, T. and Miyaushi, S., The design of planar circulators for wideband operation, *IEEE Trans. Microwave Theory Tech.*, Vol. MTT-28, pp. 210–215, March 1980.

Miyoshi, T., Yamaguchi, S., and Goto, S., Ferrite planar circuits in microwave integrated circuits, *IEEE Trans. Microwave Theory Tech.*, Vol. MTT-25, pp. 595–600, July 1977.

Mokari-Bolhassan, M. E. and Ku, W. H., Transfer function approximations for a new class of bandpass distributed network structures, *IEEE Microwave Theory Tech.*, Vol. MTT-25, pp. 837–847, 1977.

Montgomery, C. G., Dicke, R. H., and Purcell, E. M., *Principles of Microwave Circuits*, McGraw-Hill, New York, 1948.

Nagao, T., Broad-band operation of stripline Y circulators, *IEEE Trans. Microwave Theory Tech.*, Vol. MTT-25, pp. 1042–1047, December 1977.

Nagao, T., Double circulation frequency operation of stripline *Y*-junction circulators, *IEEE Trans. Microwave Theory Tech.,* Vol. MTT-25, pp. 181–189, March 1977.

Nagata, H. and Konishi, Y., New measuring method of eigen values of non-reciprocal three ports network (in Japanese), *Tech. Rept. of IEICE*, MW 95-108, pp. 79–84, 1995.

Nahito, Y. and Tanaka, N., Broad-banding and changing operation frequency of circulator, *IEEE Trans. Microwave Theory Tech.*, Vol. MTT-19, No. 4, pp. 367–372, 1971.

Nakajima, M., A simplified formula for the characteristic impedance of the microstrip lines, *Int. J. Electron.*, Vol. 44, pp. 61–64, 1978.

Nalbandian, V. and Steenaart, W., Discontinuities in symmetric striplines due to impedance steps and their compensations, *IEEE Trans. Microwave Theory Tech.*, Vol. MTT-20, pp. 573–578, 1972.

Nisbet, W. T. and Helszajn, J., Characterization of planar wye shaped resonators for use in circulator hardware, *IEEE International Conference on Microwave Theory and Techniques*, Washington DC 1980.

Nisbet, W. T. and Helszajn, J., Microstrip synthesized from waveguide equivalent, *MSN*, pp. 117–122, January 1981.

Nisbet, W. T. and Helszajn, J., Mode charts for microstrip resonators on dielectric and magnetic substrates using a traverse resonance method, *Proc. IEE H.*, pp. 69–76, March 1979.

Ogasawara, N. and Noguchi, T., Modal analysis of dielectric resonator of the normal triangular cross-section, presented at the 1974 *Annual National Convention of the Institute of Electrical Engineers*, Japan, 28 March 1974 (in Japanese).

Okada, F. and Ohwi, K., Design of a high-power CW Y-junction waveguide circulator, *IEEE Trans. Microwave Theory Tech.*, Vol. MTT-26, No. 5, pp. 364–369, May 1978.

Okamura S. and Pagai, T., VHF and UHF-band stacked junction circulators, *IEEE Trans. Microwave Theory Tech.*, Vol. MTT-17, No. 12, pp. 1151–1152, 1969.

Okoshi, T. and Miyoshi, T., The planar circuit—an approach to microwave integrated circuitry, *IEEE Trans. Microwave Theory Tech.*, Vol. MTT-20, pp. 245–252, April 1972.

Osborn, J. A., Demagnetising factors of the general ellipsoid, *Phys. Rev.*, Vol. 67, pp. 351–357, June 1945.

Owens, R. P., Accurate analytic determination of quasi static microstrip line parameters, *Radio Electron. Engineers*, Vol. 46, pp. 360–364, 1976.

Oxley, C. H., Brazil, T. J., Purcell, J. J., and Genmer, R., Design and performance of J band ferrite microstrip circulators, *Proc. IEE*, Vol. 125, No. 8, August 1978.

Pippin, J. E., Microwave ferrite devices—1968, *Microwave J.*, pp. 29–45, 1968.

Pivit, E., Circulators from lumped elements, *Telefunken*, Vol. 38, No. 2, pp. 206–213, 1965.

Polder, D., On the theory of ferromagnetic resonance, *Philos. Mag*, Vol. 40, pp. 100–115, 1949.

Poole, K. M. and Tien, P. K., *Proc. Inst. Radio Engineers*, Vol. 44, p. 1387, 1958.

Pucci, De Santis F., Novel type of M.I.C. symmetrical 3-port circulator, *Electron. Lett.*, Vol. 8, No. 1, pp. 12–13, January 1972.

Rado, G., Theory of microwave permeability tensor and Faraday effect in non-saturated ferromagnetic materials, *Phys. Rev.*, Vol. 89, pp. 529, 1953.

Reeder, T. M. and Sperry, W. R., Broad-band coupling to high-Q resonant loads, *IEEE Trans. Micrwave Theory Tech.*, Vol. MTT-20, pp. 453–458, July 1972.

Riblet, G. and Hansson, B., The use of symmetry to simplify the integral equation method with application to triangular resonators, *IEEE Trans. Microwave Theory Tech.*, Vol. MTT-30, pp. 1219–1223, 1982.

Riblet, G., Measurement of the equivalent admittance of 3-port circulators, *IEEE Trans. Microwave Theory Tech.*, Vol. MTT-25, pp. 401–405, May 1977.

Riblet, G., Technique for broad-banding above resonance circulator junctions without the use of external matching networks, *IEEE Trans. Microwave Theory Tech.*, Vol. MTT-28, pp. 125–129, February 1980.

Riblet, G. P., An eigenadmittance condition applicable to symmetrical four-port circulators and hybrids, *IEEE. Trans. Microwave Theory Tech.*, Vol. MTT-26, pp. 275–279, April 1978.

Riblet, G. P., Techniques for broadbanding above resonance circulator junctions without the use of external matching networks, *IEEE Trans. Microwave Theory Tech.*, Vol. MTT-28, pp. 125–129, February 1980.

Richardson, J. K., An approximate method of calculating Z of a symmetrical stripline, *IEEE. Trans. Microwave Theory Tech.*, Vol. MTT-15, pp. 130–131, 1967.

Robants, R. W., and Lahey, J. E., Ferrites for VHF frequencies, *IEEE Trans. Veh. Commun.*, Vol. VC-14, pp. 117–121, 1965.

Salay, S. J. and Peppiatt, H. J., An accurate junction circulator design procedure, *IEEE Trans. Microwave Theory Tech.*, Vol. MTT-20, pp. 192–193, February 1972.

Salay, S. J. and Peppiatt, H. J., Input impedance behavior of stripline circulator, *IEEE Trans. Microwave Theory Tech.*, Vol. MTT-19, pp. 109–110, 1971.

Schaug-Patterson, T., Novel design of a 3-port circulator, Norwegian Defence Research Establishment, January 1958.

Schelkunoff, S. A., *Electromagnetic Waves*, Van Nostrand, New York, 1943, pp. 393.

Schloemann, E., Broad-band stripline circulators based on YIG and Li-ferrite single crystals, *IEEE Trans. Microwave Theory Tech.*, Vol. MTT-34, pp. 1394–1400, December 1986.

Schloemann, E., Microwave behaviour of partly magnetized ferrites, *J. Appl. Phys.*, Vol. 41, pp. 202–214, January 1970.

Schloemann, E. and Blight, R. E., Broadband stripline circulator, *MTT-S International Microwave Symposium Digest* 86.1, pp. 739–742, 1986.

Schlomann, E. and Jones, R. V., Ferromagnetic resonance in polycrystalline ferrites with hexagonal crystal structure, *J. Appl. Phys. Suppl.*, Vol. 30, p. 177S, April 1959.

Schneider, M. V. Microstrip lines for microwave integrated circuits, *Bell Syst. Tech. J.*, Vol. 48, pp. 1421–1444, 1969.

Schwartz, E., Broadband matching of resonant circuits and circulators, *IEEE Trans. Microwave Theory Tech.* Vol. MTT-16, pp. 158–165, 1968.

Sharma, A. K. and Hoefer, W. J. R., Spectral domain analysis of hexagonal microstrip resonator, *IEEE Trans. Microwave Theory Tech.*, Vol. MTT-30, pp. 825–828, 1982.

Silvester, P., A general high-order finite element waveguide analysis program, *IEEE Trans. Microwave Theory Tech.*, Vol., MTT- 17, pp. 204–210, April 1969.

Silvester, P., Finite element solution of the homogeneous waveguide problem, *Alta Frequenza*, Vol. 38, pp. 593–600, May 1969.

Silvester, P., High order polynomial triangular finite elements for potential problems, *Int. J. Eng. Sci.*, Vol. 7, pp. 849–861, 1969.

Silvester, R., Finite element analysis of planar microwave networks, *IEEE Trans. Microwave Theory Tech.*, Vol. MTT-21, pp. 104–108, February 1973.

Simon, J. W., Broadband strip-transmission line Y junction circulators, *IEEE Trans. Microwave Theory Tech.*, Vol. MTT-13, pp. 335–345, 1965.

Skomal, E. P. and Medina, M. A., *J. Appl. Phys. Suppl.*, Vol. 30, No. 4, pp. 1615, 1959.

Smit, J. and Wijn, H. P. J., *Ferrites*, John Wiley & Sons, Hoboken, NJ, 1959.

Solbach, K., Equivalent circuit representation for the *E*-plane circulator, *IEEE Trans. Microwave Theory Tech.*, Vol. MTT-30, pp. 806–809, 1982.

Stoner, E. C., The demagnetizing factors of the general ellipsoid, *Philos. Mag.*, Vol. 36 pp. 803–821, 1945.

Suhl, W. The non-linear behaviour of ferrites at high microwave signal levels, *Proc. IRE*, Vol. 44, pp. 1270–1284, October 1956.

Tellengen, B. D. H., The gyrator, a new electric network element, *Philips Res. Rep.*, Vol. 3, pp. 81–101, 1948.

Treuhaft, M. A., Network properties of circulators based on the scattering concept, *Proc. IRE*, Vol. 44, pp. 1394–1402, 1956.

Tucker, R. S., Gain-bandwidth limitations of microwave transistor amplifiers, *IEEE Trans. Microwave Theory Tech.*, Vol. MTT-21, pp. 322–327, 1973.

Von Aulock, W. H. and Fay, C. E., *Linear Ferrite Devices for Microwave Applications*, Academic, Press, New York, 1968.

Vrehen, Q. H. F., Beljers, G. H., and de Lau, J. G. M., Microwave properties of fine-grain Ni and Mg ferrites, *IEEE Trans. Magnetics*, Vol. MAG-5, pp. 617–621, 1969.

Waldron, R. A. Electromagnetic fields in ferrite ellipsoids, *Br. J. Appl. Phys.*, Vol. 10, pp. 20–22, 1959,

Watkins, J., Circulator resonant structures on microstrip, *Electron. Tech. Lett.*, Vol. 5, pp. 524–525, 1969.

Webb, J. P., Absorbing boundary conditions for the finite element analysis of planar devices, *IEEE Trans. Microwave Theory Tech.*, Vol. MTT-38, No. 9, pp. 1328–1332, September 1990.

Webster, A. S., Remotely tuned circulator, *Electron. Lett.*, Vol. 6, pp. 377–379, 1970.

Wendel, R., Effective computer aided design of stripline Y-circulators, *AEA Int. J. Electron. Commun.*, Vol. 51, pp. 103–110, 1997.

Wheeler, H. A., Transmission-line properties of parallel strips by a conformal mapping approximation, *IEEE Trans. Microwave Theory Tech.*, Vol. MTT-12, pp. 280–288, May 1964.

Whiting, K., Design data for UHF circulators, *IEEE Trans. Microwave Theory Tech.* Vol. MTT-15, pp. 195–198, March 1967.

Wolff, I. and Knoppik, N. Rectangular and circular disk capacitors and resonators, *IEEE Trans. Microwave Theory Tech.*, Vol. MTT-20, pp. 857–864, 1974.

Wu, Y. S. and Rosenbaum, F. J., Wideband operation of microstrip circulators, *IEEE Trans. Microwave Theory Tech.* Vol. MTT-22, pp. 849–856, October 1974.

Yoshida, S., X-circulator, *Proc. Inst. Radio Engineers*, Vol. 47, pp. 1150, 1959.

Youla, D. C., A new theory of broad-band matching, *IEE Trans. Circuit Theory*, Vol. CT-11, pp. 30–50, March 1964.

Index